TWO-DIMENSIONAL NANOMATERIALS FOR WATER PURIFICATION

Micro and Nano Technologies

TWO-DIMENSIONAL NANOMATERIALS FOR WATER PURIFICATION

Advances and Applications

Edited by

EDISON ANG HUIXIANG

Nanyang Technological University, Singapore, Singapore

ELSEVIER

Elsevier
Radarweg 29, PO Box 211, 1000 AE Amsterdam, Netherlands
125 London Wall, London EC2Y 5AS, United Kingdom
50 Hampshire Street, 5th Floor, Cambridge, MA 02139, United States

ISBN: 978-0-443-44530-9

For Information on all Elsevier publications
visit our website at https://www.elsevier.com/books-and-journals

Publisher: Matthew Deans
Acquisitions Editor: Ana Claudia Garcia
Editorial Project Manager: Vasowati Shome
Production Project Manager: Paul Prasad Chandramohan
Cover Designer: Raman Kumar

Typeset by MPS Limited, Chennai, India

Contents

5. Application of computational methodologies in 2D nanomaterials

İdil Bali, Tuğba Baysal, Melih Doğancı, Muhammed Sarıkaya, Mütesir Temel, Hasan Topaçlı, Eyüp Yazkan and Sadiye Velioğlu

6. Advances and insights: 2D nanomaterials in membrane filtration for water purification

Meilan Pan, Jiaming Hu, Yuxuan Wei, Yueping Bao, Qiancheng Xia and Guandao Gao

7. Exploration of 2D nanomaterials in membrane distillation for water purification

Soheil Zarghami and Toraj Mohammadi

8. Advancements in 2D nanomaterials for efficient solar-driven water evaporation

Syazwani Mohd Zaki, Flora Serati, Sa'adah Ahmad Sowi and Nadiah Mokhtar

9. Harnessing the power of 2D nanomaterials for photocatalytic water treatment

Mario J. Muñoz-Batista, Manuel Peñas-Garzón, Daily Rodriguez-Padrón and Rafael Rodríguez-Solís

10. Antibacterial mechanisms of 2D nanomaterials: pathways to safer water purification

Hamyal Awais, Usman Liaqat, Afira Khan, Mudassar Azam, Waheed Miran and Sajjad Miran

11. Impact of 2D nanomaterials on adsorption performances

Sameeha Adil, Sufiyan Syed, Wajahat Khan Mohammed, Ashly Thomas, Murchana Changmai and Swarnalatha Rajaguru

12. Advances of 2D nanomaterials in capacitive deionization

Youfang Zhang and Jun Jin

List of contributors

Sameeha Adil Department of Chemical Engineering, BITS Pilani, Dubai, United Arab Emirates

Sa'adah Ahmad Sowi Department of Manufacturing and Materials Engineering, International Islamic University Malaysia, Kuala Lumpur, Malaysia

Junyi Ai Key Laboratory of the Three Gorges Reservoir Region's Eco-Environment, Ministry of Education, College of Environment and Ecology, Chongqing University, Chongqing, P.R. China

Hamyal Awais School of Chemical and Materials Engineering, National University of Sciences and Technology (NUST), Islamabad, Pakistan

Mudassar Azam Institute of Chemical Engineering and Technology, University of the Punjab, Lahore, Pakistan

Negin Karimzadeh Bajgiran Chemical and Petroleum Engineering Department, University of Tabriz, Tabriz, Iran

İdil Bali Institute of Nanotechnology, Institute of Nanotechnology, Gebze Technical University, Kocaeli, Gebze, Türkiye

Yueping Bao College of Environmental Science and Engineering, Ministry of Education Key Laboratory of Pollution Processes and Environmental Criteria, Tianjin Key Laboratory of Environmental Remediation and Pollution Control, Nankai University, Tianjin, P.R. China

Tuğba Baysal Institute of Nanotechnology, Institute of Nanotechnology, Gebze Technical University, Kocaeli, Gebze, Türkiye

Murchana Changmai Department of Chemical Engineering, BITS Pilani, Dubai, United Arab Emirates

Melih Doğancı Institute of Nanotechnology, Institute of Nanotechnology, Gebze Technical University, Kocaeli, Gebze, Türkiye

Hamid Erfan-Niya Chemical and Petroleum Engineering Department, University of Tabriz, Tabriz, Iran

Guandao Gao School of Environment, State Key Laboratory of Water Pollution Control and Green Resource Recycling, Nanjing University, Nanjing, P.R. China

Qingxiang Guo Key Laboratory of the Three Gorges Reservoir Region's Eco-Environment, Ministry of Education, College of Environment and Ecology, Chongqing University, Chongqing, P.R. China

Le Han Key Laboratory of the Three Gorges Reservoir Region's Eco-Environment, Ministry of Education, College of Environment and Ecology, Chongqing University, Chongqing, P.R. China

Ahasun Habib Hridoy Department of Chemical Engineering, Texas Tech University, Lubbock, TX, United States

Jiaming Hu College of Environmental Science and Engineering, Ministry of Education Key Laboratory of Pollution Processes and Environmental Criteria, Tianjin Key Laboratory of Environmental Remediation and Pollution Control, Nankai University, Tianjin, P.R. China; College of Environment, Zhejiang University of Technology, Hangzhou, P.R. China

Jun Jin Faculty of Materials Science and Chemistry, China University of Geosciences, Wuhan, Hubei, P.R. China

Afira Khan School of Chemical and Materials Engineering, National University of Sciences and Technology (NUST), Islamabad, Pakistan

Usman Liaqat School of Chemical and Materials Engineering, National University of Sciences and Technology (NUST), Islamabad, Pakistan

Sima Majidi Chemical and Petroleum Engineering Department, University of Tabriz, Tabriz, Iran

Sajjad Miran Department of Mechanical Engineering Technology, University of Gujrat, Gujrat, Pakistan

Waheed Miran School of Chemical and Materials Engineering, National University of Sciences and Technology (NUST), Islamabad, Pakistan; SDU Biotechnology, Department of Green Technology, University of Southern Denmark, Odense, Denmark

Toraj Mohammadi Center of Excellence for Membrane Research and Technology, Department of Chemical, Petroleum and Gas Engineering, Iran University of Science and Technology (IUST), Narmak, Tehran, Iran

Wajahat Khan Mohammed Department of Chemical Engineering, BITS Pilani, Dubai, United Arab Emirates

Syazwani Mohd Zaki Department of Manufacturing and Materials Engineering, International Islamic University Malaysia, Kuala Lumpur, Malaysia

Nadiah Mokhtar Faculty of Civil Engineering Technology, Universiti Malaysia Pahang Al-Sultan Abdullah, Pekan, Pahang, Malaysia

Mario J. Muñoz-Batista Department of Chemical Engineering, University of Granada, Granada, Spain

Meilan Pan College of Environmental Science and Engineering, Ministry of Education Key Laboratory of Pollution Processes and Environmental Criteria, Tianjin Key Laboratory of Environmental Remediation and Pollution Control, Nankai University, Tianjin, P.R. China

Manuel Peñas-Garzón Department of Chemical Engineering, University of Málaga, Málagg, Spain

Yue Qing Key Laboratory of the Three Gorges Reservoir Region's Eco-Environment, Ministry of Education, College of Environment and Ecology, Chongqing University, Chongqing, P.R. China

Swarnalatha Rajaguru Department of Electrical & Electronics Engineering, BITS Pilani, Dubai, United Arab Emirates

Daily Rodriguez-Padrón Section of Chemistry for the Technology (ChemTech), Department of Industrial Engineering, University of Padov, Padova (PD), Italy

Rafael Rodríguez-Solís Department of Chemical Engineering, University of Granada, Granada, Spain

Muhammed Sarıkaya Institute of Nanotechnology, Institute of Nanotechnology, Gebze Technical University, Kocaeli, Gebze, Türkiye

Flora Serati Department of Manufacturing and Materials Engineering, International Islamic University Malaysia, Kuala Lumpur, Malaysia; Department of Electrical and Computer Engineering, International Islamic University Malaysia, Kuala Lumpur, Malaysia

Xiaohui Song Department of Materials Science and Engineering, Hefei University of Technology, Hefei, P.R. China

Sufiyan Syed Department of Chemical Engineering, BITS Pilani, Dubai, United Arab Emirates

Mütesir Temel Chemical Engineering Division, Chalmers University of Technology, Gothenburg, Chalmers, Sweden

Ashly Thomas Department of Chemical Engineering, BITS Pilani, Dubai, United Arab Emirates

Hasan Topaçlı Institute of Nanotechnology, Institute of Nanotechnology, Gebze Technical University, Kocaeli, Gebze, Türkiye

Sadiye Velioğlu Institute of Nanotechnology, Institute of Nanotechnology, Gebze Technical University, Kocaeli, Gebze, Türkiye; Department of Chemical Engineering, Istanbul Technical University, Istanbul, Maslak, Türkiye

Ge'er Wang Key Laboratory of the Three Gorges Reservoir Region's Eco-Environment, Ministry of Education, College of

Environment and Ecology, Chongqing University, Chongqing, P.R. China

Yuxuan Wei College of Environment, Zhejiang University of Technology, Hangzhou, P.R. China

Qiancheng Xia School of Environment, State Key Laboratory of Water Pollution Control and Green Resource Recycling, Nanjing University, Nanjing, P.R. China

Eyüp Yazkan Institute of Nanotechnology, Institute of Nanotechnology, Gebze Technical University, Kocaeli, Gebze, Türkiye

Yuanyuan Yu Key Laboratory of the Three Gorges Reservoir Region's Eco-Environment, Ministry of Education, College of Environment and Ecology, Chongqing University, Chongqing, P.R. China

Soheil Zarghami Department of Chemical Engineering, Isfahan University of Technology, Isfahan, Iran

Minxiang Zeng Department of Chemical Engineering, Texas Tech University, Lubbock, TX, United States

Xingyu Zhang School of Mathematics, Statistics and Mechanics, Beijing University of Technology, Beijing, P.R. China

Youfang Zhang Hubei Key Laboratory of Polymer Materials, Ministry of Education Key Laboratory for Green Preparation and Application of Functional Materials, School of Materials Science and Engineering, Hubei University, Wuhan, Hubei, P.R. China

About the editor

Dr. Edison Ang Huixiang
Natural Sciences and Science Education, National Institute of Education/Nanyang Technological University, Singapore, Singapore

Dr. Edison Ang Huixiang is currently an assistant professor at the National Institute of Education/Nanyang Technological University. He serves on the editorial boards of *eScience*, *Carbon Energy*, *Chemical Engineering Journal*, *Journal of Energy Chemistry*, *Materials Horizons*, and *Energy & Environmental Materials* and holds the position of associate editor for Scientific Reports and Frontiers in Chemistry. Dr. Ang has been honored with several prestigious awards, including the ISE-Elsevier Prize for Applied Electrochemistry, the GCDEM Young Scientists Award, the PIERS Young Scientists Award, the AIChE SLS Young Faculty Award, the NIE/NTU Excellence in Research Award, the SNIC Early Career Research Award, and the Sabic Young Professional Award. His research focuses on leveraging nanotechnology, especially two-dimensional nanomaterials and additive manufacturing, to advance energy storage, membrane technology, catalysis, and sensing applications.

Preface

Access to clean water is a fundamental human need, yet global water scarcity and contamination challenges continue to intensify. Innovative solutions are essential to address these pressing issues, and recent advancements in two-dimensional (2D) nanomaterials have opened transformative avenues for water purification. *Two-Dimensional Nanomaterials for Water Purification: Advances and Applications* provides a comprehensive exploration of the capabilities, mechanisms, and practical applications of 2D nanomaterials in advancing water treatment technologies.

This book consolidates state-of-the-art research and development in 2D nanomaterials for water purification, offering an integrated resource that bridges fundamental science with practical implementation. It addresses a critical need in the field: while significant progress has been made, research remains scattered across specialized studies. By unifying these insights, this volume provides a coherent understanding of how 2D nanomaterials can enhance filtration membranes, membrane distillation, solar-driven evaporation, photocatalytic treatment, antibacterial strategies, adsorption, and capacitive deionization.

The text begins with an overview of these applications, followed by in-depth discussions on the fundamental separation mechanisms, material characterization techniques, and design configurations for diverse purification technologies. Subsequent chapters examine computational modeling approaches, advanced fabrication methods, and comparative performance analyses, emphasizing sustainability and efficiency in water treatment. The content is structured to benefit academic researchers, students, industrial engineers, technologists, and policymakers, providing actionable insights that translate laboratory research into real-world solutions.

This book also prioritizes diversity and inclusion in both content and authorship. The contributing experts represent a wide range of cultural, geographic, and academic backgrounds, ensuring that the perspectives and case studies reflect global water challenges. By embracing this diversity, this book fosters cross-disciplinary dialog and innovation, presenting solutions that are adaptable to varying environmental, technological, and socioeconomic contexts.

Key features of the book include the following:

1. Comprehensive coverage of applications: Offering a holistic view of 2D nanomaterials across multiple water purification strategies.
2. Fundamental and advanced insights: Combining foundational principles with cutting-edge developments to serve both newcomers and experts.
3. Practical design and implementation guidelines: Providing actionable knowledge for translating research into effective, real-world water treatment solutions.

It is our hope that this volume will serve as a definitive reference, guiding researchers, engineers, and practitioners in leveraging 2D nanomaterials to advance sustainable and effective water purification technologies. We extend our sincere gratitude to all contributors for their pioneering work and to our readers for their commitment to advancing the science and practice of clean water solutions. May this book inspire further research, innovation, and global collaboration in the pursuit of safe and accessible water for all.

CHAPTER

1

Applications of 2D nanomaterials for water purification technologies

Ahasun Habib Hridoy and Minxiang Zeng

Department of Chemical Engineering, Texas Tech University, Lubbock, TX, United States

1.1 Introduction to water purification technologies

With global population growth, ensuring access to clean and safe water has become one of the most pressing global challenges (Chen et al., 2018a). This growing demand is further complicated by water pollution, which originates from both natural sources and human activities. Contaminants such as pesticides (Zhao et al., 2018c), pharmaceuticals (Snyder et al., 2003), volatile organic compounds (VOCs) (Godri Pollitt et al., 2019), pathogenic microorganisms (Wu et al., 2019b), and toxic heavy metals and metalloids (Fu and Wang, 2011; Bolisetty et al., 2019) are responsible for numerous health issues, including waterborne diseases and even fatalities. Nearly 29% of the world's population did not have access to a safe freshwater source in 2015 alone, and each year, polluted water causes almost half a million fatalities from waterborne diseases (Prüss-Ustün et al., 2014). Although water pollution is often associated with underdeveloped, water-scarce regions, similar issues have also emerged in developed countries. Traditional water treatment methods such as coagulation, flocculation, filtration, electrolysis, and biological treatments (aerobic and anaerobic) have been extensively used. But these methods often involve large, expensive systems and labor-intensive processes that limit their wide deployment (Shannon et al., 2008). There is thus a critical need for new water treatment technologies that are not only efficient but also biocompatible, such as replacing chlorine (Cl_2) disinfection to avoid carcinogenic by-product formation (Bartram, 2003).

Nanotechnology offers a promising route toward this goal. Various nanomaterials have already demonstrated impressive capabilities in water purification, particularly due to their large surface areas. These include zero-dimensional (0D) nanoparticles of metals and metal oxides such as Au, Ag, Fe_3O_4, and TiO_2 (Wang et al., 2018b; Xu et al., 2019a; Liu et al., 2018a; Xu et al., 2019b; Zhao et al., 2016; Zhao et al., 2018b; Zhou and Fang, 2015; MiarAlipour et al., 2018), one-dimensional (1D) nanostructures such as titanate nanowires (Yin et al., 2017), natural clays (Paul et al., 2024) and carbon nanotubes (Dehghani et al., 2016), and three-dimensional (3D) macrostructures derived from these building blocks

Two-Dimensional Nanomaterials for Water Purification
DOI: https://doi.org/10.1016/B978-0-443-44530-9.00001-8

(Shehzad et al., 2016). However, the use of these low-dimensional nanomaterials has been limited by several practical issues: (1) they tend to aggregate or destabilize in solution, reducing their effective surface area (Zeng et al., 2021), (2) residual nanomaterials left in treated water can pose environmental and health risks; and (3) their nanoscale dimensions can hinder bulk device integration in scalable applications (Zeng et al., 2021).

As water purification via nanotechnology highly depends on the surface properties of these nanomaterials, 2D nanomaterials (2DMs) with exceptionally high surface-to-volume ratios (Chimene et al., 2015) offer unique advantages in terms of reduced material use, rapid processing, and highly efficient treatment (Tan et al., 2017a; Anichini et al., 2018). Their planar structure also imparts unique anisotropic physical and chemical properties that surpass those of 1D or 3D counterparts (Chimene et al., 2015). By controlling thickness, lateral size, and surface functionalities, 2DMs have enabled notable advances across multiple treatment platforms. Currently, nanomaterials have been used from atomically thin graphene-based membranes (Wei et al., 2017; Surwade et al., 2015), high-capacity oil absorbents (Ge et al., 2017; Zeng et al., 2018), and ultrafast photocatalysts under visible-light (Haque et al., 2018; Liu et al., 2016b; Zeng et al., 2019b), to responsive and self-healing functional structures (Oh et al., 2017; Zhang et al., 2014). 2D nanomaterials can be synthesized via top-down exfoliation of layered bulk crystals or bottom-up chemical routes using small-molecule precursors (Nicolosi et al., 2013). Black phosphorus (BP), transition metal dichalcogenides (TMDs), transition metal oxides (TMOs), graphene (Gr), graphitic carbon nitride (g-C_3N_4), hexagonal boron nitride (h-BN), MXenes, metal–organic frameworks (MOFs), covalent organic frameworks (COFs), and 2D nano clays are some of the prominent 2D nanomaterials used in water treatment. Fig. 1.1 illustrates their structural diversity.

Despite their promise, very few 2DM-based technologies have transitioned beyond the lab to commercial-scale applications (Anichini et al., 2018; Luo et al., 2016; Wang and Mi, 2017b; Zheng et al., 2016). To bridge this gap, it is crucial to identify the key limitations impeding their practical deployment. More importantly, strategic directions must be identified and prioritized to unlock their full potential in real-world water treatment and monitoring.

1.2 Applications in filtration membranes

At the core of modern water treatment lies the challenge of effectively removing unwanted substances, whether salts, organic pollutants, or microbial contaminants from water. Filtration membranes provide a direct and efficient route to tackle this challenge. These membranes, which function as physical barriers, can preferentially let water molecules pass while keeping out undesirable substances according to their size, charge, or other physicochemical interactions (Biesheuvel et al., 2020; Zeng et al., 2018; Liu et al., 2016a; Fathizadeh et al., 2017). However, a crucial limitation still exists in membrane design: the trade-off between water permeability and selectivity. Achieving high permeability often compromises selectivity and vice versa. Traditional membrane materials have limitations in breaking this inverse relationship. This is where 2D nanoparticles show immense promise. As a result of their atomic-level thickness, often just a few atoms, they present an exciting platform for next-generation membranes capable of delivering both high filtrate flux and superior selectivity.

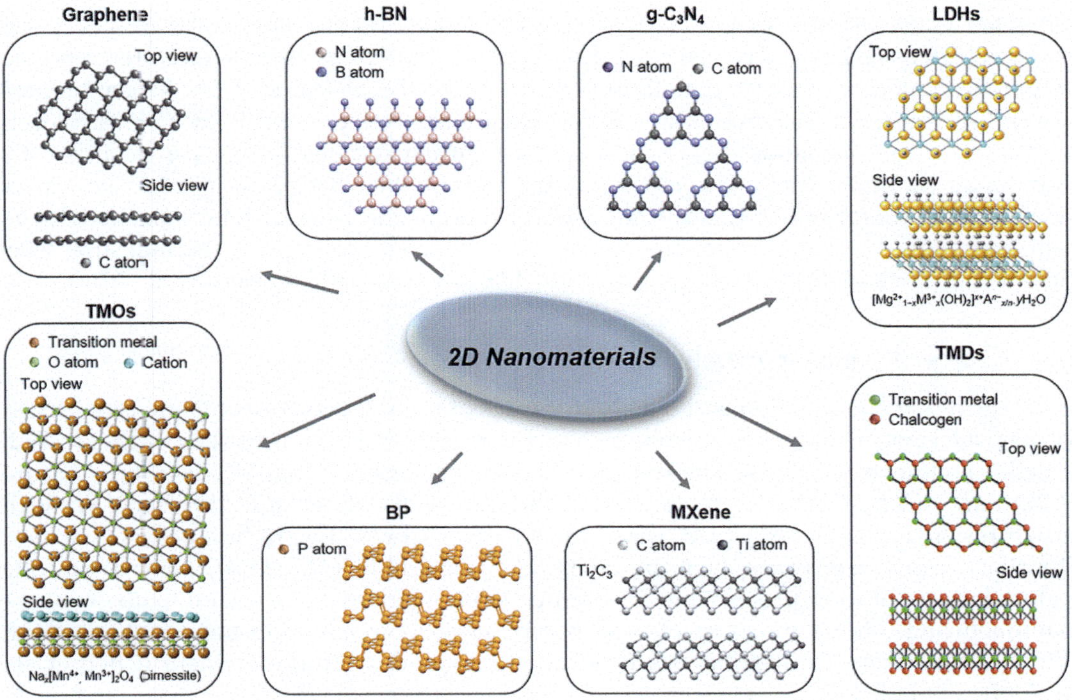

FIGURE 1.1 Various categories of 2D nanomaterials and their nano-structural schematics (Zeng et al., 2021). Source: *Reprinted with permission from Zeng, M., Chen, M., Huang, D., Lei, S., Zhang, X., Wang, L., Cheng, Z., 2021. Engineered two-dimensional nanomaterials: an emerging paradigm for water purification and monitoring.* Materials Horizons 8;*758–802. https://doi.org/10.1039/D0MH01358G. Copyright 2019, The Royal Society of Chemistry.*

1.2.1 Designing 2D material-based membranes

Two main design strategies have emerged in the fabrication of 2D material-based membranes: (1) nanoporous monolayers, where a single layer of 2D material is perforated with subnanometer pores; and (2) multilayered lamellar assemblies, which stack 2D sheets to form tunable interlayer channels for molecular sieving. Graphene oxide (GO), in particular, has become a leading material in this space. Its lamellar structure enables the formation of tunable nanochannels, allowing for selective ion and molecule separation through a combination of size-exclusion and electrostatic interactions (Song et al., 2018). What sets GO apart is its ability to maintain high ion rejection efficiency while allowing rapid water transport (Anand et al., 2018). This dual functionality makes GO-based membranes highly attractive for nanofiltration and reverse osmosis applications.

1.2.2 Nano porous monolayers

From a theoretical standpoint, nanoporous monolayers stand out for their exceptional separation capabilities. Under a pressure gradient, a nanoporous graphene sheet can act as a molecular "mesh," allowing only molecules of water to pass through whilst excluding salt ions and other contaminants. Molecular dynamics simulations predict impressive results: water fluxes as high as 66 L cm^{-2} day^{-1} MPa^{-1} and salt rejection rates of up to 99% have been reported for graphene monolayers with optimized pore structures (Nicolaï et al., 2014). When the pore size and surface functional groups are carefully controlled, water permeability can reach levels several orders of magnitude higher than those of current commercial reverse osmosis (RO) membranes (Li et al., 2016a; Hu and Mi, 2013). Despite their theoretical appeal, monolayer membranes face a major bottleneck in scalability. Fabricating large-area, defect-free monolayers with uniformly distributed nanopores remains a substantial challenge at the industrial level (Surwade et al., 2015; Cohen-Tanugi et al., 2016; O'Hern et al., 2015; O'Hern et al., 2014).

1.2.3 Stacked lamellar membranes

To address the fabrication and scalability issues of monolayers, researchers have turned to stacked nanosheet-based membranes, which assemble 2D materials into layered, lamellar architectures. These stacked membranes leverage the highly anisotropic nature of 2D nanosheets, where water can travel through well-defined interlayer spaces. Such structures are typically fabricated using scalable and cost-effective techniques like vacuum filtration, spray coating, or layer-by-layer (LbL) assembly (Mi, 2014). Since many 2D materials are solution-processable in nature, these methods work well with low-cost manufacturing. Liquid-phase exfoliation can be used to create nanosheets in high yield, allowing membrane fabrication at scales suitable for industrial deployment (Chen et al., 2017; Nie et al., 2020).

1.2.4 Material choices: GO, MXene, MoS_2, and composites

Several 2D materials have demonstrated exceptional performance in filtration applications: GO composites with metal oxide composites (e.g., TiO_2, ZnO) exhibit enhanced antifouling properties, superior mechanical strength, and extended operational lifetimes (Wang et al., 2017). These improvements arise from the synergistic effects between the functional groups on GO and the catalytic/adsorptive characteristics of the metal oxides. MXene-based membranes, alternatively, provide unique advantages through their surface charge modulation and hydration layer affinity, which enable highly selective nanofiltration and ultrafiltration processes (Bao et al., 2026). Their metallic electrical conductivity also opens possibilities for active filtration (e.g., electro-assisted separation). Molybdenum disulfide (MoS_2) and tungsten disulfide (WS_2) nanosheets, when composited into polymeric membranes, enhance water flux and reduce fouling (Li et al., 2019). Their unique surfaces, atomic-scale thickness, and flexibility allow for molecular sieving with high perm-selectivity (Lere Keshebo et al., 2020; Zeng et al., 2017).

1.2.5 Current challenges and potentials

While 2D materials have clearly opened a new frontier in membrane filtration, the key to their adoption lies in balancing performance with scalability. Monolayer membranes promise unmatched water flux and selectivity but are difficult to produce reliably on a large scale. However, 2D nanosheet-based stacked lamellar membranes offer a compelling compromise, combining high-performance with practical manufacturability. Going forward, advances in defect engineering, composite design, and roll-to-roll manufacturing will play a vital role in transitioning these technologies from lab to market. The tunability of 2D materials at the molecular level makes them uniquely suited for targeted water purification, whether it is desalination, heavy metal removal, or emerging contaminant cleanup.

1.3 Applications in membrane distillation

Membrane distillation (MD) has garnered growing attention for its efficiency in desalinating ultra-saline water. Unlike conventional pressure-driven filtration, MD operates via a thermally activated separation mechanism, making it particularly suited for harsh feed waters where reverse osmosis (RO) becomes inefficient or unstable. Besides desalination, MD has also been studied for use in wastewater treatment, pharmaceutical separations, food processing, and biomedical systems.

1.3.1 Transport mechanism in membrane distillation and role of 2D nanomaterials

MD is a nonisothermal, thermally driven separation process in which a hot saline feed and a cold permeate stream are separated by a thin hydrophobic membrane. Due to the temperature difference between the two sides, a vapor pressure gradient forms, allowing only the volatile component (typically water) to pass through the membrane as vapor. The hydrophobic membrane acts as a barrier to the liquid-phase, enabling the vapor to travel through its pores and condense as pure distillate on the permeate side (Ray et al., 2018). MD membranes typically operate at temperatures of 50°C–90°C and require lower pressure compared to RO systems, making them compatible with solar energy and other low-temperature heat sources (Bhadra et al., 2016; Kullab and Martin, 2011; Gude, 2015; Khayet, 2013; Qtaishat and Banat, 2013). However, several operational challenges remain, particularly membrane wetting, flux decline, and fouling, which compromise long-term performance and reliability. To overcome these issues, recent research has turned toward incorporating next-generation 2D nanomaterials into MD membranes. Materials such as metal oxides, carbon-based nanomaterials, and metal–organic frameworks (MOFs) have been studied for their ability to enhance key membrane properties. These include increased liquid entry pressure (LEPw), improved permeability and porosity, enhanced hydrophobicity, better chemical and thermal stability, stronger mechanical durability, etc.

1.3.2 2D nanomaterials in enhancing MD performance

2D nanomaterials introduce multiple pathways for vapor transport, which help improve both flux and selectivity. These pathways include: (1) in-plane nanopores, formed by defects in the nanosheet structure; (2) interplanar slit-like channels, the spaces between stacked 2D sheets that guide water vapor flow; and (3) voids from disordered stacking, where irregular layer assembly creates tortuous channels for vapor transport.

1.3.2.1 *Metal-organic frameworks*

MOFs are unique crystalline materials with exceptional porosity that are formed from metal ions bonded by organic ligands. Their unique physical and chemical properties, namely tunable pore size, large surface area, and high pore volume, make them ideal as fillers in mixed-matrix membranes (MMMs). MOFs show strong affinity toward polymer chains, allowing them to integrate seamlessly into membrane matrices (Furukawa et al., 2013). The tailorable structure of MOFs allows for application-specific membrane design, enabling control over both selectivity and permeability (Lin et al., 2018b). Their presence can alter the membrane architecture, often resulting in higher water flux and mechanical stability (Kadhom and Deng, 2018).

1.3.2.2 *Carbon-based nanomaterials*

Materials like GO, reduced graphene oxide (rGO), and carbon nanotubes (CNTs) represent a superior class of nanomaterials because of their high surface area-to-volume ratio, excellent mechanical strength, and in-built fouling resistance. Their integration into MD membranes improves water transport, reinforces structural durability, and extends operational life (Das et al., 2014).

1.3.3 Current challenges and potentials

While the incorporation of 2D nanomaterials has advanced MD membrane performance considerably, several critical challenges and research gaps still need to be addressed: (1) Nanoparticle Aggregation: During membrane fabrication, NPs may aggregate, leading to uneven distribution and compromised performance. (2) Cost optimization: Employing cost-effective nanomaterials is essential for translating lab-scale advancements into scalable, commercial technologies. (3) Surface roughness versus fouling: While surface roughness enhances hydrophobicity, it can also increase fouling (Drioli et al., 2015). Achieving the right balance is crucial. (4) Particle leaching: There is a risk of nanoparticle washout or delamination during long-term use, especially under dynamic flow conditions or thermal cycling. Despite these challenges, the overall effect of nanoparticle incorporation is positive. Benefits include improved super-hydrophobicity, reduced fouling tendencies, increased porosity and vapor transport pathways, and enhanced mechanical and thermal stability. Together, these attributes enable the design of robust, antiwetting, and high-flux membranes suitable for prolonged MD operations in real-world conditions. Membrane distillation integrates thermal separation and membrane technology, offering a sustainable route for clean water generation. With the strategic integration of 2D nanomaterials ranging from metallic nanoparticles to carbon-based sheets and MOFs, the next-generation of MD membranes is focused on delivering higher efficiency, longer lifespan, and broader applicability than ever before (Leaper et al., 2021). As research continues, interdisciplinary approaches combining materials science, chemical engineering, and nanotechnology will be critical to unlocking the full capabilities of 2D nanomaterials in membrane distillation systems.

1.4 Applications in solar evaporation

Using solar energy to produce freshwater through photothermal evaporation offers an environmentally friendly alternative to traditional desalination. Because of their large surface area and tunable photothermal conversion characteristics, two-dimensional nanomaterials have emerged as promising candidates for this use. These characteristics allow 2D materials to efficiently convert sunlight into localized heat, driving rapid evaporation of water even under low solar intensities.

1.4.1 Principles of solar evaporation in desalination membranes

Unlike size-exclusion membranes used in nanofiltration or reverse osmosis, solar evaporation membranes operate based on the difference in enthalpy of evaporation between water and contaminants. Four key parts make up a conventional solar desalination setup: a freshwater collection unit, a condenser, a seawater reservoir, and a solar-absorbing membrane. The most crucial of these is the membrane for solar evaporation. It serves the dual role of absorbing solar energy and converting it to thermal energy to generate water vapor from brine or wastewater. To maximize the performance of these systems, two design criteria are especially important: (1) Efficient solar absorption: this minimizes reflectance and transmittance losses across a broad-spectrum. Ideal materials should absorb nearly all incident solar radiation (Tran, 2025; Zeng et al., 2021; Zhu et al., 2018b). (2) Effective heat management: Minimizing thermal loss ensures that the generated heat is retained near the surface of evaporation. Over the years, numerous strategies have been developed to improve absorber efficiency. These include antireflection coatings (García-Vidal et al., 1997; Zhang et al., 2018b), nanostructured light-trapping layers that enhance light coupling (Zhou et al., 2016; Zhu et al., 2018b), and optical mode engineering to increase absorption across the spectrum (Yu et al., 2010). These innovations have pushed solar evaporation membranes closer to their theoretical efficiency limits.

1.4.2 Role of 2D nanomaterials in solar evaporation

Carbon-based 2D nanomaterials, such as graphene, rGO, and MXenes, have become focal points in solar desalination research. Their broad absorption bands, excellent thermal conductivity, and photothermal efficiency make them especially suited for this task (Zhang et al., 2017; Lei et al., 2019; Wu et al., 2019a,b; Fathizadeh et al., 2017). Graphene and MXenes, often referred to as "black 2D materials," exhibit strong photon-harvesting properties. Their optical behavior is driven by π-band transitions, which make them effective absorbers of visible and near-infrared light (Liu et al., 2017). However, one restriction of classic graphitic materials is their 5%–10% light reflection at the air-dielectric interface (Klinkova et al., 2014). To overcome this, researchers have designed nanostructured and porous variants of carbon materials. For example, porous membranes can trap and circulate light within their internal cavities via resonant recirculation, dramatically increasing light-matter interaction (Linic et al., 2015; Li et al., 2021; Kim et al., 2015; Sun et al., 2015b; Lin et al., 2017b). Additionally, plasmonic resonance is another powerful tool for enhancing solar absorption. When the frequency of incident light matches that of delocalized surface electrons, a collective excitation is triggered on the surface of metallic nanoparticles (Li et al., 2021; Hua et al., 2016; Li et al., 2014). This interaction generates hot electrons, which quickly relax via electron–electron scattering, leading to localized heating of the material (Zhu et al., 2016; Ding et al., 2017a; Wang et al., 2017; Zhao et al., 2007). These design strategies enhance the photothermal efficiency of 2D material-based membranes, resulting in faster water evaporation and higher freshwater output.

1.4.3 Recent examples of 2D materials in solar evaporation

One of the impactful structural innovations is the vertically aligned architecture of 2D materials. In contrast to the more conventional lamellar membranes used in filtration, vertically aligned graphene sheet membranes (VA-GSMs) promote better water transport and light-trapping. This design was shown to improve solar desalination (Zhang et al., 2017). Compared to randomly aligned or lamellar graphene sheets, the VA-GSM achieved a vapor generation rate of 1.6 and 6.3 $kg/m^2 \cdot h$ under 1 and 4 sun illumination, respectively, and solar-to-vapor conversion efficiencies of 86.5% and 94.2% (Zhang et al., 2017). Another notable example is the multiscale plasmonic nanostructure developed by Wang et al. (2018a). They constructed a solar desalination membrane by combining GO with plasmonic Ag/Au nano cubes to form plasmonic colloidosomes. The coupling between the nanoparticles produced high light absorption and heat generation. The result: a 92% efficiency at 10 kW/m^2 and a steam generation rate of 12.96 $kg/m^2 \cdot h$ (Wang et al., 2018a). These examples demonstrate the real-world potential of combining 2D nanomaterials with advanced nanophotonic architectures to create high-efficiency, high-throughput solar evaporation systems.

1.4.4 Current challenges and potentials

Despite rapid progress, several technical barriers continue to limit the real-world adoption of 2D material-based solar evaporation membranes: (1) Salt accumulation: Over time, salt tends to accumulate on the membrane surface during desalination, lowering both light absorption and supply of water, and resulting in a decline in evaporation efficiency (Zhou et al., 2016; Li et al., 2016), (2) Stability in saline environments: Prolonged exposure to high-salinity water can degrade some materials, especially under intense sunlight or cyclic wetting/drying conditions. To address these issues, recent studies have developed hydrophobic surface modifications. For example, Zhao et al. created an MXene-based membrane with a tailored hydrophobic surface that minimized salt buildup and maintained long-term stability in saline water (Zhao et al., 2018). Going forward, future research should focus on: Salt-repellent membrane designs, Self-cleaning surface chemistries, Scalable and roll-to-roll fabrication methods, and integration with solar concentrators or hybrid systems.

By tackling these challenges, 2D material-enabled solar desalination systems could offer a scalable, low-cost solution to global water shortages.

1.5 Applications in photocatalytic water treatment

In recent decades, increasing urbanization and improvements in living standards have intensified the need for safe drinking water and effective sanitation. As a result, there is growing interest in developing environmentally friendly and efficient methods for water disinfection. Conventional disinfection techniques, such as chlorination and ozonation, while widely used, are not without limitations. Chlorine and ozone, as powerful oxidants, can lead to the formation of harmful by-products (Antonopoulou et al., 2014) and are ineffective against certain resistant pathogens (Zhi et al., 2016). Likewise, ultraviolet (UV) irradiation, a physical approach, has its drawbacks: high energy consumption, costly operation, and limited functionality against some microbes (Gilboa and Friedler, 2008). These limitations underscore the urgent need for sustainable disinfection technologies that operate under a broad-spectrum, energy-efficient, and low-maintenance (Mauter et al., 2018). Among such methods, photocatalysis has emerged as one of the most promising solutions, particularly due to its potential to harness solar energy for water purification (Chen et al., 2010; Kärkäs et al., 2014).

1.5.1 Principle of photocatalysis

Photocatalysis operates through light-induced redox reactions facilitated by a semiconductor catalyst. When a photocatalyst absorbs light energy, electrons are excited from the valence band (VB) to the conduction band (CB), leaving behind positively charged holes in the VB. These charge carriers, electrons and holes, are key to driving reactions such as water splitting, degradation of pollutants, and CO_2 reduction (Turchi and Ollis, 1990; Zhao et al., 2019). However, one of the main limitations is the rapid recombination of these charge carriers, which hampers the efficiency of photocatalytic processes (Wang et al., 2014a).

1.5.2 Role of 2D nanomaterials in photocatalysis

2D nanomaterials are increasingly being explored for photocatalytic applications due to their distinctive structural and electronic features. Compared to their 0D and 1D counterparts, 2D nanostructures exhibit advantages such as enhanced architecture modulation, superior charge separation, increased light absorption, and higher tunability. Their large lateral dimensions provide numerous active sites for pollutant adsorption and catalytic reactions (Sun et al., 2015a). Furthermore, their planar surfaces, especially those rich in functional groups, facilitate the integration of guest materials, which is beneficial for constructing multifunctional hybrid architectures. Their ultrathin thickness ensures shorter charge diffusion paths, thereby reducing recombination losses and enhancing photocatalytic performance (Deng et al., 2016). Additionally, 2D layers often expose a higher fraction of low-coordinated atoms, improving light-harvesting efficiency (Di et al., 2018). By adjusting the number and thickness of layers, it is possible to tailor the bandgap and light absorption characteristics (Su et al., 2018). Structural modifications such as doping, metal deposition, or forming composites are often employed to further improve their photocatalytic efficacy (Sun et al., 2015a; Yu et al., 2014; Liu et al., 2014a). Driven by these unique properties, researchers have developed a variety of high-performance photocatalysts based on 2D nanomaterials (Su et al., 2018). These materials have been applied for biological disinfection, degradation of organic contaminants, and removal of inorganic pollutants.

1.5.3 Photocatalysts for biodisinfection

Pathogens like bacteria and viruses in drinking water present significant health hazards, especially in developing countries. Several 2D nanostructured photocatalysts have been engineered through techniques such as bandgap tuning and nano-compositing to address these threats (Yu et al., 2019b; Ma et al., 2019; Wu et al., 2019a; Sarkar et al., 2018). These materials typically generate reactive oxygen species (ROS) upon light activation, which can effectively inactivate microbial cells (Lam et al., 2018). For instance, TiO_2–Bi_2WO_6 nanosheets (TBWO), synthesized via a two-step hydrothermal method, demonstrated excellent visible-light photocatalytic activity. The incorporation of Bi_2WO_6 expanded the light absorption range, enabling efficient rupture and deactivation of *E. coli* cells (Jia et al., 2016).

1.5.4 Photocatalysts for organic pollutant removal

Unlike microbial disinfection, which primarily targets biological structures, the degradation of organic contaminants requires the complete mineralization of complex molecules. This demands the generation of a substantial amount of ROS (Sahu et al., 2018). The electronic band structure of the photocatalyst plays a critical role in defining the redox potential of the generated charge carriers (Loeb et al., 2019). For example, monolayer or few-layer MoS_2 exhibits a larger bandgap compared to its bulk form, allowing for improved redox activity despite a narrower range of solar absorption (Wang 2017). To enhance light absorption and catalytic activity, strategies such as heterojunction formation, element doping, and nanoscale size control are often employed (Gong et al., 2013). Researchers have also developed composites by incorporating 0D nanoparticles and 1D nanostructures (nanorods, nanotubes) with 2D nanosheets (Ran et al., 2014; Li et al., 2016c; Bi et al., 2015; Ebina et al., 2002; Xing et al., 2016; Tian et al., 2015; Yang et al., 2014b). A notable example includes metal nanoparticle intercalation, such as Pt, Au, RuO_2, and $Rh(OH)_3$, between oxide layers, resulting in synergistic effects that boost photocatalytic efficiency (Ebina et al., 2002; Hata et al., 2008; Hata et al., 2007). Oshima et al. further demonstrated that Pt nanoclusters intercalated into $KCa_2Nb_3O_{10}$ nanosheets enhanced catalytic activity due to the well-dispersed co-catalysts (Oshima et al., 2015).

1.5.5 Photocatalysts for inorganic pollutant removal

Some inorganic pollutants can be effectively neutralized or converted to nontoxic forms via photocatalysis. For instance, Prakash et al. developed In_2S_3/CuS nanosheets capable of converting toxic H_2S into thiosulfate ($S_2O_3^{2-}$) with concurrent hydrogen production (Prakash et al., 2018). Similarly, exfoliated monolayer g-C_3N_4 exhibited remarkable performance in removing total ammonia nitrogen (TAN). Its bandgap conveniently spans the redox potential required for ammonia oxidation, making it thermodynamically favorable. Moreover, the exfoliated variant outperformed bulk g-C_3N_4, likely due to increased active surface area and shorter diffusion lengths (Wang et al., 2014b).

1.5.6 Current challenges and potentials

Despite their potential, 2D photocatalysts face practical challenges, particularly particle aggregation, which reduces their active surface area and accelerates sedimentation in water. This aggregation occurs due to oxidation and desorption of surface ligands, as well as diminished electrostatic repulsion, which can arise from charge depletion (Haque et al., 2018; Luo et al., 2016). According to Stokes' law, sedimentation rate increases with the square of particle diameter, so aggregated particles are quickly removed from the treatment medium. To address this, researchers are focusing on surface engineering and controlled self-assembly. For example, Park et al. demonstrated that controlled aggregation of dye/TiO_2 composites improved Cr(VI) reduction by enhancing charge transfer and reducing recombination (Park et al., 2012). CdSe/CdS quantum dots that self-assembled with a Pt co-catalyst are another example, greatly increasing electron transfer and reaching an internal quantum efficiency of 65% (Li et al., 2017). Thus through tailored synthesis, structural optimization, and compositing strategies, the field of 2D nanomaterial-based photocatalysis continues to evolve toward achieving more efficient and practical water treatment solutions.

1.6 Applications in antibacterial agents

One critical phase of wastewater treatment involves disinfection or deactivation of pathogenic microorganisms to protect public health and natural ecosystems. Traditional disinfection methods, such as ultraviolet (UV) irradiation, chlorination, and ozonation, are widely used but face considerable drawbacks, including environmental toxicity, generation of harmful by-products, high energy consumption, and significant operational costs (Miao et al., 2019; Anastasi et al., 2013; Gilboa and Friedler, 2008; Krasner et al., 2006). Moreover, using natural enzymes like peroxidase for antimicrobial applications is impractical due to their instability, high production costs, and complex purification processes (Fan et al., 2021; Vatansever et al., 2013; Tang et al., 2021). To overcome these limitations, 2D nanomaterials such as GO, MoS_2, and MXenes have garnered substantial attention due to their intrinsic antimicrobial properties, high surface area, structural tunability, and favorable physicochemical characteristics (Miao et al., 2019; Hossain et al., 2014; Arabi Shamsabadi et al., 2018).

1.6.1 Mechanism of 2D nanomaterials as antibacterial agents

2D nanomaterials possess unique attributes, including a large specific surface area, exceptional stability, and minimal cytotoxicity, which make them suitable candidates for antibacterial applications (Fan et al., 2021). One major antibacterial mechanism involves photo-induced production of reactive oxygen species (ROS) under near-infrared (NIR) light, which generates oxidative stress in microbial cells (Miao et al., 2019; Hossain et al., 2014; Li et al., 2008). GO, in particular, is a biocompatible 2D nanomaterial that effectively interacts with bacterial and viral surfaces through hydrogen bonding and electrostatic interactions. Its high surface area supports bacterial adhesion, while its interaction can be enhanced by incorporating antimicrobial metals like silver (Ag) or copper (Cu) nanoparticles. The antibacterial efficacy of GO is influenced by factors such as flake size and surface roughness (Miao et al., 2019; Vi et al., 2018; María and Id, 2018; Yun et al., 2020).

1.6.2 2D nanomaterials in water disinfection as antibacterial agents

In a comprehensive review, Zou et al. outlined the antibacterial mechanisms of graphene-derived nanomaterials, citing their strong potential for antimicrobial applications (Zou et al., 2016). More recently, titanium carbide-based MXenes ($Ti_3C_2T_x$) have demonstrated strong antibacterial activity against pathogens such as *Bacillus subtilis* and *Escherichia coli* (Rasool et al., 2017). Their inherent resistance to biofouling makes them promising candidates for integration into membrane systems. The antimicrobial behavior of MXenes is attributed to their anionic surface chemistry, direct contact-based killing due to their hydrophilic structure, and strong hydrogen bonding interactions between their oxygen-containing groups and bacterial membranes, which disrupt nutrient transport (Ihsanullah, 2020). Furthermore, their unique layered atomic structure enhances their antibacterial performance (Jastrzębska et al., 2019). Graphene and GO, even without surface functionalization, have shown inherent antimicrobial activity (Liu et al., 2011). Water-dispersible GO and rGO nanosheets can suppress the growth of E. coli with negligible cytotoxicity (Hu et al., 2010). GO can impair microbial metabolic activity by disrupting oxygen uptake, thereby reducing biological oxygen demand (BOD) and ultimately leading to bacterial cell death (Ahmed and Rodrigues, 2013). The bactericidal action of graphene is also associated with its sharp nanosheet edges that physically pierce microbial membranes, causing leakage of cellular content and eventual lysis (Upadhyay et al., 2014). Moreover, graphene materials can induce oxidative damage to critical biomolecules such as proteins, lipids, and nucleic acids (Upadhyay et al., 2014). The combined effects of membrane disruption and oxidative stress account for the potent antibacterial activity observed in both GO and rGO systems.

1.6.3 Current challenges and potentials

Despite their great potential as next-generation antibacterial agents, 2D materials still face a number of challenges that must be overcome before they can be widely used. These include the adaptability of 2D nanomaterials for diverse environmental surfaces, long-term biosafety, and a comprehensive understanding of their antimicrobial mechanism. Research into scalable synthesis methods, functionalization for target-specific antimicrobial effects, and rigorous environmental risk assessments is essential to bridge the gap between laboratory success and real-world application. Overall, the unique physical and chemical properties of 2D materials continue to inspire innovation in water disinfection strategies.

1.7 Applications in adsorption materials

Adsorption is one of the most effective and widely used techniques in water purification due to its simplicity, low-cost, and regeneration ability. The process is also beneficial in mitigating oil spills and removing heavy metals. With suitable nanomaterial design, they offer high removal efficiency, involve minimal energy use, and produce fewer harmful by-products compared to other methods (Ihsanullah, 2021; Zubair et al., 2018). Among various adsorbents, 2D nanomaterials have gained substantial attention for their ability to adsorb contaminants, owing to their high surface area and effective surface/interface interactions. The unique properties of these materials make them ideal candidates for contaminant removal in water treatment.

1.7.1 Mechanisms of 2D nanomaterial-based adsorbents

The adsorption mechanisms of 2D nanomaterials include a variety of interactions, such as hydrophobic interactions, π–π stacking, hydrogen bonding, ion/ligand exchange, Lewis acid/base interactions, and electrostatic interactions (Zeng et al., 2021). Graphene, one of the most studied 2D nanomaterials, consists of a single layer of carbon atoms bonded in a honeycomb lattice (Geim and Novoselov, 2007). Its high surface area allows for the efficient adsorption of pollutants, and the possibility of functionalizing graphene to create composites enhances its adsorption capacity for water purification (Dreyer et al., 2010). To enhance its effectiveness in water purification, graphene is commonly converted into GO through oxidative treatment, often using the Hummers method (Han et al., 2013). The transformation of graphene into GO introduces a variety of hydrophilic oxygen-containing functional groups (e.g., hydroxyl, epoxide) across the graphene surface (Gao et al., 2011). This modification increases the hydrophilicity of GO, improving water flow and enabling the formation of membranes with high permeability and flux (Sweetman et al., 2017). New advancements in 2D material-based adsorbents are intended to tackle issues with heavy metal removal, oil spill mitigation, and the adsorption of organic contaminants that are hydrophilic.

1.7.2 Oil spill adsorbents

Oil spills pose a significant threat to marine ecosystems, causing severe environmental damage and loss of valuable resources (Jernelöv, 2010; Peterson et al., 2003). Efficient and eco-friendly oil adsorbents are required to address this issue. 2D nanomaterials, with their high surface area and tunable surface properties, have proven to be promising candidates for fast and efficient oil spill cleanup. rGO nanosheets have been employed in the design of oil-absorbent sponges. Ge et al. developed a graphene-wrapped sponge (GWS), which showed effective performance in cleaning up viscous crude oil (Ge et al., 2017). Often, surface functionalization is required for increased hydrophobicity. With an oil contact angle (CA) of 0° and a water CA of 162°, Jayaramulu et al. reported developing a fluorinated GO with exceptional hydrophobicity. The sorption selectivity and oil absorption capability were greatly enhanced by this change (Jayaramulu et al., 2016). Additionally, 2D covalent organic frameworks (COFs) have been used as coatings for oil/water separation membranes, with superhydrophobic COFs demonstrating excellent water repellency and oil absorption capacity (Mullangi et al., 2017). Other 2D nanomaterials, such as hexagonal boron nitride (h-BN), have also shown promise for oil/water separation. Self-assembled h-BN nanosheets were used to form superhydrophobic composites capable of separating water/oil emulsions with increased efficiency, even under high-flux conditions (Yu et al., 2015). The large surface area and customizable surface chemistry of 2D nanomaterials enable the creation of hierarchical structures, which have also been successfully applied in energy storage and catalysis (Hou et al., 2015; Cai et al., 2014; Rahman et al., 2024).

1.7.3 Hydrophilic organic pollutant adsorbents

Detergents, fertilizers, insecticides, and medications are examples of hydrophilic organic pollutants that are common water contaminants (Ali, 2012). The adsorbent materials must be engineered with characteristics including hydrogen bonding, van der Waals forces, and electrostatic attraction to efficiently remove these pollutants. BUC-17, a water-stable graphene-like metal–organic framework (MOF) developed by Li et al., demonstrated exceptional adsorption capabilities at ambient temperature when used to adsorb colors from simulated wastewater (Li et al., 2017a). In addition, MoS_2 nanosheets have also been explored as adsorbents for various organic dyes, including rhodamine B (RhB), methylene blue (MB), and Congo red (CR), demonstrating a great capacity for dye removal (Massey et al., 2016; Wang et al., 2014c). Covalent bonding, electrostatic attraction, or van der Waals interactions are the usual causes of 2D nanomaterials' high adsorption efficiency (Massey et al., 2016; Wang et al., 2014c). For example, the negative charge of MoS_2 enhances the adsorption of cationic species like RhB more effectively than anionic molecules, emphasizing the role of electrostatic interactions in pollutant removal (Wang et al., 2014c).

1.7.4 Heavy metal adsorbents

The removal of heavy metals from wastewater is essential for environmental protection since they are persistent contaminants (Duffus, 2002; Zhu et al., 2007). 2D nanomaterials, with their high surface area and layered structure, are particularly suited for adsorbing heavy metals such as Cd^{2+}, Cr^{3+}, Hg^{2+}, and Pb^{2+}. Transition metal dichalcogenides (TMDs), such as MoS_2, have been used to efficiently remove heavy metals from water. The interaction between soft Lewis bases (e.g., MoS_2) and soft Lewis acids (e.g., metal ions) enhances the adsorption process (Wang and Mi, 2017b). Nonchalcogenide 2D nanosheets, such as hexagonal boron nitride (h-BN), have also been explored for their potential to adsorb heavy metals. Li et al. synthesized porous BN whiskers and activated BN sheets that successfully removed heavy metals from water, demonstrating the versatility of 2D materials in environmental decontamination (Li et al., 2013). Similarly, 3D carbon/BN structures have been developed for enhanced heavy metal removal (Liu et al., 2018b).

1.7.5 Anion and oxyanion adsorbents

Many toxic substances, including heavy metals, form complex anions in water, and specialized 2D adsorbents are required to remove these charged species. For example, arsenic in water is highly toxic, and its removal is essential for human health (Ali, 2012). To remove anions such as arsenate (As(V)), fluoride (F^-), and phosphate (PO_4^{3-}), 2D materials with large surface areas and tunable surface chemistry are designed to optimize the adsorption process (Arefin et al., 2023). Layered double hydroxides (LDHs), a class of 2D clays, have been widely used for anion removal due to their ion-exchange properties. For instance, LDH composites have been shown to efficiently remove As(V) from contaminated groundwater, achieving high adsorption capacities and removing up to 90 mg/g of As(V) (Das et al., 2006). This efficiency is partly due to the trivalent cations in the LDH structure, which enhance the material's ability to attract and adsorb anions (Das et al., 2006).

1.7.6 Current challenges and potentials

Despite the promising findings, using 2D nanomaterials for adsorption applications still presents a number of difficulties. One major issue is the sedimentation rate of nanoparticles, which can affect their efficiency. While slow sedimentation may result in higher costs for collecting used adsorbents from water, fast sedimentation causes the premature removal of adsorbents before contaminant removal takes place efficiently. Furthermore, understanding the behavior of 2D nanomaterials in colloidal dispersions and optimizing their nano-adsorbent dimensions are crucial steps toward practical applications, especially in environments where stirring is not feasible, such as lakes or offshore conditions. Overall, while the chemical and physical diversity of 2D nanomaterials holds great potential for pollutant adsorption, continued research is needed to optimize their properties, understand their behavior at the molecular level, and fine-tune their surface chemistry for targeted pollutant removal.

1.8 Applications in capacitive deionization

Capacitive deionization (CDI) has emerged as a promising electro-sorption-based desalination method, distinguished by its environmental compatibility, efficiency, and relatively low operational cost. Its core advantage lies in the reversibility of the electrochemical process, and the CDI technology, in many cases, has shown increased efficiency compared to other contemporary technologies.

1.8.1 Mechanism of capacitive deionization

The CDI process is governed by the electric double layer (EDL) theory, involving ion adsorption and desorption at the electrode surface (Xing et al., 2020). During operation, a low external voltage (<1.8 V) is applied across two electrically conductive parallel electrodes. This conductivity is essential for forming EDLs, allowing efficient charge transfer and ion accumulation (Biesheuvel and van der Wal, 2010). The applied voltage generates an electric field that drives ions from the saline feed solution toward oppositely charged electrodes, where they are stored within the EDL and effectively removed from the water (Choi et al., 2019). Once saturation is reached, the voltage is either turned off or reversed, releasing the adsorbed ions and regenerating the electrodes for subsequent cycles (Huang et al., 2013; Suss et al., 2015; Zhao et al., 2013a).

1.8.2 Role of 2D nanomaterials in capacitive deionization

2D nanomaterials are being actively explored for CDI electrode design due to their ultrathin structures, high specific surface areas, and tunable surface functionalities. These attributes contribute to enhanced ion transport, high electrosorption capacity, and improved cycling stability (Geise et al., 2013; Zhang et al., 2015; Qian et al., 2015). Among them, MXenes ($Ti_3C_2T_x$) stand out due to their high volumetric capacitance (~1500 F cm^{-3}), conductivity, and hydrophilic interlayer spaces that enable efficient ion diffusion and high desalination rates (Zhang et al., 2015). Their lamellar architecture also offers flexibility in tuning interlayer spacing for targeted ion sieving, making them effective in both flow-by and flow-through CDI configurations. Graphene derivatives, especially rGO, are valued for their high surface area and robust chemical stability. Incorporating pseudocapacitive elements or heteroatom dopants into graphene matrices has been shown to improve charge storage and ion selectivity (Younes et al., 2022; Zhao et al., 2011; Kandambeth et al., 2019). Likewise, MoS_2 nanosheets integrated into composite systems offer additional redox-active sites and structural pathways that facilitate efficient ion adsorption (Cai et al., 2021).

1.8.3 Current challenges and potentials

Despite rapid advancements, scaling CDI for real-world applications is hindered by several persistent challenges. Among them is the co-ion expulsion effect, which reduces salt adsorption capacity (SAC) and accelerates electrode degradation. The integration of ion-exchange membranes (IEMs), particularly anion exchange membranes (AEMs), can alleviate this issue, but concerns remain regarding their long-term chemical stability, ease of fabrication, and recyclability (Tian et al., 2014). While MXenes offer excellent SAC values (~130 mg g^{-1}) owing to their unique electronic and structural features, their commercial deployment is constrained by high production costs and poor stability. Other variables, including feedwater composition, dopant incorporation, operational temperature, and long-term cycling performance, further influence CDI system efficiency. Another critical issue is the inconsistency in experimental protocols and performance metrics. Variability in feed concentration, applied voltage, and lack of standardized reporting make it difficult to benchmark different materials. Harmonizing these parameters, such as SAC, charge efficiency, and cycling durability, is essential for transparent comparison and effective material screening (Tian et al., 2014). There is also an increasing call for environmentally responsible approaches in CDI development. This includes adopting green solvents and biodegradable crosslinkers for IEM fabrication to minimize ecological impact. Additionally, efforts should focus on improving membrane fouling resistance and chemical durability under continuous operation. For CDI to become a sustainable and scalable desalination technology, future research must prioritize high-performance materials that are both cost-effective and eco-friendly.

1.9 Conclusion and perspectives

2DMs hold significant promise for water treatment; however, several challenges must be addressed to enable their adoption in real-world applications. Though we covered a number of 2DMs applications in this chapter, such as filtration membranes, membrane distillation, solar evaporation, photocatalysis, antibacterial applications, adsorption, and capacitive deionization, this is only a brief overview of the rapidly expanding field. Table 1.1 summarizes a range of 2DMs that have already demonstrated success in various water purification and monitoring technologies.

One of the primary concerns with many 2DMs is their relatively high production cost compared to traditional materials. To make 2DMs more applicable in water treatment, it's essential to develop synthesis methods that are both scalable and cost-effective. For example, top-down solution-based exfoliation methods offer a promising route; these rely on inexpensive, naturally layered precursor crystals like graphite and can be scaled up using techniques such as ultrasonication or shear force to produce large quantities of 2DMs efficiently (Cai et al., 2018; Stafford et al., 2018; Varrla et al., 2015). Furthermore, the overall sustainability of 2DMs, both in terms of how they are synthesized and applied, needs to be critically evaluated. For instance, sustainable synthesis should prioritize minimal energy consumption and the use of safer, less hazardous chemical processes to limit environmental and ecological damage (Huang et al., 2011; Liu et al., 2015b; Zhang et al., 2010). Since the goal of water purification is to remove contaminants, the materials used should themselves be safe and nontoxic. Ensuring the biocompatibility of 2DMs is essential to prevent potential recontamination. Environmentally safe materials, such as layered double hydroxides (LDHs), should be prioritized for this reason (Chimene et al., 2015). In the near future, advancing next-generation water purification technologies will focus on the design of nanomaterials that are not only highly functional but also scalable, safe, and sustainable for real-world deployment.

From the point of view of fabrication and application, it is essential that 2DMs maintain their structural integrity and functional performance over long-term use. Any degradation, leaching, or release of secondary pollutants must be avoided to ensure sustainability in real-world water treatment applications. This demands the development of 2DMs with high stability for continuous usage in multiple cycles of operation. Specifically, for solar desalination and energetic water transport, sustaining well-defined nanochannels with long-term operational stability is critical. To improve throughput, advanced self-assembly techniques and guided aggregation strategies that enable the precise formation of interconnected nanochannels could offer major advantages (Fatima et al., 2022). For technologies such as adsorption, photocatalysis, and CDI, it is vital to develop smart nanostructures that can respond selectively and adaptively to various pollutants. Many pristine 2D nanocrystals lack strong, selective interactions with contaminants. In this context, defect engineering emerges as a promising strategy to enhance their selectivity and performance in pollutant capture. In the coming future, the next-generation of water purification technologies will rely on multifunctional, scalable, and environmentally safe materials designed for precision-targeted applications. Moreover, robotic (Masubuchi et al., 2018) and additive manufacturing, 2D/3D printing technologies, particularly direct ink writing (DIW) and extrusion-based approaches, are currently revolutionizing the design of

TABLE 1.1 Overview of 2D nanomaterials and their applications in water purification technologies (As of 2025).

	Gr	TMDs/TMOs	BP	Mxene	h-BN	g-C3N4	LDH	MOF/COF
Filtration	+ Sun et al. (2020)	+ Sun et al. (2014)		+ Ding et al. (2017b)	+ Yu et al. (2015)	+ Zhao et al. (2016), Wang et al. (2017a)	+ Zeng et al. (2019a)	+ Dai et al. (2019), Zhang et al. (2018a), Shi et al. (2019)
Solar evaporation	+ Tian et al. (2020)	+ Ghim et al. (2018)		+Zhao et al., 2019		+Su et al. (2019)		+ Ma, Qarony, et al. (2019)
Adsorption	+ Bao et al. (2020)	+ Tian and Chowdhury (2019), Krasian et al. (2019)	+ Yuan et al. (2020)	+Xie et al., 2019	+ Lei et al. (2015)	+Cai et al. (2017)	+ Clark et al. (2019), Zhou et al. (2018), Rahmanian et al. (2018)	+ Sarker et al. (2018)
Photocatalysis	+ Zou et al. (2020), Kuang et al. (2020)	+ Luo et al. (2018), Li et al. (2019a), Guo et al. (2019)	+Wang et al. (2019b), Hu et al. (2019)	+ Cui et al. (2020), Shahzad et al. (2018)	+Sheng et al. (2019b)	+ EL-Sheshtawy et al. (2019), Sheng et al. (2019a), Li et al. (2020), Moreira et al. (2019), Liu et al. (2019)	+ Darie et al. (2019), Mohamed et al. (2018), Zhao et al. (2018a)	+ Pi et al. (2018), Yu et al. (2019a)
Antibacterial agents	+ Miao et al. (2019)	+ Bang et al. (2016), Masimukku et al. (2018), Krishnamoorthy et al. (2013), Yin et al. (2018)	+ Tan et al. (2018), Huang et al. (2018), Xiong et al. (2018)	+ Ihsanullah (2020), Jastrzębska et al. (2019)		+ Zhang et al. (2024)	+ Zhao et al. (2013)b, Moaty et al. (2016)	

Membrane distillation	+ Seraj et al. (2022)	+ Frappa et al. (2023)	+Mustakeem et al. (2022), Zhang et al. (2022a)	+ Frappa et al. (2023)
CDI	+Zhang et al. (2015)	+ Bentalib et al. (2020), Liu et al. (2020), Zhou et al. (2020)	+ Zhang et al. (2022b)	+ Weng et al. (2021), Li et al. (2019b)

BP, black phosphorus; *TMDs*, transition metal dichalcogenides; *TMOs*, transition metal oxides; *Gr*, graphene; g-C_3N_4, graphitic carbon nitride; *h-BN*, hexagonal boron nitride; *MOFs*, metal–organic frameworks; COFs, covalent organic frameworks.

water purification systems by enabling precise control over pore geometry, channel architecture, and device morphology (Elsersawy et al., 2025; Tijing et al., 2020). These advancements have facilitated the fabrication of customized sorbents, membranes, and modules with enhanced performance and durability. Integrating these approaches with intelligent material design strategies will enable the creation of programmable, adaptive, and high-throughput devices tailored for adsorption, photocatalysis, filtration membranes, solar evaporation, antibacterial materials, and CDI applications.

Moving forward, the design of next-generation smart 2DMs for water purification will increasingly rely on establishing fundamental structure–processing–performance relationships. To achieve this, innovative theoretical data-driven approaches such as machine learning and model-based simulations are emerging as essential tools. With rapid advancements in algorithmic development, these techniques are now playing a significant role in materials science and engineering of the nanocomposites (Lu et al., 2024; Cao et al., 2024; Lu et al., 2018; Kitchin, 2018; Zeng et al., 2019b). Methods like data mining, neural networks, and machine learning have shown strong potential in predicting material properties and structures by learning from existing datasets, significantly reducing the need for time-consuming experiments and complex simulations (Zeng et al., 2019b). In the context of 2DMs, machine learning can accelerate the discovery and optimization of materials by predicting key characteristics such as adsorption capacity, structural stability, and interaction with pollutants, especially in cases where traditional testing is resource-intensive. Moreover, integrating artificial intelligence with nanotechnology is paving the way for the intelligent design of high-performance nanocomposites tailored for water purification (Nandipati et al., 2024; Wang et al., 2021). In the future, research directions will heavily emphasize developing these nanocomposites with the integration of self-learning, predictive AI-based tools with nanoscale material engineering to develop adaptive and intelligent 2DMs for water purification processes.

Furthermore, many water treatment applications operate under harsh conditions, such as exposure to strong chemicals, extreme pH levels, high-salinity, and sustained humidity. So, critical research is needed on enhancing the stability and recyclability of 2DMs under these conditions. Although recent advances in device passivation and encapsulation have improved the nanoscale environmental stability of some 2DMs against moisture and oxygen exposure (Abate et al., 2018), truly long-term durability under severe operational conditions still remains a challenge. Addressing these issues will require not only innovations in surface chemistry and compositional nanostructuring but also a deeper fundamental understanding of the structure–property–performance relationships governing 2DMs. To bridge the current gap between theoretically predicted and experimentally synthesized 2DMs, more studies are needed to identify materials with enhanced thermal, electrical, and catalytic properties that are specifically suited for environmental remediation. Experimental validation of these predictions is essential to accelerate the development of materials with superior photocatalytic, electrocatalytic, and adsorptive performance. It is equally important to design multifunctional 2DMs capable of removing multiple types of pollutants simultaneously. Such multifunctionality could significantly improve the efficiency and practicality of water purification systems, offering a more sustainable and effective solution for both industrial and domestic applications.

Acknowledgments

The authors thank Nishat Paul, Naimul Arefin, and Brandon Dumars for the review and proofreading of this chapter.

AI disclosure

During the preparation of this work, the authors used OpenAI ChatGPT Version 4o in order to paraphrase and fix grammatical errors. After using this tool/service, the authors reviewed and edited the content as needed and take full responsibility for the content of the publication.

References

Abate, Y., Akinwande, D., Gamage, S., Wang, H., Snure, M., Poudel, N., et al., 2018. Recent progress on stability and passivation of black phosphorus. Advanced Materials 30 (29). Available at https://doi.org/10.1002/adma.201704749, http://onlinelibrary.wiley.com/journal/10.1002/(ISSN)1521-4095.

Ahmed, F., Rodrigues, D.F., 2013. Investigation of acute effects of graphene oxide on wastewater microbial community: a case study. Journal of Hazardous Materials 256-257, 33–39. Available at https://doi.org/10.1016/j.jhazmat.2013.03.064, www.elsevier.com/locate/jhazmat.

Ali, I., 2012. New generation adsorbents for water treatment. Chemical Reviews 112 (10), 5073–5091. Available at https://doi.org/10.1021/cr300133d.

Anand, Anisha, Unnikrishnan, Binesh, Mao, Ju-Yi, Lin, Han-Jia, Huang, Chih-Ching, 2018. Graphene-based nanofiltration membranes for improving salt rejection, water flux and antifouling–A review. Desalination 429, 119–133. Available at https://doi.org/10.1016/j.desal.2017.12.012.

Anastasi, E.M., Wohlsen, T.D., Stratton, H.M., Katouli, M., 2013. Survival of *Escherichia coli* in two sewage treatment plants using UV irradiation and chlorination for disinfection. Water Research 47 (17), 6670–6679. Available at https://doi.org/10.1016/j.watres.2013.09.008.

Anichini, C., Czepa, W., Pakulski, D., Aliprandi, A., Ciesielski, A., Samorì, P., 2018. Chemical sensing with 2D materials. Chemical Society Reviews 47 (13), 4860–4908. Available at https://doi.org/10.1039/c8cs00417j, http://pubs.rsc.org/en/journals/journal/cs.

Antonopoulou, M., Evgenidou, E., Lambropoulou, D., Konstantinou, I., 2014. A review on advanced oxidation processes for the removal of taste and odor compounds from aqueous media. Water Research 53, 215–234. Available at https://doi.org/10.1016/j.watres.2014.01.028.

Arabi Shamsabadi, A., Sharifian, M., Anasori, B., Soroush, M., 2018. Antimicrobial mode-of-action of colloidal Ti3C2Tx MXene nanosheets. ACS Sustainable Chemistry and Engineering 6 (12), 16586–16596. Available at https://doi.org/10.1021/acssuschemeng.8b03823, http://pubs.acs.org/journal/ascecg.

Arefin, N., Podolak, G., Lewis-Sandy, J., Zeng, M., 2023. Printing and patterning of clay-based nanocomposites for healthcare and sustainable applications. MRS Communications 13 (6), 980–993. Available at https://doi.org/10.1557/s43579-023-00449-9, https://www.springer.com/journal/43579.

Bang, G.S., Cho, S., Son, N., Shim, G.W., Cho, B.K., Choi, S.Y., 2016. DNA-assisted exfoliation of tungsten dichalcogenides and their antibacterial effect. ACS Applied Materials and Interfaces 8 (3), 1943–1950. Available at https://doi.org/10.1021/acsami.5b10136, http://pubs.acs.org/journal/aamick.

Bao, S., Yang, W., Wang, Y., Yu, Y., Sun, Y., 2020. One-pot synthesis of magnetic graphene oxide composites as an efficient and recoverable adsorbent for Cd(II) and Pb(II) removal from aqueous solution. Journal of Hazardous Materials 381, 120914. Available at https://doi.org/10.1016/j.jhazmat.2019.120914.

Bao, Luo, Li, Xindong, Huang, Lijinhong, Zhu, Qinyan, Yu, Siwei, Zhong, Dehua, Yan, Jianhui, Jia, Jianghui, Dang, Weiqi, Xu, Guangyu, 2026. MXene-PA double-layer Janus nanofiltration membrane for synergistic, efficient, and stable recovery of rare earth ions. Journal of Membrane Science 741, 124999. Available at https://doi.org/10.1016/j.memsci.2025.124999.

Bartram, J., 2003. Heterotrophic plate counts and drinking-water safety : the significance of HPCs for water quality and human health. 256.

Bentalib, A., Pan, Y., Yao, L., Peng, Z., 2020. Properties of amorphous iron phosphate in pseudocapacitive sodium ion removal for water desalination. RSC Advances 10 (29), 16875–16880. Available at https://doi.org/10.1039/d0ra02010a, http://pubs.rsc.org/en/journals/journal/ra.

Bhadra, Madhuleena, Roy, Sagar, Mitra, Somenath, 2016. Desalination across a graphene oxide membrane via direct contact membrane distillation. Desalination 378, 37–43. Available at https://doi.org/10.1016/j.desal.2015.09.026.

Bi, W., Li, X., Zhang, L., Jin, T., Zhang, L., Zhang, Q., et al., 2015. Molecular co-catalyst accelerating hole transfer for enhanced photocatalytic H 2 evolution. Nature Communications 6. Available at https://doi.org/10.1038/ncomms9647, http://www.nature.com/ncomms/index.html.

Biesheuvel, P.M., van der Wal, A., 2010. Membrane capacitive deionization. Journal of Membrane Science 346 (2), 256–262. Available at https://doi.org/10.1016/j.memsci.2009.09.043.

Biesheuvel, P.M., Zhang, L., Gasquet, P., Blankert, B., Elimelech, M., van der Meer, W.G.J., 2020. Ion selectivity in brackish water desalination by reverse osmosis: theory, measurements, and implications. Environmental Science & Technology Letters 7 (1), 42–47. Available at https://doi.org/10.1021/acs.estlett.9b00686.

Bolisetty, S., Peydayesh, M., Mezzenga, R., 2019. Sustainable technologies for water purification from heavy metals: review and analysis. Chemical Society Reviews 48 (2), 463–487. Available at https://doi.org/10.1039/c8cs00493e, http://pubs.rsc.org/en/journals/journal/cs.

Cai, S., Zhang, D., Shi, L., Xu, J., Zhang, L., Huang, L., et al., 2014. Porous Ni-Mn oxide nanosheets in situ formed on nickel foam as 3D hierarchical monolith de-NOx catalysts. Nanoscale 6 (13), 7346–7353. Available at https://doi.org/10.1039/c4nr00475b, http://www.rsc.org/publishing/journals/NR/Index.asp.

Cai, X., He, J., Chen, L., Chen, K., Li, Y., Zhang, K., et al., 2017. A 2D-g-C3N4 nanosheet as an eco-friendly adsorbent for various environmental pollutants in water. Chemosphere 171, 192–201. Available at https://doi.org/10.1016/j.chemosphere.2016.12.073, www.elsevier.com/locate/chemosphere.

Cai, X., Luo, Y., Liu, B., Cheng, H.M., 2018. Preparation of 2D material dispersions and their applications. Chemical Society Reviews 47 (16), 6224–6266. Available at https://doi.org/10.1039/c8cs00254a, http://pubs.rsc.org/en/journals/journal/cs.

Cai, Yanmeng, Zhang, Wen, Fang, Rongli, Zhao, Dongdong, Wang, Yue, Wang, Jixiao, 2021. Well-dispersed few-layered MoS2 connected with robust 3D conductive architecture for rapid capacitive deionization process and its specific ion selectivity. Desalination 520, 115325. Available at https://doi.org/10.1016/j.desal.2021.115325.

Cao, Z., Barati Farimani, O., Ock, J., Barati Farimani, A., 2024. Machine learning in membrane design: from property prediction to AI-guided optimization. Nano Letters 24 (10), 2953–2960. Available at https://doi.org/10.1021/acs.nanolett.3c05137, http://pubs.acs.org/journal/nalefd.

Chen, B., Han, M.Y., Peng, K., Zhou, S.L., Shao, L., Wu, X.F., et al., 2018a. Global land-water nexus: agricultural land and freshwater use embodied in worldwide supply chains. Science of the Total Environment 613-614, 931–943. Available at https://doi.org/10.1016/j.scitotenv.2017.09.138.

Chen, L., Shi, G., Shen, J., Peng, B., Zhang, B., Wang, Y., et al., 2017. Ion sieving in graphene oxide membranes via cationic control of interlayer spacing. Nature 550 (7676), 380–383. Available at https://doi.org/10.1038/nature24044, http://www.nature.com/nature/index.html.

Chen, X., Shen, S., Guo, L., Mao, S.S., 2010. Semiconductor-based photocatalytic hydrogen generation. Chemical Reviews 110 (11), 6503–6570. Available at https://doi.org/10.1021/cr1001645.

Chimene, D., Alge, D.L., Gaharwar, A.K., 2015. Two-dimensional nanomaterials for biomedical applications: emerging trends and future prospects. Advanced Materials 27 (45), 7261–7284. Available at https://doi.org/10.1002/adma.201502422, http://www3.interscience.wiley.com/journal/119030556/issue.

Choi, J., Dorji, P., Shon, H.K., Hong, S., 2019. Applications of capacitive deionization: desalination, softening, selective removal, and energy efficiency. Desalination 449, 118–130. Available at https://doi.org/10.1016/j.desal.2018.10.013, https://www.sciencedirect.com/science/journal/.

Clark, I., Smith, J., Gomes, R.L., Lester, E., 2019. Continuous synthesis of Zn2Al-CO3 layered double hydroxides for the adsorption of reactive dyes from water. Journal of Environmental Chemical Engineering 7 (3), 103175. Available at https://doi.org/10.1016/j.jece.2019.103175.

Cohen-Tanugi, D., Lin, L.C., Grossman, J.C., 2016. Multilayer nanoporous graphene membranes for water desalination. Nano Letters 16 (2), 1027–1033. Available at https://doi.org/10.1021/acs.nanolett.5b04089, http://pubs.acs.org/journal/nalefd.

Cui, C., Guo, R., Xiao, H., Ren, E., Song, Q., Xiang, C., et al., 2020. Bi2WO6/Nb2CTx MXene hybrid nanosheets with enhanced visible-light-driven photocatalytic activity for organic pollutants degradation. Applied Surface Science 505, 144595. Available at https://doi.org/10.1016/j.apsusc.2019.144595.

Dai, R., Zhang, X., Liu, M., Wu, Z., Wang, Z., 2019. Porous metal organic framework CuBDC nanosheet incorporated thin-film nanocomposite membrane for high-performance forward osmosis. Journal of Membrane Science 573, 46–54. Available at https://doi.org/10.1016/j.memsci.2018.11.075, www.elsevier.com/locate/memsci.

Darie, M., Seftel, E.M., Mertens, M., Ciocarlan, R.G., Cool, P., Carja, G., 2019. Harvesting solar light on a tandem of Pt or Pt-Ag nanoparticles on layered double hydroxides photocatalysts for p-nitrophenol degradation in water. Applied Clay Science 182. Available at https://doi.org/10.1016/j.clay.2019.105250, http://www.elsevier.com/inca/publications/store/5/0/3/3/2/2/index.htt.

Das, J., Patra, B.S., Baliarsingh, N., Parida, K.M., 2006. Adsorption of phosphate by layered double hydroxides in aqueous solutions. Applied Clay Science 32 (3-4), 252–260. Available at https://doi.org/10.1016/j.clay.2006.02.005.

Das, R., Ali, M.E., Hamid, S.B.A., Ramakrishna, S., Chowdhury, Z.Z., 2014. Carbon nanotube membranes for water purification: a bright future in water desalination. Desalination 336 (1), 97–109. Available at https://doi.org/10.1016/j.desal.2013.12.026.

Dehghani, M.H., Mostofi, M., Alimohammadi, M., McKay, G., Yetilmezsoy, K., Albadarin, A.B., et al., 2016. High-performance removal of toxic phenol by single-walled and multi-walled carbon nanotubes: kinetics, adsorption, mechanism and optimization studies. Journal of Industrial and Engineering Chemistry 35, 63–74. Available at https://doi.org/10.1016/j.jiec.2015.12.010, http:www.sciencedirect.com/science/journal/1226086X.

Deng, D., Novoselov, K.S., Fu, Q., Zheng, N., Tian, Z., Bao, X., 2016. Catalysis with two-dimensional materials and their heterostructures. Nature Nanotechnology 11 (3), 218–230. Available at https://doi.org/10.1038/nnano.2015.340, http://www.nature.com/nnano/index.html.

Di, J., Xiong, J., Li, H., Liu, Z., 2018. Ultrathin 2D photocatalysts: electronic-structure tailoring, hybridization, and applications. Advanced Materials 30 (1). Available at https://doi.org/10.1002/adma.201704548, http://www3.interscience.wiley.com/journal/119030556/issue.

Ding, D., Huang, W., Song, C., Yan, M., Guo, C., Liu, S., 2017a. Non-stoichiometric MoO3-x quantum dots as a light-harvesting material for interfacial water evaporation. Chemical Communications 53 (50), 6744–6747. Available at https://doi.org/10.1039/c7cc01427a, http://pubs.rsc.org/en/journals/journal/cc.

Ding, L., Wei, Y., Wang, Y., Chen, H., Caro, J., Wang, H., 2017b. A two-dimensional lamellar membrane: MXene nanosheet stacks. Angewandte Chemie - International Edition 56 (7), 1825–1829. Available at https://doi.org/10.1002/anie.201609306, http://onlinelibrary.wiley.com/journal/10.1002/(ISSN)1521-3773.

Dreyer, D.R., Park, S., Bielawski, C.W., Ruoff, R.S., 2010. The chemistry of graphene oxide. Chemical Society Reviews 39 (1), 228–240. Available at https://doi.org/10.1039/b917103g.

Drioli, E., Ali, A., Macedonio, F., 2015. Membrane distillation: recent developments and perspectives. Desalination 356, 56–84. Available at https://doi.org/10.1016/j.desal.2014.10.028.

Duffus, J.H., 2002. "heavy metals" - A meaningless term? (IUPAC technical report). Pure and Applied Chemistry 74 (5), 793–807. Available at https://doi.org/10.1351/pac200274050793, http://www.degruyter.com/view/j/pac.

EL-Sheshtawy, H.S., El-Hosainy, H.M., Shoueir, K.R., El-Mehasseb, I.M., El-Kemary, M., 2019. Facile immobilization of Ag nanoparticles on g-C 3 N 4 /V 2 O 5 surface for enhancement of post-illumination, catalytic, and photocatalytic activity removal of organic and inorganic pollutants. Applied Surface Science 467-468, 268–276. Available at https://doi.org/10.1016/j.apsusc.2018.10.109, http://www.journals.elsevier.com/applied-surface-science/.

Ebina, Y., Sasaki, T., Harada, M., Watanabe, M., 2002. Restacked perovskite nanosheets and their Pt-loaded materials as photocatalysts. Chemistry of Materials 14 (10), 4390–4395. Available at https://doi.org/10.1021/cm020622e.

Elsersawy, R., Donyadari, R., Khondoker, M.A.H., 2025. Extrusion 3D-printed kaolinite ceramic filters for water applications. Journal of Manufacturing and Materials Processing 9 (8). Available at https://doi.org/10.3390/jmmp9080278, www.mdpi.com/journal/jmmp.

Fan, X., Yang, F., Nie, C., Ma, L., Cheng, C., Haag, R., 2021. Biocatalytic nanomaterials: a new pathway for bacterial disinfection. Advanced Materials 33 (33). Available at https://doi.org/10.1002/adma.202100637.

Fathizadeh, M., Xu, W.L., Zhou, F., Yoon, Y., Yu, M., 2017. Graphene oxide: a novel 2-dimensional material in membrane separation for water purification. Advanced Materials Interfaces 4 (5). Available at https://doi.org/10.1002/admi.201600918, http://onlinelibrary.wiley.com/journal/10.1002/(ISSN)2196-7350.

Fatima, J., Shah, A.N., Tahir, M.B., Mehmood, T., Shah, A.A., Tanveer, M., et al., 2022. Tunable 2D nanomaterials; their key roles and mechanisms in water purification and monitoring. Frontiers in Environmental Science 10. Available at https://doi.org/10.3389/fenvs.2022.766743, journal.frontiersin.org/journal/environmental-science.

Frappa, M., Alessandro, F., Macedonio, F., Drioli, E., 2023. Two-dimensional materials: from discovery to application in membrane distillation/crystallization processes. Chemistry (Weinheim an der Bergstrasse, Germany) 5 (4), 2205–2228. Available at https://doi.org/10.3390/chemistry5040148.

Fu, F., Wang, Q., 2011. Removal of heavy metal ions from wastewaters: a review. Journal of Environmental Management 92 (3), 407–418. Available at https://doi.org/10.1016/j.jenvman.2010.11.011, https://www.sciencedirect.com/journal/journal-of-environmental-management.

Furukawa, H., Cordova, K.E., O'Keeffe, M., Yaghi, O.M., 2013. The chemistry and applications of metal-organic frameworks. Science (New York, N.Y.) 341 (6149). Available at https://doi.org/10.1126/science.1230444, http://www.sciencemag.org/content/341/6149/1230444.full.pdf.

Gao, W., Majumder, M., Alemany, L.B., Narayanan, T.N., Ibarra, M.A., Pradhan, B.K., et al., 2011. Engineered graphite oxide materials for application in water purification. ACS Applied Materials and Interfaces 3 (6), 1821–1826. Available at https://doi.org/10.1021/am200300u.

García-Vidal, F.J., Pitarke, J.M., Pendry, J.B., 1997. Effective Medium Theory of the Optical Properties of Aligned Carbon Nanotubes. Physical Review Letters 78 (22), 4289–4292. Available at https://doi.org/10.1103/physrevlett.78.4289.

Ge, J., Shi, L.A., Wang, Y.C., Zhao, H.Y., Yao, H.B., Zhu, Y.B., et al., 2017. Joule-heated graphene-wrapped sponge enables fast clean-up of viscous crude-oil spill. Nature Nanotechnology 12 (5), 434–440. Available at https://doi.org/10.1038/nnano.2017.33, http://www.nature.com/nnano/index.html.

Geim, A.K., Novoselov, K.S., 2007. The rise of graphene. Nature Materials 6 (3), 183–191. Available at https://doi.org/10.1038/nmat1849.

Geise, G.M., Hickner, M.A., Logan, B.E., 2013. Ionic resistance and permselectivity tradeoffs in anion exchange membranes. ACS Applied Materials and Interfaces 5 (20), 10294–10301. Available at https://doi.org/10.1021/am403207w, http://pubs.acs.org/journal/aamick.

Ghim, D., Jiang, Q., Cao, S.S., Singamaneni, S., Jun, Y.S., 2018. Mechanically interlocked 1T/2H phases of MoS2 nanosheets for solar thermal water purification. Nano Energy 53, 949–957. Available at https://doi.org/10.1016/j.nanoen.2018.09.038, http://www.journals.elsevier.com/nano-energy/.

Gilboa, Y., Friedler, E., 2008. UV disinfection of RBC-treated light greywater effluent: kinetics, survival and regrowth of selected microorganisms. Water Research 42 (4-5), 1043–1050. Available at https://doi.org/10.1016/j.watres.2007.09.027, www.elsevier.com/locate/watres.

Godri Pollitt, K.J., Kim, J.H., Peccia, J., Elimelech, M., Zhang, Y., Charkoftaki, G., et al., 2019. 1,4-Dioxane as an emerging water contaminant: state of the science and evaluation of research needs. Science of the Total Environment 690, 853–866. Available at https://doi.org/10.1016/j.scitotenv.2019.06.443, www.elsevier.com/locate/scitotenv.

Gong, Y., Liu, Z., Lupini, A.R., Shi, G., Lin, J., Najmaei, S., et al., 2013. Band gap engineering and layer-by-layer mapping of selenium-doped molybdenum disulfide. Nano Letters 14 (2), 442–449.

Gude, V.G., 2015. Energy storage for desalination processes powered by renewable energy and waste heat sources. Applied Energy 137, 877–898. Available at https://doi.org/10.1016/j.apenergy.2014.06.061, https://www.journals.elsevier.com/applied-energy.

Guo, M., Xing, Z., Zhao, T., Li, Z., Yang, S., Zhou, W., 2019. WS2 quantum dots/MoS2@WO3-x core-shell hierarchical dual Z-scheme tandem heterojunctions with wide-spectrum response and enhanced photocatalytic performance. Applied Catalysis B: Environmental 257, 117913. Available at https://doi.org/10.1016/j.apcatb.2019.117913.

Han, Y., Xu, Z., Gao, C., 2013. Ultrathin graphene nanofiltration membrane for water purification. Advanced Functional Materials 23 (29), 3693–3700. Available at https://doi.org/10.1002/adfm.201202601.

Haque, F., Daeneke, T., Kalantar-zadeh, K., Ou, J.Z., 2018. Two-dimensional transition metal oxide and chalcogenide-based photocatalysts. Nano-Micro Letters 10 (2). Available at https://doi.org/10.1007/s40820-017-0176-y, http://www.springer.com/engineering/journal/40820.

Hata, H., Kubo, S., Kobayashi, Y., Mallouk, T.E., 2007. Intercalation of well-dispersed gold nanoparticles into layered oxide nanosheets through intercalation of a polyamine. Journal of the American Chemical Society 129 (11), 3064–3065. Available at https://doi.org/10.1021/ja068272a.

Hata, H., Kobayashi, Y., Bojan, V., Youngblood, W.J., Mallouk, T.E., 2008. Direct deposition of trivalent rhodium hydroxide nanoparticles onto a semiconducting layered calcium niobate for photocatalytic hydrogen evolution. Nano Letters 8 (3), 794–799. Available at https://doi.org/10.1021/nl072571w.

Hossain, F., Perales-Perez, O.J., Hwang, S., Román, F., 2014. Antimicrobial nanomaterials as water disinfectant: applications, limitations and future perspectives. Science of the Total Environment 466-467, 1047–1059. Available at https://doi.org/10.1016/j.scitotenv.2013.08.009.

Hou, J., Cao, C., Idrees, F., Ma, X., 2015. Hierarchical porous nitrogen-doped carbon nanosheets derived from silk for ultrahigh-capacity battery anodes and supercapacitors. ACS Nano 9 (3), 2556–2564. Available at https://doi.org/10.1021/nn506394r, http://pubs.acs.org/journal/ancac3.

Hu, J., Chen, D., Mo, Z., Li, N., Xu, Q., Li, H., et al., 2019. Z-scheme 2D/2D heterojunction of black phosphorus/monolayer Bi 2 WO 6 nanosheets with enhanced photocatalytic activities. Angewandte Chemie 131 (7), 2095–2099. Available at https://doi.org/10.1002/ange.201813417.

Hu, M., Mi, B., 2013. Enabling graphene oxide nanosheets as water separation membranes. Environmental Science and Technology 47 (8), 3715–3723. Available at https://doi.org/10.1021/es400571g.

Hu, W., Peng, C., Luo, W., Lv, M., Li, X., Li, D., et al., 2010. Graphene-based antibacterial paper. ACS Nano 4 (7), 4317–4323. Available at https://doi.org/10.1021/nn101097v.

Hua, Z., Li, B., Li, L., Yin, X., Chen, K., Wang, W., 2016. Designing a novel photothermal material of hierarchical microstructured copper phosphate for solar evaporation enhancement. Journal of Physical Chemistry C 121 (1), 60–69.

Huang, L., Liu, Y., Ji, L.C., Xie, Y.Q., Wang, T., Shi, W.Z., 2011. Pulsed laser assisted reduction of graphene oxide. Carbon 49 (7), 2431–2436. Available at https://doi.org/10.1016/j.carbon.2011.01.067.

Huang, W., Zhang, Y., Bao, S., Song, S., 2013. Desalination by capacitive deionization with carbon-based materials as electrode: a review. Surface Review and Letters 20 (6). Available at https://doi.org/10.1142/S0218625X13300050.

Huang, X.W., Wei, J.J., Zhang, M.Y., Zhang, X.L., Yin, X.F., Lu, C.H., et al., 2018. Water-based black phosphorus hybrid nanosheets as a moldable platform for wound healing applications. ACS Applied Materials and Interfaces 10 (41), 35495–35502. Available at https://doi.org/10.1021/acsami.8b12523, http://pubs.acs.org/journal/aamick.

Ihsanullah, I., 2020. MXenes (two-dimensional metal carbides) as emerging nanomaterials for water purification: progress, challenges and prospects. Chemical Engineering Journal 388. Available at https://doi.org/10.1016/j.cej.2020.124340, www.elsevier.com/inca/publications/store/6/0/1/2/7/3/index.htt.

Ihsanullah, I., 2021. Boron nitride-based materials for water purification: progress and outlook. Chemosphere 263. Available at https://doi.org/10.1016/j.chemosphere.2020.127970, www.elsevier.com/locate/chemosphere.

Jastrzębska, A.M., Karwowska, E., Wojciechowski, T., Ziemkowska, W., Rozmysłowska, A., Chlubny, L., et al., 2019. The atomic structure of Ti 2 C and Ti 3 C 2 MXenes is responsible for their antibacterial activity toward E. coli bacteria. Journal of Materials Engineering and Performance 28 (3), 1272–1277. Available at https://doi.org/10.1007/s11665-018-3223-z, http://www.springerlink.com/content/1059-9495/.

Jayaramulu, K., Datta, K.K.R., Rösler, C., Petr, M., Otyepka, M., Zboril, R., et al., 2016. Biomimetic superhydrophobic/superoleophilic highly fluorinated graphene oxide and ZIF-8 composites for oil-water separation. Angewandte Chemie - International Edition 55 (3), 1178–1182. Available at https://doi.org/10.1002/anie.201507692, http://onlinelibrary.wiley.com/journal/10.1002/(ISSN)1521-3773.

Jernelöv, A., 2010. How to defend against future oil spills. Nature 466 (7303), 182–183. Available at https://doi.org/10.1038/466182a.

Jia, Y., Zhan, S., Ma, S., Zhou, Q., 2016. Fabrication of TiO2-Bi2WO6 binanosheet for enhanced solar photocatalytic disinfection of E. coli: insights on the mechanism. ACS Applied Materials and Interfaces 8 (11), 6841–6851. Available at https://doi.org/10.1021/acsami.6b00004, http://pubs.acs.org/journal/aamick.

Kadhom, M., Deng, B., 2018. Metal-organic frameworks (MOFs) in water filtration membranes for desalination and other applications. Applied Materials Today 11, 219–230. Available at https://doi.org/10.1016/j.apmt.2018.02.008, http://www.journals.elsevier.com/applied-materials-today/.

Kandambeth, S., Dey, K., Banerjee, R., 2019. Covalent organic frameworks: chemistry beyond the structure. Journal of the American Chemical Society 141 (5), 1807–1822. Available at https://doi.org/10.1021/jacs.8b10334, http://pubs.acs.org/journal/jacsat.

Khayet, M., 2013. Solar desalination by membrane distillation: dispersion in energy consumption analysis and water production costs (a review). Desalination 308, 89–101. Available at https://doi.org/10.1016/j.desal.2012.07.010.

Kim, J., Baik, S.S., Ryu, S.H., Sohn, Y., Park, S., Park, B.G., et al., 2015. Observation of tunable band gap and anisotropic Dirac semimetal state in black phosphorus. Science (New York, N.Y.) 349 (6249), 723–726. Available at https://doi.org/10.1126/science.aaa6486, http://www.sciencemag.org/content/349/6249/723.full.pdf.

Kitchin, J.R., 2018. Machine learning in catalysis. Nature Catalysis 1 (4), 230–232. Available at https://doi.org/10.1038/s41929-018-0056-y, https://www.nature.com/natcatal/.

Klinkova, Anna, Choueiri, Rachelle M., Kumacheva, Eugenia, 2014. Self-assembled plasmonic nanostructures. Chemical Society Reviews 43 (11), 3976. Available at https://doi.org/10.1039/c3cs60341e.

Krasian, T., Punyodom, W., Worajittiphon, P., 2019. A hybrid of 2D materials (MoS2 and WS2) as an effective performance enhancer for poly(lactic acid) fibrous mats in oil adsorption and oil/water separation. Chemical Engineering Journal 369, 563–575. Available at https://doi.org/10.1016/j.cej.2019.03.092, www.elsevier.com/inca/publications/store/6/0/1/2/7/3/index.htt.

Krasner, S.W., Weinberg, H.S., Richardson, S.D., Pastor, S.J., Chinn, R., Sclimenti, M.J., et al., 2006. Occurrence of a new generation of disinfection byproducts. Environmental Science and Technology 40 (23), 7175–7185. Available at https://doi.org/10.1021/es060353j.

Krishnamoorthy, K., Veerapandian, M., Yun, K., Kim, S.J., 2013. New function of molybdenum trioxide nanoplates: toxicity towards pathogenic bacteria through membrane stress. Colloids and Surfaces B: Biointerfaces 112, 521–524. Available at https://doi.org/10.1016/j.colsurfb.2013.08.026.

Kuang, P., Sayed, M., Fan, J., Cheng, B., Yu, J., 2020. 3D Graphene-based H2-production photocatalyst and electrocatalyst. Advanced Energy Materials 10 (14). Available at https://doi.org/10.1002/aenm.201903802, http://onlinelibrary.wiley.com/journal/10.1002/(ISSN)1614-6840.

Kullab, A., Martin, A., 2011. Membrane distillation and applications for water purification in thermal cogeneration plants. Separation and Purification Technology 76 (3), 231–237. Available at https://doi.org/10.1016/j.seppur.2010.09.028.

Kärkäs, M.D., Verho, O., Johnston, E.V., Åkermark, B., 2014. Artificial photosynthesis: molecular systems for catalytic water oxidation. Chemical Reviews 114 (24), 11863–12001. Available at https://doi.org/10.1021/cr400572f, http://pubs.acs.org/journal/chreay.

Lam, S.M., Quek, J.A., Sin, J.C., 2018. Mechanistic investigation of visible light responsive Ag/ZnO micro/nanoflowers for enhanced photocatalytic performance and antibacterial activity. Journal of Photochemistry and Photobiology A: Chemistry 353, 171–184. Available at https://doi.org/10.1016/j.jphotochem.2017.11.021, http://www.elsevier.com/locate/jphotochem.

Leaper, Sebastian, Abdel-Karim, Ahmed, Gorgojo, Patricia, 2021. The use of carbon nanomaterials in membrane distillation membranes: a review. Frontiers of Chemical Science and Engineering 15 (4), 755–774. Available at https://doi.org/10.1007/s11705-020-1993-y.

Lei, W., Liu, D., Chen, Y., 2015. Highly crumpled boron nitride nanosheets as adsorbents: scalable solvent-less production. Advanced Materials Interfaces 2 (3). Available at https://doi.org/10.1002/admi.201400529.

Lei, Shijun, Zeng, Minxiang, Huang, Dali, Wang, Ling, Zhang, Lecheng, Xi, Beidou, Ma, Wenchao, Chen, Guanyi, Cheng, Zhengdong, 2019. Synergistic High-flux Oil–Saltwater Separation and Membrane Desalination with Carbon Quantum Dots Functionalized Membrane. ACS Sustainable Chemistry & Engineering 7 (16), 13708–13716. Available at https://doi.org/10.1021/acssuschemeng.9b01073.

Lere Keshebo, Degu, Hu, Chien-Chieh, Kassa, Shewaye Temesgen, Hung, Wei-Song, Feng Wang, Chih, Huang, Shu-Hsien, Lee, Kueir-Rarn, Yih Lai, Juin, 2020. Solvent-Exfoliated 2D WS_2/Polyethersulfone Antifouling Mixed Matrix Ultrafiltration Membrane for Water Treatment. ACS Applied Polymer Materials 2 (11), 5039–5047. Available at https://doi.org/10.1021/acsapm.0c00871.

Li, B., Wang, Q., Zou, R., Liu, X., Xu, K., Li, W., et al., 2014. Cu7.2S4 nanocrystals: a novel photothermal agent with a 56.7% photothermal conversion efficiency for photothermal therapy of cancer cells. Nanoscale 6 (6), 3274–3282. Available at https://doi.org/10.1039/c3nr06242b, http://pubs.rsc.org/en/journals/journal/nr.

Li, Xiuqiang, Xu, Weichao, Tang, Mingyao, Zhou, Lin, Zhu, Bin, Zhu, Shining, Zhu, Jia, 2016. Graphene oxide-based efficient and scalable solar desalination under one sun with a confined 2D water path. Proceedings of the National Academy of Sciences 113 (49), 13953–13958. Available at https://doi.org/10.1073/pnas.1613031113.

Li, F., Tang, M., Li, T., Zhang, L., Hu, C., 2020. Two-dimensional graphene/g-C3N4 in-plane hybrid heterostructure for enhanced photocatalytic activity with surface-adsorbed pollutants assistant. Applied Catalysis B: Environment and Energy 268, 118397. Available at https://doi.org/10.1016/j.apcatb.2019.118397.

Li, Zhengtong, Xu, Xingtao, Sheng, Xinran, Lin, Peng, Tang, Jing, Pan, Likun, Kaneti, Yusuf Valentino, Yang, Tao, Yamauchi, Yusuke, 2021. Solar-Powered Sustainable Water Production: State-of-the-Art Technologies for Sunlight–Energy–Water Nexus. ACS Nano 15 (8), 12535–12566. Available at https://doi.org/10.1021/acsnano.1c01590.

Li, J.J., Wang, C C., Fu, H.F., Cui, J.R., Xu, P., Guo, J., et al., 2017a. High-performance adsorption and separation of anionic dyes in water using a chemically stable graphene-like metal-organic framework. Dalton Transactions 46 (31), 10197–10201. Available at https://doi.org/10.1039/c7dt02208e, http://www.rsc.org/Publishing/Journals.

Li, Yi, Yang, Shishi, Zhang, Kaisong, Van der Bruggen, Bart, 2019. Thin film nanocomposite reverse osmosis membrane modified by two dimensional laminar MoS2 with improved desalination performance and fouling-resistant characteristics. Desalination 454, 48–58. Available at https://doi.org/10.1016/j.desal.2018.12.016.

Li, J., Lin, J., Xu, X., Zhang, X., Xue, Y., Mi, J., et al., 2013. Porous boron nitride with a high surface area: hydrogen storage and water treatment. Nanotechnology 24 (15), 155603. Available at https://doi.org/10.1088/0957-4484/24/15/155603.

Li, Jie, Lin, Jing, Xu, Xuewen, Zhang, Xinghua, Xue, Yanming, Mi, Jiao, Mo, Zhaojun, Fan, Ying, Hu, Long, Yang, Xiaojing, Zhang, Jun, Meng, Fanbin, Yuan, Songdong, Tang, Chengchun, 2013. Porous boron nitride with a high surface area: hydrogen storage and water treatment. Nanotechnology 24 (15), 155603. Available at https://doi.org/10.1088/0957-4484/24/15/155603.

Li, Q., Mahendra, S., Lyon, D.Y., Brunet, L., Liga, M.V., Li, D., et al., 2008. Antimicrobial nanomaterials for water disinfection and microbial control: potential applications and implications. Water Research 42 (18), 4591–4602. Available at https://doi.org/10.1016/j.watres.2008.08.015, www.elsevier.com/locate/watres.

Li, S., Zhao, Z., Yu, D., Zhao, J.Z., Su, Y., Liu, Y., et al., 2019a. Few-layer transition metal dichalcogenides (MoS2, WS2, and WSe2) for water splitting and degradation of organic pollutants: Understanding the piezocatalytic effect. Nano Energy 66. Available at https://doi.org/10.1016/j.nanoen.2019.104083, http://www.journals.elsevier.com/nano-energy/.

Li, Y., Ding, Z., Zhang, X., Li, J., Liu, X., Lu, T., et al., 2019b. Novel hybrid capacitive deionization constructed by a redox-active covalent organic framework and its derived porous carbon for highly efficient desalination. Journal of Materials Chemistry A 7 (44), 25305–25313. Available at https://doi.org/10.1039/c9ta07344b, http://pubs.rsc.org/en/journals/journal/ta.

Li, W., Yang, Y., Weber, J.K., Zhang, G., Zhou, R., 2016a. Tunable, strain-controlled nanoporous MoS2 filter for water desalination. ACS Nano 10 (2), 1829–1835. Available at https://doi.org/10.1021/acsnano.5b05250, http://pubs.acs.org/journal/ancac3.

Li, X., Bi, W., Zhang, L., Tao, S., Chu, W., Zhang, Q., et al., 2016c. Single-atom Pt as co-catalyst for enhanced photocatalytic H2 evolution. Advanced Materials 28 (12), 2427–2431. Available at https://doi.org/10.1002/adma.201505281, http://www3.interscience.wiley.com/journal/119030556/issue.

Li, Xu-Bing, Gao, Yu-Ji, Wang, Yang, Zhan, Fei, Zhang, Xiao-Yi, Kong, Qing-Yu, Zhao, Ning-Jiu, Guo, Qing, Wu, Hao-Lin, Li, Zhi-Jun, Tao, Ye, Zhang, Jian-Ping, Chen, Bin, Tung, Chen-Ho, Wu, Li-Zhu, 2017. Self-Assembled Framework Enhances Electronic Communication of Ultrasmall-Sized Nanoparticles for Exceptional Solar Hydrogen Evolution. Journal of the American Chemical Society 139 (13), 4789–4796. Available at https://doi.org/10.1021/jacs.6b12976.

Lin, H., Wang, X., Yu, L., Chen, Y., Shi, J., 2017b. Two-dimensional ultrathin MXene ceramic nanosheets for photothermal conversion. Nano Letters 17 (1), 384–391. Available at https://doi.org/10.1021/acs.nanolett.6b04339, http://pubs.acs.org/journal/nalefd.

Lin, R., Villacorta Hernandez, B., Ge, L., Zhu, Z., 2018b. Metal organic framework based mixed matrix membranes: an overview on filler/polymer interfaces. Journal of Materials Chemistry A 6 (2), 293–312. Available at https://doi.org/10.1039/c7ta07294e, http://pubs.rsc.org/en/journals/journal/ta.

Linic, S., Aslam, U., Boerigter, C., Morabito, M., 2015. Photochemical transformations on plasmonic metal nanoparticles. Nature Materials 14 (6), 567–576. Available at https://doi.org/10.1038/nmat4281, http://www.nature com/nmat/.

Liu, Shaobin, Zeng, Tingying Helen, Hofmann, Mario, Burcombe, Ehdi, Wei, Jun, Jiang, Rongrong, Kong, Jing, Chen, Yuan, 2011. Antibacterial Activity of Graphite, Graphite Oxide, Graphene Oxide, and Reduced Graphene Oxide: Membrane and Oxidative Stress. ACS Nano 5 (9), 6971–6980. Available at https://doi.org/10.1021/nn202451x.

Liu, A., Wang, C.Z., Chu, C., Chu, H.Y., Chen, X., Du, A.F., et al., 2018a. Adsorption performance toward organic pollutants, odour control and anti-microbial activities of one Ag-based coordination polymer. Journal of Environmental Chemical Engineering 6 (4), 4961–4969. Available at https://doi.org/10.1016/j.jece.2018.07.035, http://www.journals.elsevier.com/journal-of-environmental-chemical-engineering/.

Liu, Z., Fang, Y., Jia, H., Wang, C., Song, Q., Li, L., et al., 2018b. Novel multifunctional cheese-like 3D carbon-BN as a highly efficient adsorbent for water purification. Scientific Reports 8 (1). Available at https://doi.org/10.1038/s41598-018-19541-5, www.nature.com/srep/index.html.

Zhao, J., Yang, Y., Yang, C., Tian, Y., Han, Y., Liu, J., Yin, X., Que, W., et al., 2018. A hydrophobic surface enabled salt-blocking 2D Ti3C2 MXene membrane for efficient and stable solar desalination. Journal of Materials Chemistry A 6 (33), 16196–16204. Available at https://doi.org/10.1016/j.memsci.2017.11.065https://doi.org/10.1039/C8TA05569F.

Liu, Kong, D., Hsu, P.C., Yuan, H., Lee, H.W., Liu, Y., et al., 2016a. Rapid water disinfection using vertically aligned MoS2 nanofilms and visible light. Nature Nanotechnology 11 (12), 1098–1104. Available at https://doi.org/10.1038/nnano.2016.138, http://www.nature.com/nnano/index.html.

Liu, G., Jin, W., Xu, N., 2016b. Two-dimensional-material membranes: a new family of high-performance separation membranes. Angewandte Chemie - International Edition 55 (43), 13384–13397. Available at https://doi.org/10.1002/anie.201600438, http://onlinelibrary.wiley.com/journal/10.1002/(ISSN)1521-3773.

Liu, Guohua, Xu, Jinliang, Wang, Kaiying, 2017. Solar water evaporation by black photothermal sheets. Nano Energy 41, 269–284. Available at https://doi.org/10.1016/j.nanoen.2017.09.005.

Liu, H.Y., Liang, C., Niu, C.G., Huang, D.W., Du, Y.B., Guo, H., et al., 2019. Facile assembly of g-C 3 N 4 /Ag 2 CO 3 /graphene oxide with a novel dual Z-scheme system for enhanced photocatalytic pollutant degradation. Applied Surface Science 475, 421–434. Available at https://doi.org/10.1016/j.apsusc.2019.01.018, http://www.journals.elsevier.com/applied-surface-science/.

Liu, X., Chen, T., Chu, H., Niu, L., Sun, Z., Pan, L., et al., 2015b. Fe2O3-reduced graphene oxide composites synthesized via microwave-assisted method for sodium ion batteries. Electrochimica Acta 166, 12–16. Available at https://doi.org/10.1016/j.electacta.2015.03.081, http://www.journals.elsevier.com/electrochimica-acta/.

Liu, Q., Wu, D., Zhou, Y., Su, H., Wang, R., Zhang, C., et al., 2014a. Single-crystalline, ultrathin ZnGa2O4 nanosheet scaffolds to promote photocatalytic activity in CO2 reduction into methane. ACS Applied Materials and Interfaces 6 (4), 2356–2361. Available at https://doi.org/10.1021/am404572g.

Liu, Z., Ma, W., Li, H., 2020. Elucidating the capacitive desalination behavior of Na:XCoO2: the significance of electrochemical pre-activation. Nanoscale 12 (14), 7586–7594. Available at https://doi.org/10.1039/d0nr00248h, http://pubs.rsc.org/en/journals/journal/nr.

Loeb, S.K., Alvarez, P.J.J., Brame, J.A., Cates, E.L., Choi, W., Crittenden, J., et al., 2019. The technology horizon for photocatalytic water treatment: sunrise or sunset? Environmental Science and Technology 53 (6), 2937–2947. Available at https://doi.org/10.1021/acs.est.8b05041, http://pubs.acs.org/journal/esthag.

Lu, B., Xia, Y., Ren, Y., Xie, M., Zhou, L., Vinai, G., et al., 2024. When machine learning meets 2D materials: a review. Advanced Science 11 (13). Available at https://doi.org/10.1002/advs.202305277, http://onlinelibrary.wiley.com/journal/10.1002/(ISSN)2198-3844.

Lu, S., Zhou, Q., Ouyang, Y., Guo, Y., Li, Q., Wang, J., 2018. Accelerated discovery of stable lead-free hybrid organic-inorganic perovskites via machine learning. Nature Communications 9 (1). Available at https://doi.org/10.1038/s41467-018-05761-w, http://www.nature.com/ncomms/index.html.

Luo, B., Liu, G., Wang, L., 2016. Recent advances in 2D materials for photocatalysis. Nanoscale 8 (13), 6904–6920. Available at https://doi.org/10.1039/c6nr00546b, http://pubs.rsc.org/en/journals/journal/nr.

Luo, S., Dong, S., Lu, C., Yu, C., Ou, Y., Luo, L., et al., 2018. Rational and green synthesis of novel two-dimensional WS2/MoS2 heterojunction via direct exfoliation in ethanol-water targeting advanced visible-light-responsive photocatalytic performance. Journal of Colloid and Interface Science 513, 389–399. Available at https://doi.org/10.1016/j.jcis.2017.11.044.

Ma, S., Zhan, S., Xia, Y., Wang, P., Hou, Q., Zhou, Q., 2019. Enhanced photocatalytic bactericidal performance and mechanism with novel Ag/ZnO/g-C 3 N 4 composite under visible light. Catalysis Today 330, 179–188. Available at https://doi.org/10.1016/j.cattod.2018.04.014, http://www.sciencedirect.com/science/journal/09205861.

Ma, S., Qarony, W., Hossain, M.I., Yip, C.T., Tsang, Y.H., 2019. Metal-organic framework derived porous carbon of light trapping structures for efficient solar steam generation. Solar Energy Materials and Solar Cells 196, 36–42. Available at https://doi.org/10.1016/j.solmat.2019.02.035, http://www.sciencedirect.com/science/journal/09270248/100.

María, A., Id, D.-P., 2018. Antibacterial activity of nanomaterials. Nanomaterials 2018 8, 359.

Masimukku, S., Hu, Y.C., Lin, Z.H., Chan, S.W., Chou, T.M., Wu, J.M., 2018. High efficient degradation of dye molecules by PDMS embedded abundant single-layer tungsten disulfide and their antibacterial performance. Nano Energy 46, 338–346. Available at https://doi.org/10.1016/j.nanoen.2018.02.008, http://www.journals.elsevier.com/nano-energy/.

Massey, A.T., Gusain, R., Kumari, S., Khatri, O.P., 2016. Hierarchical microspheres of MoS2 nanosheets: efficient and regenerative adsorbent for removal of water-soluble dyes. Industrial and Engineering Chemistry Research 55 (26), 7124–7131. Available at https://doi.org/10.1021/acs.iecr.6b01115, http://pubs.acs.org/journal/iecred.

Masubuchi, S., Morimoto, M., Morikawa, S., Onodera, M., Asakawa, Y., Watanabe, K., et al., 2018. Autonomous robotic searching and assembly of two-dimensional crystals to build van der Waals superlattices. Nature Communications 9 (1). Available at https://doi.org/10.1038/s41467-018-03723-w, http://www.nature.com/ncomms/index.html.

Mauter, M.S., Zucker, I., Perreault, F., Werber, J.R., Kim, J.H., Elimelech, M., 2018. The role of nanotechnology in tackling global water challenges. Nature Sustainability 1 (4), 166–175. Available at https://doi.org/10.1038/s41893-018-0046-8, www.nature.com/natsustain/.

Mi, B., 2014. Graphene oxide membranes for ionic and molecular sieving. Science (New York, N.Y.) 343 (6172), 740–742. Available at https://doi.org/10.1126/science.1250247.

Miao, H., Teng, Z., Wang, C., Chong, H., Wang, G., 2019. Recent progress in two-dimensional antimicrobial nanomaterials. Chemistry - A European Journal 25 (4), 929–944. Available at https://doi.org/10.1002/chem.201801983, http://onlinelibrary.wiley.com/journal/10.1002/(ISSN)1521-3765.

MiarAlipour, S., Friedmann, D., Scott, J., Amal, R., 2018. TiO2/porous adsorbents: recent advances and novel applications. Journal of Hazardous Materials 341, 404–423. Available at https://doi.org/10.1016/j.jhazmat.2017.07.070, www.elsevier.com/locate/jhazmat.

Moaty, S.A.A., Farghali, A.A., Khaled, R., 2016. Preparation, characterization and antimicrobial applications of Zn-Fe LDH against MRSA. Materials Science and Engineering C 68, 184–193. Available at https://doi.org/10.1016/j.msec.2016.05.110.

Mohamed, F., Abukhadra, M.R., Shaban, M., 2018. Removal of safranin dye from water using polypyrrole nanofiber/Zn-Fe layered double hydroxide nanocomposite (Ppy NF/Zn-Fe LDH) of enhanced adsorption and photocatalytic properties. Science of the Total Environment 640-641, 352–363. Available at https://doi.org/10.1016/j.scitotenv.2018.05.316, www.elsevier.com/locate/scitotenv.

Moreira, N.F.F., Sampaio, M.J., Ribeiro, A.R., Silva, C.G., Faria, J.L., Silva, A.M.T., 2019. Metal-free g-C3N4 photocatalysis of organic micropollutants in urban wastewater under visible light. Applied Catalysis B: Environmental 248, 184–192. Available at https://doi.org/10.1016/j.apcatb.2019.02.001, www.elsevier.com/inca/publications/store/5/2/3/0/6/6/index.htt.

Mullangi, D., Shalini, S., Nandi, S., Choksi, B., Vaidhyanathan, R., 2017. Super-hydrophobic covalent organic frameworks for chemical resistant coatings and hydrophobic paper and textile composites. Journal of Materials Chemistry A 5 (18), 8376–8384. Available at https://doi.org/10.1039/c7ta01302g, http://pubs.rsc.org/en/journals/journal/ta.

Mustakeem, M., El-Demellawi, J.K., Obaid, M., Ming, F., Alshareef, H.N., Ghaffour, N., 2022. MXene-coated membranes for autonomous solar-driven desalination. ACS Applied Materials and Interfaces 14 (4), 5265–5274. Available at https://doi.org/10.1021/acsami.1c20653, http://pubs.acs.org/journal/aamick.

Nandipati, M., Fatoki, O., Desai, S., 2024. Bridging nanomanufacturing and artificial intelligence—a comprehensive review. Materials 17 (7), 1621. Available at https://doi.org/10.3390/ma17071621.

Nicolaï, A., Sumpter, B.G., Meunier, V., 2014. Tunable water desalination across graphene oxide framework membranes. Physical Chemistry Chemical Physics 16 (18), 8646. Available at https://doi.org/10.1039/c4cp01051e.

Nicolosi, V., Chhowalla, M., Kanatzidis, M.G., Strano, M.S., Coleman, J.N., 2013. Liquid exfoliation of layered materials. Science (New York, N.Y.) 340 (6139). Available at https://doi.org/10.1126/science.1226419, http://www.sciencemag.org/content/340/6139/1226419.full.pdf.

Nie, L., Goh, K., Wang, Y., Lee, J., Huang, Y., Karahan, H.E.E., et al., 2020. Realizing small-flake graphene oxide membranes for ultrafast size-dependent organic solvent nanofiltration. Science Advances 6 (17). Available at https://doi.org/10.1126/sciadv.aaz9184, https://advances.sciencemag.org/content/advances/6/17/EAAZ9184.full.pdf.

O'Hern, S.C., Boutilier, M.S.H., Idrobo, J.C., Song, Y., Kong, J., Laoui, T., et al., 2014. Selective ionic transport through tunable subnanometer pores in single-layer graphene membranes. Nano Letters 14 (3), 1234–1241. Available at https://doi.org/10.1021/nl404118f.

O'Hern, S.C., Jang, D., Bose, S., Idrobo, J.C., Song, Y., Laoui, T., et al., 2015. Nanofiltration across defect-sealed nanoporous monolayer graphene. Nano Letters 15 (5), 3254–3260. Available at https://doi.org/10.1021/acs.nanolett.5b00456, http://pubs.acs.org/journal/nalefd.

Oh, Y., Armstrong, D.L., Finnerty, C., Zheng, S., Hu, M., Torrents, A., et al., 2017. Understanding the pH-responsive behavior of graphene oxide membrane in removing ions and organic micropollulants. Journal of Membrane Science 541, 235–243. Available at https://doi.org/10.1016/j.memsci.2017.07.005, www.elsevier.com/locate/memsci.

Oshima, T., Lu, D., Ishitani, O., Maeda, K., 2015. Intercalation of highly dispersed metal nanoclusters into a layered metal oxide for photocatalytic overall water splitting. Angewandte Chemie 127 (9), 2736–2740. Available at https://doi.org/10.1002/ange.201411494.

Park, Y., Kim, W., Monllor-Satoca, D., Tachikawa, T., Majima, T., Choi, W., 2012. Role of interparticle charge transfers in agglomerated photocatalyst nanoparticles: demonstration in aqueous suspension of dye-sensitized TiO2. Journal of Physical Chemistry Letters 4 (1), 189–194.

Paul, N., Zhang, L., Lei, S., Huang, D., Wang, L., Cheng, Z., et al., 2024. Ligand-directed shape reconfiguration in inorganic materials. Small 20 (4). Available at https://doi.org/10.1002/smll.202305460, http://onlinelibrary.wiley.com/journal/10.1002/(ISSN)1613-6829.

Peterson, C.H., Rice, S.D., Short, J.W., Esler, D., Bodkin, J.L., Ballachey, B.E., et al., 2003. Long-term ecosystem response to the Exxon Valdez oil spill. Science 302 (5653), 2082–2086. Available at https://doi.org/10.1126/science.1084282.

Pi, Y., Li, X., Xia, Q., Wu, J., Li, Y., Xiao, J., et al., 2018. Adsorptive and photocatalytic removal of persistent organic pollutants (POPs) in water by metal-organic frameworks (MOFs). Chemical Engineering Journal 337, 351–371. Available at https://doi.org/10.1016/j.cej.2017.12.092, www.elsevier.com/inca/publications/store/6/0/1/2/7/3/index.htt.

Prakash, A., Dan, M., Yu, S., Wei, S., Li, Y., Wang, F., et al., 2018. In2S3/CuS nanosheet composite: an excellent visible light photocatalyst for H2 production from H2S. Solar Energy Materials and Solar Cells 180, 205–212. Available at https://doi.org/10.1016/j.solmat.2018.03.011, http://www.sciencedirect.com/science/journal/09270248/100.

Prüss-Ustün, Annette, Bartram, Jamie, Clasen, Thomas, Colford, John M., Cumming, Oliver, Curtis, Valerie, Bonjour, Sophie, Dangour, Alan D., De France, Jennifer, Fewtrell, Lorna, Freeman, Matthew C., Gordon, Bruce, Hunter, Paul R., Johnston, Richard B., Mathers, Colin, Mäusezahl, Daniel, Medlicott, Kate, Neira, Maria, Stocks, Meredith, Wolf, Jennyfer, Cairncross, Sandy, 2014. Burden of disease from inadequate water, sanitation and hygiene in low- and middle-income settings: a retrospective analysis of data from 145 countries. Tropical Medicine & International Health 19 (8), 894–905. Available at https://doi.org/10.1111/tmi.12329.

Qian, B., Wang, G., Ling, Z., Dong, Q., Wu, T., Zhang, X., et al., 2015. Sulfonated graphene as cation-selective coating: a new strategy for high-performance membrane capacitive deionization. Advanced Materials Interfaces 2 (16). Available at https://doi.org/10.1002/admi.201500372, http://onlinelibrary.wiley.com/journal/10.1002/(ISSN)2196-7350.

Qtaishat, M.R., Banat, F., 2013. Desalination by solar powered membrane distillation systems. Desalination 308, 186–197. Available at https://doi.org/10.1016/j.desal.2012.01.021.

Rahman, M., Mobin, I.H.A., Alam, I., Alam, Z., Hridoy, A.H., Khan, M.M.R., et al., 2024. Conductive polymer nanocomposites-based electrode materials for supercapacitors: synthesis, characterization, and electrochemical performance. Nanostructured Materials for Energy Storage I-IV. wiley, Bangladesh, pp. 1219–1255. Available at https://www.wiley.com/en-us/Nanostructured+Materials+for+Energy+Storage%2C+4+Volumes-p-9783527838868.

Rahmanian, O., Dinari, M., Abdolmaleki, M.K., 2018. Carbon quantum dots/layered double hydroxide hybrid for fast and efficient decontamination of Cd(II): the adsorption kinetics and isotherms. Applied Surface Science 428, 272–279. Available at https://doi.org/10.1016/j.apsusc.2017.09.152, http://www.journals.elsevier.com/applied-surface-science/.

Ran, J., Zhang, J., Yu, J., Jaroniec, M., Zhang Qiao, S., 2014. Earth-abundant cocatalysts for semiconductor-based photocatalytic water splitting. Chemical Society Reviews 43 (22), 7787–7812. Available at https://doi.org/10.1039/c3cs60425j.

Rasool, K., Mahmoud, K.A., Johnson, D.J., Helal, M., Berdiyorov, G.R., Gogotsi, Y., 2017. Efficient antibacterial membrane based on two-dimensional Ti3C2Tx (MXene) nanosheets. Scientific Reports 7 (1). Available at https://doi.org/10.1038/s41598-017-01714-3, www.nature.com/srep/index.html.

Ray, S.S., Chen, S.S., Sangeetha, D., Chang, H.M., Thanh, C.N.D., Le, Q.H., et al., 2018. Developments in forward osmosis and membrane distillation for desalination of waters. Environmental Chemistry Letters 16 (4), 1247–1265. Available at https://doi.org/10.1007/s10311-018-0750-7, https://www.springer.com/journal/10311.

Sahu, S.P., Qanbarzadeh, M., Ateia, M., Torkzadeh, H., Maroli, A.S., Cates, E.L., 2018. Rapid degradation and mineralization of perfluorooctanoic acid by a new petitjeanite Bi3O(OH)(PO4)2 microparticle ultraviolet photocatalyst. Environmental Science and Technology Letters 5 (8), 533–538. Available at https://doi.org/10.1021/acs.estlett.8b00395, http://pubs.acs.org/page/estlcu/about.html.

Sarkar, D., Mondal, B., Som, A., Ravindran, S.J., Jana, S.K., Manju, C.K., et al., 2018. Holey MoS2 nanosheets with photocatalytic metal rich edges by ambient electrospray deposition for solar water disinfection. Global Challenges 2 (12). Available at https://doi.org/10.1002/gch2.201800052, https://onlinelibrary.wiley.com/doi/epdf/10.1002/gch2.201800052.

Sarker, M., Song, J.Y., Jhung, S.H., 2018. Adsorptive removal of anti-inflammatory drugs from water using graphene oxide/metal-organic framework composites. Chemical Engineering Journal 335, 74–81. Available at https://doi.org/10.1016/j.cej.2017.10.138, www.elsevier.com/inca/publications/store/6/0/1/2/7/3/index.htt.

Seraj, S., Mohammadi, T., Tofighy, M.A., 2022. Graphene-based membranes for membrane distillation applications: a review. Journal of Environmental Chemical Engineering 10 (3). Available at https://doi.org/10.1016/j.jece.2022.107974, http://www.journals.elsevier.com/journal-of-environmental-chemical-engineering/.

Shahzad, A., Rasool, K., Nawaz, M., Miran, W., Jang, J., Moztahida, M., et al., 2018. Heterostructural TiO2/Ti3C2Tx (MXene) for photocatalytic degradation of antiepileptic drug carbamazepine. Chemical Engineering Journal 349, 748–755. Available at https://doi.org/10.1016/j.cej.2018.05.148, www.elsevier.com/inca/publications/store/6/0/1/2/7/3/index.htt.

Shannon, Mark A., Bohn, Paul W., Elimelech, Menachem, Georgiadis, John G., Mariñas, Benito J., Mayes, Anne M., 2008. Science and technology for water purification in the coming decades. Nature 452 (7185), 301–310. Available at https://doi.org/10.1038/nature06599.

Shehzad, K., Xu, Y., Gao, C., Duan, X., 2016. Three-dimensional macro-structures of two-dimensional nanomaterials. Chemical Society Reviews 45 (20), 5541–5588. Available at https://doi.org/10.1039/c6cs00218h, http://pubs.rsc.org/en/journals/journal/cs.

Sheng, Y., Wei, Z., Miao, H., Yao, W., Li, H., Zhu, Y., 2019a. Enhanced organic pollutant photodegradation via adsorption/photocatalysis synergy using a 3D g-C3N4/TiO2 free-separation photocatalyst. Chemical Engineering Journal 370, 287–294. Available at https://doi.org/10.1016/j.cej.2019.03.197, www.elsevier.com/inca/publications/store/6/0/1/2/7/3/index.htt.

Sheng, Y., Yang, J., Wang, F., Liu, L., Liu, H., Yan, C., et al., 2019b. Sol-gel synthesized hexagonal boron nitride/titania nanocomposites with enhanced photocatalytic activity. Applied Surface Science 465, 154–163. Available at https://doi.org/10.1016/j.apsusc.2018.09.137, http://www.journals.elsevier.com/applied-surface-science/.

Shi, X., Xiao, A., Zhang, C., Wang, Y., 2019. Growing covalent organic frameworks on porous substrates for molecule-sieving membranes with pores tunable from ultra- to nanofiltration. Journal of Membrane Science 576, 116–122. Available at https://doi.org/10.1016/j.memsci.2019.01.034, www.elsevier.com/locate/memsci.

Snyder, S.A., Westerhoff, P., Yoon, Y., Sedlak, D.L., 2003. Pharmaceuticals, personal care products, and endocrine disruptors in water: implications for the water industry. Environmental Engineering Science 20 (5), 449–469. Available at https://doi.org/10.1089/109287503768335931, www.liebertonline.com/ees.

Song, Jun-ho, Yu, Hye-Weon, Ham, Moon-Ho, Kim, In S., 2018. Tunable Ion Sieving of Graphene Membranes through the Control of Nitrogen-Bonding Configuration. Nano Letters 18 (9), 5506–5513. Available at https://doi.org/10.1021/acs.nanolett.8b01904.

Stafford, J., Patapas, A., Uzo, N., Matar, O.K., Petit, C., 2018. Towards scale-up of graphene production via nonoxidizing liquid exfoliation methods. AIChE Journal 64 (9), 3246–3276. Available at https://doi.org/10.1002/aic.16174, http://onlinelibrary.wiley.com/journal/10.1002/(ISSN)1547-5905.

Su, H., Zhou, J., Miao, L., Shi, J., Gu, Y., Wang, P., et al., 2019. A hybrid hydrogel with protonated g-C3N4 and graphene oxide as an efficient absorber for solar steam evaporation. Sustainable Materials and Technologies 20, e00095. Available at https://doi.org/10.1016/j.susmat.2019.e00095.

Su, T., Shao, Q., Qin, Z., Guo, Z., Wu, Z., 2018. Role of interfaces in two-dimensional photocatalyst for water splitting. ACS Catalysis 8 (3), 2253–2276. Available at https://doi.org/10.1021/acscatal.7b03437, http://pubs.acs.org/page/accacs/about.html.

Sun, J., Qian, X., Wang, Z., Zeng, F., Bai, H., Li, N., 2020. Tailoring the microstructure of poly(vinyl alcohol)-intercalated graphene oxide membranes for enhanced desalination performance of high-salinity water by pervaporation. Journal of Membrane Science 599, 117838. Available at https://doi.org/10.1016/j.memsci.2020.117838.

Sun, L., Ying, Y., Huang, H., Song, Z., Mao, Y., Xu, Z., et al., 2014. Ultrafast molecule separation through layered WS2 nanosheet membranes. ACS Nano 8 (6), 6304–6311. Available at https://doi.org/10.1021/nn501786m, http://pubs.acs.org/journal/ancac3.

Sun, Y., Gao, S., Lei, F., Xie, Y., 2015a. Atomically-thin two-dimensional sheets for understanding active sites in catalysis. Chemical Society Reviews 44 (3), 623–636. Available at https://doi.org/10.1039/c4cs00236a, http://pubs.rsc.org/en/journals/journal/cs.

Sun, Z., Xie, H., Tang, S., Yu, X.F., Guo, Z., Shao, J., et al., 2015b. Ultrasmall black phosphorus quantum dots: synthesis and use as photothermal agents. Angewandte Chemie - International Edition 54 (39), 11526–11530. Available at https://doi.org/10.1002/anie.201506154, http://onlinelibrary.wiley.com/journal/10.1002/(ISSN)1521-3773.

Surwade, S.P., Smirnov, S.N., Vlassiouk, I.V., Unocic, R.R., Veith, G.M., Dai, S., et al., 2015. Water desalination using nanoporous single-layer graphene. Nature Nanotechnology 10 (5), 459–464. Available at https://doi.org/10.1038/nnano.2015.37, http://www.nature.com/nnano/index.html.

Surwade, Sumedh P., Smirnov, Sergei N., Vlassiouk, Ivan V., Unocic, Raymond R., Veith, Gabriel M., Dai, Sheng, Mahurin, Shannon M., 2015. Water desalination using nanoporous single-layer graphene. Nature Nanotechnology 10 (5), 459–464. Available at https://doi.org/10.1038/nnano.2015.37.

Suss, M.E., Porada, S., Sun, X., Biesheuvel, P.M., Yoon, J., Presser, V., 2015. Water desalination via capacitive deionization: what is it and what can we expect from it? Energy & Environmental Science 8 (8), 2296–2319. Available at https://doi.org/10.1039/C5EE00519A.

Sweetman, M., May, S., Mebberson, N., Pendleton, P., Vasilev, K., Plush, S., et al., 2017. Activated carbon, carbon nanotubes and graphene: materials and composites for advanced water purification. C 3 (2), 18. Available at https://doi.org/10.3390/c3020018.

Tan, C., Cao, X., Wu, X.J., He, Q., Yang, J., Zhang, X., et al., 2017a. Recent advances in ultrathin two-dimensional nanomaterials. Chemical Reviews 117 (9), 6225–6331. Available at https://doi.org/10.1021/acs.chemrev.6b00558, http://pubs.acs.org/journal/chreay.

Tan, L., Li, J., Liu, X., Cui, Z., Yang, X., Yeung, K.W.K., et al., 2018. In situ disinfection through photoinspired radical oxygen species storage and thermal-triggered release from black phosphorous with strengthened chemical stability. Small (Weinheim an der Bergstrasse, Germany) 14 (9). Available at https://doi.org/10.1002/smll.201703197, http://onlinelibrary.wiley.com/journal/10.1002/(ISSN)1613-6829.

Tang, Q., Cao, S., Ma, T., Xiang, X., Luo, H., Borovskikh, P., et al., 2021. Engineering biofunctional enzyme-mimics for catalytic therapeutics and diagnostics. Advanced Functional Materials 31 (7). Available at https://doi.org/10.1002/adfm.202007475, http://onlinelibrary.wiley.com/journal/10.1002/(ISSN)1616-3028.

Tian, G., Liu, L., Meng, Q., Cao, B., 2014. Preparation and characterization of cross-linked quaternised polyvinyl alcohol membrane/activated carbon composite electrode for membrane capacitive deionization. Desalination 354, 107–115. Available at https://doi.org/10.1016/j.desal.2014.09.024.

Tian, J., Hao, P., Wei, N., Cui, H., Liu, H., 2015. 3D Bi2MoO6 nanosheet/TiO2 nanobelt heterostructure: enhanced photocatalytic activities and photoelectochemistry performance. ACS Catalysis 5 (8), 4530–4536. Available at https://doi.org/10.1021/acscatal.5b00560, http://pubs.acs.org/page/accacs/about.html.

Tian, J., Huang, X., Wu, W., 2020. Graphene-based stand-alone networks for efficient solar steam generation. Industrial and Engineering Chemistry Research 59 (3), 1135–1141. Available at https://doi.org/10.1021/acs.iecr.9b03523, http://pubs.acs.org/journal/iecred.

Tian, Y., Chowdhury, I., 2019. Adsorption of selected pharmaceutical and personal care products with molybdenum disulfide and tungsten disulfide nanomaterials. Environmental Engineering Science 36 (3), 305–315. Available at https://doi.org/10.1089/ees.2018.0313, www.liebertonline.com/ees.

Tijing, L.D., Dizon, J.R.C., Ibrahim, I., Nisay, A.R.N., Shon, H.K., Advincula, R.C., 2020. 3D printing for membrane separation, desalination and water treatment. Applied Materials Today 18. Available at https://doi.org/10.1016/j.apmt.2019.100486, http://www.journals.elsevier.com/applied-materials-today/.

Tran, Q.T., Mikhailova, I.A., 2025. Cost-effective graphene-based 2D solar evaporator for high-efficiency water purification and desalination. Solar Energy 298, 113736. Available at https://doi.org/10.1016/j.solener.2025.113736.

Turchi, C.S., Ollis, D.F., 1990. Photocatalytic degradation of organic water contaminants: mechanisms involving hydroxyl radical attack. Journal of Catalysis 122 (1), 178–192. Available at https://doi.org/10.1016/0021-9517(90)90269-P.

Upadhyay, R.K., Soin, N., Roy, S.S., 2014. Role of graphene/metal oxide composites as photocatalysts, adsorbents and disinfectants in water treatment: a review. RSC Advances 4 (8), 3823–3851. Available at https://doi.org/10.1039/c3ra45013a.

Varrla, E., Backes, C., Paton, K.R., Harvey, A., Gholamvand, Z., McCauley, J., et al., 2015. Large-scale production of size-controlled MoS2 nanosheets by shear exfoliation. Chemistry of Materials 27 (3), 1129–1139. Available at https://doi.org/10.1021/cm5044864, http://pubs.acs.org/journal/cmatex.

Vatansever, F., de Melo, W.C.M.A., Avci, P., Vecchio, D., Sadasivam, M., Gupta, A., et al., 2013. Antimicrobial strategies centered around reactive oxygen species - bactericidal antibiotics, photodynamic therapy, and beyond. FEMS Microbiology Reviews 37 (6), 955–989. Available at https://doi.org/10.1111/1574-6976.12026, http://femsre.oxfordjournals.org/.

Vi, T.T.T., Kumar, S.R., Rout, B., Liu, C.H., Wong, C.B., Chang, C.W., et al., 2018. The preparation of graphene oxide-silver nanocomposites: the effect of silver loads on gram-positive and gram-negative antibacterial activities. Nanomaterials 8 (3). Available at https://doi.org/10.3390/nano8030163, http://www.mdpi.com/2079-4991/8/3/163/pdf.

Wang, Zhongying, Mi, Baoxia, 2017. Environmental Applications of 2D Molybdenum Disulfide (MoS_2) Nanosheets. Environmental Science & Technology 51 (15), 8229–8244. Available at https://doi.org/10.1021/acs.est.7b01466.

Wang, D., Zhang, Y., Jin, B., Wang, Y., Zhang, M., 2017. Quasi-distributed optical fiber sensor for liquid-level measurement. IEEE Photonics Journal 9 (6). Available at https://doi.org/10.1109/JPHOT.2017.2776245, http://www.ieee.org.

Wang, Jing, Wang, Yuanming, Zhu, Junyong, Zhang, Yatao, Liu, Jindun, Van der Bruggen, Bart, 2017. Construction of TiO2@graphene oxide incorporated antifouling nanofiltration membrane with elevated filtration performance. Journal of Membrane Science 533, 279–288. Available at https://doi.org/10.1016/j.memsci.2017.03.040.

Wang, X., Xiang, Y., Zhou, B., Zhang, Y., Wu, J., Hu, R., et al., 2019b. Enhanced photocatalytic performance of Ag/TiO2 nanohybrid sensitized by black phosphorus nanosheets in visible and near-infrared light. Journal of Colloid and Interface Science 534, 1–11. Available at https://doi.org/10.1016/j.jcis.2018.09.013, http://www.elsevier.com/inca/publications/store/6/2/2/8/6/1/index.htt.

Wang, H., Zhang, L., Chen, Z., Hu, J., Li, S., Wang, Z., et al., 2014a. Semiconductor heterojunction photocatalysts: Design, construction, and photocatalytic performances. Chemical Society Reviews 43 (15), 5234–5244. Available at https://doi.org/10.1039/c4cs00126e, http://pubs.rsc.org/en/journals/journal/cs.

Wang, H., Su, Y., Zhao, H., Yu, H., Chen, S., Zhang, Y., et al., 2014b. Photocatalytic oxidation of aqueous ammonia using atomic single layer graphitic-C 3 N 4. Environmental Science & Technology 48 (20), 11984–11990. Available at https://doi.org/10.1021/es503073z.

Wang, X., Ding, J., Yao, S., Wu, X., Feng, Q., Wang, Z., et al., 2014c. High supercapacitor and adsorption behaviors of flower-like MoS 2 nanostructures. Journal of Materials Chemistry A 2 (38), 15958–15963. Available at https://doi.org/10.1039/c4ta03044c.

Wang, M., Zhang, J., Wang, P., Li, C., Xu, X., Jin, Y., 2018a. Bifunctional plasmonic colloidosome/graphene oxide-based floating membranes for recyclable high-efficiency solar-driven clean water generation. Nano Research 11 (7), 3854–3863. Available at https://doi.org/10.1007/s12274-017-1959-7, http://www.springer.com/materials/nanotechnology/journal/12274.

Wang, S., Niu, H., Cai, Y., Cao, D., 2018b. Multifunctional Au NPs-polydopamine-polyvinylidene fluoride membrane chips as probe for enrichment and rapid detection of organic contaminants. Talanta 181, 340–345. Available at https://doi.org/10.1016/j.talanta.2018.01.038, https://www.journals.elsevier.com/talanta.

Wang, Y., Li, L., Wei, Y., Xue, J., Chen, H., Ding, L., et al., 2017a. Water transport with ultralow friction through partially exfoliated g-C3N4 nanosheet membranes with self-supporting spacers. Angewandte Chemie - International Edition 56 (31), 8974–8980. Available at https://doi.org/10.1002/anie.201701288, http://onlinelibrary.wiley.com/journal/10.1002/(ISSN)1521-3773.

Wang, Z., Mi, B., 2017b. Environmental applications of 2D molybdenum disulfide (MoS2) nanosheets. Environmental Science and Technology 51 (15), 8229–8244. Available at https://doi.org/10.1021/acs.est.7b01466, http://pubs.acs.org/journal/esthag.

Wang, Y., Cao, Z., Barati Farimani, A., 2021. Efficient water desalination with graphene nanopores obtained using artificial intelligence. npj 2D Materials and Applications 5 (1). Available at https://doi.org/10.1038/s41699-021-00246-9.

Wei, G., Quan, X., Chen, S., Yu, H., 2017. Superpermeable atomic-thin graphene membranes with high selectivity. ACS Nano 11 (2), 1920–1926. Available at https://doi.org/10.1021/acsnano.6b08000, http://pubs.acs.org/journal/ancac3.

Weng, J.-ze, Wang, S.-yong, Zhang, P.-xin, Li, C.-ping, Wang, G., 2021. A review of metal-organic framework-derived carbon electrode materials for capacitive deionization. New Carbon Materials 36 (1), 117–132. Available at https://doi.org/10.1016/s1872-5805(21)60009-4.

Wu, B., Li, Y., Su, K., Tan, L., Liu, X., Cui, Z., et al., 2019a. The enhanced photocatalytic properties of MnO2/g-C3N4 heterostructure for rapid sterilization under visible light. Journal of Hazardous Materials 377, 227–236. Available at https://doi.org/10.1016/j.jhazmat.2019.05.074, www.elsevier.com/locate/jhazmat.

Wu, L., Ning, D., Zhang, B., Li, Y., Zhang, P., Shan, X., et al., 2019b. Global diversity and biogeography of bacterial communities in wastewater treatment plants. Nature Microbiology 4 (7), 1183–1195. Available at https://doi.org/10.1038/s41564-019-0426-5, www.nature.com/nmicrobiol/.

Xie, Xiuqiang, Chen, Chi, Zhang, Nan, Tang, Zi-Rong, Jiang, Jianjun, Xu, Yi-Jun, 2019. Microstructure and surface control of MXene films for water purification. Nature Sustainability 2 (9), 856–862. Available at https://doi.org/10.1038/s41893-019-0373-4.

Xing, W., Liang, J., Tang, W., He, D., Yan, M., Wang, X., et al., 2020. Versatile applications of capacitive deionization (CDI)-based technologies. Desalination 482, 114390. Available at https://doi.org/10.1016/j.desal.2020.114390.

Xing, Z., Zong, X., Zhu, Y., Chen, Z.G., Bai, Y., Wang, L., 2016. A nanohybrid of CdTe@CdS nanocrystals and titania nanosheets with p-n nanojunctions for improved visible light-driven hydrogen production. Catalysis Today 264, 229–235. Available at https://doi.org/10.1016/j.cattod.2015.08.007, http://www.sciencedirect.com/science/journal/09205861.

Xiong, Z., Zhang, X., Zhang, S., Lei, L., Ma, W., Li, D., et al., 2018. Bacterial toxicity of exfoliated black phosphorus nanosheets. Ecotoxicology and Environmental Safety 161, 507–514. Available at https://doi.org/10.1016/j.ecoenv.2018.06.008, http://www.elsevier.com/inca/publications/store/6/2/2/8/1/9/index.htt.

Xu, L., Suo, H., Liu, R., Liu, H., Qiu, H., 2019a. Design of GO-Ag-functionalized Fe3O4@CS composite for magnetic adsorption of rhodamine B. RSC Advances 9 (52), 30125–30133. Available at https://doi.org/10.1039/c9ra04897a, http://pubs.rsc.org/en/journals/journal/ra.

Xu, L., Suo, H., Wang, J., Cheng, F., Liu, H., Qiu, H., 2019b. Magnetic graphene oxide decorated with chitosan and Au nanoparticles: synthesis, characterization and application for detection of trace rhodamine B. Analytical Methods 11 (30), 3837–3843. Available at https://doi.org/10.1039/c9ay01287g, http://pubs.rsc.org/en/journals/journal/ay.

Yang, X., Xue, H., Xu, J., Huang, X., Zhang, J., Tang, Y.B., et al., 2014b. Synthesis of porous ZnS:Ag2S nanosheets by ion exchange for photocatalytic H2 generation. ACS Applied Materials and Interfaces 6 (12), 9078–9084. Available at https://doi.org/10.1021/am5020953, http://pubs.acs.org/journal/aamick.

Yin, Ling, Wang, Pengyi, Wen, Tao, Yu, Shujun, Wang, Xiangxue, Hayat, Tasawar, Alsaedi, Ahmed, Wang, Xiangke, 2017. Synthesis of layered titanate nanowires at low temperature and their application in efficient removal of U (VI). Environmental Pollution 226, 125–134. Available at https://doi.org/10.1016/j.envpol.2017.03.078.

Yin, Q., Tan, L., Lang, Q., Ke, X., Bai, L., Guo, K., et al., 2018. Plasmonic molybdenum oxide nanosheets supported silver nanocubes for enhanced near-infrared antibacterial activity: synergism of photothermal effect, silver release and photocatalytic reactions. Applied Catalysis B: Environmental 224, 671–680. Available at https://doi.org/10.1016/j.apcatb.2017.11.024, www.elsevier.com/inca/publications/store/5/2/3/0/6/6/index.htt.

Yu, D., Li, L., Wu, M., Crittenden, J.C., 2019a. Enhanced photocatalytic ozonation of organic pollutants using an iron-based metal-organic framework. Applied Catalysis B: Environmental 251, 66–75. Available at https://doi.org/10.1016/j.apcatb.2019.03.050, www.elsevier.com/inca/publications/store/5/2/3/0/6/6/index.htt.

Yu, P., Zhou, X., Yan, Y., Li, Z., Zheng, T., 2019b. Enhanced visible-light-driven photocatalytic disinfection using AgBr-modified g-C 3 N 4 composite and its mechanism. Colloids and Surfaces B: Biointerfaces 179, 170–179. Available at https://doi.org/10.1016/j.colsurfb.2019.03.074, www.elsevier.com/locate/colsurfb.

Yu, J., Low, J., Xiao, W., Zhou, P., Jaroniec, M., 2014. Enhanced photocatalytic CO2-reduction activity of anatase TiO2 by coexposed {001} and {101} facets. Journal of the American Chemical Society 136 (25), 8839–8842. Available at https://doi.org/10.1021/ja5044787, http://pubs.acs.org/journal/jacsat.

Yu, Y., Chen, H., Liu, Y., Craig, V.S.J., Wang, C., Li, L.H., et al., 2015. Superhydrophobic and superoleophilic porous boron nitride nanosheet/polyvinylidene fluoride composite material for oil-polluted water cleanup. Advanced Materials Interfaces 2 (1). Available at https://doi.org/10.1002/admi.201400267, http://onlinelibrary.wiley.com/journal/10.1002/(ISSN)2196-7350.

Yuan, Y., Niu, B., Yu, Q., Guo, X., Guo, Z., Wen, J., et al., 2020. Photoinduced multiple effects to enhance uranium extraction from natural seawater by black phosphorus nanosheets. Angewandte Chemie 132 (3), 1236–1243. Available at https://doi.org/10.1002/ange.201913644.

Yun, T., Jeong, G.H., Padmajan Sasikala, S., Kim, S.O., 2020. 2D graphene oxide liquid crystal for real-world applications: energy, environment, and antimicrobial. APL Materials 8 (7). Available at https://doi.org/10.1063/5.0012465, http://scitation.aip.org/content/aip/journal/aplmater.

Zeng, H., Yu, Z., Peng, Y., Zhu, L., 2019a. Environmentally friendly electrostatically driven self-assembled LDH/GO/PVDF composite membrane for water treatment. Applied Clay Science 183, 105322. Available at https://doi.org/10.1016/j.clay.2019.105322.

Zeng, M., Yuan, S., Huang, D., Cheng, Z., 2019b. Accelerated design of catalytic water-cleaning nanomotors via machine learning. ACS Applied Materials and Interfaces 11 (43), 40099–40106. Available at https://doi.org/10.1021/acsami.9b14792, http://pubs.acs.org/journal/aamick.

Zeng, M., Peng, B., Ybanez, C., Tan, N.W., Deeb, E.A., Bordovsky, E., et al., 2017. High-flux underwater superoleophobic hybrid membranes for effective oil-water separation from oil-contaminated water. RSC Advances 7 (15), 9051–9056. Available at https://doi.org/10.1039/c6ra27124c, http://pubs.rsc.org/en/journals/journalissues.

Zeng, M., Echols, I., Wang, P., Lei, S., Luo, J., Peng, B., et al., 2018. Highly biocompatible, underwater superhydrophilic and multifunctional biopolymer membrane for efficient oil-water separation and aqueous pollutant removal. ACS Sustainable Chemistry and Engineering 6 (3), 3879–3887. Available at https://doi.org/10.1021/acssuschemeng.7b04219, http://pubs.acs.org/journal/ascecg.

Zeng, M., Chen, M., Huang, D., Lei, S., Zhang, X., Wang, L., et al., 2021. Engineered two-dimensional nanomaterials: an emerging paradigm for water purification and monitoring. Materials Horizons 8 (3), 758–802. Available at https://doi.org/10.1039/d0mh01358g, http://pubs.rsc.org/en/journals/journal/mh.

Zeng, Minxiang, Chen, Mingfeng, Huang, Dali, Lei, Shijun, Zhang, Xuan, Wang, Ling, Cheng, Zhengdong, 2021. Engineered two-dimensional nanomaterials: an emerging paradigm for water purification and monitoring. Materials Horizons 8 (3), 758–802. Available at https://doi.org/10.1039/d0mh01358g.

Zhang, B., Wong, P.W., Guo, J., Zhou, Y., Wang, Y., Sun, J., et al., 2022a. Transforming Ti3C2Tx MXene's intrinsic hydrophilicity into superhydrophobicity for efficient photothermal membrane desalination. Nature Communications 13 (1). Available at https://doi.org/10.1038/s41467-022-31028-6, http://www.nature.com/ncomms/index.html.

Zhang, B., Boretti, A., Castelletto, S., 2022b. Mxene pseudocapacitive electrode material for capacitive deionization. Chemical Engineering Journal 435, 134959. Available at https://doi.org/10.1016/j.cej.2022.134959.

Zhang, H., Hou, J., Hu, Y., Wang, P., Ou, R., Jiang, L., et al., 2018a. Ultrafast selective transport of alkali metal ions in metal organic frameworks with subnanometer pores. Science Advances 4 (2). Available at https://doi.org/10.1126/sciadv.aaq0066, http://advances.sciencemag.org/content/advances/4/2/eaaq0066.full.pdf.

Zhang, Q., Xu, W., Wang, X., 2018b. Carbon nanocomposites with high photothermal conversion efficiency. Science China Materials 61 (7), 905–914. Available at https://doi.org/10.1007/s40843-018-9250-x, http://www.springer.com/journal/40843/about.

Zhang, J., Yang, H., Shen, G., Cheng, P., Zhang, J., Guo, S., 2010. Reduction of graphene oxide vial-ascorbic acid. Chemical Communications 46 (7), 1112–1114. Available at https://doi.org/10.1039/b917705a, http://pubs.rsc.org/en/journals/journal/cc.

Zhang, M., Feng, G., Song, Z., Zhou, Y.P., Chao, H.Y., Yuan, D., et al., 2014. Two-dimensional metal-organic framework with wide channels and responsive turn-on fluorescence for the chemical sensing of volatile organic compounds. Journal of the American Chemical Society 136 (20), 7241–7244. Available at https://doi.org/10.1021/ja502643p, http://pubs.acs.org/journal/jacsat.

Zhang, P., Li, J., Lv, L., Zhao, Y., Qu, L., 2017. Vertically aligned graphene sheets membrane for highly efficient solar thermal generation of clean water. ACS Nano 11 (5), 5087–5093. Available at https://doi.org/10.1021/acsnano.7b01965, http://pubs.acs.org/journal/ancac3.

Zhang, Panpan, Li, Jing, Lv, Lingxiao, Zhao, Yang, Qu, Liangti, 2017. Vertically Aligned Graphene Sheets Membrane for Highly Efficient Solar Thermal Generation of Clean Water. ACS Nano 11 (5), 5087–5093. Available at https://doi.org/10.1021/acsnano.7b01965.

Zhang, X., Wu, X., Zhang, J., Xu, H., Yu, X., 2024. Recent progress in graphitic carbon nitride-based materials for antibacterial applications: synthesis, mechanistic insights, and utilization. Microstructures 4 (2). Available at https://doi.org/10.20517/microstructures.2023.77, https://f.oaes.cc/xmlpdf/a3ba1c3f-1dc3-4982-a8fc-195924983367/microstructures4077.pdf.

Zhang, Y., Zou, L., Wimalasiri, Y., Lee, J.Y., Chun, Y., 2015. Reduced graphene oxide/polyaniline conductive anion exchange membranes in capacitive deionisation process. Electrochimica Acta 182, 383–390. Available at https://doi.org/10.1016/j.electacta.2015.09.128, http://www.journals.elsevier.com/electrochimica-acta/.

Zhao, G., Jiang, L., He, Y., Li, J., Dong, H., Wang, X., et al., 2011. Sulfonated graphene for persistent aromatic pollutant management. Advanced Materials 23 (34), 3959–3963. Available at https://doi.org/10.1002/adma.201101007.

Zhao, G., Li, C., Wu, X., Yu, J., Jiang, X., Hu, W., et al., 2018a. Reduced graphene oxide modified NiFe-calcinated layered double hydroxides for enhanced photocatalytic removal of methylene blue. Applied Surface Science 434, 251–259. Available at https://doi.org/10.1016/j.apsusc.2017.10.181, http://www.journals.elsevier.com/applied-surface-science/.

Zhao, J., Niu, Y., Ren, B., Chen, H., Zhang, S., Jin, J., et al., 2018b. Synthesis of Schiff base functionalized superparamagnetic Fe3O4 composites for effective removal of Pb(II) and Cd(II) from aqueous solution. Chemical Engineering Journal 347, 574–584. Available at https://doi.org/10.1016/j.cej.2018.04.151, www.elsevier.com/inca/publications/store/6/0/1/2/7/3/index.htt.

Zhao, P., Alvarez, P.J.J., Li, X., Pan, C., 2018c. Development of an analytical method for pesticide residues in berries with dispersive solid phase extraction using multiwalled carbon nanotubes and primary secondary amine sorbents. Analytical Methods 10 (7), 757–766. Available at https://doi.org/10.1039/c7ay02178j, http://pubs.rsc.org/en/journals/journal/ay.

Zhao, H., Chen, S., Quan, X., Yu, H., Zhao, H., 2016. Integration of microfiltration and visible-light-driven photocatalysis on g-C3N4 nanosheet/reduced graphene oxide membrane for enhanced water treatment. Applied Catalysis B: Environmental 194, 134–140. Available at https://doi.org/10.1016/j.apcatb.2016.04.042, www.elsevier.com/inca/publications/store/5/2/3/0/6/6/index.htt.

Zhao, J., Liu, J., Li, N., Wang, W., Nan, J., Zhao, Z., et al., 2016. Highly efficient removal of bivalent heavy metals from aqueous systems by magnetic porous Fe3O4-MnO2: adsorption behavior and process study. Chemical Engineering Journal 304, 737–746. Available at https://doi.org/10.1016/j.cej.2016.07.003, www.elsevier.com/inca/publications/store/6/0/1/2/7/3/index.htt.

Zhao, L.M., Tang, D.Y., Zhang, H., Cheng, T.H., Tam, H.Y., Lu, C., 2007. Dynamics of gain-guided solitons in an all-normal-dispersion fiber laser. Optics Letters 32 (13), 1806–1808. Available at https://doi.org/10.1364/OL.32.001806.

Zhao, X., Pan, D., Chen, X., Li, R., Jiang, T., Wang, W., et al., 2019. g-C 3 N 4 photoanode for photoelectrocatalytic synergistic pollutant degradation and hydrogen evolution. Applied Surface Science 467-468, 658–665. Available at https://doi.org/10.1016/j.apsusc.2018.10.090, http://www.journals.elsevier.com/applied-surface-science/.

Zhao, Xing, Zha, Xiang-Jun, Pu, Jun-Hong, Bai, Lu, Bao, Rui-Ying, Liu, Zheng-Ying, Yang, Ming-Bo, Yang, Wei, 2019. Macroporous three-dimensional MXene architectures for highly efficient solar steam generation. Journal of Materials Chemistry A 7 (17), 10446–10455. Available at https://doi.org/10.1039/c9ta00176j.

Zhao, Y., Wang, Y., Wang, R., Wu, Y., Xu, S., Wang, J., 2013a. Performance comparison and energy consumption analysis of capacitive deionization and membrane capacitive deionization processes. Desalination 324, 127–133. Available at https://doi.org/10.1016/j.desal.2013.06.009.

Zhao, Y., Wang, C.J., Gao, W., Li, B., Wang, Q., Zheng, L., et al., 2013b. Synthesis and antimicrobial activity of ZnTi-layered double hydroxide nanosheets. Journal of Materials Chemistry B 1 (43), 5988–5994. Available at https://doi.org/10.1039/c3tb21059f.

Zheng, Z., Grünker, R., Feng, X., 2016. Synthetic two-dimensional materials: a new paradigm of membranes for ultimate separation. Advanced Materials 28 (31), 6529–6545. Available at https://doi.org/10.1002/adma.201506237, http://onlinelibrary.wiley.com/journal/10.1002/(ISSN)1521-4095.

Zhi, S., Banting, G., Li, Q., Edge, T.A., Topp, E., Sokurenko, M., et al., 2016. Evidence of naturalized stress-tolerant strains of *Escherichia coli* in municipal wastewater treatment plants. Applied and Environmental Microbiology 82 (18), 5505–5518. Available at https://doi.org/10.1128/AEM.00143-16, http://aem.asm.org/content/82/18/5505.full.pdf.

Zhou, H., Jiang, Z., Wei, S., 2018. A new hydrotalcite-like absorbent FeMnMg-LDH and its adsorption capacity for Pb2 + ions in water. Applied Clay Science 153, 29–37. Available at https://doi.org/10.1016/j.clay.2017.11.033, http://www.elsevier.com/inca/publications/store/5/0/3/3/2/2/index.htt.

Zhou, Q., Fang, Z., 2015. Graphene-modified TiO2 nanotube arrays as an adsorbent in micro-solid phase extraction for determination of carbamate pesticides in water samples. Analytica Chimica Acta 869, 43–49. Available at https://doi.org/10.1016/j.aca.2015.02.019, http://www.journals.elsevier.com/analytica-chimica-acta/.

Zhou, Lin, Tan, Yingling, Ji, Dengxin, Zhu, Bin, Zhang, Pei, Xu, Jun, Gan, Qiaoqiang, Yu, Zongfu, Zhu, Jia, 2016. Self-assembly of highly efficient, broadband plasmonic absorbers for solar steam generation. Science Advances 2 (4). Available at https://doi.org/10.1126/sciadv.1501227.

Zhou, R., Li, J., Wei, W., Li, X., Luo, M., 2020. Atomic substituents effect on boosting desalination performances of Zn-doped NaxCoO2. Desalination 496, 114695. Available at https://doi.org/10.1016/j.desal.2020.114695.

Zhu, C., Luan, Z., Wang, Y., Shan, X., 2007. Removal of cadmium from aqueous solutions by adsorption on granular red mud (GRM). Separation and Purification Technology 57 (1), 161–169. Available at https://doi.org/10.1016/j.seppur.2007.03.013.

Zhu, G., Xu, J., Zhao, W., Huang, F., 2016. Constructing black titania with unique nanocage structure for solar desalination. ACS Applied Materials and Interfaces 8 (46), 31716–31721. Available at https://doi.org/10.1021/acsami.6b11466, http://pubs.acs.org/journal/aamick.

Zhu, Liangliang, Gao, Minmin, Peh, Connor Kang Nuo, Ho, Ghim Wei, 2018. Solar-driven photothermal nanostructured materials designs and prerequisites for evaporation and catalysis applications. Materials Horizons 5 (3), 323–343. Available at https://doi.org/10.1039/c7mh01064h.

Zhu, L., Gao, M., Peh, C.K.N., Ho, G.W., 2018b. Solar-driven photothermal nanostructured materials designs and prerequisites for evaporation and catalysis applications. Materials Horizons 5 (3), 323–343. Available at https://doi.org/10.1039/c7mh01064h, http://pubs.rsc.org/en/journals/journal/mh.

Zou, R., Xu, T., Lei, X., Wu, Q., Xue, S., 2020. Novel design of porous hollow hydroxyapatite microspheres decorated by reduced graphene oxides with superior photocatalytic performance for tetracycline removal. Solid State Sciences 99, 106067. Available at https://doi.org/10.1016/j.solidstatesciences.2019.106067.

Zou, X., Zhang, L., Wang, Z., Luo, Y., 2016. Mechanisms of the antimicrobial activities of graphene materials. Journal of the American Chemical Society 138 (7), 2064–2077. Available at https://doi.org/10.1021/jacs.5b11411, http://pubs.acs.org/journal/jacsat.

Zubair, M., Jarrah, N., Ihsanullah, Khalid, A., Manzar, M.S., Kazeem, T.S., et al., 2018. Starch-NiFe-layered double hydroxide composites: efficient removal of methyl orange from aqueous phase. Journal of Molecular Liquids 249, 254–264. Available at https://doi.org/10.1016/j.molliq.2017.11.022.

CHAPTER

2

Fundamental separation mechanisms of 2D nanomaterials

Hamid Erfan-Niya, Sima Majidi, and Negin Karimzadeh Bajgiran

Chemical and Petroleum Engineering Department, University of Tabriz, Tabriz, Iran

2.1 Background

The production of purified target substances from complex mixtures is a crucial operation in chemical engineering, environmental cleanup, medical applications, and various day-to-day processes (Wang et al., 2017a). Conventional separation methods, such as cryogenic distillation, thermal desalination, adsorption, and electrodialysis, can meet basic separation needs but suffer from excessive energy consumption, operational complexity, and significant waste generation. Membrane-based separation technologies have recently gained prominence due to their lower-energy consumption, simpler operations, smaller environmental footprint, and reduced CO_2 emissions (Werber et al., 2016). All membrane operations can be performed continuously and are considered a more environmentally friendly method for purifying, recovering, and recycling liquids and gases (Zhu et al., 2018).

Conventional membranes, even with their benefits, typically face the inherent permeability-selectivity trade-off; that is, when one is increased, the other will likely decrease (Zhang et al., 2019). Otherwise, the limited tunability and empirical design of traditional membranes have limited their potential for more diverse and efficient separation applications. Thus, the desire to overcome these issues has been paving the way for newer, next-generation membrane materials, particularly ultrathin two-dimensional (2D) nanomaterials (Wang et al., 2023).

Since graphene was first isolated in 2004, 2D nanomaterials—defined as materials with thickness at the atomic or molecular scale—have proven to be useful building blocks for membrane design (Novoselov et al., 2004). The unique properties of 2D nanomaterials, including a very large specific surface area, molecular-scale precision for sieving, mechanical strength, and tunable surface chemistry, make them a promising way to address the permeability-selectivity gap (Jahangirzadeh et al., 2024; Jiang et al., 2017). In contrast to bulk materials, 2D membranes enable the rapid and selective transport of ions and molecules

Two-Dimensional Nanomaterials for Water Purification
DOI: https://doi.org/10.1016/B978-0-443-44530-9.00002-X

through interlayer channels or engineered nanopores. Graphene-based materials (for instance, graphene oxide, reduced graphene oxide) (Han et al., 2021), transition metal dichalcogenides (e.g., MoS_2, WS_2), MXenes, and porous organic frameworks (for example, MOFs and COFs) are the most studied 2D nanomaterials for separation (Bajgiran et al., 2025).

2.2 Problem statement

This chapter provides an overview of the essential separation mechanisms made possible by 2D nanomaterials. The mechanisms examined include size and shape selectivity, charge-driven separations, hydrophobic/hydrophilic interactions, and selective molecular recognition through functionalization. The chapter will focus specifically on structural features and physicochemical interactions that govern transport phenomena in the context of these types of materials. The final section of the chapter addresses significant challenges for the ongoing development of 2D membranes, while also exploring innovative new approaches to address them and lay the groundwork for future development and innovation in separation science.

2.3 Method procedure

This chapter is based on an inclusive literature analysis of 2D nanomaterials applied in separation science which summarizes the studies from 2016–2025.

2.4 Pore size and shape selectivity

Molecular sieving is a crucial mechanism that controls separation in 2D nanomaterial-based membranes through the size and shape-selective resolution of molecules and ions. Molecular sieving ultimately occurs in two architectures: (1) controllably tunable interlayer channels between the 2D sheets, and (2) purposely designed nanopores inside or across the 2D plane. In both architectures, the level of control of the porosity provides 2D materials for separations that are spatial, controlled, and energy efficient (Zhao et al., 2024a; Bondaz et al., 2023).

2.4.1 Interlayer spacing and laminar structures

Graphene oxide (GO) membranes are also an example of size-based separation enabled through tunable interlayer spacing. The layers can be tunably spaced with angstrom-level precision through the use of ion intercalation (e.g., K^+, Na^+, Ca^{2+}), pressure-driven intercalation, or simply by adjusting the deposition rate (Bondaz et al., 2023; Deng et al., 2016). For example, during vacuum filtration, membranes can be made slower by drop-wise adding single layers of GO to create membranes with significantly improved salt rejection (>95% vs. notch rates between 2.5 and 4 times those of membranes made from fast-deposited GO). The mechanism for improvement is attributable to unimodal interlayer channels and consistently narrow interlayer channels in slowly deposited membranes. GO laminates also provide >99% retention of organically dyes while relying on modest retention of ionic salts, which are characteristics driven by both size exclusion and electrostatic repulsion (Wang et al., 2019; Murali et al., 2021). Likewise, similar membrane structures can be made by dispersing zeolitic imidazolate frameworks (ZIF-L, ZIF-8) nanosheets, which create a lamellar structure with aligned apertures. The application of ZIF-L, for instance, resulted in membranes with higher selectivity and permeability than those of ZIF-8. In an order laminar tube-type membrane, they were able to quickly transport water for the membrane via pervaporation-based dehydration of ethanol while broadly excluding ethanol via size-selective sieving (Coe-Sessions et al., 2024; Qu et al., 2021).

Similarly, layered materials comprised of metal–organic frameworks (MOFs), covalent–organic frameworks (COFs), transition metal dichalcogenides (TMDs), MXenes, graphitic carbon nitride ($g\text{-}C_3N_4$), and zeolite nanosheets utilize interlayer spacing modulation to enable molecular sieving. For instance, ultrathin 2D MOF nanosheets can be assembled into homogeneous films with controlled pore sizes, which exhibit outstanding gas separation performances, with selectivity for H_2/CO_2 above 200 and permeance beyond Robeson's upper limit (Zou et al., 2025; Wu et al., 2025). COF nanosheets have also been shown to provide excellent H_2/CO_2 separation in mixed matrix membranes (MMMs) with functional laminar channels formed via covalent bonds. Similar to MOFs and COFs, TMDs like MoS_2 and WS_2 exhibit layered structures with strong van der Waals interactions, and have higher rigidity than GO, which can imply significant mechanical stabilization in the material. The TMD MoS_2, for example, has been shown to provide water permeabilities that are 3–5 times higher than those of GO membranes, without compromising their ability to reject direct dyes, while also maintaining an intolerable degree of chemical swelling over time when contacted with the aqueous phase (Côté et al., 2007; Zhu et al., 2018).

MXenes, specifically Ti_3C_2Tx, are inherently hydrophilic, negatively charged, and mech (Hoenig et al., 2024) anically strong nanosheets. These properties create unique hydrogel-like properties for high water-permeable membranes (>1000 L/min/mbar) that can reject selective molecules. They have been used with polymers and GO to provide nanosheets with fewer defects, and they will enable better performing membranes overall if properly incorporated into the film (Wu et al., 2024; Hoenig et al., 2024). $G\text{-}C_3N_4$ is a nanosheet material with intrinsic nanopores and artificial spacers, to create ordered channels in membranes to provide the chemical and mechanical flexibility needed for high-performance membranes (Thomas et al., 2008). The $g\text{-}C_3N_4$ can improve mechanical strength, resistance to swelling, and separation efficiency when used with water/alcohol mixtures, as well as in gas separations. Nanoporous zeolite nanosheets, fabricated through graphite exfoliation and gel-less secondary growth to yield ultrathin oriented and continuous films, contained microporous crystalline silica molecular sieves that achieved high selectivity and separation performance for hydrocarbon isomers (Berry, 2013; Ji et al., 2016).

Fig. 2.1 demonstrates size and shape-selective molecular sieving through the interlayer space in laminar membranes. The diagram compares several classes of 2D materials—GO, ZIF-L, MOFs, COFs, TMDs (e.g., MoS_2), MXenes, $g\text{-}C_3N_4$, and zeolite nanosheets. It also

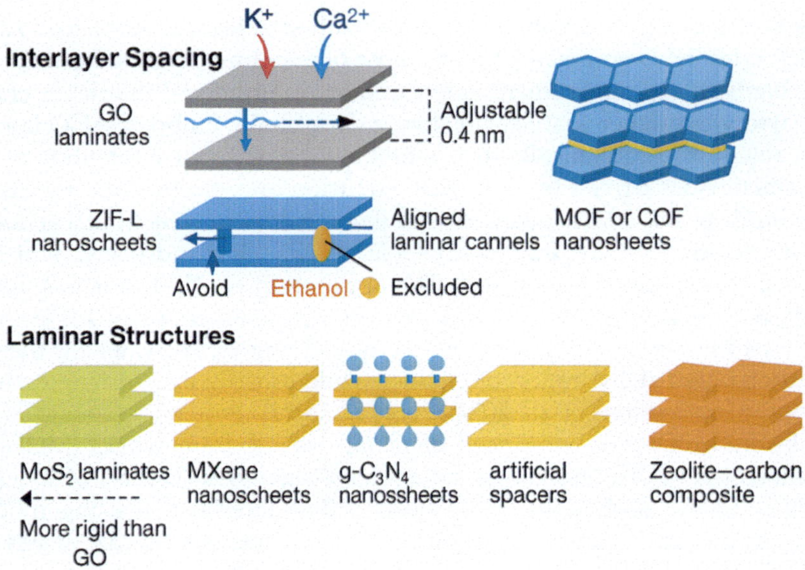

FIGURE 2.1 Schematic illustration of interlayer spacing and laminar structures in 2D material-based membranes.

described their layered stacking architecture, having tunable interlayer channels. These interlayer spaces serve as sieving pathways for H_2O while sieving out larger solutes or contaminants. As illustrated, limited interlayer distances that are narrower and more consistent, as well as inherent hydrophilicity and stiffness to the materials, can improve water permeability and solute rejection for a range of separation applications (Zhu et al., 2018).

2.4.2 Engineered nanoporous membranes

As an alternative means of achieving shape and size selectivity, the design of nanopores directly within 2D materials can be accomplished through top-down methods, such as ion irradiation and plasma etching, or through the bottom-up synthesis of porous materials, such as 2D MOFs, COFs, and polymers (Wu et al., 2025; Chen et al., 2021). Plasma-etched, oxygen-porous graphene membranes with subnanometer pores achieve nearly 100% salt rejection during desalination while allowing ultrafast water fluxes. Ion-gated nanoporous graphene (NPG) membranes coated with ionic liquids can dynamically modulate pore size in order to achieve very high selectivity for CO_2/CH_4 (~42). NPG membranes may be multilayered and either stacked in aligned layers or have their interlayer spacing tuned, thereby enabling enhancement in gas separation and water desalination (Zhang et al., 2018; Joshi et al., 2025). Hybrid membranes that incorporate porous graphene with a 3D MOF such as UIO-66 can have water permeability which is upwards of 15x the permeability of pristine reduced graphene oxide (rGO) membranes and that still retain good dye rejection, where this enhanced permeability is believed to arise from the synergy between the vertical pores of the MOF and the horizontal laminar channels of the graphene (Jang et al., 2020).

Nanoporous MoS_2 membranes with pores at the atomic scale can effectively block more than 88% of the ions while providing water flux several orders of magnitude higher than other nanoporous membranes. Favorable pore designs in MoS_2-produced membranes can effectively outperform otherwise identical nanoporous graphene membranes. Furthermore, MoS_2-based MMMs embedded into polymers have exhibited high CO_2 permeability and selectivity, which is attributed to the solution-diffusion mechanisms and higher CO_2 adsorption energy in the polymer (Dai et al., 2025; Zhao et al., 2024b).

2.4.3 Expanding applications to gas separation

Although significant research has been conducted on graphene-based membranes and their performance in water treatment applications, recent advances in GO/rGO membranes for gas separation have garnered more attention. Pristine graphene is impermeable to all molecules, but the introduction of intrinsic defects and controlled stacking can create nanochannels that enable selective gas permeation (Dai et al., 2025; Cheng et al., 2025). Kim et al. (2013) demonstrated selective gas permeation through few-layer GO membranes through abandoned gas-flow nanochannels and intrinsic defects of a few layers within the GO membranes. Li et al. (2013a) demonstrated a sharp exponential decrease in the gas permeance of GO membranes, with H_2 gas permeation over other gases, by increasing the layer thickness of the GO membranes, thus exhibiting molecular sieving. GO/polymer laminates constructed through hydrogen bonding enable the manipulation of interlayer spacing and straight diffusion pathways, resulting in excellent CO_2/N_2 selectivity, with CO_2 permeability of 100 Barrer, CO_2/N_2 selectivity of 91, and long-term stability. Different approaches to constructing GO-stacked membranes, including vacuum-spin processing, have been developed to reduce the spacing between the interlayers to < 0.4 nm to obtain optimal gas sieving performance (Tong et al., 2024a; Ding et al., 2018).

Table 2.1 summarizes much of the recent work conducted to develop 2D nanomaterials used in membranes for molecular sieving and selective separations. It organizes materials based on structural features, pore/channel size, separation mechanism, fabrication methods, and delivery of performance values. For each type of material, Table 2.1 highlights the advantages, the types of membranes that can be made, and high-level commercialization evidence. It provides an imagery landscape that specifically compares different 2D materials, including GO, MOFs, COFs, TMDs, MXenes, and others. The sizes and shapes of the pores and channels, as well as laminar or nanoporous architectures, dictate the selectivity and permeability performance when applied to gas and liquid separations. Separation processes are influenced by the fabrication method, whether it is top-down or bottom-up, in situ or at the nanoscale, which determines the final morphology, layer stability, and market entry. Most importantly, learning and understanding both high-level and specific technical characteristics are critical to the effective and efficient application of membranes towards particular separation problems, such as those related to desalination, gas purification, and organic solvents/nanofiltration. For scholars, Table 2.1 provides a practical reference for those seeking to broadly optimize membranes to incorporate thresholds and/or achieve high permeability.

TABLE 2.1 2D materials for molecular sieving membranes.

2D material type	Structure and morphology	Pore size/ channel dimensions	Separation mechanism	Fabrication methods	Key advantages	Representative performance	Applications	References
GO	Laminar stacks with tunable interlayer spacing	Adjustable from ~0.4 to 1.2 nm via intercalation	Size exclusion and electrostatic repulsion through interlayer channels	Vacuum filtration, ion intercalation, spin coating	Precise d-spacing control; high water flux; high dye rejection	Salt rejection >99%; water flux 2.5–4 times higher for slow-deposited membranes	Water desalination, dye removal	Han et al. (2013)
MOF nanosheets	Ultrathin 2D nanosheets with uniform nanopores	~1–2 nm pore size, sub-10 nm thickness	Molecular sieving through uniform pores & ultrathin layers	Top-down exfoliation, bottom-up surfactant-assisted synthesis	High selectivity; short diffusion paths; tuneable pore chemistry	H_2/CO_2 selectivity >200; permeance surpasses Robeson upper bound	Gas separation, solvent dehydration	Tripathi et al. (2021)
COF nanosheets	Crystalline, covalently linked laminar sheets	~1.5 nm pore diameter, robust covalent bonds	Size-based sieving via ordered nanopores and interlayer passages	Exfoliation, GO-assisted LbL assembly	Thermal/ chemical stability; excellent polymer compatibility	Surpassing 2008 Robeson upper bound for H_2/CO_2 separation	Gas separation, water/ethanol separation	Ciampalini et al. (2022)
Transition metal dichalcogenides (TMDs) (MoS_2, WS_2)	Layered sandwich structure: metal atom between chalcogen layers	~0.3–1.2 nm interlayer spacing; atomic scale nanopores possible	Size exclusion & selective adsorption via nanochannels and pores	Chemical exfoliation, nanostrand intercalation	High mechanical strength; stable in water; antibacterial	Water permeability 3–5 times that of GO; ion rejection ~99%; CO_2/N_2 selectivity 93	Water purification, gas separation	Tang et al. (2022)

MXenes	2D transition metal carbides/nitrides with surface terminations (O, OH, F)	Variable, depends on intercalation and restacking	Hydrophilic nanochannels and surface charge-based sieving	Selective etching (MAX phase), vacuum filtration	Hydrophilicity; high mechanical strength; excellent permeability	Water permeability >1000 LMH/bar; ~90% rejection of small molecules	Solvent-resistant nanofiltration, water treatment	Liu et al. (2022)
Zeolite nanosheets	Crystalline microporous silica frameworks	0.25–1.0 nm, unit cell thickness	Molecular sieving via highly ordered micropores	Exfoliation, gel-free secondary growth, float-coating	Ultrahigh selectivity; thermal stability	Separation factor >10,000 for xylene isomers	Hydrocarbon isomer separation, gas separation	Qi et al. (2023)
Nanoporous graphene	Graphene with subnanometer engineered pores	<1 nm pore diameter	Size exclusion & tunable pore size by ion gating	Oxygen plasma etching, ion irradiation	Ultrafast water flux; near 100% salt rejection	Water flux orders of magnitude higher than polymer membranes	Water desalination, gas separation	Hao et al. (2024)

2.5 Charge-driven separation

Charge-driven separation in 2D nanomaterials is the process where the distribution of electric charge in these atomically thin materials enables selective molecular transport and efficient energy conversion. This phenomenon relies on the distinct electronic properties of 2D materials, such as quantum confinement, high surface-to-volume ratios, and tunable band structures, which collectively enable engineered charge separation and transport (Guo et al., 2019; Liu et al., 2019; Lee et al., 2023). Charge-driven separation is crucial in 2D materials because their unique electronic and optical properties—arising from their confined, planar structure—allow precise control of charge carrier behavior. These features separate 2D materials from bulk systems and position them as leading candidates for applications in energy harvesting, storage, and advanced electronics (Wang et al., 2017b).

2.5.1 Electrostatic interactions and surface charge effects

Surface charge is important to the selective properties of 2D membrane systems. GO, MXenes, layered double hydroxides (LDHs), and other materials possess surface functional groups (–COOH, –OH, –F, etc.) that modulate the surface charge density of the materials in response to variations in pH or ionic strength (Li et al., 2018; Hao et al., 2024). For example, GO is negatively charged in neutral to basic conditions, and thus preferentially repels anions while allowing the passage of cations. This phenomenon can be advantageous for the sequestration of cationic dyes or for ion sieving applications. Janus 2D materials, which have two distinct surfaces, exhibit an asymmetric charge distribution, resulting in intrinsic electric fields. These internal fields have the ability to drive selective transport of ions or direct the direction of adsorption through different water or ion routing pathways (Chatterjee et al., 2024; Shakeelur Raheman et al., 2021). Additionally, these charge states can influence the electrostatic self-assembly and hybridization of oppositely charged 2D sheets (for example, MOF-GO) into nanoscale, stable layered heterostructures comprised of confined nanochannels, which can be useful for charge-selective separation (Wang et al., 2021).

2.5.2 Band alignment and heterojunction engineering

Band alignment in heterostructure 2D materials establishes carrier migration pathways. Type-II (staggered) heterojunctions are especially beneficial, as they spatially separate the conduction band minimum (CBM) of one material from the valence band maximum (VBM) of another material (Xiong et al., 2020). For instance, in a MoS_2/WS_2 bilayer, photoexcited electrons are confined in MoS2, while holes remain in WS_2. This leads to ultrafast charge separation (<50 fs) and minimal recombination. Architectures designed in a Z-scheme mimic the natural pathways in photosynthesis by allowing charge carriers with high redox potential to remain available while recombining with lower-energy charge carriers. With an internal electric field and differences in work function, these architectures suggest the possibility for electroactive membranes, solar-assisted photocatalytic reduction of CO_2, and photocatalytic degradation of environmental pollutants (Zhao et al., 2024b; Tang et al., 2024).

2.5.3 Built-in electric fields and polarization effects

Built-in electric fields (BEFs) are manifested in heterointerfaces, originating from discrepancies in Fermi levels, dipoles at the interfaces, and the presence of surface states. These BEFs can enable directional charge transfer, limit electron–hole pair recombination, and improve charge selectivity. For instance, CoO/CoS_2 heterostructures are examples of significant BEFs that allow for an efficient oxygen evolution reaction (OER). Some ferroelectric 2D materials, such as α-In2Se3, exhibit spontaneous polarization, generating a permanent internal field that governs ionic migration or traversal when the BEF potential exceeds the intrinsic electric potential of the material (Zhang et al., 2022). Janus structures also lend themselves to switchable out-of-plane polarizations, allowing for field-dependent selectivity due to charge flow in a certain direction. Outside of BEFs, application of an external electric field can exceed 3.5 V/nm with dual ionic gating of the surface zone of functional nanoscale channels, allowing for semiconductor-to-metal transitions or electro-gated ion transport (Yang et al., 2021).

Fig. 2.2 demonstrates how electric fields (intrinsic and external) affect charge and ion transport in state-of-the-art 2D materials. Various heterostructures, such as CoO/CoS_2, can yield intrinsic electric fields that enhance charge separation due to band alignment and interfacial dipoles. Alternatively, ferroelectric materials have built-in out-of-plane polarization that can focus ionic movements through internal electric fields. Last, Janus structures, such as molybdenum sulfide selenide (MoSSe), are capable of responding to external electric fields, which allows for switchable dipoles and tunable electronic phases—key features for future electronic and membrane-based devices.

2.5.4 Defect-induced charge trapping and surface engineering

Atomic defects, such as vacancies, dopants, and surface terminations, serve as localized charge trapping sites, modifying electronic structures and promoting separation. Sulfur or indium vacancies in MoS_2 and $ZnIn_2S_4$ nanosheets introduce mid-gap states that prolong charge carrier lifetimes and enhance CO_2 photoreduction or reactive oxygen species (ROS) generation for disinfection (Yang et al., 2021; Tawfik et al., 2019). Time-resolved

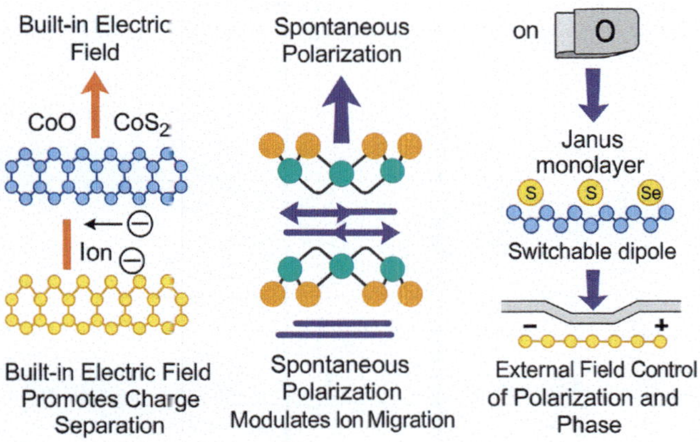

FIGURE 2.2 Schematic illustration of electric field effects in 2D material systems. The diagram shows three mechanisms: (1) Built-in electric fields at the CoO/CoS_2 heterointerface promote directional charge separation (electron and hole transport), (2) spontaneous polarization in ferroelectric 2D materials (e.g., α-In_2Se_3) facilitates ion migration by internal electric dipoles, and (3) Janus MoSSe monolayers under dual-gate ionic fields demonstrate field-controlled polarization switching and phase modulation.

TABLE 2.2 Overview of charge-driven separation mechanisms in 2D materials.

Mechanism	2D material example	Key effect	Separation target	Reference case	References
Surface charge	Graphene Oxide (GO)	Electrostatic repulsion of anions, attraction of cations	Dyes, salts, metal ions	GO laminates in NF/UF	Tawfik et al. (2019)
Janus dipoles	Janus MoS_2/WS_2	Internal field for asymmetric ion transport	Directional ion sieving	2D Janus heterostructures	Tripathi et al. (2021)
Band alignment	MoS_2/WS_2 heterostructure	Type-II junctions for spatial charge separation	Photocatalytic CO_2 reduction	Charge transfer in 50 fs	Ciampalini et al. (2022)
Built-in electric field (BEF)	CoO/CoS_2	Fermi-level induced directional transport	Oxygen evolution, redox separations	Co-based catalysis	Tang et al. (2022)
Ferroelectricity	α-In_2Se_3	Spontaneous polarization enables field control	Ion gating, switchable membranes	Field-switchable sieving	Liu et al. (2022)
Defect engineering	MoS_2, $ZnIn_2S_4$	Defect-induced charge trapping improves carrier lifetime	CO_2 photoreduction, ROS-based disinfection	S vacancies & mid-gap states	Qi et al. (2023)
External electric fields	MXene, GO composites	Electro-gated control of ion transport	Cation/anion-selective membranes	Dual ionic gating > 3.5 V/nm	Hao et al. (2024)

spectroscopy and ab initio nonadiabatic molecular dynamics reveal that hole transfer in MoS_2/WS_2 is faster than electron transfer, primarily due to a larger density of acceptor states and stronger interlayer coupling. Recombination is hindered by state localization. Electron—hole recombination, for example, occurs on the nanosecond timescale (2.2 ns), approximately 3–4 orders of magnitude slower than charge separation. Surface modification strategies, such as hydrogenation, heteroatom doping, or bandgap engineering, further enhance charge separation. Hydrogenated TiO_2 nanosheets, for instance, exhibit bandgap narrowing and enhanced visible-light absorption, leading to improved hydrogen generation by up to 3.73 mmol h^{-1} g^{-1} (Dai et al., 2024).

Table 2.2 provides an overview of important 2D nanomaterials employed in charge-driven separation processes, highlighting the contributions of surface charge, chemical functionalities, and electrostatic interactions in facilitating selective ion transport and molecular separation. Nanomaterials such as graphene oxide (GO), MoS_2, and MXenes possess surface functional groups (e.g., –COOH, –OH, –F, $–SO_3H$) that generate tunable zeta potentials, allowing for the electrostatic repulsion or attraction of specific ions. For example, negatively charged GO membranes electrostatically repel multivalent anions strongly due to both size exclusion and electrostatic interactions. In contrast, functionalization with polycationic species (e.g., PEI, chitosan) promotes cation selectivity. In the same manner, MXenes such as Ti_3C_2Tx, bearing surface terminations like –OH and –F, demonstrate tunable charge densities and show superior rejection of negatively charged dyes and heavy metal anions.

Charge-mediated separation performance is also a function of environmental conditions, specifically solution pH and ionic strength, which alter the membrane surface charge and ion hydration. Several advanced studies employ layer-by-layer (LbL) assembly or interfacial polymerization to introduce charge heterogeneity, thereby achieving ion sieving with enhanced selectivity, permeability, and antifouling properties. Therefore, by combining charge effects with structural control (e.g., interlayer spacing, pore size), 2D membranes provide excellent platforms for precise ion separations in nanofiltration, water desalination, and selective ion recovery from wastewater streams.

2.6 Hydrophobic and hydrophilic interactions

The interaction between a surface and a liquid in contact with is determined by surface forces, which can be categorized into three types. One of them is the van der Waals forces that can be created in any interface between solids and liquids. The other is the electrostatic interaction that is created between charged or polar surfaces and the liquid in contact with them, and the third is the structural forces, which are mainly hydrogen bonds (Hirasakl, 1991). Wettability is the most fundamental interaction between a surface and a liquid, in which the geometry of the liquid spreading on the surface is controlled by a balance between adhesive and cohesive forces. These interactions affect the wetting behavior and determine the hydrophilicity or hydrophobicity of the interaction. They also control the electronic structure at the liquid–solid interface and affect the frictional interaction, thereby modulating the chemical activity. However, these interactions are complicated by the consideration of surface inhomogeneities, such as roughness and contamination, as they disrupt the creation of an equilibrium configuration between the liquid and the surface, leading to non-ideal behaviors, including super slipping, superwetting, and super hydrophobicity (Li et al., 2013b; Snapp et al., 2020).

The separation mechanisms based on using 2D nanomaterials heavily rely on hydrophilic and hydrophobic interactions. Oil–water separation, gas separation, and liquid–liquid extraction are just a few of the separation applications made possible by these interactions as well as the special qualities of 2D materials, such as their large surface area and adjustable surface chemistry. In hydrophobic interactions, the nonpolar molecules or surfaces band together to decrease contact with a polar solvent like water. The hydrophilic 2D nanomaterials, like graphene oxide (GO), easily interact with water due to the existence of the oxygen-containing functional groups (epoxy, hydroxyl, and carbonyl), while hydrophobic resources like graphene can repel the polar solvent like water. The attractive forces between polar or charged molecules and water molecules are an example of the hydrophilic interactions. These interactions are driven by the tendency of water molecules to create hydrogen bonds with other polar or charged molecules (Li et al., 2023).

Based on the explanations, it is clear that various interactions between solute, solvent, and membrane, not only are the main driving forces responsible for transport through the membrane, but also play important roles in inducing molecular adsorption, repulsion, and diffusion. The structural features of the membrane along with the molecular interactions should be included in transport models that are useful in explaining membrane transport and help in the prediction of membrane performance. All these parameters are required to be analyzed precisely (Tang et al., 2024).

2.6.1 Membrane-based separations

Obviously, various interactions between solute, solvent, and the membrane have the decisive impact on molecular adsorption, exclusion, and permeation, along with the primary driving forces in charge of transport across the membrane. The structural features of the membrane associated with molecular interactions have to be incorporated into transport models to be helpful to explain membrane permeation and allow for the estimation of membrane performance (Tong et al., 2024b; Marchetti et al., 2014). In addition to the exclusive properties of 2D nanomaterials, the hydrophobic and hydrophilic interaction offers flexible and efficient separation procedures for a variety of applications. Filtering air, gases, and nonaqueous fluids, as well as separating or filtering liquids that contain hydrophobic substances, are the main applications for hydrophobic membranes. For example, 2D nanomaterials with a hydrophobic surface may be a viable option for oil and water separation (Avasarala and Bose, 2023; Majidi et al., 2023a). While the hydrophilic surface can be used to help transport water, the nanomaterial repels water, allowing oil to pass through. Additionally, they are employed in venting applications, which permit the passage of gases or air while obstructing liquids. Oil and water can be separated using hydrophobic membranes, which is crucial for industrial processes like wastewater treatment. They are also suitable for filtering various nonaqueous fluids, including organic solvents. The specific applications of the hydrophilic nanomaterials in membrane structure, which are attracted to water, include clarifying aqueous solutions (Ghadei et al., 2025; Majidi et al., 2023b).

Some of the current 2D nanomaterials used in separation include graphene and GO, hexagonal boron nitride (h-BN), MXenes, transition metal dichalcogenides (TMDs). They are used in filtration membranes for specific purposes like water purification, oil–water separation, and liquid–liquid extraction due to their tunable surface chemistry and desired interactions (Kim et al., 2024). Hydrophobic features can be indicated by graphene derivatives and molybdenum disulfide (MoS_2) nanosheets as the 2D nanomaterials. In particular, using some special modification in atomic scale can engineer the structure of MoS_2 nanosheets to make them superhydrophobic (Fang et al., 2021). The same is true for graphene, and it can be modified in a various method to make it hydrophobic. Due to their hydrophobic properties, they can be merged into polymer matrices to fabricate hybrid membranes with improved water permeability and reduced swelling, which makes them suitable for applications such as wastewater treatment and separation of special substances (Liu et al., 2018).

Graphene oxide (GO) is one of the graphene derivative 2D nanomaterial in which the oxygen-containing functional groups in its molecular structure make it hydrophilic (Pakdel et al., 2021). Other current examples are certain MXenes and LDH. LDHs are utilized in membrane separation due to their specific hydrophilicity, which increases water permeability and antifouling properties of the membranes. This modification is essential in membrane-based water purification due to the attraction of water molecules, which increases the water flux and enhances separation efficiency. Furthermore, this modification enhances the membrane's long-term stability and performance due to increase the antibiofouling features (Sajid et al., 2022). In a study by Fang et al. (2021), an organic–inorganic hybrid membrane was developed by embedding highly hydrophobic MoS_2 nanosheets into a PEBA polymer matrix for pervaporation, a separation technique for separating the pyridine as an organic compound from water. This study reported the enhanced and efficient selectivity toward pyridine over water due to the hydrophobicity of the membrane. The other study evaluated the development of a novel nanomaterial for separating oil from aqueous solution. These materials are synthesized by adding graphene to melamine sponges and have a reverse wettability feature, which means they are extremely attracting oil and oleophilic, and also superhydrophobic that repel water, causing the effective separation of oil from water.

Pandey et al. (2024) evaluated the application of Ti3C2Tx MXene-based membranes for water purification to study the hydrophilic performance of this membrane as a 2D nanomaterial. They specifically focused on how altering the membrane's hydrophilicity and surface charge affects its performance in desalination. The performance of two types of MXene-based membranes was evaluated in this study. One of them hybridized with WS2 nanosheets, and another one was modified with polyvinyl phosphonic acid (PVPA). This study reported that these modifications not only improve water permeance but also improve salt rejection by enhancing the selectivity.

2.7 Specific molecular recognition and functionalization

Due to the exclusive structural features and modifiable surface chemistry, 2D nanomaterials have attracted more attention for application in membrane-based separation. These materials afford a flexible platform for enhancing the selectivity and specificity of separation membranes in the field of functionalization and molecular recognition (Hyun et al., 2019).

2.7.1 Surface functionalization techniques

The deliberate alteration of the surface of 2D nanomaterials to comprise specific chemical agents or functional groups, active sites, or molecular structures that interact with target species in a certain way is known as functionalization. The covalent method is one of the functionalization strategies defined as the addition of chemical groups, molecules, or atoms to the surface of 2D nanomaterials through covalent bonds. This technique is suitable for 2D materials because of their large surface area and exclusive features that allow precise tuning of their chemical, mechanical, and electronic properties (Zeng et al., 2025b). Adapted applications are caused possible by covalent functionalization, which improves environmental stability, conductivity, and dispersibility. TGA, the XPS, and IR spectroscopy are the analyses that are used to verify the creation of covalent the bonds (Wang et al., 2016; Garrido et al., 2024).

Noncovalent functionalization is the other strategy which use of van der Waals interaction, hydrogen bonding, or π–π stacking, to attach recognition elements such as peptides, polymers, or small molecules, without changing the structure of the nanomaterials. In this technique, the molecules or other considered materials attached to the surface of the 2D nanomaterials without forming strong covalent bonds. In this method, the properties of 2D nanomaterials, including their interactions with other materials is modified, while their intrinsic characteristics are maintained (Zhou et al., 2019). Since the interactions in noncovalent functionalization are weaker than that of the covalent bonding and are less likely to damage the structure and electrical properties of the 2D material. These interactions may prevent the nanomaterials aggregation and so improve the dispersion of them in solvents by enhancing their processability (Georgakilas et al., 2016). The other methods for functionalizing 2D nanomaterials denoted "In Situ Growth of Receptors" that is used to offer molecular recognition for extremely selective separation. The molecules or functional groups are directly grown onto the surface of the nanomaterial during synthesis to enhance properties such as dispersion. The ability of 2D nanomaterials to interact with other molecules can be improved by adding specific receptors, which develops their applicability for a diversity of applications (Wang et al., 2024).

2.7.2 Molecular recognition mechanisms

The fundamental basis for 2D nanomaterials' molecular recognition capabilities is their Size and Shape Selectivity, which allows for accurate control over pore size or interlayer spacing to exclude or allow molecules to pass based on dimensional criteria. In molecular recognition with 2D nanomaterials, the size and shape selectivity is a decisive aspect of their application in various fields, including membrane separation. Highly selective interactions are made possible by the capability to modify these nanomaterials to interact with specific molecules according to their size and shape (Wang et al., 2024). According to the literature, these properties of the nanomaterials can be precisely controlled during synthesis, manufacturing a diversity of surface characteristics and reactivity. These characteristics of the molecule can determine its ability to fit into the binding sites or pores on the surface of the 2D nanomaterial, which can be affected by the thickness and lateral dimensions of the 2D material (Saleh and Hassan, 2023).

When discussing 2D nanomaterials, chemical affinity describes the materials' tendency to interact with particular molecules via attractive forces, resulting in molecular recognition. These interactions can be covalent or noncovalent including electrostatic interactions, π–π stacking, or van der Waals interactions. Designing 2D nanomaterial-based applications in fields like membrane separation requires an understanding of and ability to control these affinities (Zhan et al., 2022). Moreover, surface charge introduced by functionalization has an influence on ion transport via electrostatic interactions, a principle exploited in nanofiltration and ion sieving applications. A molecular recognition via charge-based selectivity by 2D nanomaterials is an approach that leverages electrostatic interactions to selectively bind target molecules. By tailoring the surface charge of 2D nanomaterials, scientists can develop their affinity for specific molecules, leading to improved sensing and separation capabilities. This approach is mainly beneficial in membrane separation and electrodialysis (Zeng et al., 2025a).

For a more detailed explanation, the surface charge modification can be presented. For instance, 2D nanomaterials, such as graphene, MXenes, and transition metal dichalcogenides (TMDs), possess unique properties like high surface area and tunable electronic characteristics. These properties can be leveraged by modifying the surface charge of the 2D nanomaterial. This can be achieved through various methods (Abdulrahman et al., 2025). The Surface functionalization is one of them used in membrane separation processes in which the molecules with specific charges (positive or negative) attach to the nanomaterial surface. Another one is doping that introduces foreign atoms into the crystal lattice of the 2D material to alter its charge distribution. Additionally, the modified surface charge can create electrostatic interactions with target molecules, which can be either attractive or repulsive. By carefully controlling the surface charge, researchers can design 2D nanomaterials that selectively bind molecules with complementary charges. For example, a positively charged 2D nanomaterial will preferentially bind to negatively charged molecules (Speranza, 2019). This is employed extensively in membrane-based separation processes to isolate specific molecules from complex mixtures. Or, totally, can be used to detect pollutants in water or air. For example, a negatively charged nanomaterial can be designed to selectively adsorb positively charged heavy metal ions (Chen et al., 2022).

In a study by Majidi et al. (2021), the separation of Cu^{2+} and Hg^{2+} as heavy metal ions from water was investigated theoretically using a functionalized γ-graphyne-1 nanosheet membrane. The membrane was functionalized with amine and carboxyl functional groups on the edge of pores. The ion-membrane and water molecule-membrane interaction energies, the radial distribution function of cations, and the hydrogen bond results confirm that the functional groups improved the membranes' performance for the cations separation. In other study (Yu et al., 2020), three oxygen-containing functional groups dominant (C-GO, H-GO, and E-GO) GO membranes were fabricated and systematically studied. Based on the results, the oxygen-contained functional groups with different hydrophilicity and electricity controlled the D-spacing of GONMs efficiently, which caused different water permeance and filtering performance of these materials. In addition, C-GONMs with abundant negative charges displayed outstanding antifouling capacity for trypan blue (dye) due to the impact of the electrostatic interaction. In the study by Liu et al. (2020), the advances and challenges of an innovative composite membrane fabricated with GO and MXene were explained. Because of the advantages of the 2D interlayer channels and hydrophilicity, a composite membrane containing GO/MXene displayed superior water flux ($71.9\ L\ m^{-2}\ h^{-1}.bar^{-1}$) in comparison with a pristine GO membrane ($6.5\ L\ m^{-2}\ h^{-1}.bar^{-1}$) under the same experimental conditions. More than 99.5% of the considered small organic molecule was separated from the water. According to other similar studies, the functional groups have a significant impact on membrane performance by improving the characteristics like conductivity, permeability, stability, and selectivity. The type and arrangement of the functional groups can change a membrane's interactions with several molecules, determining its capacity for separation or other purposes.

2.8 Summary and conclusions

In summary, 2D nanomaterials constitute a transformative class of materials with significant potential to advance molecular separation and water purification technologies due to their controllable pore architecture, tunable surface chemistry, and high transport efficiency. Their physicochemical flexibility enables adjustment of size, shape, surface charge, hydrophilicity or hydrophobicity, and interaction mechanisms after synthesis, resulting in improved separation selectivity and permeability compared to traditional membranes. Functionalization is critical for enhancing dispersibility, wettability, and molecular recognition sites, as well as refining chemical interactions such as hydrogen bonding, π–π stacking, host–guest interactions, and van der Waals forces. Despite these advantages, several challenges hinder practical implementation. Scalable production of uniform, defect-free nanosheets remains difficult, and integrating them into mechanically stable, chemically robust membrane architectures is still unresolved. Defects generated during synthesis can impair transport pathways, and although in situ growth on substrates improves stability, consistent pore structure and filtration performance require further optimization. Current demonstrations remain largely at the laboratory scale, with limited pilot-scale or industrial validation. Future progress will depend on developing reliable top-down and bottom-up fabrication strategies, incorporating hybrid reinforcement to enhance mechanical durability, and leveraging simulation-guided design for precise control of surface charge and pore geometry. Equally important are pilot-scale testing, techno-economic assessment, and long-term stability studies under realistic operational conditions. As research moves toward stimuli-responsive and adaptive membrane systems, 2D nanomaterials are poised to evolve into intelligent, energy-efficient platforms capable of addressing global water challenges and enabling next-generation separation technologies.

References

Abdulrahman, G.A.Q., Aziz, A., Qasem, N.A.A., Alazzam, A., 2025. 2D Materials: synthesis, properties, and energy-related applications. Coordination Chemistry Reviews 544, 216950. Available at https://doi.org/10.1016/j.ccr.2025.216950.

Avasarala, S., Bose, S., 2023. 2D nanochannels and huge specific surface area offer unique ways for water remediation and adsorption: assessing the strengths of hexagonal boron nitride in separation technology. Functional Composite Materials 4 (1). Available at https://doi.org/10.1186/s42252-023-00042-2.

Bajgiran, N.K., Majidi, S., Azamat, J., Erfan-Niya, H., 2025. Multiscale modeling advances in MOF-based membranes for heavy metals separation from aqueous solutions. Computational Materials Science 256, 113979. Available at https://doi.org/10.1016/j.commatsci.2025.113979.

Berry, V., 2013. Impermeability of graphene and its applications. Carbon 62, 1–10. Available at https://doi.org/10.1016/j.carbon.2013.05.052.

Bondaz, L., Ronghe, A., Li, S., Čerņevičs, K., Hao, J., Yazyev, O.V., et al., 2023. Selective photonic gasification of strained oxygen clusters on graphene for tuning pore size in the å regimejacs. Au 3 (10), 2844–2854. Available at https://doi.org/10.1021/jacsau.3c00395.

Chatterjee, S., Dey, S., Kar, A.K., 2024. Influence of FRET, charge separation, and density of defect states on the improvement of UV light-driven photocatalytic activity of PMMA capped ZnO nanomaterials. Applied Surface Science 664, 160246. Available at https://doi.org/10.1016/j.apsusc.2024.160246.

Cheng, P., Ferrell, N., Hus, S.M., Moehring, N.K., Coupin, M.J., Warner, J., et al., 2025. Protein-enabled size-selective defect-sealing of atomically thin 2d membranes for dialysis and nanoscale separations. Nano Letters 25 (1), 193–203. Available at https://doi.org/10.1021/acs.nanolett.4c04706.

Chen, X., Zhang, S., Hou, D., Duan, H., Deng, B., Zeng, Z., et al., 2021. Tunable pore size from sub-nanometer to a few nanometers in large-area graphene nanoporous atomically thin membranes. ACS Applied Materials & Interfaces 13 (25), 29926–29935. Available at https://doi.org/10.1021/acsami.1c06243.

Chen, Z., Li, Z., Chen, J., Kallem, P., Banat, F., Qiu, H., 2022. Recent advances in selective separation technologies of rare earth elements: a review. Journal of Environmental Chemical Engineering 10 (1), 107104. Available at https://doi.org/10.1016/j.jece.2021.107104.

Ciampalini, G., Fabbri, F., Menichetti, G., Buoni, L., Pace, S., Mišeikis, V., et al., 2022. Unexpected electron transport suppression in a heterostructured graphene–MoS_2 multiple field-effect transistor architecture. ACS Nano 16 (1), 1291–1300. Available at https://doi.org/10.1021/acsnano.1c09131.

Coe-Sessions, K., Davies, A.E., Dhokale, B., Wenzel, M.J., Mahmoudi Gahrouei, M., Vlastos, N., et al., 2024. functionalized graphene via a one-pot reaction enabling exact pore sizes, modifiable pore functionalization, and precision doping. Journal of the American Chemical Society 146 (48), 33056–33063. Available at https://doi.org/10.1021/jacs.4c10529.

Côté, A.P., El-Kaderi, H.M., Furukawa, H., Hunt, J.R., Yaghi, O.M., 2007. Reticular synthesis of microporous and mesoporous 2D covalent organic frameworks. Journal of the American Chemical Society 129 (43), 12914–12915. Available at https://doi.org/10.1021/ja0751781.

Dai, W., Wang, Y., Li, M., Chen, L., Yan, Q., Yu, J., et al., 2024. 2D Materials-based thermal interface materials: structure, properties, and applications. Advanced Materials 36 (37). Available at https://doi.org/10.1002/adma.202311335.

Dai, Z., Jin, P., Yuan, S., Yang, J., Agrawal, K.V., Wang, H., 2025. A molecularly engineered large-area nanoporous atomically thin graphene membrane for ion separation. Nature Communications 16 (1). Available at https://doi.org/10.1038/s41467-025-59625-1.

Deng, Q., Pan, J., Yin, X., Wang, X., Zhao, L., Kang, S.-gu, et al., 2016. Toward high permeability, selectivity and controllability of water desalination with FePc nanopores. Physical Chemistry Chemical Physics 18 (11), 8140–8147. Available at https://doi.org/10.1039/C6CP00322B.

Ding, L., Wei, Y , Li, L., Zhang, T., Wang, H., Xue, J., et al., 2018. MXene molecular sieving membranes for highly efficient gas separation. Nature Communications 9 (1). Available at https://doi.org/10.1038/s41467-017-02529-6.

Fang, L.J., Chen, J.H., Wang, J.M., Lin, W.W., Lin, X.G., Lin, Q.J., et al., 2021. Hydrophobic two-dimensional MoS_2 nanosheets embedded in a polyether copolymer block amide (PEBA) membrane for recovering pyridine from a dilute solution. ACS Omega 6 (4), 2675–2685. Available at https://doi.org/10.1021/acsomega.0c04852, pubs.acs.org/journal/acsodf.

Garrido, M., Naranjo, A., Pérez, E.M., 2024. Characterization of emerging 2D materials after chemical functionalization. Chemical Science 15 (10), 3428–3445. Available at https://doi.org/10.1039/d3sc05365b, http://pubs rsc.org/en/journals/journal/sc.

Georgakilas, V., Tiwari, J.N., Kemp, K.C., Perman, J.A., Bourlinos, A.B., Kim, K.S., et al., 2016. Noncovalent functionalization of graphene and graphene oxide for energy materials, biosensing, catalytic, and biomedical applications. Chemical Reviews 116 (9), 5464–5519. Available at https://doi.org/10.1021/acs.chemrev.5b00620.

Ghadei, S.K., Sankaran, K.J., Sakthivel, R., 2025. Hybrid nano-interfacial engineering of special-wettable surfaces for advanced oil/water separation: multifunctional strategies and future directions. Results in Surfaces and Interfaces 19. Available at https://doi.org/10.1016/j.rsurfi.2025.100526, https://www.sciencedirect.com/science/journal/26668459.

Guo, L., Zhang, K., Han, X., Zhao, Q., Wang, D., Fu, F., 2019. 2D In-Plane CuS/Bi2WO6 p-n Heterostructures with promoted visible-light-driven photo-fenton degradation performance. Nanomaterials 9 (8), 1151. Available at https://doi.org/10.3390/nano9081151.

Han, Y., Xu, Z., Gao, C., 2013. Ultrathin graphene nanofiltration membrane for water purification. Advanced Functional Materials 23 (29), 3693–3700. Available at https://doi.org/10.1002/adfm.201202601.

Han, Zy, Huang, Lj, Qu, Hj, Wang, Yx, Zhang, Zj, Rong, Ql, et al., 2021. A review of performance improvement strategies for graphene oxide-based and graphene-based membranes in water treatment. Journal of Materials Science 56 (16), 9545–9574. Available at https://doi.org/10.1007/s10853-021-05873-7, www.springer.com/journal/10853.

Hao, Y., Liu, X., Zhang, Y., Zhang, X., Li, Z., Chen, X., 2024. Fine-tuning 2D heterogeneous channels for charge-lock enhanced lithium separation from brine. Advanced Science 11 (41). Available at https://doi.org/10.1002/advs.202406535.

Hirasakl, G.J., 1991. Wettability: fundamentals and surface forces. SPE Formation Evaluation 6 (2), 217–226. Available at https://doi.org/10.2118/17367-pa.

Hoenig, E., Han, Y., Xu, K., Li, J., Wang, M., Liu, C., 2024. In situ generation of (sub) nanometer pores in MoS_2 membranes for ion-selective transport. Nature Communications 15 (1). Available at https://doi.org/10.1038/s41467-024-52109-8.

Hyun, T., Jeong, J., Chae, A., Kim, Y.K., Koh, D.-Y., 2019. 2D-enabled membranes: materials and beyond. BMC Chemical Engineering 1 (1). Available at https://doi.org/10.1186/s42480-019-0012-x.

Jahangirzadeh, M., Bajgiran, N.K., Majidi, S., Azamat, J., Erfan-Niya, H., 2024. Atomistic understanding on desalination performance of pristine graphenylene nanosheet membrane at high applied pressures. Journal of Molecular Graphics and Modelling 132, 108833. Available at https://doi.org/10.1016/j.jmgm.2024.108833.

Jang, J.S., Lee, J., Koo, W.T., Kim, D.H., Cho, H.J., Shin, H., et al., 2020. Pore-size-tuned graphene oxide membrane as a selective molecular sieving layer: toward ultraselective chemiresistors. Analytical Chemistry 92 (1), 957–965. Available at https://doi.org/10.1021/acs.analchem.9b03869, http://pubs.acs.org/journal/ancham.

Jiang, Z.-Y., Liu, H.-L., Ahmed, S.A., Hanif, S., Ren, S.-B., Xu, J.-J., et al., 2017. Insight into Ion transfer through the sub-nanometer channels in zeolitic imidazolate frameworks. Angewandte Chemie 129 (17), 4845–4849. Available at https://doi.org/10.1002/ange.201701279.

Ji, Y., Dong, H., Lin, H., Zhang, L., Hou, T., Li, Y., 2016. Heptazine-based graphitic carbon nitride as an effective hydrogen purification membrane. RSC Advances 6 (57), 52377–52383. Available at https://doi.org/10.1039/C6RA06425F.

Joshi, M., Ren, X., Lin, T., Joshi, R., 2025. Mechanistic insights into gas adsorption on 2D materials. Small 21 (7). Available at https://doi.org/10.1002/smll.202406706.

Kim, H.W., Yoon, H.W., Yoon, S.M., Yoo, B.M., Ahn, B.K., Cho, Y.H., et al., 2013. Selective gas transport through few-layered graphene and graphene oxide membranes. Science 342 (6154), 91–95. Available at https://doi.org/10.1126/science.1236098, http://www.sciencemag.org/content/342/6154/91.full.pdf.

Kim, H., Kim, S., Park, C., 2024. Advancements and challenges in ceramic membranes incorporating two-dimensional nanomaterials for semiconductor wastewater treatment: a critical review. Journal of Water Process Engineering 68, 106308. Available at https://doi.org/10.1016/j.jwpe.2024.106308.

Lee, J.-H., Jeong, S.-Y., Son, Y.-D., Lee, S.-W., 2023. Facile fabrication of TiO_2 quantum dots-anchored g-C3N4 nanosheets as 0D/2D heterojunction nanocomposite for accelerating solar-driven photocatalysis. Nanomaterials 13 (9), 1565. Available at https://doi.org/10.3390/nano13091565.

Liu, Y., Peng, J., Wang, S., Xu, M., Gao, M., Xia, T., et al., 2018. Molybdenum disulfide/graphene oxide nanocomposites show favorable lung targeting and enhanced drug loading/tumor-killing efficacy with improved biocompatibility. NPG Asia Materials 10 (1), e458. Available at https://doi.org/10.1038/am.2017.225.

Liu, D., Zhang, S., Wang, J., Peng, T., Li, R., 2019. Direct Z-scheme 2D/2D photocatalyst based on ultrathin g-C3N4 and WO_3 nanosheets for efficient visible-light-driven H_2 generation. ACS Applied Materials & Interfaces 11 (31), 27913–27923. Available at https://doi.org/10.1021/acsami.9b08329.

Liu, T., Liu, X., Graham, N., Yu, W., Sun, K., 2020. Two-dimensional MXene incorporated graphene oxide composite membrane with enhanced water purification performance. Journal of Membrane Science 593, 117431. Available at https://doi.org/10.1016/j.memsci.2019.117431.

Liu, S., Wang, J., Shao, J., Ouyang, D., Zhang, W., Liu, S., et al., 2022. Nanopatterning technologies of 2D materials for integrated electronic and optoelectronic devices. Advanced Materials 34 (52). Available at https://doi.org/10.1002/adma.202200734.

Li, J., Li, Y., Lu, Y., Wang, Y., Guo, Y., Shi, W., 2023. Preparation of 2D materials and their application in oil–water separation. Biomimetics 8 (1), 35. Available at https://doi.org/10.3390/biomimetics8010035.

Li, H., Song, Z., Zhang, X., Huang, Y., Li, S., Mao, Y., et al., 2013a. Ultrathin, molecular-sieving graphene oxide membranes for selective hydrogen separation. Science 342 (6154), 95–98. Available at https://doi.org/10.1126/science.1236686.

Li, Z., Wang, Y., Kozbial, A., Shenoy, G., Zhou, F., McGinley, R., et al., 2013b. Effect of airborne contaminants on the wettability of supported graphene and graphite. Nature Materials 12 (10), 925–931. Available at https://doi.org/10.1038/nmat3709.

Li, Q., Zhao, F., Qu, C., Shang, Q., Xu, Z., Yu, L., et al., 2018. Two-dimensional morphology enhances light-driven H_2 generation efficiency in CdS nanoplatelet-pt heterostructures. Journal of the American Chemical Society 140 (37), 11726–11734. Available at https://doi.org/10.1021/jacs.8b06100.

Majidi, S., Erfan-Niya, H., Azamat, J., Cruz-Chú, E.R., Walther, J.H., 2021. Efficient removal of heavy metals from aqueous solutions through functionalized γ-graphyne-1 membranes under external uniform electric fields: insights from molecular dynamics simulations. The Journal of Physical Chemistry B 125 (44), 12254–12263. Available at https://doi.org/10.1021/acs.jpcb.1c06617.

Majidi, S., Erfan-Niya, H., Azamat, J., Cruz-Chú, E.R., Walther, J.H., 2023a. Understanding the performance of graphdiyne membrane for the separation of nitrate ions from aqueous solution at the atomistic scale. Journal of Molecular Graphics and Modelling 118, 108337. Available at https://doi.org/10.1016/j.jmgm.2022.108337.

Majidi, S., Erfan-Niya, H., Azamat, J., Ziaei, S., Cruz-Chú, E.R., Walther, J.H., 2023b. Membrane based water treatment: insight from molecular dynamics simulations. Separation & Purification Reviews 52 (4), 336–352. Available at https://doi.org/10.1080/15422119.2022.2111263.

Marchetti, P., Jimenez Solomon, M.F., Szekely, G., Livingston, A.G., 2014. Molecular separation with organic solvent nanofiltration: a critical review. Chemical Reviews 114 (21), 10735–10806. Available at https://doi.org/10.1021/cr500006j.

Murali, A., Lokhande, G., Deo, K.A., Brokesh, A., Gaharwar, A.K., 2021. Emerging 2D nanomaterials for biomedical applications. Materials Today 50, 276–302. Available at https://doi.org/10.1016/j.mattod.2021.04.020.

Novoselov, K.S., Geim, A.K., Morozov, S.V., Jiang, D., Zhang, Y., Dubonos, S.V., et al., 2004. Electric field effect in atomically thin carbon films. Science 306 (5696), 666–669. Available at https://doi.org/10.1126/science.1102896.

Pakdel, S., Majidi, S., Azamat, J., Erfan-Niya, H., 2021. Graphene Oxide and Reduced Graphene Oxide as Nanofillers in Membrane Separation. Springer Science and Business Media LLC, pp. 113–144. https://doi.org/10.1007/978-3-030-72457-3_5.

Pandey, L., Liang, W., VahidMohammadi, A., Zhang, T., Gogotsi, Y., Wanunu, M., 2024. Hydrophilicity and surface charge modulation of $Ti_3C_2T_x$ MXene based membranes for water desalination. RSC Advances 14 (30), 21635–21643. Available at https://doi.org/10.1039/d4ra02678k.

Qi, Y., Sadi, M.A., Hu, D., Zheng, M., Wu, Z., Jiang, Y., et al., 2023. Recent progress in strain engineering on van der waals 2D materials: tunable electrical, electrochemical, magnetic, and optical properties. Advanced Materials 35 (12). Available at https://doi.org/10.1002/adma.202205714.

Qu, H., Rayabharam, A., Wu, X., Wang, P., Li, Y., Fagan, J., et al., 2021. Selective filling of n-hexane in a tight nanopore. Nature Communications 12 (1). Available at https://doi.org/10.1038/s41467-020-20587-1.

Shakeelur Raheman, A.R., Mane, R.S., Wilson, H.M., Jha, N., 2021. CdSe quantum dot/white graphene hexagonal porous boron nitride sheet (h-PBNs) heterostructure photocatalyst for solar driven H_2 production. Journal of Materials Chemistry C 9 (27), 8524–8536. Available at https://doi.org/10.1039/d1tc01556g.

Sajid, M., Sajid Jillani, S.M., Baig, N., Alhooshani, K., 2022. Layered double hydroxide-modified membranes for water treatment: recent advances and prospects. Chemosphere 287 (Pt 1), 132140. Available at https://doi.org/10.1016/j.chemosphere.2021.132140.

Saleh, H.M., Hassan, A.I., 2023. Synthesis and characterization of nanomaterials for application in cost-effective electrochemical devices. Sustainability 15 (14), 10891. Available at https://doi.org/10.3390/su151410891.

Snapp, P., Kim, J.M., Cho, C., Leem, J., Haque, M.F., Nam, S.W., 2020. Interaction of 2D materials with liquids: wettability, electrochemical properties, friction, and emerging directions. NPG Asia Materials 12 (1). Available at https://doi.org/10.1038/s41427-020-0203-1.

Speranza, G., 2019. The role of functionalization in the applications of carbon materials: an overview. C 5 (4), 84. Available at https://doi.org/10.3390/c5040084.

Tang, K., Wang, Y., Gong, C., Yin, C., Zhang, M., Wang, X., et al., 2022. Electronic and photoelectronic memristors based on 2D materials. Advanced Electronic Materials 8 (4). Available at https://doi.org/10.1002/aelm.202101099.

Tang, B., Sivan, M., Leong, J.F., Xu, Z., Zhang, Y., Li, J., et al., 2024. Solution-processable 2D materials for monolithic 3D memory-sensing-computing platforms: opportunities and challenges. npj 2D Materials and Applications 8 (1). Available at https://doi.org/10.1038/s41699-024-00508-2.

Tawfik, S.A., Isayev, O., Stampfl, C., Shapter, J., Winkler, D.A., Ford, M.J., 2019. Efficient prediction of structural and electronic properties of hybrid 2D materials using complementary DFT and machine learning approaches. Advanced Theory and Simulations 2 (1). Available at https://doi.org/10.1002/adts.201800128, www.advtheorysimul.com.

Thomas, A., Fischer, A., Goettmann, F., Antonietti, M., Müller, J.O., Schlögl, R., et al., 2008. Graphitic carbon nitride materials: variation of structure and morphology and their use as metal-free catalysts. Journal of Materials Chemistry 18 (41), 4893–4908. Available at https://doi.org/10.1039/b800274f.

Tong, Y., Dai, S., Jiang, D.E., 2024a. 2D Carbonaceous materials for molecular transport and functional interfaces: simulations and insights. Accounts of Chemical Research 57 (18), 2678–2688. Available at https://doi.org/10.1021/acs.accounts.4c00398, http://pubs.acs.org/journal/achre4.

Tong, Y.H., Luo, L.H., Jia, R., Han, R., Xu, S.J., Xu, Z.L., 2024b. Whether membranes developed for organic solvent nanofiltration (OSN) tend to be hydrophilic or hydrophobic? — a review. Heliyon 10 (2). Available at https://doi.org/10.1016/j.heliyon.2024.e24330, https://www.sciencedirect.com/science/journal/24058440.

Tripathi, M., Lee, F., Michail, A., Anestopoulos, D., McHugh, J.G., Ogilvie, S.P., et al., 2021. Structural defects modulate electronic and nanomechanical properties of 2D materials. ACS Nano 15 (2), 2520–2531. Available at https://doi.org/10.1021/acsnano.0c06701.

Wang, A., Yu, W., Huang, Z., Zhou, F., Song, J., Song, Y., et al., 2016. Covalent functionalization of reduced graphene oxide with porphyrin by means of diazonium chemistry for nonlinear optical performance. Scientific Reports 6 (1). Available at https://doi.org/10.1038/srep23325.

Wang, J., Zhu, J., Zhang, Y., Liu, J., Van der Bruggen, B., 2017a. Nanoscale tailor-made membranes for precise and rapid molecular sieve separation. Nanoscale 9 (9), 2942–2957. Available at https://doi.org/10.1039/C6NR08417F.

Wang, W., Li, G., Xia, D., An, T., Zhao, H., Wong, P.K., 2017b. Photocatalytic nanomaterials for solar-driven bacterial inactivation: recent progress and challenges. Environmental Science: Nano 4 (4), 782–799. Available at https://doi.org/10.1039/C7EN00063D.

Wang, S., Dai, S., Jiang, D.E., 2019. Continuously tunable pore size for gas separation via a bilayer nanoporous graphene membrane. ACS Applied Nano Materials 2 (1), 379–384. Available at https://doi.org/10.1021/acsanm.8b01967, https://pubs.acs.org/journal/aanmf6.

Wang, X.R., Yang, K.X., Ding, S.S., Li, Y.Y., Zhou, B.X., Huang, G.F., et al., 2021. 2D Amorphous CoO incorporated g-C_3N_4 nanotubes for improved photocatalytic performance. Physica Status Solidi - Rapid Research Letters 15 (9). Available at https://doi.org/10.1002/pssr.202100254, http://onlinelibrary.wiley.com/journal/10.1002/(ISSN)1862-6270.

Wang, Y., Niu, Z., Dai, Y., Mu, P., Li, J., 2023. Two-dimensional nanomaterial MXenes for efficient gas separation: a review. Nanoscale 15 (9), 4170–4194. Available at https://doi.org/10.1039/d2nr06625d.

Wang, Y., Zhang, X., Yue, H., 2024. Two-dimensional nanomaterials induced nano-bio interfacial effects and biomedical applications in cancer treatment. Journal of Nanobiotechnology 22 (1). Available at https://doi.org/10.1186/s12951-024-02319-5, https://jnanobiotechnology.biomedcentral.com.

Werber, J.R., Osuji, C.O., Elimelech, M., 2016. Materials for next-generation desalination and water purification membranes. Nature Reviews Materials 1. Available at https://doi.org/10.1038/natrevmats.2016.18, www.nature.com/natrevmats/.

Wu, B., Song, Z., Xiang, Y., Sun, H., Yao, H., Chen, J., 2024. Desalination performance of MoS_2 membranes with different single-pore sizes: a molecular dynamics simulation study. ACS Omega 9 (21), 22851–22857. Available at https://doi.org/10.1021/acsomega.4c01208.

Wu, Z.H., Wang, M., Liu, X.Y., Wu, J., Liu, M.M., Liu, Y., et al., 2025. Spatial size manipulation of 1D/2D channels in covalent organic framework membranes through dopamine chemistry for ion separations. Advanced Functional Materials 35 (9). Available at https://doi.org/10.1002/adfm.202416228, http://onlinelibrary.wiley.com/journal/10.1002/(ISSN)1616-3028.

Xiong, J., Di, J., Li, H., 2020. Charge steering in ultrathin 2D nanomaterials for photocatalysis. Journal of Materials Chemistry A 8 (26), 12928–12950. Available at https://doi.org/10.1039/d0ta04923a, http://pubs.rsc.org/en/journals/journal/ta.

Yang, C., Wang, G., Liu, M., Yao, F., Li, H., 2021. Mechanism, material, design, and implementation principle of two-dimensional material photodetectors. Nanomaterials 11 (10), 2688. Available at https://doi.org/10.3390/nano11102688.

Yu, H., He, Y., Xiao, G., Fan, Y., Ma, J., Gao, Y., et al., 2020. The roles of oxygen-containing functional groups in modulating water purification performance of graphene oxide-based membrane. Chemical Engineering Journal 389, 124375. Available at https://doi.org/10.1016/j.cej.2020.124375.

Zeng, X., Meng, C., Xu, Z., Li, X., Zhu, H., Li, G., 2025a. Recent advances on the positively-charged nanofiltration membranes for Mg^{2+}/Li^{+} separation through interfacial polymerization. Nanomaterials 15 (13), 967. Available at https://doi.org/10.3390/nano15130967.

Zeng, J., Zhao, Q., Xiong, Z., Zhang, S., Deng, S., Liu, D., et al., 2025b. Surface functionalization of two-dimensional nanomaterials beyond graphene: applications and ecotoxicity. Advances in Colloid and Interface Science 336, 103357. Available at https://doi.org/10.1016/j.cis.2024.103357.

Zhang, H., Quan, X., Chen, S., Fan, X., Wei, G., Yu, H., 2018. Combined effects of surface charge and pore size on co-enhanced permeability and ion selectivity through RGO-OCNT nanofiltration membranes. Environmental Science & Technology 52 (8), 4827–4834. Available at https://doi.org/10.1021/acs.est.8b00515.

Zhang, J., Li, Z., Zhan, K., Sun, R., Sheng, Z., Wang, M., et al., 2019. Two dimensional nanomaterial-based separation membranes. ELECTROPHORESIS 40 (16-17), 2029–2040. Available at https://doi.org/10.1002/elps.201800529.

Zhang, H., Zheng, Y., Yu, S., Chen, W., Yang, J., 2022. A review of advancing two-dimensional material membranes for ultrafast and highly selective liquid separation. Nanomaterials 12 (12), 2103. Available at https://doi.org/10.3390/nano12122103.

Zhan, J., Lei, Z., Zhang, Y., 2022. Non-covalent interactions of graphene surface: mechanisms and applications. Chem 8 (4), 947–979. Available at https://doi.org/10.1016/j.chempr.2021.12.015, http://www.cell.com/chem/home.

Zhao, K., Lee, W.-C., Rezaei, M., Chi, H.-Y., Li, S., Villalobos, L.F., et al., 2024a. Tuning pore size in graphene in the angstrom regime for highly selective ion–ion separation. ACS Nano. Available at https://doi.org/10.1021/acsnano.3c11068.

Zhao, C., Bao, J., Gao, Y., Yan, X., Sun, F., Cheng, H., et al., 2024b. Layer distortion engineering for spontaneous charge separation in two-dimensional perovskites. ACS Energy Letters 9 (10), 4994–5001. Available at https://doi.org/10.1021/acsenergylett.4c02099.

Zhou, Y., Fang, Y., Ramasamy, R., 2019. Non-covalent functionalization of carbon nanotubes for electrochemical biosensor development. Sensors 19 (2), 392. Available at https://doi.org/10.3390/s19020392.

Zhu, J., Hou, J., Uliana, A., Zhang, Y., Tian, M., Van der Bruggen, B., 2018. The rapid emergence of two-dimensional nanomaterials for high-performance separation membranes. Journal of Materials Chemistry A 6 (9), 3773–3792. Available at https://doi.org/10.1039/c7ta10814a.

Zou, Y., Zhang, Y., Zhang, J., He, P., Fan, Z., He, N., et al., 2025. Cooperative steric modulation of flexibility, disorder, and pore size in two-dimensional covalent organic framework membranes for enhanced selective ion sieving. Journal of the American Chemical Society 147 (36), 32580–32590. Available at https://doi.org/10.1021/jacs.5c07154

CHAPTER

3

Designs of water purification technologies with 2D nanomaterials

Yuanyuan Yu, Qingxiang Guo, Ge'er Wang, Junyi Ai, Yue Qing, and Le Han

Key Laboratory of the Three Gorges Reservoir Region's Eco-Environment, Ministry of Education, College of Environment and Ecology, Chongqing University, Chongqing, P.R. China

3.1 Introduction

With the rapid growth of the global population and accelerated urbanization, the challenges of water scarcity and pollution are becoming increasingly severe (He et al., 2021). Industrial effluents, agricultural runoff, and domestic sewage are widely contaminated to some degree with harmful constituents such as heavy metal ions, organic pollutants, microorganisms, and excessive salinity, posing serious threats to ecosystems and human health (Wang et al., 2024a). To this end, the development of efficient, ecofriendly, and low-energy water purification technologies has emerged as a critical focus in the field of water treatment.

Two-dimensional nanomaterials (2D NMs), characterized by atomic- or nanoscale thickness, layered structures, and pronounced anisotropy, have attracted extensive attention in recent years for water purification applications (Pang et al., 2021). Typical 2D NMs include graphene oxide (GO) (Zhang et al., 2021c), reduced graphene oxide (rGO) (Xing et al., 2023), MXenes (Ding et al., 2020), transition metal sulfides (e.g., MoS_2) (Wu et al., 2023b), layered double hydroxides (LDHs) (Gao et al., 2024), and black phosphorus (BP) (Kwon et al., 2016). These materials possess several distinct advantages (Wan et al., 2016): (1) an ultrahigh specific surface area offering abundant adsorption and reaction sites; (2) excellent electrical and thermal conductivity enabling efficient mass and energy transfer; (3) abundant surface functional groups that impart tunable hydrophilicity, selectivity, and reactivity; and (4) controllable interlayer spacing and porous structures suited for molecular sieving, ion rejection, and fluid regulation. Compared with conventional functional materials, 2D NMs demonstrate superior performance in pollutant rejection, selective adsorption, photothermal conversion, electron transport, antifouling, and structural stability, thereby opening new avenues for the design of next-generation high-performance water purification technologies.

Two-Dimensional Nanomaterials for Water Purification
DOI: https://doi.org/10.1016/B978-0-443-44530-9.00004-3

Focusing on "System Design of 2D NMs-Based Water Purification Technologies," this chapter systematically reviews the structural configurations, key components, and functional mechanisms of 2D NMs in diverse water treatment systems, with particular emphasis on their structure–function synergism in membrane separation, thermal-driven processes, photocatalysis, electrochemical treatment, and adsorption pathways. The chapter aims to provide readers with an integrated perspective across materials science and water treatment engineering, offering a systematic understanding of how 2D NMs serve as critical building blocks within the "configuration–component–function" triad of water purification system design, and to propose new strategies and tools for researchers engaged in the development of efficient purification systems.

3.2 Configurations, components, and functions of filtration membranes for water

Membrane separation technology has become a core unit operation in modern water treatment. By employing selective barriers, it enables efficient separation of solutes from solvents and is widely deployed in drinking water purification, municipal wastewater treatment, and industrial water reclamation (Xu et al., 2025). Common membrane types include microfiltration (MF), ultrafiltration (UF), nanofiltration (NF), and reverse osmosis (RO), which mainly differ in pore size and sieving mechanism. Nevertheless, polymeric membranes face inherent trade-offs between flux and selectivity, as well as vulnerability to fouling and limited mechanical robustness (Qin et al., 2025). In recent years, two-dimensional nanomaterials (2D NMs), such as graphene oxide (GO), MXene, and transition-metal chalcogenides, which have shown great promise for membrane design owing to their layered architectures, high specific surface areas, and rich surface functionalities.

3.2.1 Typical configurations and components of water filtration membranes

A membrane module is an integrated system comprising functional layers and hydraulic channels, and its architecture directly shapes mass-transfer pathways, fouling control, and modular integration potential. Depending on membrane format and operating environment, modules can be categorized as shown in Fig. 3.1.

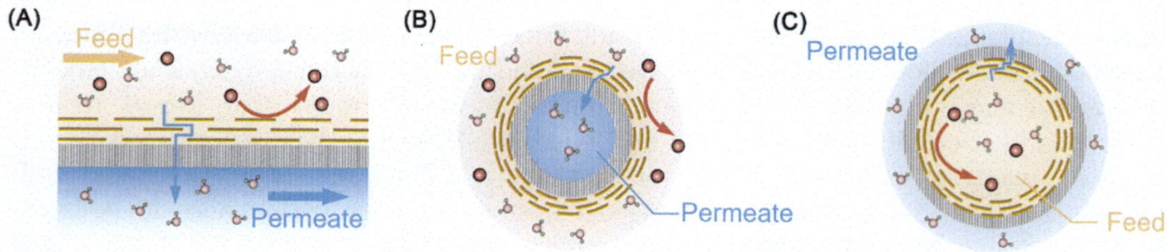

FIGURE 3.1 Representative configurations and components of 2D-NM-enabled filtration membranes. (A) Flat-sheet, (B) hollow fiber, and (C) tubular/spiral-wound.

Flat-sheet membranes are commonly used in laboratory studies, decentralized treatment systems, and early-stage industrial processes. Membrane sheets are laid against a support plate, yielding a simple structure that is easy to maintain. This format is suitable for directly integrating 2D NMs to form active surface layers or self-supporting films, for example, GO layer-by-layer stacks and MXene nanosheet films prepared by vacuum filtration or interfacial self-assembly (Yu et al., 2025; see Fig. 3.1A). However, such laminates may suffer from interlayer collapse and hydraulic failure under long-term operation (Zheng et al., 2017). Crosslinkers or inorganic backings (e.g., TiO_2, Al_2O_3) can improve mechanical stability, but often at the expense of water flux-another manifestation of the strength-permeance trade-off (Zhang et al., 2022).

Hollow-fiber membranes provide extremely high surface area, compact packing, and compatibility with high-throughput continuous operation; they are therefore prevalent in wastewater treatment and membrane bioreactors (MBRs). Because 2D NMs are nanoscale and processable, they are commonly integrated on inner/outer fiber surfaces via coating, chemical infiltration, or interfacial bonding (see Fig. 3.1B; Su et al., 2020). Yet achieving uniform deposition on high-curvature fibers and maintaining long-term adhesion remain challenging. Initiator layers such as polydopamine (PDA) can strengthen adhesion, but may perturb the original pore structure (Alves et al., 2024).

The spiral-wound format is the most mature architecture in seawater desalination and RO/NF industries. It consists of stacked membrane leaves and spacers rolled into a compact element, combining high area density with energy-recovery compatibility and standardized packaging (Oron et al., 2024). 2D NMs are typically incorporated in two ways (Zhang et al., 2019): (1) intercalated into the polyamide active layer to construct selective nanochannels; or (2) applied as surface coatings to endow hydrophilicity, antifouling, or antibiofouling functions for long-term stability (Fig. 3.1C). However, the sealed construction and high-shear environment can lead to interfacial instability and coating delamination, calling for improved mechanical adhesion and chemical durability.

Tubular modules employ rigid ceramics or high-strength polymers as supports, with the membrane layer formed on the inner or outer tube wall (Neto et al., 2020). They are suitable for viscous effluents and organic-solvent recovery, offering high pressure and corrosion resistance (Merlet et al., 2020). The large-diameter channels and accessible housings facilitate uniform, large-area deposition or interfacial compounding of 2D NMs (Fig. 3.1C). 2D NMs can act as direct filtration layers or be coupled with ceramic substrates to enhance antifouling and wetting control (Yang et al., 2019). Given the harsh conditions, however, stringent stability with respect to pH, temperature, and adhesion is required.

In summary, different membrane architectures target distinct scenarios and performance priorities. Through interface tuning, functional enhancement, and structural support, 2D NMs can boost flux and mechanics in flat-sheet membranes, improve wettability and fouling resistance in hollow fibers, and deliver multifunctional, robust interfaces in spiral-wound and tubular modules, together enabling efficient and durable operation.

3.2.2 Optimization strategies of 2D nanomaterials for water-filtration membrane systems

In high-performance separation systems, different membrane layers synergistically provide separation, mechanical support, transport, and antifouling (Pang et al., 2021).

Leveraging atomic-scale thickness, large surface area, and rich surface chemistry, 2D NMs have become key enhancers across layers. Researchers now deploy them from the active layer to the support, interlayers, and surface coatings to impart precise sieving, interfacial reinforcement, rapid transport, and fouling resistance (Fig. 3.2). This multidimensional "function–architecture– interface" codesign improves performance while introducing new challenges in structural optimization and engineering.

In the core active layer, applications of 2D NMs are most extensive and critical. Layer-by-layer stacks or intercalated assemblies build well-ordered 2D nanochannels for efficient rejection of dyes, heavy metal ions, and small organic solutes (Fig. 3.2A). For instance, GO enables ångström-scale control of interlayer spacing to selectively sieve molecules and even ions (Wang et al., 2021b). MXenes, enriched with hydrophilic groups and charged sites, show excellent exclusion of multivalent/hydrated ions (Xu et al., 2023a). Yet lamellar structures tend to swell and delaminate under hydration; crosslinking, ion

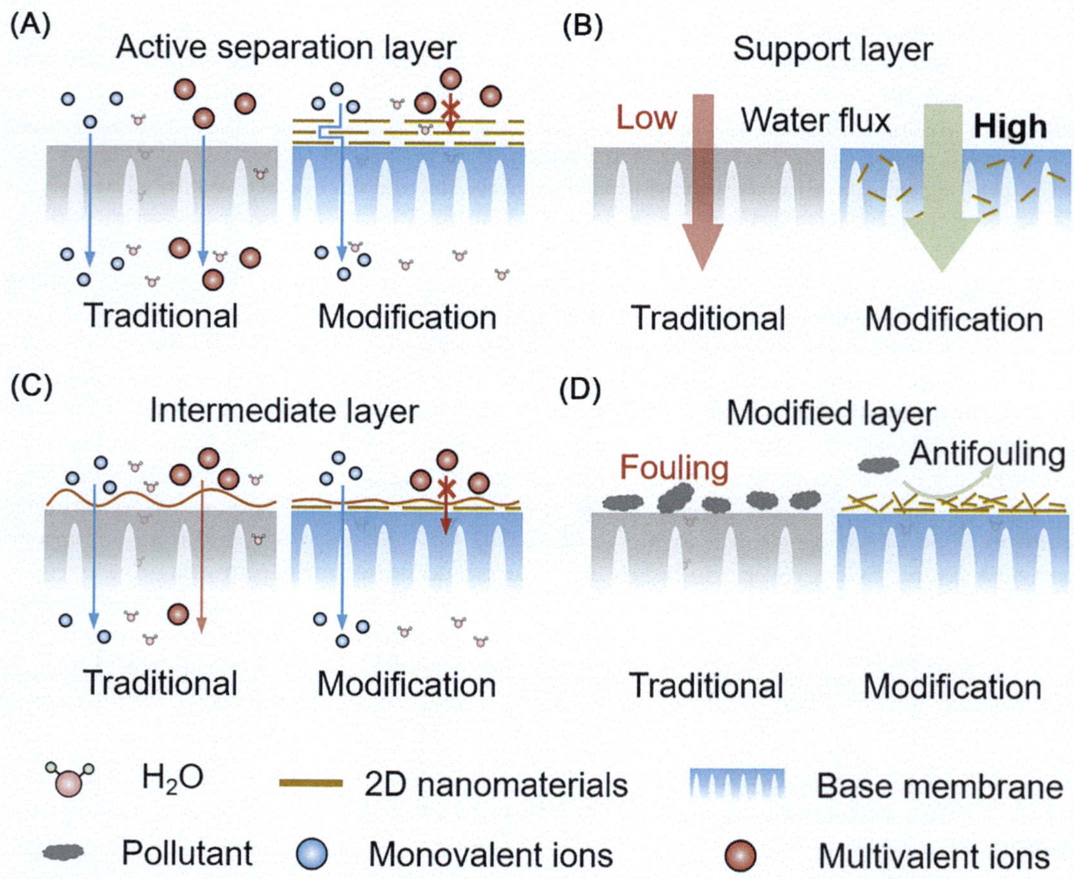

FIGURE 3.2 Schematic of optimization strategies using 2D NMs in water-filtration membranes. (A) Active selective layer, (B) support, (C) interlayer, and (D) surface modification.

bridging, or pillarization with nanofillers are needed to improve stability and pressure tolerance (Zheng et al., 2017).

As the mechanical backbone, the support's wettability, roughness, and interfacial adhesion govern active-layer formation and service life (Fig. 3.2B). Using MXene nanosheets as composite supports can substantially increase the hydrophilicity and affinity of hydrophobic substrates (e.g., PVDF), yielding more uniform and stable composites (Ostovar et al., 2024). Excessive 2D-NM loading, however, may reduce permeability or introduce defects; balancing support strength and mass transfer remains a design challenge.

In multilayer composites, 2D-NM-based interlayers or flow-guiding layers are introduced to optimize water/solute pathways and hydrodynamics (Fig. 3.2C). Porous, ordered layers of GO, MXene, or their hybrids (e.g., MXene/CNT) create "microchannels + conductive pathways" that sustain high water flux while mitigating foulant accumulation (Yang et al., 2022a). These intermediate structures demand tight control over thickness and porosity: too thick increases hydraulic resistance, too thin provides insufficient support; fabrication complexity also challenges scale-up (Wang et al., 2022d).

As shown in Fig. 3.2D, 2D NMs are also used as outer surface functional coatings to improve antifouling, antibacterial behavior, and stimuli responsiveness. Coatings formed by solution casting, self-assembly, or interfacial polymerization (e.g., with GO, MXene, MoS_2) impart super-hydrophilicity or bioinertness that suppresses organic and microbial adhesion (Kaneda et al., 2019). However, shear stresses and cleaning processes can erode coatings; a balance between adhesion and function retention is essential.

Notably, membrane design has moved beyond single-component 2D NMs. As summarized in Table 3.1, more studies couple 2D NMs with other functional nanomaterials to achieve multiple enhancements. For example, GO combines with Ag, TiO_2, ZnO, or polymeric modifiers to provide antibacterial, photocatalytic, and selective-sieving functions. In such coupled systems, 2D NMs serve not only as the core channel builders but also as interfacial reinforcers and dispersion scaffolds that raise synergy and stabilize performance. This "2D platform with functional component" paradigm is becoming mainstream, while placing higher demands on controllable interfacial engineering and multifunctional integration.

In short, 2D NMs offer compelling opportunities across membrane layers, from precise separation to interfacial reinforcement, from guided transport to surface antifouling, constituting a new design paradigm. Yet layer-specific constraints on material properties, assembly, and operating environments remain, including limited structural stability, possible detachment, and integration complexity.

3.2.3 Underlying mechanisms of 2D nanomaterials for enhanced water-filtration performance

Integrating 2D NMs yields multiscale synergies by engineering nanochannel structure, surface charge, and interfacial behavior-improving size sieving, charge repulsion, water permeance, antifouling, and structural stability.

First, 2D NMs active layers provide dual selectivity via size sieving and electrostatic repulsion. GO laminates form nanochannels with an average spacing of around 0.6–0.8 nm

TABLE 3.1 Summary of 2D NMs applications in water filtration membranes.

Type of membrane	2D nanomaterials	Application	Target separation material	Separation mechanism	Separation efficiency[a]	References
Flat-sheet	*MXene*	Salt/water separation	NaCl, KCl, LiCl, $CaCl_2$, and $MgCl_2$	Constructing low-resistance and hydrophilic nanochannels	Water flux increased by 3 times, with NaCl rejection rate >90 %	Ding et al. (2020)
Flat-sheet	MXene-EDTA	Monovalent/Divalent ion separation	Na^+, K^+, Li^+, $Ca^{2+,}$ and Mg^{2+}	Size sieving ↑ Mass transfer resistance ↑	The K^+/Mg^{2+} selectivity was increased by 11 times	Xu et al. (2023a)
Flat-sheet	$Ti_3C_2T_x$ MXene	Oil/water separation	Oil-in-water emulsions, e.g., gasoline/water, petroleum ether/water, cyclohexane/water	Hydrophilic ↑ surface energy ↓ Water molecule transfer resistance ↓	Water flux increased by 4 times; Superhydrophilicity (contact angle < 5 °); Antifouling (flux recovery ratio > 98%)	Zeng et al. (2024)
Flat-sheet	GO	Heavy metal ion rejection Dye rejection	Dye, Pb^{2+}, Cu^{2+}, Cd^{2+}, Cr^{2+}	Size sieving ↑ Charge repulsion ↑ Hydrophilic ↑	Water flux increased by 28-35 times, with dye and heavy metal ion rejection > 95%	Zhang et al. (2017)
Flat-sheet	GO-NH_2	Heavy metal ion rejection	Co^{2+}	Size sieving ↑ Chemical adsorption ↑	98%	Fang et al. (2014)
Flat-sheet	GO	Heavy metal ion rejection	Pb^{2+} Ni^{2+} Zn^{2+} Cd^{2+}	Charge repulsion ↑	96%, 96%, 97%, 91%	Dai et al. (2021)
Flat-sheet	rGO-Ag	Heavy metal ion rejection	Cr^{6+}	Size sieving ↑	100%	Han et al. (2022)
Flat-sheet	GO-Ag	Heavy metal ion rejection	Cr^{3+} Pb^{2+} Cu^{2+}	Size sieving ↑ Charge repulsion ↑	97%, 100 %, 100%	Han et al. (2022)
Flat-sheet	GO-MnO_2	Heavy metal ion rejection	Ni^{2+} Cu^{2+}	Charge repulsion ↑ Complexation ↑	67.4%, 87.1%	Ibrahim et al. (2022)
Flat-sheet	GO/SAA	Heavy metal ion rejection	Ni^{2+} Pb^{2+}	Charge repulsion ↑	88%, 92%	Chandio et al. (2021)
Flat-sheet	rGO-K^+	Heavy metal ion rejection	Cr^{3+} Fe^{3+}	Charge repulsion ↑	99.4%, 99.8%	Yang et al. (2021)

Flat-sheet	GO-UiO-66	Heavy metal ion rejection	Cu^{2+} Pb^{2+}	Size sieving ↑	99.40%	He et al. (2020b)
Flat-sheet	CW-GO	Dye rejection	MB RB	Size sieving ↑	>90%	Zhang et al. (2021b)
Flat-sheet	ZIF-8-GO	Dye rejection	MB	Charge repulsion ↑	100%	Zhang et al. (2021c)
Flat-sheet	GO-g-C_3N_4	Dye rejection	MB MO	Size sieving ↑	100%, 60%	Wu et al. (2021)
Flat-sheet	GO-TBO	Dye rejection	MO CR	Size sieving ↑ Charge repulsion ↑	100%	Wang et al. (2021b)
Flat-sheet	GO	Dye rejection	RB MB	Size sieving ↑	90%	Zheng et al. (2020)
Flat-sheet	GO-Fe	Dye rejection	MB MO	Size sieving ↑ Charge repulsion ↑	99.5%, 90%	Li et al. (2021)
Flat-sheet	PANI/GO	Dye rejection	MO Allura	Size Sieving ↑ Charge Repulsion ↑	95%, 98%	Chen et al. (2021)
Flat-sheet	GO	Dye rejection	MO CVr Afs AR AB	Size sieving ↑ Charge repulsion ↑	43%, 9%, 95%, 95%, 100%	Nie et al. (no date)
Flat-sheet	UiO-66-$(COOH)_2$/GO	Dye rejection	MB	Size sieving ↑ Charge repulsion ↑	95.10%	Zhang et al. (2018a)
Flat-sheet	UiO-66-rGO	Dye rejection	MB	Transfer channel density ↑	98.70%	Guan et al. (2017)
Flat-sheet	MXene-MOF	Dye rejection	MB	Transfer channel density ↑	85.30%	Zhang et al. (2023a)
Flat-sheet	UiO-66-rGO	Dye rejection	RhB	Transfer channel density ↑	95%	Guan et al. (2017)
Flat-sheet	GO/MIP-202(Zr)	Dye rejection	RhB	Size sieving ↑ Charge repulsion ↑	89.40%	Fei et al. (2023)
Flat-sheet	GO-ZIF-7	Dye rejection	HA	Hydrophilic ↑	94%	Jafarian et al. (2023)

(Continued)

TABLE 3.1 (Continued)

Type of membrane	2D nanomaterials	Application	Target separation material	Separation mechanism	Separation efficiency[a]	References
Spiral-wound	GO	Desalination	NaCl + microorganism	Hydrophilic ↑ size sieving ↑	Rejection rate remained unchanged; flux increased by 80%	Chae et al. (2015)
Spiral-wound	GO	Desalination	NaCl + BSA	Hydrophilic ↑ Size Sieving ↑	Flux increased by 39%	Ali et al. (2016)
Spiral-wound	rGO-TiO_2	Desalination	NaCl + BSA	Hydrophilic ↑ Size Sieving ↑	Desalination rate of 94.45%; flux increased by 21%	Safarpour et al. (2015)
Spiral-wound	GO-ZnO	Desalination	NaCl	Hydrophilic ↑ Size Sieving ↑	99.40%	Rajakumaran et al. (2019)
Spiral-wound	GO-ZnO	Desalination	NaCl	Hydrophilic ↑ Size Sieving ↑	96.30%	Rajakumaran et al. (2019)
Spiral-wound	GO-ZnO	Desalination	NaCl	Hydrophilic ↑ Size Sieving ↑	100%	Rajakumaran et al. (2019)
Hollow fiber	TiO_2/WO_3-GO	Photocatalytic degradation	Crude oil	Hydrophilic ↑ Photocatalysis ↑	TOC removal rate of 98.62%	Anjum et al. (2019)

[a] *Unless otherwise specified, all data refer to rejection rates.*

for tailored rejection according to hydrated radius (Zheng et al., 2017). $Ti_3C_2T_x$ MXenes carry abundant negative groups (e.g., -OH, -O) that repel multivalent cations (e.g., Ca^{2+}, Pb^{2+}) while interfering less with monovalent ions, enabling refined ionic selectivity (Jang et al., 2022).

Second, water permeance is markedly enhanced. GO and MXene sheets are inherently hydrophilic and permit low-friction transport along interlayer planes, forming low-resistance pathways for rapid permeation (Ren et al., 2015). Cross-linked GO with functional components (e.g., diamines or carbon-chain crosslinkers) can further increase flux by >30% (Zhou et al., 2023).

Third, antifouling and self-cleaning benefits arise from 2D NMs' low surface energy, superhydrophilicity, and ease of functionalization, thereby reducing organic adsorption and biofilm attachment. For instance, GO-based coatings with small amounts of PDA maintain stable fluxes with flux recovery ratios exceeding 90% (Abounahia et al., 2022). MXene surfaces can also couple with photocatalytic or electrostimulated responses to enable online self-cleaning (Xu et al., 2023b).

Finally, structural stability is crucial. Thin lamellae lacking internal pillars may delaminate or collapse under pressure or shear. MXene/carbon nanotube (CNT) composites create pillarized frameworks that preserve 80-85 % of initial performance under continuous pressurization and solvent exposure(Wang et al., 2020). Multipoint anchoring and surface crosslinking strengthen MXene–polymer interfaces, improving oxidation and chemical-aging resistance (Mu et al., 2024).

Overall, when systematically integrated across the architecture, 2D NMs deliver additive and cooperative gains: nanochannels enforce size/charge selectivity, connected water pathways boost flux, super-hydrophilic layers limit foulant accumulation, and robust interfaces extend service life, together yielding efficient, stable, and fouling-resistant membranes.

3.2.4 Summary and outlook

This section organizes design strategies and structural integration routes for 2D-NM-enabled water-filtration membranes around the triad of architecture, components, and functions. Thanks to ultrathin thickness, tunable surface chemistry, and high processability, 2D NMs can be co-designed across the active layer, support, interlayers, and surface modifications. Via size sieving, charge control, wetting regulation, and mechanical reinforcement, they significantly improve separation efficiency, water flux, antifouling, and stability. Challenges include limited interfacial compatibility between 2D NMs and polymeric supports, delamination/collapse under pressure or shear in laminated structures, and insufficient long-term validation of smart, multifunctional membranes under complex waters.

Future research should concentrate on three areas: (1) developing green, low-cost, and scalable sources and modification technologies for 2D NMs; (2) building composite membrane modules that combine multifunctional responsiveness with interfacial stability, for example, self-cleaning membranes integrating photothermal, electric-field, and pH responses; and (3) exploring the integrated synergistic effects of 2D NMs across multistage water-treatment processes to advance membrane systems from "structural optimization" toward "intelligent water-treatment platforms."

3.3 Configurations, components, and functions of membrane distillation systems

Membrane distillation (MD) is a thermally driven membrane separation process in which a vapor-pressure gradient transports water vapor from the hot feed side to the cold condensation side, achieving selective liquid–vapor–liquid separation (Hu et al., 2025). MD shows broad promise for high-salinity wastewater treatment, seawater desalination, and ultrapure-water production. Unlike pressure-driven membrane processes, MD typically operates at 40°C– 90°C, accommodates highly concentrated feeds, and theoretically delivers 100% rejection of nonvolatile solutes (Liao et al., 2023). However, conventional hydrophobic polymer membranes (e.g., PVDF, PTFE) suffer from insufficient thermal stability, susceptibility to pore wetting, and fouling accumulation, which constrain long-term operation and scale-up. To overcome these challenges, researchers have incorporated 2D NMs (e.g., MXenes, GO) to improve thermal conductivity, enable controllable surface architectures, and integrate interfacial functionalities, advancing MD membranes from single hydrophobic layers toward multifunctional, hierarchical designs.

3.3.1 Configurations and components of typical membrane distillation systems

MD systems can be categorized by the condenser-side configuration and the mode of vapor transport into several architectures, including direct contact MD (DCMD), air gap MD (AGMD), vacuum MD (VMD), and sweeping gas MD (SGMD) (Kakkar et al., 2025). Each configuration offers distinct strengths and limitations in heat-transfer efficiency, energy consumption, operational complexity, and application scope (see Fig. 3.3).

DCMD is the most common and structurally simplest design. As illustrated in Fig. 3.3A, the feed and the coolant flow along opposite sides of the membrane and directly contact the membrane surfaces, yielding good vapor mass-transfer performance and a compact footprint, well-suited to laboratory studies and small-scale integrated systems. However, substantial convective and conductive heat losses lead to low overall thermal efficiency. Incorporating 2D NMs into the membrane can create high-thermal-conductivity, low-thermal-resistance pathways, effectively alleviating temperature polarization (TP) and enhancing both heat utilization and interfacial vapor transport (Gao et al., 2021).

AGMD introduces a static air gap between the membrane and the condensing plate to reduce conductive heat loss (Fig. 3.3B), thereby improving thermal efficiency and making it a preferred option for low-grade heat sources (e.g., solar energy). However, the high thermal resistance of the air gap limits condensation efficiency and can cause delayed condensation and thermal hysteresis. Constructing 2D NMs composite layers can provide dual functionalities-enhanced infrared radiative heat transfer and photothermal absorption-thereby elevating the membrane-surface temperature and promoting vapor transport across the gap (Zhou, 2022).

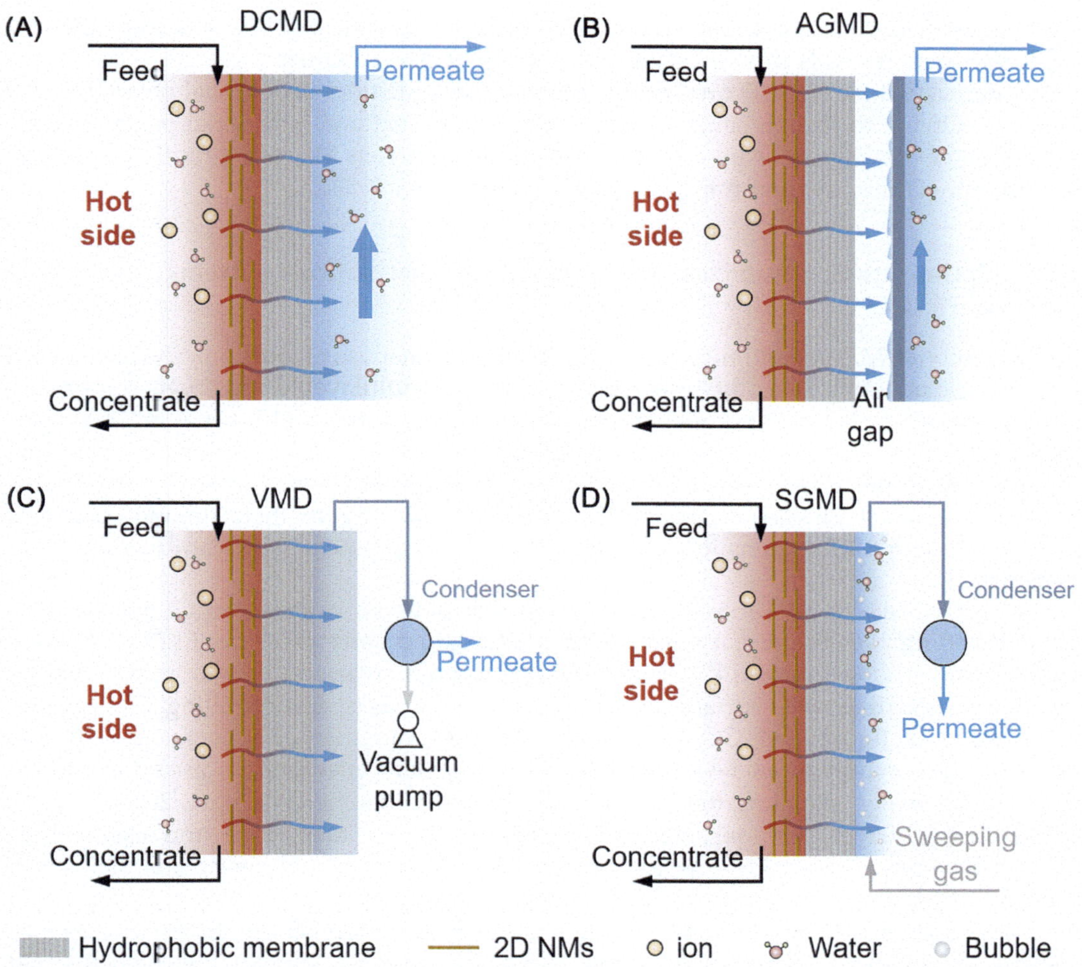

FIGURE 3.3 Schematic illustration of typical membrane distillation configurations and components based on 2D NMs. (A) Direct contact membrane distillation (DCMD), (B) air gap membrane distillation (AGMD), (C) vacuum membrane distillation (VMD), and (D) sweeping gas membrane distillation (SGMD).

VMD applies a vacuum on the permeate side to lower the transmembrane pressure and rapidly remove permeated vapor (Fig. 3.3C), offering very high flux potential and suitability for concentrated brines and high-purity water recovery. This configuration demands membranes with strong compressive resistance and thermal stability. Incorporation of 2D NMs enables composite membranes with improved mechanical strength and thermal conductivity for robust VMD operation.

SGMD uses an inert sweep gas to carry vapor away from the membrane surface, with external condensation, thereby avoiding in-module condenser fouling (Fig. 3.3D). It is appropriate for highly corrosive feeds but requires tight control of gas flow and careful coupling with the condensation system. The antifouling functionality and high thermal conductivity of 2D NMs provide stable support for SGMD modules.

In summary, different MD architectures are suited to distinct operational objectives and scenarios. Owing to their tunable structures and multifunctional interfacial properties, 2D NMs open new avenues for performance enhancement across MD configurations, particularly in antiwetting, thermal transport, and structural robustness.

3.3.2 Optimization strategies for membrane distillation systems using 2D nanomaterials

Conventional MD membranes are typically homogeneous, of only microscale porous hydrophobic material. Recently, researchers introduced multilayered constructs comprising a surface-selective layer, a porous support, intermediate functional layers, and a surface functionalization coating, probably inspired by the achievements in water-permeable pressure-driven membranes. Each layer plays a critical role in sustaining overall membrane performance, yet each also faces specific challenges. Owing to their unique structural attributes and multifunctionality, 2D NMs show strong potential for addressing these challenges.

The surface-selective layer is the key region that governs vapor-selective permeability and antiwetting performance; it must exhibit a high liquid entry pressure (LEP) and strong hydrophobicity. Conventional PVDF or PTFE membranes are susceptible to liquid penetration due to heterogeneous surface energy or pore-size defects. Incorporating 2D nanomaterials such as fluorinated GO (F-GO) and MXenes can markedly increase LEP (see Fig. 3.4A). For example, an F-GO-modified polypropylene membrane raised the LEP to 250 kPa, effectively extending the membrane's operating lifetime (Wu et al., 2023a).

The support layer provides mechanical stability to the membrane and assists heat transfer; however, the intrinsically low thermal conductivity of polymeric materials often leads to TP, thereby diminishing MD efficiency. Incorporating 2D NMs, as shown in Fig. 3.4B, such as rGO and MXenes, into the support can establish low-thermal-resistance heat-transfer pathways while maintaining mechanical strength, thereby reducing thermal losses and alleviating TP (Liu et al., 2025a).

The intermediate functional layer is commonly incorporated into multilayer membrane architectures to provide thermal isolation, regulate heat transfer, or buffer fouling (see Fig. 3.4C). 2D NMs like g-C_3N_4 nanosheets can impart photothermal responsiveness to the membrane, facilitating operation in solar-driven, low-temperature MD systems (Xu et al., 2023c).

The surface functional-modification layer is intended to enhance antifouling performance or impart self-cleaning capability. Introducing 2D NMs such as $Ti_3C_2T_x$ MXene and g-C_3N_4 onto the membrane surface can create functional interfaces with photocatalytic and/or antimicrobial activity (Liu et al., 2025b), which facilitate the degradation of adsorbed contaminants or inhibit biofilm formation, thereby extending service life and stabilizing permeate flux (see Fig. 3.4D).

Table 3.2 summarizes the applications of 2D NMs in membrane distillation. Through synergistic, multilayer architectural design and the judicious incorporation of 2D NMs, MD

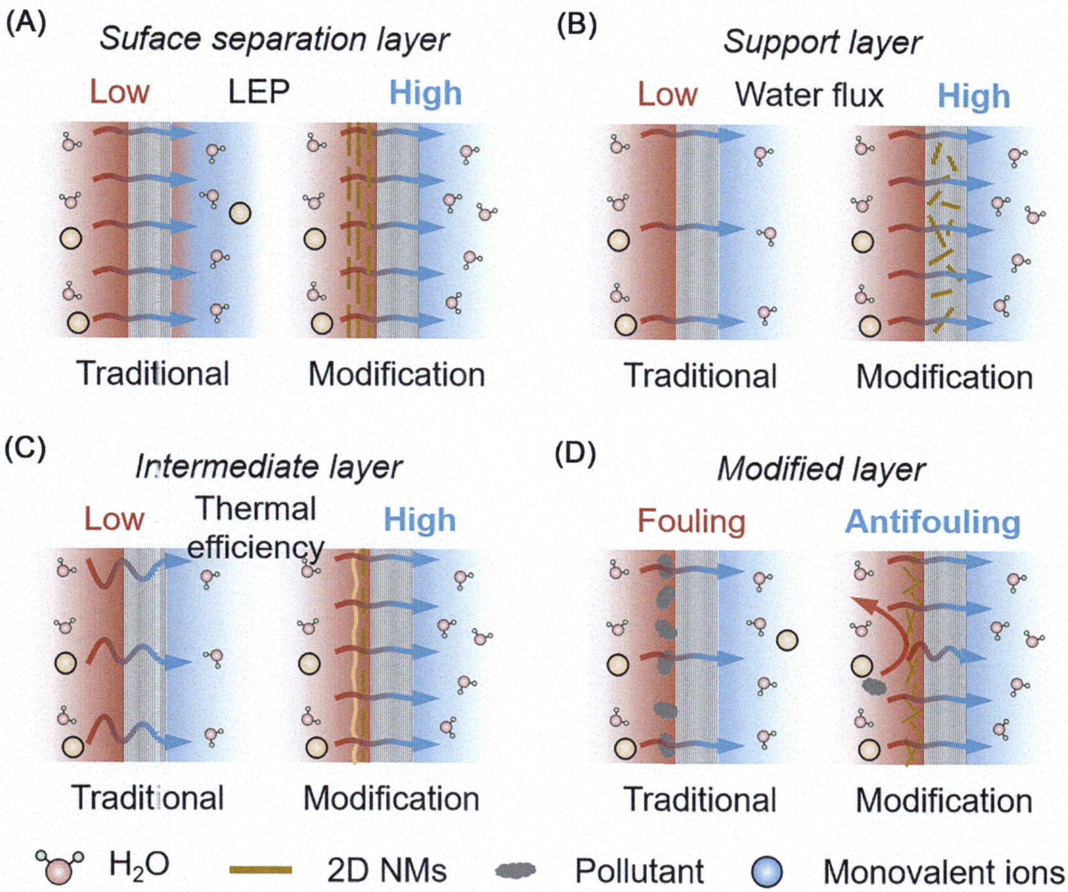

FIGURE 3.4 Schematic illustration of optimization strategies for membrane distillation systems using 2D NMs. (A) Surface-selective layer, (B) support layer, (C) intermediate layer, and (D) surface-modification layer.

systems can achieve comprehensive enhancements in structural robustness, thermal management, wetting resistance, and antifouling performance. This 2D NMs-centric, functionally integrated design paradigm has emerged as a mainstream strategy for elevating the multidimensional performance of MD membranes.

3.3.3 Underlying mechanisms of 2D nanomaterials for enhanced membrane distillation

The performance enhancement imparted by 2D NMs in MD derives from their unique physical structures, chemical compositions, and interfacial regulation capabilities, which act concurrently on multiple core mechanisms, including mass transfer, heat transfer, wetting control, and antifouling, to yield coupled, synergistic improvements.

TABLE 3.2 Summary of the application of 2D NMs in membrane distillation systems.

Type of MD	2D NMs	Application	Separation material	Separation efficiency	Mechanism	References
AGMD	GO	Treatment of complex wastewater containing humic acid and paraffin oil	NaCl Humic acid Paraffin oil Triton X-100	Stable operation for 72 h; flux decreased by 0.1%; removal efficiency >99.9%; FDR as low as 0.0021%/h	Reduced surface roughness and contact angle; suppressed foulant adsorption; Improved antifouling and antiwetting performance.	Alberto et al. (2022)
AGMD	hBN	Treatment of oil/organic mixtures	NaCl Humic acid Paraffin oil Triton X-100	Flux decreased by 1.9 %; removal efficiency >99.9 % FDR=0.0274 %/h	Polar B–N bonds enhanced selective water channels; inhibited foulant deposition.	Srivastava and Murthy (2021)
AGMD	MoS_2	Treatment of complex oily wastewater	NaCl Humic acid Paraffin oil Triton X-100	Flux decreased by 18.1 %; Removal efficiency >99.9 % FDR=0.25 %/h	Hydrophobic materials enhanced oil adsorption; caused slight wetting over time.	Skuse et al. (2024)
AGMD	WS_2	Antifouling against oil and surfactants	NaCl Humic acid Paraffin oil Triton X-100	Stable operation for 72 h; flux decreased by 16.9%; removal efficiency >99.9 %; FDR as low as 0.24 %/h	WS_2 enhanced membrane hydrophobicity; high surface roughness led to slight foulant deposition.	Skuse et al. (2024)
DCMD	MXene	Treatment of SDS-containing saline water and landfill leachate	SDS NaCl	>99 % removal at 50–200 mg/L SDS; stable flux (11.5 LMH) over 24 h; pristine PTFE membrane rapidly wetted; membrane flux decline <3 %	MXene nanosheets formed 5.5 Å sub-nanochannels; constructed partial dehydration energy barrier; blocked large SDS molecules; enabled fast, ordered monolayer water transport; base membrane maintained high evaporation area.	Jiang et al. (2022)

DCMD	Graphene GO rGO	Seawater desalination; treatment of high-salinity wastewater; concentration of RO brine; Brackish water purification	NaCl SDS HA	rGO membrane maintained high flux (149 L/m^2·h) for 66 h in DCMD with SDS fouling resistance GO-modified membrane exhibited low flux decline and high salt rejection in hypersaline water. Monolayer graphene membrane achieved high water flux and near-complete salt rejection	Physical sieving nanochannels regulated water/ion transport; surface polarity contrast suppressed organic adsorption; enhanced surface hydrophobicity and antiwetting ability.	Jia et al. (2022)
DCMD	Graphene	Desalination; treatment of dye wastewater; small-scale drinking water treatment	NaCl dye acid	High water flux (33 L/m^2·h); rejection >99.99 %	Selective water vapor transport.	Maqbool et al. (2025)

3.3.3.1 *Construction of low-resistance vapor pathways*

Layered, plate-like 2D materials such as GO and MXenes can form lamellar, ordered, or expanded micro/nanochannels within the membrane, thereby shortening the migration path of water-vapor molecules and increasing their diffusivity. For example, a PVDF membrane incorporating an MXene layer achieved a more than 30 % increase in vapor flux relative to the unmodified membrane, while maintaining a salt rejection of up to 99.5% (Zhou et al., 2023).

3.3.3.2 *Enhanced thermal conduction*

High-thermal-conductivity 2D NMs (e.g., GO, MXenes) can efficiently deliver heat from the hot side to the membrane exterior interface, thereby reducing energy losses associated with TP, stabilizing the interfacial temperature gradient, and enabling continuous, efficient vapor mass transfer (Yadav et al., 2022).

3.3.3.3 *Enhanced antiwetting performance*

By constructing micro-/nanohierarchical roughness and introducing low-surface-energy moieties (e.g., -CF_3 groups, F-GO), 2D NMs can substantially increase the LEP, stabilize the vapor-liquid meniscus, and suppress liquid breakthrough. A fluorinated-MXene-based dual-superhydrophobic layer raised the LEP to >320 kPa, significantly delaying wetting failure (Huang et al., 2020).

3.3.3.4 *Fouling control and self-cleaning mechanisms*

Certain 2D NMs (e.g., MXenes) exhibit strong photocatalytic and/or photothermal effects, enabling in situ degradation of adherent foulants during operation ("on-stream self-cleaning")(Ray et al., 2024), thereby reducing cleaning frequency and extending membrane lifetime.

3.3.3.5 *Interfacial stability and structural reinforcement*

Under high vacuum (VMD) or high flow velocity (SGMD) conditions, membranes are prone to interlayer delamination and mechanical fatigue. Synergistically integrating 2D NMs with functional components such as PDA and CNTs to form composite supports can markedly enhance compressive and shear strength. An MXene/CNT cross-linked membrane maintained >85% of its initial flux after 72 h at a negative pressure of 0.08 MPa, exhibiting interfacial stability superior to membranes composed of a single material.

3.3.4 Summary and outlook

In summary, the configuration of MD membranes is evolving from single hydrophobic structures toward multilayer, synergistically functional architectures. Distinct layers within the system assume complementary roles in separation, wetting resistance, heat transfer, mechanical support, and fouling control. Leveraging their high thermal conductivity, tunable wettability, surface functional, and structural diversity, 2D NMs provide an effective platform to realize these functions. In particular, 2D NMs show pronounced advantages in constructing hierarchically rough top layers, high-thermal-conductivity supports, multifunctional intermediate layers, and self-cleaning surface layers.

Notably, contemporary MD research increasingly adopts "2D NMs + auxiliary functional materials" cointegration strategies, for example, combining GO or MXenes with PDA, metal oxides, or polymeric interfacial modifiers, which simultaneously enhance performance and mitigate limitations associated with single-component materials (e.g., mechanical robustness, stability, or interfacial compatibility). This interface-design-driven, multifunctional, composite-integration approach has emerged as a mainstream pathway for improving the comprehensive performance of MD membranes. Looking ahead, achieving precise spatial placement and controllable fabrication of 2D NMs within membrane architectures, together with further expansion of their interfacial responsiveness and multifunctional integration, will be key to advancing MD systems toward higher efficiency, stability, and intelligent operation.

3.4 Configurations, components, and functions of solar evaporation systems

Solar-driven evaporation (SDE) is a clean water treatment technology that uses solar radiation to heat water and separate the solvent from solutes through evaporation (Chu et al., no date). This approach is particularly suitable for seawater desalination, wastewater reclamation, and the utilization of low-grade heat sources. Compared with conventional thermal distillation and membrane-based separation, SDE offers high energy efficiency, environmental compatibility, and operational simplicity, making it well suited for decentralized water treatment in water-scarce regions. A typical SDE system consists of three core components: a solar absorber layer, a thermal conduction layer, and an evaporation interface (Huang et al., no date). In recent years, 2D NMs such as graphene, MXenes, and black phosphorus have been widely explored in SDE owing to their superior photothermal conversion performance and high thermal conductivity. Incorporating 2D NMs into SDE designs can effectively enhance optical absorption, increase evaporation rates, and improve fouling resistance and operational stability.

3.4.1 Configurations and components of typical solar-driven evaporation systems

SDE systems can be classified into three types according to the thermal coupling between the absorbing interface and the water: direct-contact, indirect-heating, and isolated (Fig. 3.5). This mechanism-oriented taxonomy helps elucidate the key roles of 2D NMs in different architectures and informs pathways for functional integration.

As shown in Fig. 3.5A, the direct-contact system laminates the photothermal absorber directly onto the water interface, where solar heating of the interfacial layer drives prompt evaporation. Its advantages include structural simplicity and intuitive vapor generation; however, heat readily dissipates into the bulk water, limiting efficiency. By implementing localized thermal management with an MXene/rGO sponge, an evaporation rate of 2.35 kg m^{-2} h^{-1} has been achieved (Liu et al., 2023a). This design concentrates solar

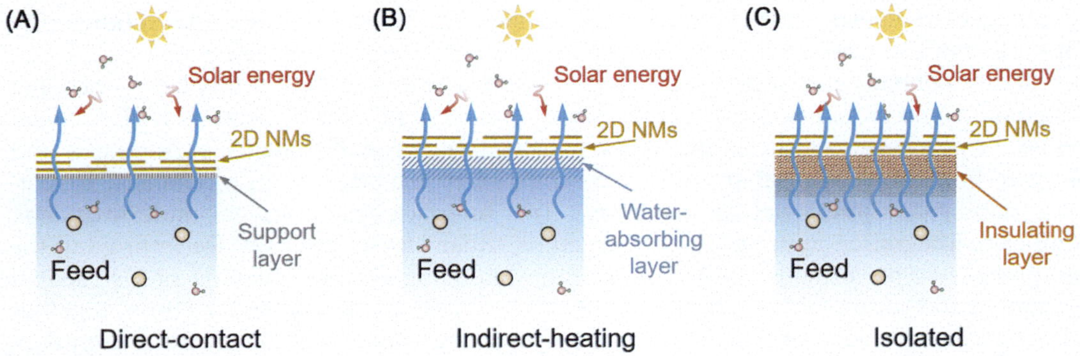

FIGURE 3.5 Schematic illustration of the configurations and components of typical solar-driven evaporation systems based on 2D NMs. (A) Direct contact, (B) indirect heating, and (C) isolated architectures.

absorption at the interface and mitigates thermal losses, exemplifying the interfacial heat-transfer advantage of the direct-contact architecture.

In the indirect-heating configuration, the photothermal layer is decoupled from the water and positioned above the surface; heat is delivered to the evaporation interface through thermally conductive supports or capillary wicking pathways (Fig. 3.5B). This arrangement effectively suppresses heat diffusion and improves energy utilization. Typical material setups integrate graphene, MXenes, or CNTs with porous substrates or foam scaffolds. By employing a GO/CNT composite capillary architecture to heat the interfacial region indirectly, one can not only enhance thermal utilization but also localize evaporation sites, thereby avoiding bulk-water heating and markedly boosting efficiency.

In the isolated configuration, the photothermal interface is physically separated from the feed water. Capillary wicks lift water from the lower reservoir to the absorber, where evaporation occurs, thereby spatially decoupling heat from contaminants (Fig. 3.5C). This design is well-suited for high-salinity or heavily polluted feeds. Finnerty et al. (2021) reported a 3D GO architecture that achieved an evaporation rate of up to 34.7 $kg\ m^{-2}\ h^{-1}$ on the isolated interface, approximately 100-fold higher than conventional evaporation ponds, highlighting both the efficacy of the isolated configuration and its fouling-isolation capability. Composite architectures not only accelerate evaporation but also prevent contaminant back-diffusion.

3.4.2 Optimization strategies for solar-driven evaporation systems using 2D nanomaterials

In SDE systems, interfacial architecture and material configuration directly govern evaporation efficiency and operational stability. To achieve efficient photothermal conversion and continuous vapor release, these systems are typically built as multilayer assemblies, comprising a solar absorber, an evaporation interface, a supporting layer, and thermal transport channels (often capillary), which must operate synergistically. Owing to their distinctive photothermal, electrothermal, and interfacial-regulation functionalities, 2D NMs offer new avenues for performance enhancement in SDE systems (Fig. 3.6).

The primary function of the absorber layer is to efficiently convert solar irradiation into heat and locally raise the temperature at the evaporation interface. Accordingly, an ideal absorber should exhibit high broadband absorptance, low radiative (thermal) losses, and robust thermal stability. 2D NMs such as graphene, MXenes, and BP are widely employed in this layer owing to their superior photothermal properties (see Fig. 3.6A). Monolayer graphene absorbs approximately 2.3% of incident light; when stacked into multilayers, it offers excellent broadband absorption (Bonaccorso et al., 2010). MXenes, by virtue of their high electron density and surface polarization, show strong absorption in the visible and near-infrared, and when composited with polymers can yield absorber layers with reduced thermal losses and high photothermal efficiency (Wen et al., 2025).

The evaporation interface governs water-molecule transport and phase change, thereby determining the vapor generation rate and its spatial uniformity. An ideal interface should feature strong hydrophilicity, unobstructed vapor pathways, and uniform heat distribution. However, conventional interfaces often experience local overheating, insufficient liquid backflow, or contaminant deposition. As shown in Fig. 3.6B, 2D NMs can optimize both the physical and chemical characteristics of the interface through structural engineering and regulation of surface functional groups. Surface-modified graphene or MXenes can be tuned to hydrophilic or hydrophobic states to enable rapid liquid spreading or expedited vapor release, thus improving evaporation efficiency (Jia et al., 2016).

The supporting layer serves as a load-bearing scaffold in solar-driven evaporation systems and, via thermal conduits, efficiently delivers heat to the evaporation interface. This structure should combine mechanical robustness, high thermal conductivity, appropriate porosity, and resistance to water-induced degradation. However, conventional porous sponge substrates (e.g., polyurethane) exhibit low thermal conductivity and loose architectures, leading to substantial thermal losses (see Fig. 3.6C). 2D NMs such as graphene promote uniform heat distribution and porous, suppresse local heat accumulation, and enhance long-term operational stability (Zhang et al., 2024).

3.4.3 Functional performance enhancements of solar-driven evaporation systems enabled by 2D nanomaterials

In solar-driven evaporation systems, the incorporation of 2D NMs delivers not only structural innovations but also substantive performance gains. Table 3.3 summarizes the

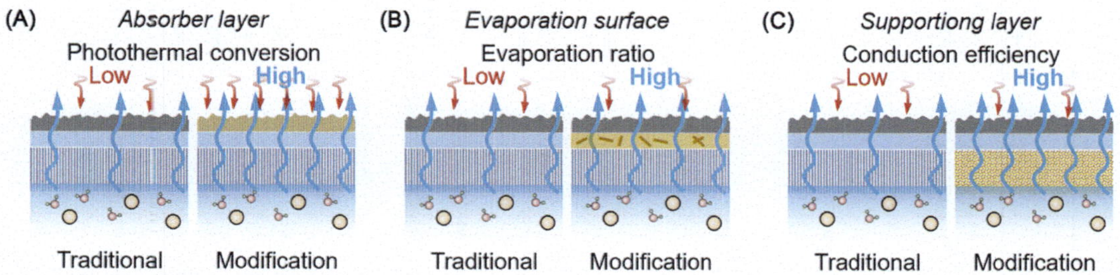

FIGURE 3.6 Schematic of optimization strategies for SDE systems using 2D NMs. (A) Absorber layer, (B) evaporation interface, and (C) supporting layer with thermal transport channels.

TABLE 3.3 Summary of the application of 2D NMs in solar evaporation systems.

2D NMs	Application	Separation material	Efficiency	Mechanism	References
rGO	Seawater desalination; water purification	Na^+, Mg^{2+}, Ca^{2+}, etc.; original	Evaporation rate of 3.81–8.34 kg/m^2/h under 1 sun; average evaporation rate of 6.61 kg/m^2/h over 10-day continuous testing.	High thermal conductivity supports enhanced energy transfer between water and the evaporation interface	Wang et al. (2021a)
rGO MXene	Seawater desalination; wastewater purification	Na^+, Mg^{2+}, Ca^{2+}, etc.; original; heavy metal ions	Evaporative water yield improved by 12.2%– 73.8% in passive systems; up to 285 % improvement in active systems; graphite and similar materials showed the best performance; over 100% enhancement with external condensers; some materials achieved 91% –94% energy conversion efficiency.	Enhanced photothermal conversion; improved localized thermal management; accelerated water vapor diffusion; controlled salt ion migration	Peng et al. (2021)
Graphene HGO RGO	Seawater desalination; wastewater purification	Na^+, Mg^{2+}, Ca^{2+}, etc.; original	Evaporation efficiency reached 70%–85%	Enhanced photothermal conversion; improved localized thermal management; efficient capillary transport; interfacial evaporation enhancement	Liu et al. (2017)
MoS_2-Mo_5N_6	Seawater desalination; wastewater purification	Na^+, Mg^{2+}, Ca^{2+}, etc.; original	Evaporation rate of 2.31 kg/m^2/h under 1 sun; evaporation efficiency up to 106.6 %; no performance decline over 20 cycling tests; measured seawater evaporation rate of 2.15 kg/m^2/h; water resistance increased to 1.3 MΩ (effective desalination).	Heterostructures enhanced broadband light absorption; heterointerfaces enabled localized thermal management and reduced heat loss	Wang et al. (2022b)

SWNT-MoS_2	Seawater desalination wastewater purification	Na^+, Mg^{2+}, Ca^{2+}, etc.; original	Evaporation rate of 6.6 kg/m²/h under 5 kW/m² irradiation; evaporation efficiency of 91.5 %; evaporation rates up to 15.6 kg/m²/h under vacuum conditions; no significant performance decline after 20 consecutive evaporation cycles; desalinated water quality met WHO standards.	Enhanced solar light absorption; localized heating concentrated thermal energy at the evaporation interface, reducing heat dissipation to the bulk water; ultrathin porous structure accelerated vapor escape and reduced evaporation resistance; low thermal conductivity suppressed conductive heat loss	Yang et al. (2018a)
Graphene RGO	Seawater desalination; wastewater purification; clean water production	Na^+, Mg^{2+}, Ca^{2+}, etc.; original heavy metal ions	2D systems exhibit evaporation rates of 1.0–1.5 kg/m²/h; 3D systems achieve up to 16.1 kg/m²/day; multistage systems reduce ion concentrations by 3–4 orders of magnitude after desalination; desalinated water meets WHO drinking water standards	Multidimensional structural design enhanced solar energy harvesting; localized interfacial heating suppressed heat dissipation to bulk water; porous insulating layer reduced conductive and convective heat loss; expanded evaporation area increased water vapor release	Yin et al. (2017)
Xenes (phosphorene, antimonene, tellurene, borophene) MXenes (Ti_3C_2、Nb_2C、Ta_4C_3) TMDs（MoS_2、WS_2） TMOs（WO_x、TiO_2）	Seawater desalination; heavy metal wastewater treatment; organic pollutant removal; clean water production	Na^+, Mg^{2+}, Ca^{2+}, etc.; original; heavy metal ions	2D material systems typically achieve 80%–100% evaporation efficiency; maximum evaporation rates up to 15.6 kg/m²/h; desalinated water quality meets WHO drinking water standards.	Broadband absorption enhancement by 2D materials; localized interfacial heating for concentrated energy; low thermal conductivity support for effective insulation; porous structure facilitating vapor escape; capillary-driven water supply for continuous replenishment	Zhang et al. (2020)
Graphene RGO	Seawater desalination; wastewater purification; clean water production	Na^+, Mg^{2+}, Ca^{2+}, etc.; original; heavy metal ions	Optimized systems achieve evaporation efficiency above 90%; certain structures enable continuous evaporation	Broadband absorption improves solar energy utilization; localized heating minimizes heat diffusion to bulk water; low thermal	Tao et al. (2018)

(Continued)

TABLE 3.3 (Continued)

2D NMs	Application	Separation material	Efficiency	Mechanism	References
			for over 10 hours without salt crystallization; desalinated water quality meets WHO drinking water standards.	conductivity support enables effective thermal insulation; porous structure accelerates water vapor release 1D/2D capillary channels sustain continuous evaporation	
Graphene RGO	Seawater desalination; wastewater purification; energy recovery	Na^+, Mg^{2+}, Ca^{2+}, etc. Original Heavy metal ions	Evaporation efficiency up to 90 %; desalinated water meets WHO drinking water standards.	2D materials enhanced broadband light absorption; localized thermal management concentrated energy for evaporation; porous structure with capillary water supply enabled continuous evaporation; low thermal conductivity insulating substrate reduced heat dissipation	Jiang et al. (2016)

applications of 2D NMs in SDE; their distinctive physicochemical properties provide marked advantages in photothermal conversion, water–vapor transport, antifouling capability, and environmental adaptability. In the following, we systematically elucidate the mechanisms by which 2D NMs enhance overall SDE performance across four dimensions: photothermal properties, evaporation kinetics, operational stability, and environmental tolerance.

3.4.3.1 *Enhanced photothermal conversion*

2D NMs exhibit strong optical absorption, enabling efficient transduction of solar radiation into heat. Materials such as graphene and MXenes absorb sunlight and convert it to thermal energy, thereby increasing the evaporation rate. Studies have reported photothermal conversion efficiencies exceeding 90% for graphene-based composite membranes in solar-driven evaporation, leading to markedly accelerated vapor generation(Chen et al., 2025).

3.4.3.2 *Increased evaporation rate*

Incorporating 2D NMs into solar-driven evaporation systems enhances both vapor generation and transport. Beyond concentrating heat, 2D NMs enlarge the effective evaporation area, thereby accelerating vapor production. Under identical conditions, MXene-enhanced systems exhibit higher evaporation rate than their conventional counterparts (Zhang et al., 2025a).

3.4.3.3 *Enhanced thermal and chemical stability*

2D NMs intrinsically possess high thermal stability and chemical resilience, enabling sustained operation of solar-driven evaporation systems under high-temperature and high-salinity conditions. Graphene and MXenes retain their thermal conductivity and structural integrity at elevated temperatures; accordingly, they are ideal choices for SDE systems that must withstand prolonged exposure to heat or saline environments.

3.4.4 Summary and outlook

This section reviews the configuration designs and key components of SDE systems, systematically outlining integration strategies and functional implementation pathways for 2D NMs. Leveraging high photothermal conversion efficiency, excellent thermal conductivity, and antifouling characteristics, 2D NMs can be coengineered across multiple structural units—the absorber, evaporation interface, and supporting layer—to markedly increase vapor yield, extend service life, and reduce energy consumption and operation-and-maintenance costs. Nevertheless, several challenges remain: insufficient interfacial adhesion between 2D NMs and substrates, which jeopardizes long-term stability; attenuated fouling and scaling suppression in high-salinity and organics-rich feeds; and a lack of systematic assessment of performance-degradation mechanisms under large-area, long-duration operation.

Future research should prioritize the following directions: (1) Develop composite membrane modules that combine high evaporation efficiency with self-cleaning functionality to enable multifunctional integration and robust fouling suppression. (2) Establish green, low-cost, and scalable methods for the preparation and functional modification of 2D NMs to accelerate engineering deployment. (3) Pursue tight integration with solar-thermal utilization and other renewable-energy systems to create high-efficiency, low-carbon solar-evaporation platforms.

3.5 Configurations, components, and functions of photocatalytic degradation systems for wastewater treatment

Photocatalytic degradation is a water treatment technology that employs photocatalysts under light irradiation to promote the decomposition of organic pollutants(Hou et al., 2024). This technique offers advantages such as high efficiency, low cost, and environmental friendliness, making it particularly suitable for the treatment of recalcitrant organic wastewater, dye effluents, and pesticide-contaminated water. The degradation process typically relies on reactive oxygen species (ROS) such as $\cdot OH$ and $\cdot O_2^-$, generated on the catalyst surface, to oxidize pollutants(Zhuang et al., 2025). With increasing demands for enhanced photocatalytic performance, the limitations of conventional photocatalysts (e.g., TiO_2, ZnO) in degradation efficiency, light absorption range, and stability have become apparent(Maaz et al., 2025). In recent years, 2D NMs such as graphene, MXenes, and transition metal sulfides have emerged as promising candidates for photocatalytic degradation systems due to their excellent light absorption capacity, efficient electron transport properties, and tunable surface structures(Gao et al., 2025). These features enable 2D NMs to significantly enhance photocatalytic efficiency, broaden the light response spectrum, and improve system stability (Wang et al., 2017c).

3.5.1 Configurations and components of typical photocatalytic degradation systems

The design of photocatalytic reactor configurations directly determines light intensity distribution, catalyst–pollutant interaction efficiency, as well as catalyst separation and recovery performance. Depending on the catalyst state and reactor architecture, photocatalytic systems are generally categorized into slurry photoreactors, fluidized-bed photoreactors, and fixed-bed photoreactors (Fig. 3.7). In different configurations, 2D NMs can act either as photocatalysts or as catalyst supports to enhance light absorption efficiency and electron transfer.

In a slurry photoreactor, photocatalysts are dispersed as fine particles in the contaminated water (Fig. 3.7A). This configuration provides abundant catalyst–pollutant contact opportunities due to the large specific surface area, while also facilitating exposure of catalysts to light. However, challenges remain, including difficulty in catalyst recovery, light scattering, and particle aggregation. The introduction of 2D NMs can improve catalyst dispersion in aqueous media. Moreover, graphene and MXenes, with their excellent conductivity and light absorption properties, can act as electron transport pathways in photocatalyst

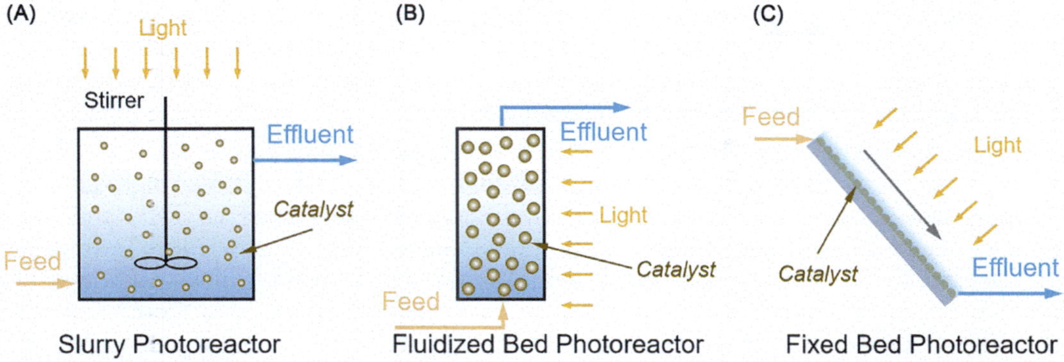

FIGURE 3.7 Schematic diagram of a typical photocatalytic degradation system configuration and components. (A) Suspension photoreactor, (B) fluidized-bed photoreactor, and (C) fixed-bed photoreactor.

composites, suppressing electron–hole recombination and enhancing quantum efficiency (Wang et al., 2025).

In a fluidized-bed photoreactor, the flow rate is controlled to maintain photocatalyst particles in a semi-suspended state (Fig. 3.7B). This configuration combines efficient mass transfer with moderate particle controllability and facilitates catalyst separation and reuse, making it suitable for pilot-scale or continuous-flow treatment. However, the system's stability depends on the density and morphology of the particles. To enhance photocatalytic efficiency and structural robustness, 2D NMs have been composited with supports such as alumina and silica (Kusworo et al., 2024). Studies using MXenes or graphene to coat photocatalyst particles have demonstrated extended light absorption and improved charge transfer pathways, suppressing electron–hole recombination (Rathod et al., 2025).

In a fixed-bed photoreactor, photocatalysts are immobilized on inert substrates or membrane surfaces, and contaminated water flows through the reactor for degradation (Fig. 3.7C). This design facilitates catalyst recovery and minimizes leaching, offering stable operation for continuous treatment of low-concentration pollutants. However, its limited surface area and light-shielding effects restrict reaction efficiency. To overcome these drawbacks, researchers have integrated 2D NMs into photocatalytic membranes or supports, creating high-surface-area, light-responsive architectures. For example, graphene/TiO_2 composite membranes exhibit outstanding antifouling properties and photocatalytic performance, making them suitable for long-term degradation of organic pollutants (Wu et al., 2025).

3.5.2 Optimization strategies of 2D nanomaterials for photocatalytic degradation systems

In photocatalytic degradation systems, the structural design and multifunctional integration of catalysts directly affect light absorption efficiency, charge separation capability, and degradation stability. To achieve efficient pollutant degradation, photocatalysts are generally composed of multiple functional regions, including light-absorbing matrices, charge-carrier separation channels, active reaction surfaces, and dispersive supports. Owing to their unique optoelectronic properties, surface tunability, and interfacial coupling characteristics, 2D NMs have emerged as key materials for multifunctional photocatalyst design.

3.5.2.1 *Light absorption enhancement layer design*

The core function of a photocatalyst is to efficiently absorb light and generate electron–hole pairs. An ideal absorption layer should feature broad spectral response, high quantum efficiency, and robust stability. However, traditional catalysts such as TiO_2 and ZnO primarily respond to ultraviolet light, limiting their solar utilization efficiency. To improve light harvesting, 2D NMs such as graphene, MXenes, and transition metal sulfides (e.g., MoS_2) have been introduced into absorption layer design. For instance, rGO combined with TiO_2 effectively extends absorption into the visible region, promotes electron transfer, and enhances RhB degradation efficiency (Alwan et al., 2022).

3.5.2.2 *Charge carrier separation channel design*

The rapid recombination of photogenerated electron–hole pairs is a major limitation of photocatalytic efficiency. Constructing effective charge migration channels is therefore essential for enhancing carrier separation. With their high electron mobility and favorable interfacial contact, 2D NMs can function as electron acceptors or conductive fillers within photocatalyst structures. TiO_2 loaded with rGO forms TiO_2/rGO composites, which significantly reduce electron–hole recombination and accelerate degradation, achieving enhancement in phenol-4-sulfonic acid degradation efficiency (Shen et al., 2021).

3.5.2.3 *Catalytic surface design and active site regulation*

Since photocatalytic reactions occur primarily at catalyst surfaces, the surface architecture and functional groups directly determine pollutant adsorption and catalytic activity. The rich functionalities and tunable configurations of 2D NMs provide opportunities for constructing highly active catalytic interfaces. The carboxyl and hydroxyl groups on GO can act as adsorption and reactive sites for organic molecules, while promoting degradation through surface electron transfer (Çinar et al., 2020).

3.5.2.4 *Catalyst stability and recyclability design*

In practical wastewater treatment, catalyst stability and recyclability are critical for long-term operation. The mechanical flexibility and interfacial assembly properties of 2D NMs allow the fabrication of magnetic composites or immobilized structures that facilitate catalyst recovery. Fe_3O_4@SiO_2@TiO_2-Co/rGO ternary composites can be rapidly separated magnetically and reused, maintaining over 90% degradation efficiency after five cycles (Fu et al., 2019).

Recent studies reveal that single-component photocatalysts often struggle to simultaneously achieve broad spectral response, efficient charge transport, and high stability. Therefore multifunctional synergistic composites have become a key pathway for performance optimization. Benefiting from their excellent interfacial tunability and integration potential, 2D NMs are frequently combined with traditional semiconductors (e.g., TiO_2, g-C_3N_4, ZnO), metal nanoparticles (e.g., Ag, Fe), or conductive carbon materials (e.g., CNTs) to construct heterojunctions, Z-scheme, or S-scheme architectures for efficient charge transfer. Table 3.4 summarizes the photocatalytic performance of 2D NMs combined with other

TABLE 3.4 Summary of applications and performance of 2D NM-based photocatalysts.

Catalyst	Pollutant	Light	Efficiency	References
GO–ZnO	MB	Sunlight	84.00%	Al-Rawashdeh et al. (2020)
GO–ZnO–Ag	MB	Sunlight	100%	Al-Rawashdeh et al. (2020)
CuO–ZnO–Fe_2O_3/rGO	MB	Low-pressure bulb (200 W)	87.00%	Fatima et al. (2020)
ZnO–CdO–RGO	Bisphenol A	UV light	98.00%	Kumar S. Kumar et al. (2022)
ZnO TiO_2/rGO	MB	UV light	99.00%	Viet et al. (2021)
RGO/TiO_2	Bisphenol	Visible light	98.00%	Ramesh et al. (2021)
ZrO_2–RGO	Nitrophenol	Visible light	98.00%	Anjaneyulu et al. (2022)
TiO_2/ZnO/RGO	MB	Illumination (300 W)	99.60%	Nguyen et al. (2020)
	RhB	Illumination (300 W)	99.20%	Nguyen et al. (2020)
	MO	Illumination (300 W)	99.40%	Nguyen et al. (2020)
Ni/rGO	Phenols	UV–Vis (150 W)	85.70%	Çinar et al. (2020)
Ag_3PO_4/GO film	Norfloxacin	Visible light (250 W)	83.60%	Ji et al. (2020)
K-doped g-C_3N_4	Naphthalene	Visible light (200 W)	82.20%	Deebansok et al. (2021)
α-Fe_2O_3/g-C_3N_4	Hg(II)	Visible light (400 W)	90.00%	Kadi et al. (2020b)
CuO/g-C_3N_4	Hg(II)	Visible light (150 W)	100%	Kadi et al. (2020c)
$CoFe_2O_4$/g- C_3N_4	Hg(II)	Visible light (300 W)	100%	Kadi et al. (2020a)
g- C_3N_4@Co-TiO_2	Escherichia coli	Visible light (300 W)	6-log inactivation	Song et al. (2020)
SnO_2–MoS_2	MB	Visible light (300 W)	96.40%	Ni et al. (2019)
	RhB	Visible light (300 W)	93.10%	Ni et al. (2019)
TiO_2/CuO/MCM-41 MoS_2	Tetracycline	UV light (125 W)	70.50%	Khanmohammadi et al. (2021)
GO–MnO_2	Reactive black 5	UV light	85.00%	Saroyan et al. (2019)
CdS/ZnO/GO	MB	UV light	99.00%	Khan et al. (2012)

(Continued)

TABLE 3.4 (Continued)

Catalyst	Pollutant	Light	Efficiency	References
Sn-WO_3/g-C_3N_4	RhB	UV–Vis	99.00%	Ahmed et al. (2019)
CeO_2/Ti_3C_2Tx	TC	Xe lamp (350 W)	80.20%	Ahmed et al. (2019)
MoS_2/Ti_3C_2Tx	RAN	UV light (25 W)	88.40%	Zou et al. (2021)
ZnO/Ti_3C_2Tx	RhB	UV light	97.50%	Khadidja et al. (2021)
CNTs-Ti_3C_2Tx	RhB	Visible light	75.00%	Thirumal et al. (2022)
Nb_2O_5/Nb_2CTx	RhB	Xe lamp (500 W)	98.50%	Cui et al. (2021)
Nb_2O_5/Nb_2CTx	TC	Xe lamp (500 W)	91.20%	Cui et al. (2021)
ZnO/Nb_2C	MB	Xe lamp (500 W)	62.62%	Zhou et al. (2022)
ZnO/V_2C	MB	Xe lamp (500 W)	99.53%	Zhou et al. (2022)
$NiCo_2S_4$@Ti_3C_2Tx	RhB	Xe lamp (250 W)	~100%	Vigneshwaran et al. (2021)
TiO_2/Ti_3C_2	MO	Xe lamp (375 W)	99.60%	Li et al. (2022a)
Bi_2WO_6/Nb_2CTx	RhB	Xe lamp (500 W)	99.80%	Cui et al. (2020)
Bi_2WO_6/Nb_2CTx	TC	Xe lamp (500 W)	83.10%	Cui et al. (2020)
Ag_2WO_4/Ti_3C_2Tx	SFE	Xe lamp (300 W)	88.60%	Fang et al. (2019)
Ag_2WO_4/Ti_3C_2Tx	TC	Xe lamp (300 W)	62.90%	Fang et al. (2019)
Ag_2S/Ti_3C_2Tx	MB	Xe lamp (300 W)	> 98.00%	Feng et al. (2021)
$Bi_2O_2CO_3$/Ti_3C_2Tx	LEV	Xe lamp (300 W)	95.40%	Zhang et al. (2023b)
Bi_3O_4Br/Ti_3C_2	BPA	Xe lamp (300 W)	91.26%	Wang et al. (2023c)
Ti_3C_2Tx/TiO_2/BiOCl	RhB	Xe lamp (500 W)	78.36%	Liu et al. (2020)
Ti_3C_2Tx/TiO_2-black phosphorus	RhB	Xe lamp (300 W)	99.09%	Shao et al. (2020)
Ti_3C_2Tx/TiO_2-black phosphorus	TC	Xe lamp (300 W)	92.70%	Shao et al. (2020)
MIL-125-NH_2(TiO_2)/ Ti_3C_2Tx	TC	Xe lamp (300 W)	82.80%	Wu et al. (2020a)
Ag/g-C_3N_4/Ti_3C_2Tx	TC	Xe lamp (300 W)	92.10%	Huang et al. (2021)
Ag/g-C_3N_4/Ti_3C_2Tx	RhB	Xe lamp (300 W)	99.20%	Huang et al. (2021)
Ag/Ag_3PO_4/Ti_3C_2	MO	Xe lamp (300 W)	93.00%	Sun et al. (2021)
Ag/Ag_3PO_4/Ti_3C_2	Cr(VI)	Xe lamp (300 W)	61.00%	Sun et al. (2021)
TiO_2@Ti_3C_2Tx/g-C_3N_4	Aniline	Xe lamp (300 W)	76.40%	Ding et al. (2019a)
TiO_2@Ti_3C_2Tx/g-C_3N_4	RhB	Xe lamp (300 W)	99.90%	Ding et al. (2019a)
NCDs/TiO_2/Ti_3C_2Mx	TC	Xe lamp (500 W)	~ 45.00%	Peng et al. (2022)
H-Ti_3C_2/Ag/AgBr	MO	Xe lamp (300 W)	99.20%	Yang et al. (2023)
Ag-decorated g-C_3N_4/ Ti_3C_2	Aniline	Xe lamp (300 W)	81.80%	Ding et al. (2019b)

(Continued)

TABLE 3.4 (Continued)

Catalyst	Pollutant	Light	Efficiency	References
BiOBr/(001)-TiO_2/Ti_3C_2Tx	RhB	Xe lamp (500 W)	98.00%	Yang et al. (2022b)
CdS/Nb_2O_5/Nb_2C	CBZ	Xe lamp (500 W)	92.00%	Zu et al. (2020)
BN/Ti_3C_2Tx/$ZnIn_2S_4$	4-CP	Xe lamp (300 W)	90.00%	Dai et al. (2023)
g-C_3N_4/TiO_2/Ti_3C_2Tx	NO	Xe lamp (300 W)	66.30%	Hu et al. (2021)
g-C_3N_4/C-TiO_2	CIP	Xe lamp (300 W)	88.14%	Li et al. (2022b)
α-Fe_2O_3@$ZnIn_2S_4$/Ti_3C_2	BPA	Xe lamp (λ > 420 nm)	96.10%	Wang et al. (2023b)
In_2S_3/MQDs/$SrFeO_3$	SMX	Xe lamp (300 W)	98.00%	Saravanakumar et al. (2023)
In_2S_3/MQDs/$SrFeO_3$	4-CP	Xe lamp (300 W)	95.40%	Saravanakumar et al. (2023)
Ti-MOF/Ti_3C_2 QDs/$ZnIn_2S_4$	SMZ	Xe lamp (300 W)	98.00%	Liu et al. (2022)
Ti-MOF/Ti_3C_2 QDs/$ZnIn_2S_4$	TC	Xe lamp (300 W)	96.00%	Liu et al. (2022)
Graphene/TiO_2/g-C_3N_4	TC	Xe lamp (300 W)	83.50%	Wu et al. (2020b)
Graphene/TiO_2/g-C_3N_4	RhB	Xe lamp (300 W)	98.00%	Wu et al. (2020b)
TiO_2/Ti_3C_2Tx/AgI	TC	Xe lamp (350 W)	97.00%	Wu et al. (2022)
g-C_3N_4/Ti_3C_2/Bi_2MoO_6	TC	Xe lamp (300 W)	81.30%	Bao et al. (2022)
Ti,C-doped BiOBr/Ti_3C_2Tx	NOx	Xe lamp (450 W)	61.90%	Hermawan et al. (2021)
Sm-doped g-C_3N_4/Ti_3C_2	CIP	Xe lamp (300 W)	> 99.00%	Yu et al. (2021)
La- and Mn-codoped $BiFeO_3$/Ti_3C_2Tx	CR	Xe lamp (300 W)	100%	Iqbal et al. (2019)
Ti_2CTx/TiO_2/Ag	SA	Xe lamp (150 W)	97.10%	Wojciechowski et al. (2019)
Chitosan@g-C_3N_4/Ti_3C_2Tx	MB	Xe lamp (250 W)	~ 99.00%	Vigneshwaran et al. (2020)
Chitosan@g-C_3N_4/Ti_3C_2Tx	RhB	Xe lamp (250 W)	~ 98.50%	Vigneshwaran et al. (2020)
Co/$ZnTiO_3$/Ti_3C_2Tx	TC	Xe lamp (300 W)	88.43%	Park et al. (2023)
N-Ti_3C_2	MO	Hg lamp (1000 W)	98.73%	Cui et al. (2023)
PDI/g-C_3N_4/TiO_2@Ti_3C_2/PMS	ATZ	Xe lamp (300 W)	78.00%	Tang et al. (2022)
Ti_3C_2/g-C_3N_4/PMS	DCF	Xe lamp (300 W)	100%	He et al. (2020a)
Fe/Ti_3C_2/ZSM-5/PMS	Phenols	Xe lamp (300 W)	~ 99.00%	Zhang et al. (2021a)
g-C_3N_4/Ti_3C_2/$MnFe_2O_4$/PMS	NAP	Xe lamp (300 W)	99.00%	Hou et al. (2023)

(Continued)

TABLE 3.4 (Continued)

Catalyst	Pollutant	Light	Efficiency	References
$Ti_3C_2/ZnFe_2O_4$/PDS	TC	LED (120 W, 450 nm)	87.00%	Wang et al. (2023a)
$CuS@Ti_3C_2$/PMS	Orange II	LED (120 W, 450 nm)	100%	Wang et al. (2022c)

materials. Such synergistic effects not only extend light absorption ranges but also optimize interfacial charge migration and pollutant enrichment, providing a solid material and structural basis for efficient, stable, and sustainable photocatalytic degradation.

3.5.3 Mechanisms of 2D nanomaterials in enhancing the performance of photocatalytic degradation systems

The performance of photocatalytic degradation systems largely depends on the light absorption capacity of photocatalysts, the separation efficiency of photogenerated charge carriers, and their stability in complex aquatic environments. Owing to their ultrathin layered structure, large specific surface area, tunable surface chemistry, and excellent electrical properties, 2D NMs enable precise modulation of interfacial interactions, band structures, and surface reactivity, thereby providing new pathways and mechanisms to enhance photocatalytic performance.

3.5.3.1 *Enhancement of photocatalytic efficiency*

The incorporation of 2D NMs can markedly improve the photocatalytic efficiency of degradation systems. Graphene can absorb more solar radiation, facilitate the separation of photogenerated electron–hole pairs, and accelerate degradation kinetics. Studies have shown that graphene/TiO_2 composites achieve over 6.5 times higher degradation rates under UV irradiation compared with pristine TiO_2 (Yang et al., 2016).

3.5.3.2 *Extension of light response range*

Conventional photocatalysts mainly respond to UV light, which accounts for only 5% of solar energy, whereas visible and near-infrared light dominate the spectrum. By doping with metal/nonmetal elements or forming composites with other materials, 2D NMs can significantly broaden the optical response of photocatalysts, enabling more efficient utilization of solar energy and enhancing system performance (Yu et al., 2022).

In summary, 2D NMs enhance photocatalytic degradation systems by improving efficiency and broadening spectral response. Nevertheless, further investigations are required to optimize interfacial coupling between 2D NMs and photocatalysts, improve visible and near-infrared light utilization, and ensure long-term stability under real water treatment conditions.

3.5.4 Summary and outlook

This section systematically reviewed the configurations, material optimization strategies, and performance enhancement mechanisms of photocatalytic degradation systems, highlighting the integration pathways and functional roles of 2D NMs. With their strong light absorption, high electron mobility, and tunable surface chemistry, 2D NMs can serve as photocatalysts or catalyst carriers in slurry, fluidized-bed, and fixed-bed systems, significantly enhancing light utilization and charge separation efficiency. Through synergistic design of multiple functional zones, including absorption layers, charge separation channels, catalytic surfaces, and stabilizing structures, 2D NMs effectively broaden spectral response, strengthen pollutant adsorption and activation, and improve catalyst stability and recyclability in complex water environments. However, current challenges remain, including weak interfacial bonding between 2D NMs and substrates, limited utilization of visible and near-infrared light, and insufficient validation of long-term stability under real wastewater conditions. Future research should focus on developing green, low-cost, and scalable synthesis methods for 2D NMs, exploring multifunctional integration and heterostructure strategies, and conducting long-term evaluations in real water environments to advance photocatalytic degradation technology toward high efficiency, intelligence, and practical engineering applications.

3.6 Configurations, components, and functional mechanisms of antibacterial systems in wastewater treatment

Wastewater commonly contains diverse microorganisms, including bacteria, viruses, and fungi, which impair water quality and pose environmental and health risks (Frieri et al., 2017). Traditional disinfection (chemical or UV) often faces drug-resistant strains, secondary pollution, and high energy costs (Zheng et al., 2025). Consequently, efficient, green, and residue-free antibacterial technologies have become a key research direction. 2D NMs-based antibacterial systems, endowed with large surface areas, photocatalytic activity, and facile surface modification, are widely applied in wastewater treatment. They can inactivate pathogens and impart self-cleaning to extend membrane or reactor lifetimes (Kaneda et al., 2019).

3.6.1 Typical configurations of antibacterial systems

The architecture of an antibacterial system directly determines its efficacy and applicability in wastewater treatment (Fig. 3.8). According to the distribution and mechanism of the antibacterial agents, typical formats include those below.

Antibacterial coatings are the most widely used format (see Fig. 3.8A). By grafting active agents to membrane or equipment surfaces, they realize contact-killing or antiadhesion effects. This design is structurally simple and easily retrofits existing modules, pipes, and screens, but may suffer from agent leaching and limited long-term stability. 2D NMs (e.g., rGO, MXene) can form robust coatings by layer-by-layer self-assembly or interfacial

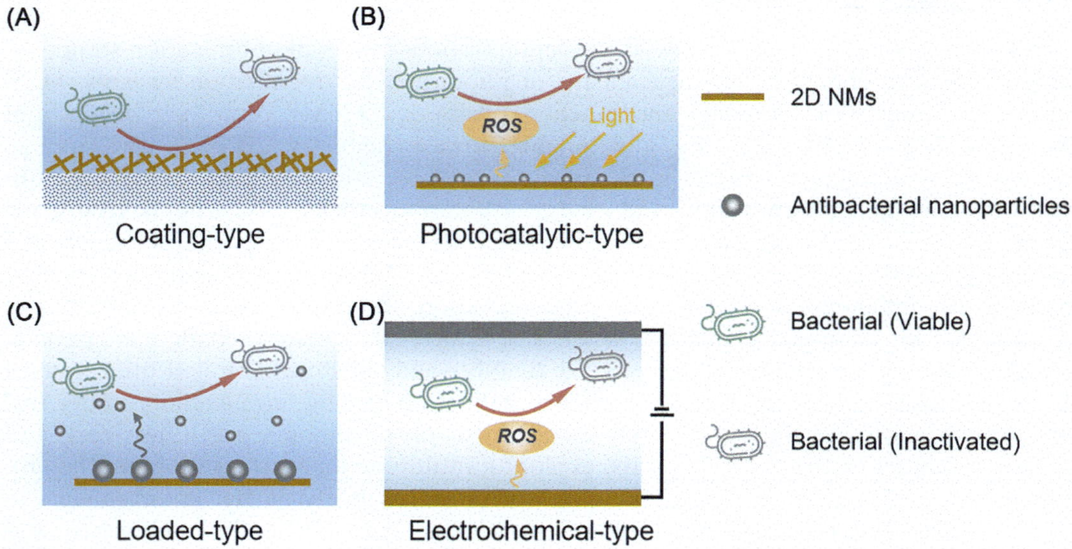

FIGURE 3.8 Schematic configurations of antibacterial systems. (A) Antibacterial coating, (B) photocatalytic, (C) carrier-loaded, and (D) electrochemical.

bonding; rGO/g-C_3N_4 composites notably enhance antibacterial and antifouling performance while maintaining hydrophilicity (Zhang et al., 2025b).

Photocatalytic antibacterial systems couple a photocatalyst with light to generate reactive species (·OH, $\cdot O_2^-$) via photo-excited electron-hole pairs, enabling noncontact inactivation (see Fig. 3.8B). 2D NMs form heterojunctions with catalysts (e.g., MXene/Ag) to promote charge separation and efficiency; under visible light, MXene/Ag can inactivate Staphylococcus aureus by >99 % (Chen et al., 2024).

Carrier-loaded systems immobilize antibacterial agents within porous supports (e.g., carbon cloth, ceramic cartridges) and rely on controlled release of metal ions (Ag^+, Cu^{2+}) or actives under flow (see Fig. 3.8C). 2D NMs are excellent carriers; MXene/Ag hybrids offer sustained release (Chen et al., 2024). This format suits low-flow, high-contamination scenarios but faces challenges of agent depletion and ion-release control during long-term operation (Yu et al., 2022).

Electrochemical antibacterial systems apply electric fields or pulsed bias to drive interfacial electron transfer, directly damaging cell membranes or generating ROS (see Fig. 3.8D). Conductive 2D NMs (MXene, rGO) are well-suited electrodes. $Ti_3C_2T_x$ electrodes can generate ROS at low voltages to rapidly inactivate strains (Zhao et al., 2020).

3.6.2 Optimization strategies of 2D nanomaterials in antibacterial systems

Rational integration of 2D NMs maximizes intrinsic antibacterial activity while enhancing system-level performance. Strategies include using 2D NMs as active agents, surface-functional coatings, composite fillers, and photosensitive components (Table 3.5). Each differs in integration, stability, applicable scenarios, and antibacterial potency.

TABLE 3.5 Summary of the application and performance of 2D NMs in antimicrobial systems.

2D NMs	Bacterial	Mechanism	Antibacterial efficiency	References
GO	*E. coli*	Physical disruption; induction of oxidative stress; modulation of interfacial charge behavior	74.90%	Yu et al. (2013)
GO	*E. coli*		99%	Mural et al. (2016)
GO	*E. coli*		Colony count reduced by nearly 2/3	Zhang et al. (2015)
rGO	*P. aeruginosa*		Effective inhibition	Lim et al. (2012)
GO-NH_2	*E. coli*		75%	Mural et al. (2014)
GO/Ag	*E. coli P.aeruginosa S. aureus*	Physical disruption; induction of oxidative stress; modulation of interfacial charge behavior; enhanced dispersion and stability of antimicrobial agents	Effective inactivation	De Faria et al. (2015)
GO/Ag	*E. coli*		> 90%	Mahmoudi et al. (2015)
GO/Ag	*E. coli*		> 99%	Soroush et al. (2015)
GO/Ag	*E. coli*		86%	Sun et al. (2015)
GO/Ag	*E. coli*		Removal efficiency up to 3-log	Musico et al. (2014)
Ti_3C_2Tx Mxene	*E. coliP. aeruginosa*	Physical disruption, photothermal effect, induction of oxidative stress, modulation of interfacial charge behavior, enhanced dispersion and stability of antimicrobial agents	67% (*E. coli*) 73 % (*P. aeruginosa*)	Rasool et al. (2017)
Ti_3C_2Tx Mxene	*E. coliP. aeruginosa*		82% (*E. coli*) 85.55% (*P. aeruginosa*)	Mayerberger et al. (2018)
Ti_3C_2Tx Mxene/Ag	*E. coliS. aureus*		99.9% (*E. coli*) 99.9% (*S. aureus*)	Chen et al. (2024)
Ti_3C_2 Mxene	*E. coli*		95%	Mayerberger et al. (2018)
LDH	*E. coliS. aureus*	Modulation of interfacial hydrophilicity/ hydrophobicity and charge behavior Zn^{2+} release	~92% (*E. coli*) > 99% (*S. aureus*)	Wang et al. (2024b)

Embedding 2D NMs directly as antibacterial actives leverages membrane-disruptive, cutting, and oxidative-stress actions (e.g., GO). Cointegrating 2D NMs with conventional agents (e.g., Ag, Cu, ZnO nanoparticles) markedly boosts antibacterial strength and suppresses resistant strains. 2D NMs serve as dispersive scaffolds that reduce dosage and toxicity while stabilizing catalytic sites (e.g., GO/Ag hybrids) (Chen et al., 2024). This is effective against multidrug-resistant bacteria but requires control of metal leaching and biocompatibility (Yu et al., 2022). As light-responsive components, 2D NMs (e.g., MXene/g-C_3N_4) generate ROS or local photothermal heating under illumination to achieve photocatalytic or thermal sterilization (Yin et al., 2025). Such systems excel in adaptive disinfection under variable influent loads, though performance depends on light intensity and incidence; integration is more complex.

Overall, diverse strategies with 2D NMs open broad routes for architecture optimization and functional integration in antibacterial systems. Future work should advance multi-mechanism hybrids tailored for complex wastewaters, aiming for efficient, green, and durable antimicrobial treatment.

3.6.3 Mechanisms by which 2D nanomaterials enhance antibacterial performance

The performance gains stem from unique physical structures, electronic properties, and interfacial activity, involving membrane disruption, oxidative stress, photothermal effects, ROS generation, and carrier-assisted dispersion.

Physical disruption: 2D NMs such as graphene, MoS_2, and BN possess sharp edges and high aspect ratios that puncture cell membranes via a "blade" effect. This passive mechanism works without external stimuli and is effective even at high cell densities, suiting biofilm-rich systems (Hu et al., 2010). Oxidative stress induction: certain 2D NMs (e.g., rGO, $Ti_3C_2T_x$) catalyze the formation of ROS, such as ·OH and $\cdot O_2^-$ on cell surfaces, damaging organelles and DNA to trigger cell death. This selective, rapid mechanism is especially effective when combined with light or peroxides. Photothermal sterilization: 2D NMs, including MXene and black phosphorus, exhibit high photothermal conversion; under solar or near-infrared light, they heat rapidly, denaturing proteins and rupturing cells (Su et al., 2025). This oxygen-independent, noncatalytic pathway is suitable for anaerobic environments. Carrier function for enhanced dispersion and stability: high surface area and dispersibility allow 2D NMs to carry Ag/Cu nanoparticles or antibiotics, controlling release profiles and spatial distribution to cut dosage while increasing efficacy (Yu et al., 2022). GO/Ag hybrids, for example, provide long-lasting activity and suppress toxic aggregation-ideal for long-running membrane systems (Soroush et al., 2015). Interfacial wetting/charge modulation to suppress adhesion: surface functional groups tune charge density and hydrophilicity. Because bacteria are generally negatively charged, rendering the surface similarly charged or super-hydrophilic repels adhesion and breaks the colonization chain. PDA-modified MXene layers enhance stability while improving antiadhesion performance (Lin et al., 2021).

In summary, 2D NMs enhance antibacterial systems via physical, chemical, photothermal, and interfacial pathways that can act cooperatively. Future efforts should explore multi-mechanism hybrids with strong responsiveness, broad adaptability, and robust stability for realistic wastewater conditions.

3.6.4 Summary and outlook

This section has outlined antibacterial system architectures, 2D-NM application strategies, and mechanistic enhancements. Owing to ultrathin, high-area, and tunable optoelectronic properties, 2D NMs can be embedded as coatings, photocatalytic/photothermal components, carriers, or active agents to realize contact killing, ROS generation, photothermal sterilization, and antiadhesion. While these strategies significantly elevate antibacterial efficiency, stability, and durability, challenges persist in long-term stability, activity decay, agent leaching, and environmental risk. Priorities include robust, low-cost, scalable synthesis/modification; realistic, long-term assessments; and environmental safety validation to enable efficient, durable, and sustainable antibacterial treatment.

However, current research still faces challenges such as insufficient long-term stability of materials in complex aquatic environments, attenuation of antibacterial activity, leaching of antibacterial agents, and potential environmental risks. Future efforts should prioritize the design of highly stable, multimechanism composite antibacterial systems based on 2D NMs; the development of green, low-cost, and scalable fabrication and modification methods; and strengthened medium- to long-term performance assessments and environmental safety validations under real wastewater conditions, to advance antibacterial systems toward higher efficiency, durability, and sustainability.

3.7 Configurations, components, and functions of water purification setup using adsorbents

Adsorption technology is a water purification method that removes contaminants from aqueous systems by transferring them onto the surface of an adsorbent through mechanisms such as physical adsorption, electrostatic interactions, or chemical coordination (Yuan et al., 2022). Owing to its operational simplicity, low energy consumption, high selectivity, and broad adaptability, it has been widely applied in the removal of various pollutants from water, including heavy metal ions, organic contaminants, dyes, and radioactive substances (Jiang et al., 2018). Traditional adsorbents, such as activated carbon, zeolites, and clays, have been extensively used but are often limited by relatively low specific surface areas, poor selectivity, and challenges in regeneration (Zhang et al., 2025c). 2D NMs, including GO, MXenes, and LDHs, have attracted significant attention as next-generation water treatment adsorbents due to their layered architectures, high specific surface areas, abundant functional groups, and tunable structures (Marouani et al., 2025). These materials exhibit outstanding adsorption capacities, rapid adsorption kinetics, and excellent regeneration performance, offering high-performance components for advanced water purification systems.

3.7.1 Configuration and components of a typical adsorption system

The configuration of an adsorption system plays a pivotal role in determining mass transfer efficiency between pollutants and adsorbents, operational stability, and overall removal performance. In practical engineering applications, adsorption systems are generally categorized into three typical configurations: fixed-bed, moving-bed, and fluidized-bed, based on the state of the adsorbent in the reactor and its hydrodynamic behavior (see Fig. 3.9). These configurations differ significantly in structural design, operational characteristics, and application scope, making them suitable for different treatment requirements. With the growing research on 2D NMs, their merits offer new strategies for enhancing the performance of conventional adsorption systems, particularly in improving adsorption capacity, regeneration efficiency, and the structural stability of composite materials.

The fixed-bed system is currently the most widely employed adsorption configuration in engineering applications such as advanced drinking water treatment and end-of-pipe industrial wastewater remediation. In this setup, the adsorbent is packed into a column reactor, and water flows unidirectionally through the bed, allowing sufficient contact between the adsorbent and the target contaminants for efficient removal (see Fig. 3.9A). Its advantages include a simple structure, stable effluent quality, and ease of modular integration. However, it also presents challenges such as high-pressure drop, bed clogging, and difficulty in adsorbent replacement. Due to their small particle size and tendency to aggregate, 2D NMs in fixed-bed systems are prone to increased pressure loss and particle leaching. To address these issues, researchers often immobilize 2D NMs onto porous supports or polymer matrices to fabricate structurally stable composite granules (Kamil et al., 2025).

The moving-bed system maintains continuous movement or periodic replacement of the adsorbent within the reactor through external mechanical or hydraulic means (see Fig. 3.9B). It is well-suited for high-throughput or continuous-operation wastewater treatment scenarios, particularly for industrial effluents with large fluctuations in contaminant concentrations. This configuration offers high adsorbent utilization, low mass transfer resistance, and convenient regeneration operations. However, it also faces engineering challenges such as particle attrition, increased energy consumption, and difficulties in solid–liquid separation. For the application of 2D NMs in moving-bed systems, the adsorbents need to be engineered into mechanically robust, shear-resistant particulate forms. This is commonly achieved through magnetic compositing (e.g., Fe_3O_4/rGO

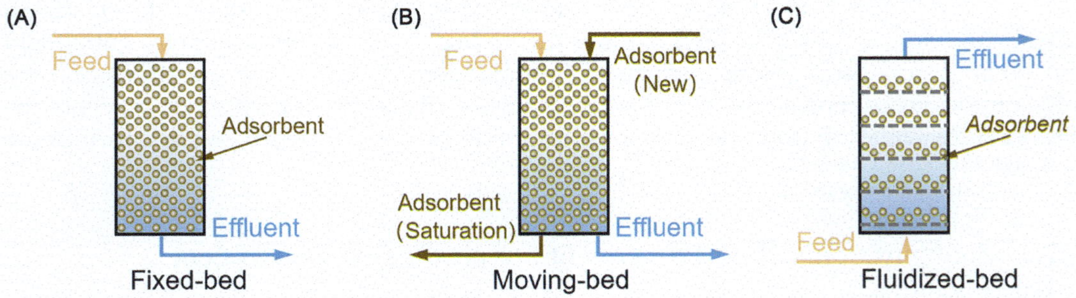

FIGURE 3.9 Schematic illustration of typical adsorption system configurations and components. (A) Fixed-bed, (B) moving-bed, and (C) fluidized-bed adsorption systems.

hybrids), polymer encapsulation, or spray granulation to enhance structural stability (Minitha et al., 2018).

The fluidized-bed configuration suspends adsorbent particles through an upward flow of water, enabling highly efficient liquid–solid contact (see Fig. 3.9C). It is commonly employed for treating wastewater with high pollutant concentrations, such as textile dyeing effluents and electroplating wastewater containing heavy metals. Its key advantages include high mass transfer efficiency, uniform bed distribution, and reduced clogging risk. However, challenges such as particle agglomeration, low separation efficiency, and high system variability remain. Due to their low density and tendency to aggregate, 2D NMs in such systems are often composited with inert supports (e.g., alumina, activated carbon) or encapsulated into spherical granules to maintain a stable suspended state.

The selection of an adsorption system configuration should be carefully aligned with water quality characteristics, treatment objectives, and operational conditions. Fixed-bed systems are well-suited for treating low-concentration, stable water sources, whereas fluidized-bed and moving-bed systems are more appropriate for high-concentration, highly variable, or continuously operated scenarios. Nevertheless, challenges remain in engineering the structural assembly, dispersion stability, and reusability of 2D NMs in practical systems. Further optimization is essential to enable their large-scale application in advanced adsorption processes.

3.7.2 Optimization strategies for adsorption systems using 2D nanomaterials

In recent years, 2D NMs have been incorporated as performance-enhancing components into various adsorbent systems (see Table 3.6). By improving structural features, surface chemistry, and interfacial stability, 2D NMs have significantly enhanced the overall performance of conventional adsorbents. The key optimization strategies can be summarized as follows:

Constructing composite adsorption interfaces to enhance adsorption capacity. 2D NMs, with their atomic-scale thickness and high specific surface areas, provide abundant active sites for pollutant capture. When integrated with conventional adsorbents, such as activated carbon, clays, or polymer matrices, 2D NMs can effectively fill structural voids and generate hierarchical pore architectures, thereby expanding the available surface area and improving surface accessibility. For example, rGO/activated carbon composites have demonstrated more than a 2.24 times higher adsorption affinities (KF) of PFOS compared with pristine activated carbon (Ahmad et al., 2025).

Introducing selective functional groups to enhance targeted removal. The surfaces of 2D NMs are rich in tunable functional groups (e.g., -OH, -COOH, $-NH_2$, -F), which can be incorporated into conventional adsorbents through in situ grafting, covalent modification, or noncovalent adsorption, thereby imparting selective affinity toward specific pollutants. Incorporating GO with surface-carboxyl modifications into a chitosan matrix to form a GO-CS composite markedly improves its coordination-based adsorption capacity for dye (Bai et al., 2025).

Enhancing dispersibility and interfacial stability to prolong service life. Conventional powdered adsorbents often suffer from agglomeration, material loss, and structural collapse during repeated use. The layered architecture and excellent interfacial compatibility of 2D

TABLE 3.6 Performance summary of 2D NM-based adsorbents.

2D NMs	Species	Experimental condition	Adsorption capacity (mg/g)	References
CTAB-GO	CR	pH=3 T=298 K T=60 min	2767	Mahmoodi et al. (2017)
rGO/ZIF-67 aerogel	CV MO	pH 3–9 T=298 K t=0–16 h	1714 426	Yang et al. (2018b)
GO/Fauasite	MB	t=5 min t=2–300 min	92.4	Kim et al. (2018)
SGO			2530	Wei et al. (2018)
rSGO	MB	T=298 K	2187	Araujo et al. (2018)
GO0123	MB	pH=5.5 T=303 K t=10 min	365	Othman et al. (2018)
GO0125			504	
GO-MNP	MB	pH=9 t=10–30 min	1428	Balkız et al. (2018)
GO-calcium alginate Gel	MB	t=60 min T=298 K 306 K	122	Balkız et al. (2018)
GO-$FeCl_3$	MB	313 K	200	Ojha and Thareja (2018)
MgO-GO	MB	pH=6–7	237	Xu et al. (2018)
PDA/RGO/HNTs	MB	T=303 K	12	Zhao et al. (2018)
	CR	T=313.15 K	11	
GO/MOF	MB	T=298 K	274	Zhao et al. (2018)
MGO	Eriochrome Black T (EBT)	t=12 h pH=2 T=303 K t=5 h	210	Khurana et al. (2018)
rGO/Cu(4)	MB	pH=6	1200	Aboelfetoh et al. (2018)
5%GO/MIL-100(Fe)	CV MB MO	t=20 min pH=4 T=298 K t=48 h	238 1231 1189	Luo and Wang (2018)
$MnFe_2O_4$-GO nanocomposite	MB	pH=7 T=RT	177	Huong et al. (2016)
rGO-MMT	MB	t=20 min pH=8 T=313 K t=4 h	227	Zhang et al. (2018b)
GO-Au	MG	pH=2	71	Naeem et al. (2018)
Bi_2O_3@GO	Ethyl Violet (EV)	T=308 K t=50 min pH=4	39	Das et al. (2018)
	RhB	T=308 K t=65 min	320	

(*Continued*)

TABLE 3.6 (Continued)

2D NMs	Species	Experimental condition	Adsorption capacity (mg/g)	References
Fe_3O4/porous graphene nanocomposite	CV	pH=7 T=RT t=0–300 min	460	Bharath et al. (2017)
Graphene	MB	pH=12	481	El Essawy et al. (2017)
Y-Fe_2O_3/N-rGO	Acid Blue 25 (AB25) R6G	T=308 K t=30–50 min pH=7 T=303 K t=120 min	460 44	Yang et al. (2017)
N-doped graphene hydrogels	Acridine orange (AO)	pH=5 T=298 K t=60-400 min	124	Jiang et al. (2017)
Fe_3O_4@SiO_2@CS-TETA-GO	MB	pH=10 T=303 K t=20 min	529	Wang et al. (2017a)
PmPD/rGO/NFO	MB	pH=7 T=RT t=5–90 min	103	Wang et al. (2017d)
	MO		136	
	CR		285	
rGO	MB	pH=7 T=RT t=60 min	145	Gupta and Khatri (2017)
	MO		244	
GO-SBA-16	MG	pH=8 and 9 T=308 K t=40 min	358	Chaudhuri et al. (2017)
	MV		536	
N/S-GHs	MG	pH=7 T=293 K t=600 min	738	Shi et al. (2016)
PPD-GO	CR	pH=3 T=298 K t=60 min	892	Wu et al. (2016)
Ti_3C_2Tx	Ba	pH=3–9 T=298 K	9.3	Fard et al. (2017)
Ti_3C_2Tx	Cu	pH=2–6 T=293–318 K	86.5	Shahzad et al. (2017)
Magnetic Ti_3C_2Tx	Hg	pH=2–9 T=288–318 K	1128	Shahzad et al. (2018)
Ti_3C_2 (OH)0.8F1 2	Cr	pH=3–6 T=298 K	62	Zou et al. (2016)
Ti_2CTx-hydratec	Th(IV)	pH=1–3.5 T=282–312 K	213	Li et al. (2019b)
Ti_3C_2Tx-10%	Cr(VI)	pH=2–13 T=RT	250	Ying et al. (2015)
Alk-Ti_2Csheet	Cd	pH=2–9 T=RT	326	Shahzad et al. (2019)
Ti_3C_2Tx-DMSO-hydated	U	pH=2–6 temp.=RT	214	Wang, Tao, Yuan, et al. (2017b)
Ti_3C_2Tx	Cs	pH=2–12 T=273–303 K	25.4	Khan et al. (2019)
Mxene-25	Pd	pH=0–5 T=293–313 K	119	Mu et al. (2019)

(Continued)

TABLE 3.6 (Continued)

2D NMs	Species	Experimental condition	Adsorption capacity (mg/g)	References
Mxene-35			164	
Mxene-45			185	
Ti_3C_2Tx	Cr	T=298 K	80	Tang et al. (2018)
Alk-MXene	Pb(I)	pH=1–7 T=293–325 K	140	Peng et al. (2014)
BN	As(III)	pH=2–12 t=6 h T=283–313 K	8.5	Bangari et al. (2020)
BN nanosheets	As(V)	pH=2–10 T=283–313 K	5.3	Bangari et al. (2019)
BN	Cu	pH=6.7 t=5 min T=298 K	19.7	Ng et al. (2020)
	Cd		15.2	
	Ni		9.47	
Hexagonal BN	Cd	pH=2–12 t=2 h T=298–328 K	~80	khan et al. (2021)
Hexagonal BN	Cd	pH=1–7 t=3 h T=303–328 K	184	Li et al. (2020)
BN	U(VI)	pH=2–10 t=24 h T=293–313 K	413	Li et al. (2019a)
BN nanosheets	Hg	pH=3–8 t=0.5 h T=298 K	625	Peng et al. (2018)
	Pb		211	
Fe_3O_4-BN	As(II)	pH=2–12 t=6 h T=298–308 K	30.3	Bangari et al. (2020)
Fe_3O_4-BN	As(V)	pH=2–10 T=283–313 K	26.3	Bangari et al. (2019)
Fe_3O_4-BN nanotubes	As(V)	pH=1–13 T=298 K	32.2	Chen et al. (2011)
Polyaniline-hexagonal BN	PO_4^{3}	pH=3–12 t=1 h T=293 K	106	Karthikeyan et al. (2020)
	NO_3		67.9	
	Cr(VI)		133	
BN microrods@Fe_3O_4	Pb	pH=5.5 t=4 h T=293 K	250	Song et al. (2018)

NMs make them ideal as structural reinforcement components in adsorption systems. Composites formed by integrating 2D NMs with natural polymers into hydrogel or sponge-type adsorbents exhibit superior hydraulic shear resistance and enhanced recyclability over multiple adsorption - desorption cycles(Taye et al., 2023).

Synergistic tuning of hydrophilicity/hydrophobicity to adapt to complex water matrices. By tailoring the surface properties of 2D NMs, including hydrophilicity, hydrophobicity, and surface charge, it is possible to precisely modulate the interfacial behavior of adsorbents, thereby meeting the capture requirements for either polar or nonpolar pollutants. For instance, incorporating ZnO/Si@rGO hybrids enables efficient oil–water separation and removal of organic compounds, thus extending the application scope of conventional adsorbents (Babitha et al., 2015).

In summary, 2D NMs not only address the performance limitations of conventional adsorbents but also expand their applicability in complex water treatment scenarios through structural synergy and interfacial functional coupling. Nonetheless, current technologies still face challenges related to large-scale fabrication, cost control, and long-term operational stability. Future research should place greater emphasis on integration with reactor systems, multifunctional composite strategies, and the development of intelligent regeneration mechanisms, thereby accelerating the engineering application of 2D NMs in advanced adsorption processes.

3.7.3 Performance enhancement mechanisms of 2D nanomaterials in adsorption systems

The application of 2D NMs in water treatment adsorption systems is not only reflected in their strategic enhancement of structure and performance but is fundamentally rooted in their unique physicochemical mechanisms at the microscale. By examining the key steps of the adsorption process, such as pollutant mass transfer, interfacial recognition, binding and capture, and desorption stability, it becomes evident that 2D NMs contribute to performance enhancement through several core mechanisms.

3.7.3.1 Highly exposed active sites for enhanced interfacial recognition and rapid capture

The atomic-scale thickness and large specific surface area of 2D NMs enable the efficient exposure of surface-active sites, such as carboxyl, hydroxyl, amino, -F, and -OH groups, facilitating the initial contact between pollutants and the adsorbent surface. This structural advantage markedly reduces diffusion pathways and increases the probability of interfacial collisions, thereby accelerating the initial adsorption rate and delivering a faster kinetic response (Kamil et al., 2025). For instance, in GO- or MXene-enhanced composite adsorbents, the synergistic effect of porous networks and layered structures promotes the rapid capture of both metal ions and organic contaminants.

3.7.3.2 Multiple interaction mechanisms for enhanced selective pollutant recognition

The surfaces of 2D NMs can engage in multiple interaction mechanisms, including electrostatic attraction, coordination bonding, hydrogen bonding, and hydrophilic/hydrophobic interactions, endowing them with high adaptability toward various classes of pollutants. GO can adsorb cations such as Pb^{2+} through a combination of electrostatic attraction and carboxyl group coordination (Velusamy et al., 2021), while efficiently capturing aromatic dyes via π–π stacking interactions (Kamil et al., 2025).

In summary, 2D NMs endow adsorbents with enhanced interfacial tunability, higher selectivity, and improved structural stability through multiscale synergistic mechanisms. The combined effects of these mechanisms not only increase adsorption rates and capacities but also provide the theoretical foundation and functional basis for the development of highly selective, regenerable, and engineering-ready adsorption systems capable of adapting to complex water quality conditions.

3.7.4 Conclusion and outlook

This section provides a systematic overview of typical adsorption system configurations, optimization strategies involving 2D NMs, and the underlying mechanisms behind their performance enhancement. With their ultrathin layered structures, high specific surface areas, and tunable functional groups, 2D NMs can be flexibly integrated into diverse adsorption configurations, including fixed-bed, moving-bed, and fluidized-bed systems, for the efficient capture of heavy metals, organic pollutants, dyes, and radioactive substances. They exhibit notable advantages in composite interface construction, selective functionalization, enhanced dispersion and stability, interfacial hydrophilicity/hydrophobicity tuning, and controllable regeneration. Mechanistically, 2D NMs facilitate rapid interfacial recognition through highly exposed active sites, strengthen pollutant-specific binding via multiple interaction pathways, achieve adsorption - reaction synergy through electron transfer and interfacial energy-level matching, and enable efficient regeneration under external stimuli, collectively imparting adsorption systems with superior removal efficiency and operational stability.

However, the practical engineering application of 2D NMs still faces significant challenges, including high production costs, limited scalable manufacturing techniques, and the need to verify their long-term stability and safety in complex water matrices. Future research should focus on developing green, low-cost, and scalable synthesis methods; achieving efficient integration with various reactor configurations; implementing multi-mechanism synergistic composite strategies; and designing intelligent, regenerable systems. Furthermore, long-term operational evaluations under real water conditions are essential to promote the large-scale deployment and sustainable application of 2D NMs in advanced adsorption-based water treatment systems.

3.8 Configurations, components, and functions of capacitive deionization systems

Capacitive deionization (CDI) is a green water treatment technology that removes ions from water through electrically driven ion migration and physical adsorption onto electrode surfaces(Shi et al., 2025). Operating typically at low voltages (<1.5 V), CDI offers the advantages of low energy consumption, regenerability, and precise electrochemical control, making it particularly suitable for the desalination and mineral content regulation of low-salinity waters such as brackish water and groundwater(Narayanan et al., 2024). As performance demands on electrodes increase, 2D NMs have emerged as highly promising candidates for CDI systems due to their excellent electrical conductivity, high specific surface area, tunable surface chemistry, and controllable pore structures(Al-Saidi et al., 2025). By constructing functionalized 2D NM-based electrodes, CDI systems can achieve significant improvements in deionization capacity, mass transfer rate, selectivity, and cycling stability, driving the technology toward higher efficiency, intelligence, and multi-functionality.

3.8.1 Configuration and components of a typical capacitive deionization system

The performance of CDI, including operational efficiency, charge utilization, and pollutant selectivity, is governed by factors such as reactor configuration, electrode material structure, and interfacial reaction characteristics. As shown in Fig. 3.10, variations in electrode arrangement, solution flow paths, and ion migration modes have led to the development of multiple CDI configurations, including conventional CDI, flow-electrode CDI (FCDI), membrane-assisted CDI (MCDI), and hybrid CDI (HCDI). Owing to their exceptional electrical conductivity, high specific surface area, and interfacial tunability, 2D NMs have been integrated into these configurations as electrode frameworks or functional layers, significantly enhancing charge storage capacity, electrochemical stability, and desalination efficiency.

Conventional CDI systems can be classified into symmetric and asymmetric configurations. A symmetric CDI system consists of two identical carbon-based electrodes (see Fig. 3.10A) that adsorb and release ions via the electric double-layer (EDL) effect. This design offers advantages such as structural simplicity and low energy consumption, making it suitable for treating low- to medium-salinity water. However, its performance is constrained by limited charge utilization and poor ion selectivity. In contrast, asymmetric CDI introduces structural or material differences between the cathode and anode to achieve functional differentiation and enhanced selectivity. This approach is often employed for targeted ion separation or improved energy storage; for example, heterostructure electrodes composed of MXene and GO can modulate selective adsorption behavior through interfacial potential differences (Alam et al., 2023). With their abundant surface functional groups and high electrochemical activity, 2D NMs offer high specific capacitance, excellent conductivity, and programmable interfacial architectures for both symmetric and asymmetric CDI systems, significantly improving salt removal efficiency and cycling stability (Al-Saidi et al., 2025).

MCDI incorporates anion and/or cation exchange membranes (AEM/CEM) in front of the electrodes to restrict co-ion diffusion and enhance ion migration efficiency, making it one of the most widely studied CDI configurations in recent years (see Fig. 3.10B). MCDI can significantly improve charge utilization and system stability, making it well-suited for applications requiring high-selectivity deionization. However, membrane resistance and instability at the membrane–electrode interface can lead to increased energy consumption and electrode polarization during operation.

FCDI employs recirculating slurry electrodes to enable continuous desalination under an applied electric field (see Fig. 3.10C). This configuration allows electrode regeneration and charge transfer to occur simultaneously, making it highly suitable for high-throughput and high-salinity water treatment. However, its operational stability can be significantly affected by slurry rheology, charge transfer resistance, and particle agglomeration.

HCDI is an emerging CDI configuration in which one electrode incorporates Faradaic reaction materials, such as transition metal oxides, Prussian blue analogs, or doped MXenes, to induce pseudocapacitive behavior, while the other electrode operates via electric double-layer adsorption. This design enables a synergistic ion removal mechanism that combines capacitive and Faradaic processes. HCDI has demonstrated remarkable performance in enhancing desalination capacity, charge utilization, and cycling life, making it particularly suitable for high-salinity treatment and resource recovery applications. Studies have shown

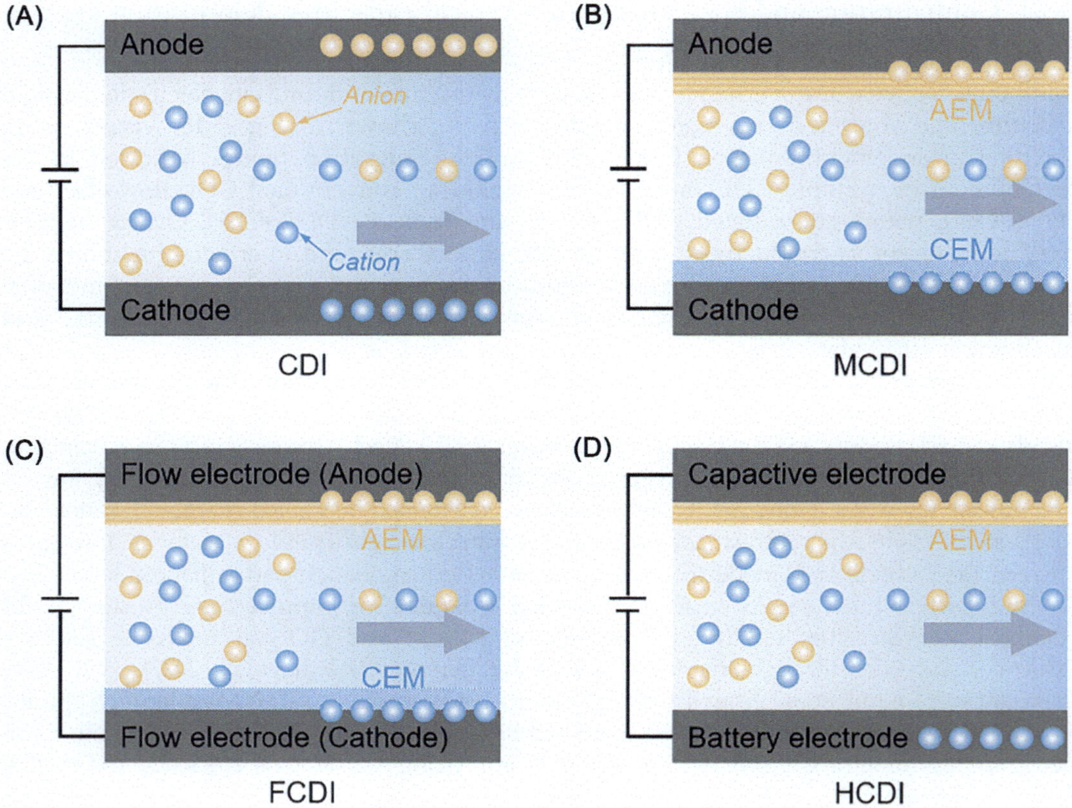

FIGURE 3.10 Schematic illustration of typical CDI system configurations and components. (A) Conventional CDI, (B) flow-electrode CDI (FCDI), (C) membrane-assisted CDI (MCDI), and (D) hybrid CDI (HCDI).

that compositing two-dimensional MXene with pseudocapacitive materials such as Fe_3O_4 yields HCDI electrodes that maintain excellent electrical conductivity while providing abundant ion intercalation sites, achieving higher Na^+ adsorption capacity and faster response rates (Wang et al., 2022a).

In summary, the evolution of CDI system configurations highlights their high adaptability to diverse water qualities and operational conditions. With their outstanding interfacial functional integration, excellent electronic conductivity, and structural versatility, 2D NMs play a pivotal role across different CDI architectures, serving as a key strategy for performance enhancement.

3.8.2 Optimization strategies for capacitive deionization electrode components using 2D nanomaterials

In conventional CDI systems, commonly used electrode materials such as activated carbon and graphite suffer from limitations including restricted capacitive storage, low electrical conductivity, and poor surface reactivity, making it difficult to meet engineering requirements for high throughput, high selectivity, and long-term operational stability. In recent years, 2D NMs have been widely incorporated into CDI systems due to their high electrical conductivity, large specific surface area, and excellent ion accessibility, becoming key material units for electrode performance enhancement. The primary optimization strategies can be summarized in the following five aspects:

Constructing multiscale conductive networks to enhance specific capacitance and electron transport efficiency. 2D NMs such as graphene and MXenes possess high electron mobility and layered stacking structures, which facilitate the formation of continuous conductive pathways within electrodes, thereby improving electron transport rates and specific capacitance. For example, compositing $Ti_3C_2T_x$ MXene with carbon nanotubes to form a three-dimensional porous network electrode can significantly enhance coupled electron/ion transport, resulting in a capacitance of 222.3 F/g and exhibiting good cyclic stability, making it well-suited for fast-cycling CDI systems (Liu et al., 2024).

Introducing ion-affinitive functional groups to enhance ion selectivity and adsorption rate. In conventional CDI electrodes, ion adsorption is primarily governed by EDL formation, lacking specific recognition and binding sites. By modifying 2D NMs with ion-affinitive functional groups, such as carboxyl, sulfonic acid, or pyrrolyl amine groups, their affinity and selective adsorption capacity toward target ions can be significantly improved. Some research has demonstrated that $Mo_{1.33}C(OH)_2$-MXene is an efficient electrode to capture Cl^- and Na^+ from sea and brackish water (Guerrero-Sanchez et al., 2022).

Optimizing pore structure and wettability to enhance ion transport efficiency and water compatibility. The CDI process relies on the rapid migration of ions through pore channels to the electrode surface for adsorption, making pore structure engineering critical to overall performance. Compositing 2D NMs with mesoporous carbon, activated carbon, or natural minerals can create hierarchical pore architectures that not only improve ion accessibility but also enhance wettability and interfacial affinity with water (Folaranmi et al., 2021).

Enhancing electrode structural stability to prolong cycling life. During long-term repeated charging–discharging, conventional electrodes often suffer from degradation issues such as mechanical swelling, delamination, or pore blockage. The flexible layered structure and interfacial assembly capability of 2D NMs offer a viable pathway for constructing electrodes with antiswelling properties and structural robustness (Morshedi Dehaghi et al., 2025).

Table 3.7 summarizes the applications and performance of 2D NMs in CDI systems. As performance-enhancing components, 2D NMs systematically improve conventional electrode materials through multidimensional strategies, including conductivity optimization, enhanced ion selectivity, improved structural stability, and interfacial modulation, thereby boosting the overall operational performance of CDI systems. Nevertheless, their application still faces challenges such as high raw material costs, difficulties in large-scale production, and insufficient long-term stability.

3.8.3 Underlying mechanisms of 2D nanomaterials for enhanced capacitive deionization system performance

CDI systems, the incorporation of 2D NMs not only optimizes electrode structure, electrochemical stability, and macroscopic mass transfer behavior, but also delivers deeper

TABLE 3.7 Summary of applications and performance of 2D NMs in capacitive deionization.

2D NMs	Application	Pollutions	Efficiency	Mechanism	References
MXene	Desalination	Inorganic ionic species	117 ± 4.7 mg/cm(Pang et al., 2021) salt removal in 5000 mg/L NaCl; stable over 24 CDI cycles	MXene materials offer high conductivity and surface area, providing more adsorption sites and faster electron transport.	Solangi et al. (2023)
MoS_2	Wastewater treatment	Heavy metal ions, such as Pb(II), Cd(II), and Hg(II) Organic compounds, such as tetracycline and other organic pollutants	1T-MoS_2 electrode: 65.1 mg/cm(Pang et al., 2021) desalination capacity (CDI); MnMoS_2 electrode: 24.5 mg/g salt removal in 500 mg/L NaCl; FeOOH/Pd/MoS_2 composite: 99.42 % Cr(VI) removal efficiency (high concentration); $Ti_3C_2T_x$/1T-MoS_2/CNT heterostructure: Stable performance after 144 cycles (49.64 → 43.73 mg/g; 90.32 % → 81.92 %); O-doped MoS_2: ~100 % rare earth ion recovery from low-concentration solution.	Layered structure provides abundant active sites; surface S atoms enable chemical adsorption of metal ions; tunable interlayer spacing allows ion intercalation; composite formation enhances removal via synergistic effects.	Zhu et al. (2025)
Graphene/GO/RGO	Brackish water reuse	Ionic pollutants in brackish water, including various cations such as Na^+, K^+, Ca^{2+}, Mg^{2+}, Ni^{2+}, Cu^{2+}, and Pb^{2+}	GAC electrode: electro sorption efficiency increased from 1.6 % to 3.0 % (vs. AC: 2.7 % → 1.6 %); divalent cation selectivity: 27-65 % for GAC, significantly higher than AC; monovalent ion removal: low for both electrodes (<9.3 %).	Electric double layers form on the electrode surfaces, attracting and storing ions from water for removal.	Chong et al. (2018)

NPC/rGO	Desalination	Salts in water, especially Na^+ and Cl^-	Desalination capacity: 39.34 mg/g at 1.2 V in 1000 mg/L NaCl; Cycling stability: Minimal degradation after 50 cycles.	Micro-/macroporous structure: More adsorption sites and faster ion diffusion; EDL formation: Ion electrosorption via electric double layers under applied field; N,P codoping: enhanced electronegativity, hydrophilicity, and ion adsorption capacity;	Guo et al. (2021)
Amine-functionalized ZIF-67@MXene（A-ZIF@MXene）	Desalination Brackish water reuse	Inorganic ionic species	A-ZIF@MXene (NaCl): SAC of 80.3 mg/g (1000 mg/L) and 60.3 mg/g (500 mg/L), outperforming ZIF@MXene and ZIF-67; A-ZIF@MXene (divalent ions): SAC of 84.1 mg/g (Ca^{2+}), 87.3 mg/g (Mg^{2+}), 70.3 mg/g (K^+) in 5 mM solutions; cycling stability: A-ZIF@MXene retained 99% after 50 cycles; ZIF@MXene 84 %; ZIF-67 75% (25 cycles).	MXene: high conductivity and active sites enable rapid ion adsorption and enrichment; pseudocapacitance: fast surface redox reactions enhance salt adsorption capacity and rate; micro-/macropores: Improve ion storage and diffusion for better mass transfer; amine-functionalized ZIF-67/MXene: enhanced ion adsorption via amine–ion interactions; MXene surface groups (–O, –F, –OH): promote additional ion adsorption via surface reactions.	Molaei Aghdam et al. (2023)
MoS_2	Desalination	Na^+ and Cl^-	Ion removal capacity: 8.81 mg/g and 16.51 mg/cm^3 at 1.2 V in 400 mM NaCl; superior performance at 50 mM vs. carbon materials; cycling stability: fully regenerated over 20 cycles in 400 mg/L NaCl; outperformed bulk MoS_2 in durability.	Interlayer spacing: enlarged MoS_2 spacing enables ion intercalation and adsorption; surface charge: negative MoS_2 surface attracts cations via Coulombic interactions; surface area: large CV-derived area ensures strong electrosorption capacity.	Xing et al. (2017)

(*Continued*)

TABLE 3.7 (Continued)

2D NMs	Application	Pollutions	Efficiency	Mechanism	References
T-MoS_2	Desalination	Na^+ and Cl^-	Desalination capacity: T-MoS_2 reached 24.6 mg/g at 0.8 V (≈2.8× MoS_2); max capacity 35 mg/g; Na^+ selectivity: higher capacity than MoS_2 across varying Na^+ concentrations; cycling stability: specific capacitance remained nearly unchanged after 20 cycles.	Surface charge: negative MoS_2 surface attracts cations via Coulombic forces; surface area: large CV-derived area supports strong ion electro sorption; conductivity: low resistance facilitates ion transport in the electrode.	Jia et al. (2018)

advantages arising from their microscale physicochemical mechanisms, such as electron transfer, interfacial regulation, and reaction coupling. By examining the core CDI processes, including ion migration, interfacial adsorption, charge storage, and desorption/dissociation, it becomes evident that 2D NMs play a functional and reinforcing role through several key mechanisms outlined below.

3.8.3.1 *High-specific surface area and accessible channel structures for enhanced charge storage and ion accessibility*

2D NMs inherently possess extremely high specific surface areas and atomically thin layered structures, enabling the formation of continuous and accessible ion transport channels within electrodes. This greatly increases the probability of electrolyte ions contacting the electrode surface and accelerates interfacial response rates. Such structural advantages manifest in the electrosorption stage as higher deionization rates and improved instantaneous desalination efficiency.

3.8.3.2 *Surface functional group-driven ion-selective adsorption mechanism*

The surfaces of 2D NMs are rich in polar functional groups such as -OH, -F, and -COOH, which, in CDI systems, can form stable associations with specific target ions through electrostatic attraction, complexation, or ion-exchange interactions (Chen et al., 2020). This mechanism enables 2D NMs to deliver not only high ion storage capacity but also pronounced selectivity toward particular ions.

3.8.3.3 *Structural stability for enhanced long-term operational reliability*

CDI systems require electrodes with high durability under prolonged charge–discharge cycling and strong resistance to mechanical and electrochemical stress. 2D NMs can improve overall electrode robustness through strategies such as interlaced layer stacking, conductive scaffold construction, and flexible interfacial assembly, thereby ensuring stable performance during long-term operation (Liu et al., 2023b).

In summary, the performance enhancement of CDI systems by 2D NMs stems not merely from structural assembly or physical reinforcement, but fundamentally from their functional advantages in multiple mechanistic dimensions, including electron transfer, ion recognition, energy-level matching, and interfacial polarization. The synergistic interplay of these microscopic mechanisms provides both the theoretical foundation and functional platform for achieving high capacity, high selectivity, and high stability in CDI systems, while also offering clear guidance for the future design of advanced electro-adsorption materials.

3.8.4 Summary and outlook

This section has systematically reviewed the typical configurations of CDI systems, the optimization strategies for electrode components using 2D NMs, and the mechanisms underlying their performance enhancement, highlighting the pivotal role of these materials in improving deionization capacity, mass transfer efficiency, selectivity, and cycling stability. With their ultrathin layered structures, high electron mobility, tunable functional groups, and excellent interfacial regulation capabilities, 2D NMs can be flexibly integrated into various CDI configurations, including conventional CDI, MCDI, FCDI, and HCDI, serving both as electrode frameworks to enhance electron/ion transport and as functional layers or interfacial modifiers to improve ion selectivity and mitigate fouling.

From an optimization perspective, 2D NMs enable systematic improvements to conventional electrode materials through the construction of multiscale conductive networks, incorporation of ion-affinitive functional groups, tailoring of pore structures and wettability, enhancement of structural stability, and development of interfacial synergy systems. Mechanistically, their high specific surface area and accessible channel structures strengthen charge storage and ion accessibility; surface functional groups confer selective ion-binding capabilities; high conductivity promotes electron migration and interfacial polarization; interlayer space regulation supports stable ion intercalation; flexible architectures improve cycling durability; and excellent electro-responsiveness facilitates precise, controllable regeneration.

Although 2D NMs have significantly enhanced the overall performance of CDI systems, their large-scale application still faces challenges such as high raw material costs, insufficient scalable fabrication processes, and limited verification of long-term stability and safety in complex water matrices. Future research should focus on developing green, low-cost, and scalable synthesis routes; achieving high-level integration with reactor configurations; designing materials with multimechanism synergistic functions; and conducting long-term operational and environmental safety assessments under real water conditions.

3.9 Conclusion and perspectives

This chapter systematically reviews the configuration design, key components, and functional mechanisms of 2D nanomaterials (2D NMs) in various water purification systems (filtration membranes, membrane distillation, solar evaporation, photocatalytic degradation, antibacterial treatment, adsorption purification, capacitive deionization). Although these systems differ in working principles, application scenarios, and engineering requirements, they share commonalities in structural design logic and material integration strategies, primarily manifested in the following aspects:

- Functional escalation from material to system level: Initially studied as standalone materials, 2D NMs offer new possibilities for water treatment due to their exceptional specific surface area, conductivity, interfacial activity, and surface functionalization potential. Across purification systems, they are integrated as active layers, adsorption layers, antibacterial coatings, thermal conduction channels, or electrode scaffolds, ultimately evolving into key functional units of complete systems.
- Configuration-driven performance optimization: The structural evolution of systems (planar → stacked → hollow → 3D architectures) aligns well with the lamellar/platelet characteristics of 2D NMs. Performance optimization at the system level is achieved through interfacial engineering, interlayer spacing design, and fluid channel construction.
- Multifunctional synergy as a design paradigm: 2D NMs (e.g., GO, MXene, MoS_2, LDH, black phosphorus) not only exhibit individual functions (adsorption, conductivity,

photothermal conversion, catalysis) but also enable integrated multifunctionality through material hybridization and interface engineering (e.g., antifouling + photocatalysis, thermal conduction + antiwetting, adsorption + membrane separation), addressing complex water treatment demands.

- Despite their promise, scaling up 2D NM-based water purification technologies from lab-scale research to industrial applications faces critical challenges:
- Material cost and scalable production: Most high-performance 2D NMs (e.g., MXene, black phosphorus) still rely on expensive or complex synthesis routes with low yields and batch-to-batch variations. Developing green, low-energy, and scalable synthesis/post-processing methods is essential for industrialization.
- Stability and long-term reliability: Some 2D NMs suffer from oxidation, delamination, or structural degradation in aqueous or oxidative environments, compromising their stability in continuous operation. Enhanced material durability and interfacial robustness are needed for real-world systems facing pH fluctuations, organic fouling, and microbial interference.
- Structure–function coupling optimization: The integration of 2D NMs with porous supports, polymers, or metal components requires balanced structural integrity and functional synergy. Key challenges include maintaining efficient mass transfer (membrane flux, evaporation efficiency, charge migration pathways) while preserving material performance.
- Insufficient system integration and process evaluation: Current research predominantly focuses on lab-scale material testing, lacking comprehensive assessments of system-level metrics (total energy consumption, carbon footprint, lifecycle analysis, economic viability) for practical water treatment applications.

To address these challenges, future development directions for 2D NM-based water purification systems include:

- Multifunctional integrated platforms: Future systems will transcend single-function operation, evolving toward unified platforms combining adsorption, electrocatalysis, membrane separation, antibacterial action, and photodriven processes for one-step removal of complex contaminants. Example: MXene/GO hybrid membranes integrating nanofiltration, conductivity, and photocatalysis for simultaneous ion separation and organic degradation.
- Stimuli-responsive smart systems: Environmentally adaptive systems will be developed using 2D NMs responsive to pH, temperature, light, or electric fields. Examples include photothermal GO evaporators with sunlight-regulated evaporation rates and electro-responsive CDI systems with tunable ion selectivity.
- Green and sustainable 2D NMs: Eco-friendly synthesis strategies will employ natural precursors (plant-derived carbons), aqueous exfoliation, room-temperature processing, and low-toxicity modification to produce cost-effective materials. Emerging trends include bio-reduced graphene, chitosan-modified MXene, and waste-derived LDH.
- Modular engineering and lifecycle assessment (LCA) integration: Engineering applications must incorporate modular designs (spiral-wound membrane modules, mobile desalination units), operational stability, and recycling protocols, complemented by

LCA and techno-economic analysis (TEA) to evaluate environmental and economic feasibility.

2D NMs are transitioning from "functional materials" to "system components" and "technological platforms," becoming integral to water purification system design. By adopting a holistic material–structure–function–system perspective, these advanced materials may overcome current performance bottlenecks and environmental challenges, ushering water purification technology into an era of high efficiency, intelligence, and sustainability.

References

Aboelfetoh, E.F., Elhelaly, A.A., Gemeay, A.H., 2018. Synergistic effect of Cu(II) in the one-pot synthesis of reduced graphene oxide (rGO/CuxO) nanohybrids as adsorbents for cationic and anionic dyes. Journal of Environmental Chemical Engineering 6 (1), 623–634. Available from. Available at https://doi.org/10.1016/j.jece.2017.12.047, http://www.journals.elsevier.com/journal-of-environmental-chemical-engineering/.

Abounahia, N., Qiblawey, H., Zaidi, S.J., 2022. Progress for co-incorporation of polydopamine and nanoparticles for improving membranes performance. Membranes 12 (7) Available from. Available at https://doi.org/10.3390/membranes12070675, https://www.mdpi.com/2077-0375/12/7/675/pdf?version=1656600477.

Ahmed, K.E., Kuo, D.H., Zeleke, M.A., Zelekew, O.A., Abay, A.K., 2019. Synthesis of Sn-WO3/g-C3N4 composites with surface activated oxygen for visible light degradation of dyes. Journal of Photochemistry and Photobiology A: Chemistry 369, 133–141. Available from. Available at https://doi.org/10.1016/j.jphotochem.2018.10.027, http://www.elsevier.com/locate/jphotochem.

Ahmad, R., Liu, X., Ilyas, H.N., Hanphaiboon, P., Wang, W., Noman, M., et al., 2025. One-step synthesis of reduced graphene oxide/activated carbon composite for efficient removal of per- and polyfluoroalkyl substances from drinking water: adsorption mechanisms and DFT study. Separation and Purification Technology 367, 132797. Available at https://doi.org/10.1016/j.seppur.2025.132797.

Alam, R., Faheem, M., He, Y., Su, X., Zou, L., 2023. Ion selective electrosorption by two pseudocapacitive intercalating nanocomposite electrodes. Desalination 566, 116923. Available at https://doi.org/10.1016/j.desal.2023.116923.

Alberto, M., Skuse, C., Tamaddondar, M., Gorgojo, P., 2022. Immobilized graphene oxide-based membranes for improved pore wetting resistance in membrane distillation. Desalination 537, 115898. Available at https://doi.org/10.1016/j.desal.2022.115898.

Ali, M.E.A., Wang, L., Wang, X., Feng, X., 2016. Thin film composite membranes embedded with graphene oxide for water desalination. Desalination 386, 67–76. Available at https://doi.org/10.1016/j.desal.2016.02.034.

Al-Rawashdeh, N.A.F., Allabadi, O., Aljarrah, M.T., 2020. Photocatalytic activity of graphene oxide/zinc oxide nanocomposites with embedded metal nanoparticles for the degradation of organic dyes. American Chemical Society, Jordan ACS Omega 5 (43), 28046–28055. Available from. Available at https://doi.org/10.1021/acsomega.0c03608, pubs.acs.org/journal/acsodf.

Al-Saidi, S., Kyaw, H.H., Myint, M.T.Z., Al-Hajri, R., Al-Abri, M., 2025. Enhanced desalination of produced water using graphene oxide-coated activated carbon cloth via capacitive deionization. Desalination 603 Available from. Available at https://doi.org/10.1016/j.desal.2025.118668, https://www.sciencedirect.com/science/journal/00119164.

Alves, Bd.S.G., Barbosa, R.F., Santo, A.C.Wd.E., Habert, A.C., Borges, C.P., da Fonseca, F.V., 2024. Dynamic deposition of PDA on a hollow fiber ceramic membrane for oily water treatment. ACS Omega 9 (31), 34046–34055. Available from. Available at https://doi.org/10.1021/acsomega.4c04643, pubs.acs.org/journal/acsodf.

Alwan, S.H., Salem, K.H., Alshamsi, H.A., 2022. Visible light-driven photocatalytic degradation of Rhodamine B dye onto TiO2/rGO nanocomposites. Materials Today Communications 33 Available from. Available at https://doi.org/10.1016/j.mtcomm.2022.104558, http://www.journals.elsevier.com/materials-today-communications/.

Anjaneyulu, R.B., Rao, K.S., Madhavi, K., Mahender, C., Gupta, C.H.L.D.S.N., Shekhar, B., 2022. Efficient photocatalytic degradation of o-nitrophenol using ZrO_2–RGO nanocomposite: Hydrothermal approach. Research Square [Preprint]. Available at https://doi.org/10.21203/rs.3.rs-2189071/v1.

Anjum, M., Miandad, R., Waqas, M., Gehany, F., Barakat, M.A., 2019. Remediation of wastewater using various nano-materials. Arabian Journal of Chemistry 12 (8), 4897–4919. Available from. Available at https://doi.org/10.1016/j.arabjc.2016.10.004, http://colleges.ksu.edu.sa/Arabic%20Colleges/CollegeOfScience/ChemicalDept/AJC/default.aspx (ScienceDirect http://www.sciencedirect.com/science/journal/18785352).

Araujo, C.M.B.D., Assis Filho, R.B.D., Baptisttella, A.M.S., Nascimento, G.F.O.D., Costa, G.R.B.D., Carvalho, M.N., et al., 2018. Systematic study of graphene oxide production using factorial design techniques and its application to the adsorptive removal of methylene blue dye in aqueous medium. Materials Research Express 5 (6) Available from. Available at https://doi.org/10.1088/2053-1591/aacb51, http://iopscience.iop.org/article/10.1088/2053-1591/aacb51/pdf.

Babitha, K.B., Jani Matilda, J., Peer Mohamed, A., Ananthakumar, S., 2015. Catalytically engineered reduced graphene oxide/ZnO hybrid nanocomposites for the adsorption, photoactivity and selective oil pick-up from aqueous media. RSC Advances 5 (62), 50223–50233. Available at https://doi.org/10.1039/C5RA04850H.

Bai, Y., Gan, X., Jing, Z., Fan, P., 2025. Adsorption of methyl orange using modified chitosan/graphene oxide gel beads in wastewater. Langmuir: The ACS Journal of Surfaces and Colloids 41 (34), 22723–22737. Available from. Available at https://doi.org/10.1021/acs.langmuir.5c01431, http://pubs.acs.org/journal/langd5.

Balkız, G., Pingo E., Kahya, N., Kaygusuz, H., Bedia Erim, F., 2018. Graphene oxide/alginate quasi-cryogels for removal of methylene blue. Water, Air, and Soil Pollution 229 (4) Available from. Available at https://doi.org/10.1007/s11270-018-3790-5, http://www.kluweronline.com/issn/0049-6979/.

Bangari, R.S., Yadav, V.K., Singh, J.K., Sinha, N., 2020. Fe3O4-functionalized boron nitride nanosheets as novel adsorbents for removal of arsenic(III) from contaminated water. ACS Omega 5 (18), 10301–10314. Available from. Available at https://doi.org/10.1021/acsomega.9b04295, pubs.acs.org/journal/acsodf.

Bangari, R.S., Singh, A.K., Namsani, S., Singh, J.K., Sinha, N., 2019. Magnetite-coated boron nitride nanosheets for the removal of arsenic(V) from water. ACS Applied Materials and Interfaces 11 (21), 19017–19028. Available from. Available at https://doi.org/10.1021/acsami.8b22401, http://pubs.acs.org/journal/aamick.

Bao, Y., Liu, Y., Pan, J., Chen, P., Liu, X., Li, Y., et al., 2022. Constructing 2D layered PCN/Ti3C2/Bi2MoO6 heterojunction with MXene as charge mediator for enhanced photocatalytic performance. Applied Surface Science 589, 152883. Available at https://doi.org/10.1016/j.apsusc.2022.152883.

Bharath, G., Alhseinat, E., Ponpandian, N., Khan, M.A., Siddiqui, M.R., Ahmed, F., et al., 2017. Development of adsorption and electrosorption techniques for removal of organic and inorganic pollutants from wastewater using novel magnetite/porous graphene-based nanocomposites. Separation and Purification Technology 188, 206–218. Available from. Available at https://doi.org/10.1016/j.seppur.2017.07.024, http://www.journals.elsevier.com/separation-and-purification-technology/.

Bonaccorso, F., Sun, Z., Hasan, T., Ferrari, A.C., 2010. Graphene photonics and optoelectronics. Nature Photonics 4 (9), 611–622. Available at https://doi.org/10.1038/nphoton.2010.186.

Chae, H.R., Lee, J., Lee, C.H., Kim, I.C., Park, P.K., 2015. Graphene oxide-embedded thin-film composite reverse osmosis membrane with high flux, anti-biofouling, and chlorine resistance. Journal of Membrane Science 483, 128–135. Available from. Available at https://doi.org/10.1016/j.memsci.2015.02.045, www.elsevier.com/locate/memsci.

Chandio, I., Janjhi, F.A., Memon, A.A., Memon, S., Ali, Z., Thebo, K.H., et al., 2021. Ultrafast ionic and molecular sieving through graphene oxide based composite membranes. Desalination 500 Available from. Available at https://doi.org/10.1016/j.desal.2020.114848, https://www.journals.elsevier.com/desalination.

Chaudhuri, H., Dash, S., Gupta, R., Pathak, D.D., Sarkar, A., 2017. Room-temperature in-situ design and use of graphene oxide-SBA-16 composite for water remediation and reusable heterogeneous catalysis. Chemistry Select 2 (5), 1835–1842. Available from. Available at https://doi.org/10.1002/slct.201601817, http://onlinelibrary.wiley.com/journal/10.1002/(ISSN)2365-6549.

Chen, B., Feng, A., Deng, R., Liu, K., Yu, Y., Song, L., 2020. MXene as a cation-selective cathode material for asymmetric capacitive deionization. ACS Applied Materials & Interfaces 12 (12), 13750–13758. Available at https://doi.org/10.1021/acsami.9b19684.

Chen, C., Wang, B., Xu, J., Fei, L., Raza, S., Li, B., et al., 2024. Recent advancement in emerging mxene-based photocatalytic membrane for revolutionizing wastewater treatment. Small (Weinheim an der Bergstrasse,

Germany) 20 (37) Available from. Available at https://doi.org/10.1002/smll.202311427, http://onlinelibrary.wiley.com/journal/10.1002/(ISSN)1613-6829.

Chen, Q., Kim, J., Choi, M., Jeon, S., 2025. Advancing solar energy applications with graphene: the potential of minimally oxidized graphene. Nano Convergence 12 (1) Available from. Available at https://doi.org/10.1186/s40580-025-00498-x, https://nanoconvergencejournal.springeropen.com.

Chen, R., Zhi, C., Yang, H., Bando, Y., Zhang, Z., Sugiur, N., et al., 2011. Arsenic (V) adsorption on Fe3O4 nanoparticle-coated boron nitride nanotubes. Journal of Colloid and Interface Science 359 (1), 261–268. Available at https://doi.org/10.1016/j.jcis.2011.02.071.

Chen, T., Butt, F.S., Zhang, M., Wei, X., Lewis, A., Radacsi, N., et al., 2021. Ultra-permeable zeolitic imidazolate frameworks-intercalated graphene oxide membranes for unprecedented ultrafast molecular separation. Chemical Engineering Journal 419 Available from. Available at https://doi.org/10.1016/j.cej.2021.129507, www.elsevier.com/inca/publications/store/6/0/1/2/7/3/index.htt.

Chong, L.G., Chen, P.A., Huang, J.Y., Huang, H.L., Wang, H.P., 2018. Capacitive deionization of a RO brackish water by AC/graphene composite electrodes. Chemosphere 191, 296–301. Available from. Available at https://doi.org/10.1016/j.chemosphere.2017.10.064, www.elsevier.com/locate/chemosphere.

Chu, A., Zhang, S., Jin, J. Recent progress on solar-driven interfacial evaporation for resource recovery and pollutant removal. Advanced Materials, 2505656.

Çinar, B., Kerimoğlu, I., Tönbül, B., Demirbüken, A., Dursun, S., Kaya, I.C., et al., 2020. Hydrothermal/electrospinning synthesis of CuO plate-like particles/TiO2 fibers heterostructures for high-efficiency photocatalytic degradation of organic dyes and phenolic pollutants. Materials Science in Semiconductor Processing 109, 104919. Available at https://doi.org/10.1016/j.mssp.2020.104919.

Cui, C., Guo, R., Xiao, H., Ren, E., Song, Q., Xiang, C., et al., 2020. Bi2WO6/Nb2CTx MXene hybrid nanosheets with enhanced visible-light-driven photocatalytic activity for organic pollutants degradation. Applied Surface Science 505, 144595. Available at https://doi.org/10.1016/j.apsusc.2019.144595.

Cui, C., Guo, R., Ren, E., Xiao, H., Lai, X., Qin, Q., et al., 2021. Facile hydrothermal synthesis of rod-like Nb2O5/Nb2CTx composites for visible-light driven photocatalytic degradation of organic pollutants. Environmental Research 193, 110587. Available at https://doi.org/10.1016/j.envres.2020.110587.

Cui, L., Wen, J., Deng, Q., Du, X., Tang, T., Li, M., et al., 2023. Improving the photocatalytic activity of Ti3C2 MXene by surface modification of N doped. Materials 16 (7) Available from. Available at https://doi.org/10.3390/ma16072836, http://www.mdpi.com/journal/materials.

Dai, F., Zhou, F., Chen, J., Liang, S., Chen, L., Fang, H., 2021. Ultrahigh water permeation with a high multivalent metal ion rejection rate through graphene oxide membranes. Journal of Materials Chemistry A 9 (17), 10672–10677. Available from. Available at https://doi.org/10.1039/d1ta00647a, http://pubs.rsc.org/en/journals/journal/ta.

Dai, M., He, Z., Cao, W., Zhang, J., Chen, W., Jin, Q., et al., 2023. Rational construction of S-scheme BN/MXene/ZnIn2S4 heterojunction with interface engineering for efficient photocatalytic hydrogen production and chlorophenols degradation. Separation and Purification Technology 309, 123004. Available at https://doi.org/10.1016/j.seppur.2022.123004.

Das, T.R., Patra, S., Madhuri, R., Sharma, P.K., 2018. Bismuth oxide decorated graphene oxide nanocomposites synthesized via sonochemical assisted hydrothermal method for adsorption of cationic organic dyes. Journal of Colloid and Interface Science 509, 82–93. Available from. Available at https://doi.org/10.1016/j.jcis.2017.08.102, http://www.elsevier.com/inca/publications/store/6/2/2/8/6/1/index.htt.

De Faria, A.F., Perreault, F., Shaulsky, E., Arias Chavez, L.H., Elimelech, M., 2015. Antimicrobial electrospun biopolymer nanofiber mats functionalized with graphene oxide-silver nanocomposites. ACS Applied Materials and Interfaces 7 (23), 12751–12759. Available from. Available at https://doi.org/10.1021/acsami.5b01639, http://pubs.acs.org/journal/aamick.

Deebansok, S., Amornsakchai, T., Sae-ear, P., Siriphannon, P., Smith, S.M., 2021. Sphere-like and flake-like ZnO immobilized on pineapple leaf fibers as easy-to-recover photocatalyst for the degradation of congo red. Journal of Environmental Chemical Engineering 9 (2), 104746. Available at https://doi.org/10.1016/j.jece.2020.104746.

Ding, X., Li, Y., Li, C., Wang, W., Wang, L., Feng, L., et al., 2019a. 2D visible-light-driven TiO2@Ti3C2/g-C3N4 ternary heterostructure for high photocatalytic activity. Journal of Materials Science 54 (13), 9385–9396. Available from. Available at https://doi.org/10.1007/s10853-018-03289-4, https://link.springer.com/content/pdf/10.1007/s10853-018-03289-4.pdf.

Ding, X., Li, C., Wang, L., Feng, L., Han, D., Wang, W., 2019b. Fabrication of hierarchical g-C 3 N 4 /MXene-AgNPs nanocomposites with enhanced photocatalytic performances. Materials Letters 247, 174–177. Available from. Available at https://doi.org/10.1016/j.matlet.2019.02.125, http://www.journals.elsevier.com/materials-letters/.

Ding, L., Li, L., Liu, Y., Wu, Y., Lu, Z., Deng, J., et al., 2020. Effective ion sieving with Ti3C2T x MXene membranes for production of drinking water from seawater. Nature Sustainability 3 (4), 296–302. Available from. Available at https://doi.org/10.1038/s41893-020-0474-0, www.nature.com/natsustain/.

El Essawy, N.A., Ali, S.M., Farag, H.A., Konsowa, A.H., Elnouby, M., Hamad, H.A., 2017. Green synthesis of graphene from recycled PET bottle wastes for use in the adsorption of dyes in aqueous solution. Ecotoxicology and Environmental Safety 145, 57–68. Available from. Available at https://doi.org/10.1016/j.ecoenv.2017.07.014, http://www.elsevier.com/inca/publications/store/6/2/2/8/1/9/index.htt.

Fang, F., Kong, L., Huang, J., Wu, S., Zhang, K., Wang, X., et al., 2014. Removal of cobalt ions from aqueous solution by an amination graphene oxide nanocomposite. Journal of Hazardous Materials 270, 1–10. Available at https://doi.org/10.1016/j.jhazmat.2014.01.031.

Fang, Y., Cao, Y., Chen, Q., 2019. Synthesis of an Ag2WO4/Ti3C2 Schottky composite by electrostatic traction and its photocatalytic activity. Ceramics International 45 (17), 22298–22307. Available from. Available at https://doi.org/10.1016/j.ceramint.2019.07.256, https://www.journals.elsevier.com/ceramics-international.

Fard, A.K., Mckay, G., Chamoun, R., Rhadfi, T., Preud'Homme, H., Atieh, M.A., 2017. Barium removal from synthetic natural and produced water using MXene as two dimensional (2-D) nanosheet adsorbent. Chemical Engineering Journal 317, 331–342. Available from. Available at https://doi.org/10.1016/j.cej.2017.02.090, www.elsevier.com/inca/publications/store/6/0/1/2/7/3/index.htt.

Fatima, R., Warsi, M.F., Zulfiqar, S., Ragab, S.A., Shakir, I., Sarwar, M.I., 2020. Nanocrystalline transition metal oxides and their composites with reduced graphene oxide and carbon nanotubes for photocatalytic applications. Ceramics International 46 (10), 16480–16492. Available from. Available at https://doi.org/10.1016/j.ceramint.2020.03.213, https://www.journals.elsevier.com/ceramics-international.

Fei, L., Chen, C., Shen, L., Zhang, Y., Wang, B., Xu, J., et al., 2023. Graphene oxide assisted assembly of superhydrophilic MOF-based membrane with 2D/3D hybrid nanochannels for enhanced water purification. Chemical Engineering Journal 460, 141694. Available at https://doi.org/10.1016/j.cej.2023.141694.

Feng, X., Yu, Z., Sun, Y., Shan, M., Long, R., Li, X., 2021. 3D MXene/Ag2S material as Schottky junction catalyst with stable and enhanced photocatalytic activity and photocorrosion resistance. Separation and Purification Technology 266, 118606. Available at https://doi.org/10.1016/j.seppur.2021.118606.

Finnerty, C.T.K., Menon, A.K., Conway, K.M., Lee, D., Nelson, M., Urban, J.J., et al., 2021. Interfacial Solar Evaporation by a 3D graphene oxide stalk for highly concentrated brine treatment. Environmental Science and Technology 55 (22), 15435–15445. Available from. Available at https://doi.org/10.1021/acs.est.1c04010, http://pubs.acs.org/journal/esthag.

Folaranmi, G., Bechelany, M., Sistat, P., Cretin, M., Zaviska, F., 2021. Activated carbon blended with reduced graphene oxide nanoflakes for capacitive deionization. Nanomaterials 11 (5), 1090. Available at https://doi.org/10.3390/nano11051090.

Frieri, M., Kumar, K., Boutin, A., 2017. Antibiotic resistance. Journal of Infection and Public Health 10 (4), 369–378. Available from. Available at https://doi.org/10.1016/j.jiph.2016.08.007, http://www.elsevier.com/wps/find/journaldescription.cws_home/716388/descriptio.

Fu, C., Liu, X., Wang, Y., Li, L., Zhang, Z., 2019. Preparation and characterization of Fe3O4@SiO2@TiO2-Co/rGO magnetic visible light photocatalyst for water treatment. RSC Advances 9 (35), 20256–20265. Available from. Available at https://doi.org/10.1039/c9ra04002a, http://pubs.rsc.org/en/journals/journal/ra.

Gao, J., Cai, M., Nie, Z., Zhang, J., Chen, Y., 2021. Superwetting PVDF membrane prepared by in situ extraction of metal ions for highly efficient oil/water mixture and emulsion separation. Separation and Purification Technology 275 Available from. Available at https://doi.org/10.1016/j.seppur.2021.119174, http://www.journals.elsevier.com/separation-and-purification-technology/.

Gao, J., Wu, J., Chen, S.B., Chung, T.S., 2024. Thin-film composite (TFC) membranes incorporated with hydrophilic additives and layered double hydroxides (LDHs) for effective brackish water (BW) desalination. Journal of Membrane Science 712 Available from. Available at https://doi.org/10.1016/j.memsci.2024.123228, https://www.sciencedirect.com/science/journal/03767388.

Gao, W., Pan, X., Wu, J., Guan, T., Cai, F., Han, J., et al., 2025. Synergistic degradation mechanism of roxithromycin by the combination of TiO2/graphene oxide/polypropylene fiber photocatalytic network and microalgae. Chemical Engineering Journal 519, 165218. Available at https://doi.org/10.1016/j.cej.2025.165218.

Guan, K., Zhao, D., Zhang, M., Shen, J., Zhou, G., Liu, G., et al., 2017. 3D nanoporous crystals enabled 2D channels in graphene membrane with enhanced water purification performance. Journal of Membrane Science 542, 41–51. Available from. Available at https://doi.org/10.1016/j.memsci.2017.07.055, www.elsevier.com/locate/memsci.

Guerrero-Sanchez, J., Muñoz-Pizza, D.M., Moreno-Armenta, M.G., Takeuchi, N., 2022. Atomic-scale understanding of the Na and Cl trapping on the Mo1.33C(OH)2-MXene. Scientific Reports 12 (1) Available from. Available at https://doi.org/10.1038/s41598-022-12177-6, www.nature.com/srep/index.html.

Guo, J., Xu, X., Hill, J.P., Wang, L., Dang, J., Kang, Y., et al., 2021. Graphene-carbon 2D heterostructures with hierarchically-porous P,N-doped layered architecture for capacitive deionization. Chemical Science 12 (30), 10334–10340. Available from. Available at https://doi.org/10.1039/d1sc00915j, http://pubs.rsc.org/en/journals/journal/sc.

Gupta, K., Khatri, O.P., 2017. Reduced graphene oxide as an effective adsorbent for removal of malachite green dye: plausible adsorption pathways. Journal of Colloid and Interface Science 501, 11–21. Available from. Available at https://doi.org/10.1016/j.jcis.2017.04.035, http://www.elsevier.com/inca/publications/store/6/2/2/8/6/1/index.htt.

Han, S., Li, W., Xi, H., Yuan, R., Long, J., Xu, C., 2022. Plasma-assisted in-situ preparation of graphene-Ag nanofiltration membranes for efficient removal of heavy metal ions. Journal of Hazardous Materials 423 (Pt A), 127012. Available at https://doi.org/10.1016/j.jhazmat.2021.127012.

He, J., Yang, J., Jiang, F., Liu, P., Zhu, M., 2020a. Photo-assisted peroxymonosulfate activation via 2D/2D heterostructure of Ti3C2/g-C3N4 for degradation of diclofenac. Chemosphere 258, 127339. Available at https://doi.org/10.1016/j.chemosphere.2020.127339.

He, M., Wang, L., Zhang, Z., Zhang, Y., Zhu, J., Wang, X., et al., 2020b. Stable forward osmosis nanocomposite membrane doped with sulfonated graphene oxide@metal–organic frameworks for heavy metal removal. ACS Applied Materials & Interfaces 12 (51), 57102–57116. Available at https://doi.org/10.1021/acsami.0c17405.

He, C., Liu, Z., Wu, J., Pan, X., Fang, Z., Li, J., et al., 2021. Future global urban water scarcity and potential solutions. Nature Communications 12 (1) Available from. Available at https://doi.org/10.1038/s41467-021-25026-3, http://www.nature.com/ncomms/index.html.

Hermawan, A., Hasegawa, T., Asakura, Y., Yin, S., 2021. Enhanced visible-light-induced photocatalytic NOx degradation over (Ti,C)-BiOBr/Ti3C2Tx MXene nanocomposites: role of Ti and C doping. Separation and Purification Technology 270, 118815. Available at https://doi.org/10.1016/j.seppur.2021.118815.

Hou, Q., Wang, M., Li, T., Hou, Y., Xuan, K., Hao, Y., 2023. Peroxymonosulfate-assisted photocatalysis by a novel Ti3C2-based heterojunction catalyst (g-C3N4/Ti3C2/MnFe2O4) for enhanced degradation of naphthalene. Chemical Engineering Journal 464, 142566. Available at https://doi.org/10.1016/j.cej.2023.142566.

Hou, Y., Zhou, P., Liu, F., Tong, K., Lu, Y., Li, Z., et al., 2024. Rigid covalent organic frameworks with thiazole linkage to boost oxygen activation for photocatalytic water purification. Nature Communications 15 (1) Available from. Available at https://doi.org/10.1038/s41467-024-51878-6, https://www.nature.com/ncomms/.

Hu, W., Peng, C., Luo, W., Lv, M., Li, X., Li, D., et al., 2010. Graphene-based antibacterial paper. ACS Nano 4 (7), 4317–4323. Available at https://doi.org/10.1021/nn101097v.

Hu, J., Zhu, G., Ronen, A., Jassby, D., Li, Q., Wang, P., et al., 2025. Interfacial heating in membrane distillation: advances, optimization strategies, and industrial applications for desalination. Environmental Science and Technology 59 (22), 10750–10769. Available from. Available at https://doi.org/10.1021/acs.est.5c01447, http://pubs.acs.org/journal/esthag.

Hu, X., Wang, Y., Ling, Z., Song, H., Cai, Y., Li, Z., et al., 2021. Ternary g-C3N4/TiO2/Ti3C2 MXene S-scheme heterojunction photocatalysts for NOx removal under visible light. Applied Surface Science 556, 149817. Available at https://doi.org/10.1016/j.apsusc.2021.149817.

Huang, X., Zhu, J., Wu, D., Zhang, Q., Nie, Z., Surface-engineered solar-driven interfacial evaporation: innovations and challenges. Advanced Materials Interfaces, n/a (n/a), e00371.

Huang, M.Y., Chen, Y., Yan, X., Guo, X.J., Dong, L., Lang, W.Z., 2020. Two-dimensional montmorillonite membranes with efficient water filtration. Journal of Membrane Science 614 Available from. Available at https://doi.org/10.1016/j.memsci.2020.118540, www.elsevier.com/locate/memsci.

Huang, K., Li, C., Wang, L., Wang, W., Meng, X., 2021. Layered Ti3C2 MXene and silver co-modified g-C3N4 with enhanced visible light-driven photocatalytic activity. Chemical Engineering Journal 425, 131493. Available at https://doi.org/10.1016/j.cej.2021.131493.

Huong, P.T.L., Huy, L.T., Phan, V.N., Huy, T.Q., Nam, M.H., Lam, V.D., et al., 2016. Application of graphene oxide-MnFe2O4 magnetic nanohybrids as magnetically separable adsorbent for highly efficient removal of arsenic from water. Journal of Electronic Materials 45 (5), 2372–2380. Available at https://doi.org/10.1007/s11664-015-4314-3.

Ibrahim, Y., Wadi, V.S., Ouda, M., Naddeo, V., Banat, F., Hasan, S.W., 2022. Highly selective heavy metal ions membranes combining sulfonated polyethersulfone and self-assembled manganese oxide nanosheets on positively functionalized graphene oxide nanosheets. Chemical Engineering Journal 428 Available from. Available at https://doi.org/10.1016/j.cej.2021.131267, www.elsevier.com/inca/publications/store/6/0/1/2/7/3/index.htt.

Iqbal, M.A., Ali, S.I., Amin, F., Tariq, A., Iqbal, M.Z., Rizwan, S., 2019. La-and Mn-codoped bismuth ferrite/Ti 3 C 2 MXene composites for efficient photocatalytic degradation of congo red dye. ACS Omega 4 (5), 8661–8668. Available from. Available at https://doi.org/10.1021/acsomega.9b00493, pubs.acs.org/journal/acsodf.

Jafarian, H., Dadashi Firouzjaei, M., Aghapour Aktij, S., Aghaei, A., Pilevar Khomami, M., Elliott, M., et al., 2023. Synthesis of heterogeneous metal organic framework-graphene oxide nanocomposite membranes for water treatment. Chemical Engineering Journal 455, 140851. Available at https://doi.org/10.1016/j.cej.2022.140851.

Jang, J., Kang, Y., Jang, K., Kim, S., Chee, S.S., Kim, I.S., 2022. Ti3C2TX-ethylenediamine nanofiltration membrane for high rejection of heavy metals. Chemical Engineering Journal 437 Available from. Available at https://doi.org/10.1016/j.cej.2022.135297, www.elsevier.com/inca/publications/store/6/0/1/2/7/3/index.htt.

Ji, B., Zhao, W., Duan, J., Fu, L., Ma, L., Yang, Z., 2020. Immobilized Ag3PO4/GO on 3D nickel foam and its photocatalytic degradation of norfloxacin antibiotic under visible light. RSC Advances 10 (8), 4427–4435. Available from. Available at https://doi.org/10.1039/c9ra08678a, http://pubs.rsc.org/en/journals/journal/ra.

Jia, J., Seitz, L.C., Benck, J.D., Huo, Y., Chen, Y., Ng, J.W.D., et al., 2016. Solar water splitting by photovoltaic-electrolysis with a solar-to-hydrogen efficiency over 30%. Nature Communications 7 Available from. Available at https://doi.org/10.1038/ncomms13237, http://www.nature.com/ncomms/index.html.

Jia, F., Sun, K., Yang, B., Zhang, X., Wang, Q., Song, S., 2018. Defect-rich molybdenum disulfide as electrode for enhanced capacitive deionization from water. Desalination 446, 21–30. Available at https://doi.org/10.1016/j.desal.2018.08.024.

Jia, Y., Xu, G., An, X., Hu, Y., 2022. Robust reduced graphene oxide composite membranes for enhanced anti-wetting property in membrane distillation. Desalination 526, 115549. Available at https://doi.org/10.1016/j.desal.2022.115549.

Jiang, Q., Tian, L., Liu, K.K., Tadepalli, S., Raliya, R., Biswas, P., et al., 2016. Bilayered biofoam for highly efficient solar steam generation. Advanced Materials 28 (42), 9400–9407. Available from. Available at https://doi.org/10.1002/adma.201601819, http://www3.interscience.wiley.com/journal/119030556/issue.

Jiang, Y., Chowdhury, S., Balasubramanian, R., 2017. Nitrogen-doped graphene hydrogels as potential adsorbents and photocatalysts for environmental remediation. Chemical Engineering Journal 327, 751–763. Available from. Available at https://doi.org/10.1016/j.cej.2017.06.156, www.elsevier.com/inca/publications/store/6/0/1/2/7/3/index.htt.

Jiang, N., Shang, R., Heijman, S.G.J., Rietveld, L.C., 2018. High-silica zeolites for adsorption of organic micro-pollutants in water treatment: a review. Water Research 144, 145–161. Available from. Available at https://doi.org/10.1016/j.watres.2018.07.017, www.elsevier.com/locate/watres.

Jiang, L., Chen, L., Zhu, L., 2022. Minimize the trade-off between wetting resistance and water permeance in membrane distillation with ion-sieving coating layer. Chemical Engineering Journal 430, 133165. Available at https://doi.org/10.1016/j.cej.2021.133165.

Kadi, M.W., Mohamed, R.M., Ismail, A.A., Bahnemann, D.W., 2020a. Decoration of g-C3N4 nanosheets by mesoporous CoFe2O4 nanoparticles for promoting visible-light photocatalytic Hg(II) reduction. Colloids and

Surfaces A: Physicochemical and Engineering Aspects 603 Available from. Available at https://doi.org/10.1016/j.colsurfa.2020.125206, www.elsevier.com/locate/colsurfa.

Kadi, M.W., Mohamed, R.M., Ismail, A.A., Bahnemann, D.W., 2020b. Performance of mesoporous α-Fe2O3/g-C3N4 heterojunction for photoreduction of Hg(II) under visible light illumination. Ceramics International 46 (14), 23098–23106. Available from. Available at https://doi.org/10.1016/j.ceramint.2020.06.087, https://www.journals.elsevier.com/ceramics-international.

Kadi, M.W., Mohamed, R.M., Ismail, A.A., Bahnemann, D.W., 2020c. Soft and hard templates assisted synthesis mesoporous CuO/g-C3N4 heterostructures for highly enhanced and accelerated Hg(II) photoreduction under visible light. Journal of Colloid and Interface Science 580, 223–233. Available from. Available at https://doi.org/10.1016/j.jcis.2020.07.001, http://www.elsevier.com/inca/publications/store/6/2/2/8/6/1/index.htt.

Kakkar, R., Kaur, D.P., Raj, S., 2025. Membrane distillation for sustainable water desalination: a review of principles, materials, and applications. Water, Air, and Soil Pollution 236 (1) Available from. Available at https://doi.org/10.1007/s11270-024-07622-3, https://www.springer.com/journal/11270.

Kamil, A.M., Kumar, A.C., Maria, H.J., Thomas, S., 2025. Review: adsorption of dyes from wastewater by graphene oxide and its composites. Journal of Materials Science 60 (23), 9397–9446. Available from. Available at https://doi.org/10.1007/s10853-025-10944-0, https://www.springer.com/journal/10853.

Kaneda, M., Lu, X., Cheng, W., Zhou, X., Bernstein, R., Zhang, W., et al., 2019. Photografting graphene oxide to inert membrane materials to impart antibacterial activity. Environmental Science and Technology Letters 6 (3), 141–147. Available from. Available at https://doi.org/10.1021/acs.estlett.9b00012, http://pubs.acs.org/page/estlcu/about.html.

Karthikeyan, P., Elanchezhiyan, S.S.D., Preethi, J., Meenakshi, S., Park, C.M., 2020. Mechanistic performance of polyaniline-substituted hexagonal boron nitride composite as a highly efficient adsorbent for the removal of phosphate, nitrate, and hexavalent chromium ions from an aqueous environment. Applied Surface Science 511, 145543. Available at https://doi.org/10.1016/j.apsusc.2020.145543.

Khadidja, M.F., Fan, J., Li, S., Li, S., Cui, K., Wu, J., et al., 2021. Hierarchical ZnO/MXene composites and their photocatalytic performances. Colloids and Surfaces A: Physicochemical and Engineering Aspects 628 Available from. Available at https://doi.org/10.1016/j.colsurfa.2021.127230, www.elsevier.com/locate/colsurfa.

Khan, Z., Chetia, T.R., Vardhaman, A.K., Barpuzary, D., Sastri, C.V., Qureshi, M., 2012. Visible light assisted photocatalytic hydrogen generation and organic dye degradation by CdS-metal oxide hybrids in presence of graphene oxide. RSC Advances 2 (32), 12122–12128. Available at https://doi.org/10.1039/c2ra21596a.

Khan, A.R., Husnain, S.M., Shahzad, F., Mujtaba-Ul-Hassan, S., Mehmood, M., Ahmad, J., et al., 2019. Two-dimensional transition metal carbide (Ti3C2 T x) as an efficient adsorbent to remove cesium (Cs+). Dalton Transactions 48 (31), 11803–11812. Available from. Available at https://doi.org/10.1039/c9dt01965k, http://pubs.rsc.org/en/journals/journal/dt.

khan, F., A, N., Din, I.U., Saeed, T., Alotaibi, M.A., Alharthi, A.I., et al., 2021. Synthesis, characterization and adsorption studies of h-BN crystal for efficient removal of Cd2+ from aqueous solution. Ceramics International 47 (4), 4749–4757. Available from. Available at https://doi.org/10.1016/j.ceramint.2020.10.044, https://www.journals.elsevier.com/ceramics-international.

Khanmohammadi, M., Shahrouzi, J.R., Rahmani, F., 2021. Insights into mesoporous MCM-41-supported titania decorated with CuO nanoparticles for enhanced photodegradation of tetracycline antibiotic. Environmental Science and Pollution Research 28 (1), 862–879. Available from. Available at https://doi.org/10.1007/s11356-020-10546-0, https://link.springer.com/journal/11356.

Khurana, I., Shaw, A.K., Bharti, Khurana, J.M., Rai, P.K., 2018. Batch and dynamic adsorption of Eriochrome Black T from water on magnetic graphene oxide: experimental and theoretical studies. Journal of Environmental Chemical Engineering 6 (1), 468–477. Available from. Available at https://doi.org/10.1016/j.jece.2017.12.029, http://www.journals.elsevier.com/journal-of-environmental-chemical-engineering/.

Kim, D.W., Han, H., Kim, H., Guo, X., Tsapatsis, M., 2018. Preparation of a graphene oxide/faujasite composite adsorbent. Microporous and Mesoporous Materials 268, 243–250. Available from. Available at https://doi.org/10.1016/j.micromeso.2018.04.034, www.elsevier.com/inca/publications/store/6/0/0/7/6/0.

Kumar (S. Kumar), S., Kaushik, R.D., Purohit, L.P., 2022. ZnO-CdO nanocomposites incorporated with graphene oxide nanosheets for efficient photocatalytic degradation of bisphenol A, thymol blue and ciprofloxacin. Journal of Hazardous Materials 424 (Pt A), 127332. Available at https://doi.org/10.1016/j.jhazmat.2021.127332.

Kusworo, T.D., Purwanto, P., Jos, B., Budiyono, B., Astuti, D.A.P., Inamullah, A.M.A., et al., 2024. Photocatalytic nanohybrid UV-light-driven PVDF/GO-NiFe@SiO2 membrane coupled with bentonite adsorption and ozonation process for a sustainable textile wastewater treatment. Process Safety and Environmental Protection 190, 438–457. Available from. Available at https://doi.org/10.1016/j.psep.2024.08.048, https://www.sciencedirect.com/science/journal/09575820.

Kwon, H., Seo, S.W., Kim, T.G., Lee, E.S., Lanh, P.T., Yang, S., et al., 2016. Ultrathin and flat layer black phosphorus fabricated by reactive oxygen and water rinse. ACS Nano 10 (9), 8723–8731. Available from. Available at https://doi.org/10.1021/acsnano.6b04194, http://pubs.acs.org/journal/ancac3.

Li, L., Zhao, Y., Jin, Y., Linghu, W., Chen, C., Asiri, A.M., et al., 2019a. Efficient scavenging of uranium (VI) using porous hexagonal boron nitride by a combined process of surface adsorption and induced precipitation crystallization. Journal of Radioanalytical and Nuclear Chemistry 321 (3), 1035–1044. Available from. Available at https://doi.org/10.1007/s10967-019-06644-0, www.wkap.nl/journalhome.htm/0236-5731.

Li, S., Wang, L., Peng, J., Zhai, M., Shi, W., 2019b. Efficient thorium(IV) removal by two-dimensional Ti2CTx MXene from aqueous solution. Chemical Engineering Journal 366, 192–199. Available from. Available at https://doi.org/10.1016/j.cej.2019.02.056, www.elsevier.com/inca/publications/store/6/0/1/2/7/3/index.htt.

Li, L., Guo, X., Jin, Y., Chen, C., Asiri, A.M., Marwani, H.M., Sheng, G., et al., 2020. Distinguished Cd(II) capture with rapid and superior ability using porous hexagonal boron nitride: kinetic and thermodynamic aspects. Wuji Cailiao Xuebao/Journal of Inorganic Materials 35 (3), 284–292. Available from. Available at https://doi.org/10.15541/jim20190371, http://www.jim.org.cn/EN/article/showTenYearVolumnDetail.do?nian=2010.

Li, S., Wan, Y., Guo, S., Luo, J., 2021. Ferric ions mediated defects narrowing of graphene oxide nanofiltration membrane for robust removal of organic micropollutants. Chemical Engineering Journal 411, 128587. Available at https://doi.org/10.1016/j.cej.2021.128587.

Li, H., Sun, B., Gao, T., Li, H., Ren, Y., Zhou, G., 2022a. Ti3C2 MXene co-catalyst assembled with mesoporous TiO2 for boosting photocatalytic activity of methyl orange degradation and hydrogen production. Chinese Journal of Catalysis 43 (2), 461–471. Available from. Available at https://doi.org/10.1016/S1872-2067(21)63915-3, http://www.elsevier.com.

Li, R., Chen, A., Deng, Q., Zhong, Y., Kong, L., Yang, R., 2022b. Well-designed MXene-derived carbon-doped TiO2 coupled porous g-C3N4 to enhance the degradation of ciprofloxacin hydrochloride under visible light irradiation. Separation and Purification Technology 295, 121254. Available at https://doi.org/10.1016/j.seppur.2022.121254.

Liao, X., Lim, Y.J., Khayet, M., Liao, Y., Yao, L., Zhao, Y., et al., 2023. Applications of electrically conductive membranes in water treatment via membrane distillation: joule heating, membrane fouling/scaling/wetting mitigation and monitoring. Water Research 244 Available from. Available at https://doi.org/10.1016/j.watres.2023.120511, www.elsevier.com/locate/watres.

Lim, H.N., Huang, N.M., Loo, C.H., 2012. Facile preparation of graphene-based chitosan films: enhanced thermal, mechanical and antibacterial properties. Journal of Non-Crystalline Solids 358 (3), 525–530. Available at https://doi.org/10.1016/j.jnoncrysol.2011.11.007.

Lin, Q., Liu, Y., Zeng, G., Li, X., Wang, B., Cheng, X., et al., 2021. Bionics inspired modified two-dimensional MXene composite membrane for high-throughput dye separation. Journal of Environmental Chemical Engineering 9 (4), 105711. Available at https://doi.org/10.1016/j.jece.2021.105711.

Liu, G., Xu, J., Wang, K., 2017. Solar water evaporation by black photothermal sheets. Nano Energy 41, 269–284. Available from. Available at https://doi.org/10.1016/j.nanoen.2017.09.005, http://www.journals.elsevier.com/nano-energy/.

Liu, H., Yang, C., Jin, X., Zhong, J., Li, J., 2020. One-pot hydrothermal synthesis of MXene Ti3C2/TiO2/BiOCl ternary heterojunctions with improved separation of photoactivated carries and photocatalytic behavior toward elimination of contaminants. Colloids and Surfaces A: Physicochemical and Engineering Aspects 603, 125239. Available at https://doi.org/10.1016/j.colsurfa.2020.125239.

Liu, S., Jiang, X., Waterhouse, G.I.N., Zhang, Z.M., Yu, Lm, 2022. A novel Z-scheme NH2-MIL-125(Ti)/Ti3C2 QDs/ZnIn2S4 photocatalyst with fast interfacial electron transfer properties for visible light-driven antibiotic degradation and hydrogen evolution. Separation and Purification Technology 294 Available from. Available at https://doi.org/10.1016/j.seppur.2022.121094, http://www.journals.elsevier.com/separation-and-purification-technology/.

Liu, Z., Ding, R., Xue, F., Zhao, X., Chen, Z., Zheng, H., et al., 2023a. MXene-reduced graphene oxide sponge-based solar evaporators with integrated water-thermal management by anisotropic design. Communications Materials 4 (1) Available from. Available at https://doi.org/10.1038/s43246-023-00398-9, https://www.nature.com/commsmat/.

Liu, N., Yu, L., Liu, B., Yu, F., Li, L., Xiao, Y., et al., 2023b. Ti(3) C(2) -MXene partially derived hierarchical 1D/2D TiO(2) /Ti(3) C(2) heterostructure electrode for high-performance capacitive deionization. Advanced Science (Weinh) 10 (2), e2204041.

Liu, Z., Wang, Y., Wang, H., Lee Smith, R., Qi, X., 2024. Direct preparation of 3D monolithic carbon-structured Ti3C2-MXene electrodes for capacitive deionization with laser-driven carbonization method. Chemical Engineering Journal 484, 149664. Available at https://doi.org/10.1016/j.cej.2024.149664.

Liu, Q., He, Z., Yu, J., Zou, L., 2025a. Engineering morphological architecture of superhydrophobic/hydrophilic composite membrane for efficient photothermal membrane distillation. Desalination 593, 118240. Available at https://doi.org/10.1016/j.desal.2024.118240.

Liu, H., Yang, B., Liao, G., Huang, B., Li, J., Rodriguez, R.D., et al., 2025b. Facilitating carrier kinetics in ultrathin porous carbon nitride through shear-repair strategy for peroxymonosulfate-assisted water purification. Nature Communications 16 (1) Available from. Available at https://doi.org/10.1038/s41467-025-61185-3, https://www.nature.com/ncomms/.

Luo, S., Wang, J., 2018. MOF/graphene oxide composite as an efficient adsorbent for the removal of organic dyes from aqueous solution. Environmental Science and Pollution Research 25 (6), 5521–5528. Available from. Available at https://doi.org/10.1007/s11356-017-0932-z, http://www.springerlink.com/content/0944-1344.

Maaz, A., Ullah, U., Shah, Sh.A., Farheen, R., Dahshan, A., Khan, Z.A., 2025. Enhancing charge transport and photocatalytic properties of zinc oxide via Ca2+ doping as electron transport material in perovskite solar cells and waste water treatments. MRS Advances Available from. Available at https://doi.org/10.1557/s43580-025-01354-0, https://www.springer.com/journal/43580.

Mahmoodi, N.M., Maroofi, S.M., Mazarji, M., Nabi-Bidhendi, G., 2017. Preparation of modified reduced graphene oxide nanosheet with cationic surfactant and its dye adsorption ability from colored wastewater. Journal of Surfactants and Detergents 20 (5), 1085–1093. Available from. Available at https://doi.org/10.1007/s11743-017-1985-1, http://link.springer.com/journal/11743.

Mahmoudi, E., Ng, L.Y., Ba-Abbad, M.M., Mohammad, A.W., 2015. Novel nanohybrid polysulfone membrane embedded with silver nanoparticles on graphene oxide nanoplates. Chemical Engineering Journal 277, 1–10. Available from. Available at https://doi.org/10.1016/j.cej.2015.04.107, www.elsevier.com/inca/publications/store/6/0/1/2/7/3/index.htt.

Maqbool, F., Kumar, L., Soomro, M.I., Harijan, K., 2025. Global advancement of solar photovoltaic thermal technologies integrated with membrane distillation systems: a comprehensive review. Environmental Science and Pollution Research 32 (15), 9361–9411. Available from. Available at https://doi.org/10.1007/s11356-025-36279-6, https://www.springer.com/journal/11356.

Marouani, I., Hassan, W.H., Singh, N.S.S., Hasan, M., Abduvokhidov, A., Karimov, M., et al., 2025. Preparation of a novel reusable 2D-MXene with flower-like LDH composite for ultra-high adsorption of congo red and doxycycline: stability and environmental application. Journal of Water Process Engineering 76 Available from. Available at https://doi.org/10.1016/j.jwpe.2025.108288, https://www.sciencedirect.com/science/journal/22147144.

Mayerberger, E.A., Street, R.M., McDaniel, R.M., Barsoum, M.W., Schauer, C.L., 2018. Antibacterial properties of electrospun Ti3C2Tz (MXene)/chitosan nanofibers. RSC Advances 8 (62), 35386–35394. Available from. Available at https://doi.org/10.1039/c8ra06274a, http://pubs.rsc.org/en/journals/journal/ra.

Merlet, R.B., Pizzoccaro-Zilamy, M.A., Nijmeijer, A., Winnubst, L., 2020. Hybrid ceramic membranes for organic solvent nanofiltration: state-of-the-art and challenges. Journal of Membrane Science 599 Available from. Available at https://doi.org/10.1016/j.memsci.2020.117839, www.elsevier.com/locate/memsci.

Minitha, C.R., Martina Susan Arachy, M., Rajendra Kumar, R.T., 2018. Influence of Fe 3 O 4 nanoparticles decoration on dye adsorption and magnetic separation properties of Fe 3 O 4 /rGO nanocomposites. Separation Science and Technology 53 (14), 2159–2169. Available at https://doi.org/10.1080/01496395.2018.1446986.

Molaei Aghdam, A., Mikaeili Chahartagh, N., Delfani, E., 2023. High-efficient capacitive deionization using amine-functionalized ZIF-67@ 2D MXene: toward ultrahigh desalination performance. Advanced Materials

Technologies 8 (18) Available from. Available at https://doi.org/10.1002/admt.202300628, http://onlinelibrary.wiley.com/journal/10.1002/(ISSN)2365-709X.

Dehaghi, M., Aberoumand, Sundararaj, 2025. A review on multifunctional polymer-MXene hybrid materials for electronic applications. Molecules (Basel, Switzerland) 30 (9).

Mu, W., Du, S., Li, X., Yu, Q., Wei, H., Yang, Y., et al., 2019. Removal of radioactive palladium based on novel 2D titanium carbides. Chemical Engineering Journal 358, 283–290. Available from. Available at https://doi.org/10.1016/j.cej.2018.10.010, www.elsevier.com/inca/publications/store/6/0/1/2/7/3/index.htt.

Mu, M., Graham, N., Yu, W., Sun, K., Xu, X., Liu, T., 2024. Optimal cross-linking of MXene-based membranes for high rejection and low adsorption with long-term stability for water treatment. Journal of Membrane Science 701, 122738. Available at https://doi.org/10.1016/j.memsci.2024.122738.

Mural, P.K.S., Banerjee, A., Rana, M.S., Shukla, A., Padmanabhan, B., Bhadra, S., et al., 2014. Polyolefin based antibacterial membranes derived from PE/PEO blends compatibilized with amine terminated graphene oxide and maleated PE. Journal of Materials Chemistry A 2 (41), 17635–17648. Available from. Available at https://doi.org/10.1039/c4ta03997a, http://pubs.rsc.org/en/journals/journalissues/ta.

Mural, P.K.S., Jain, S., Kumar, S., Madras, G., Bose, S., 2016. Unimpeded permeation of water through biocidal graphene oxide sheets anchored on to 3D porous polyolefinic membranes. Nanoscale 8 (15), 8048–8057. Available from. Available at https://doi.org/10.1039/c6nr01356b, http://www.rsc.org/publishing/journals/NR/Index.asp.

Musico, Y.L.F., Santos, C.M., Dalida, M.L.P., Rodrigues, D.F., 2014. Surface modification of membrane filters using graphene and graphene oxide-based nanomaterials for bacterial inactivation and removal. ACS Sustainable Chemistry and Engineering 2, 1559–1565. Available at https://doi.org/10.1021/sc500044p, http://pubs.acs.org/journal/ascecg.

Naeem, H., Ajmal, M., Muntha, S., Ambreen, J., Siddiq, M., 2018. Synthesis and characterization of graphene oxide sheets integrated with gold nanoparticles and their applications to adsorptive removal and catalytic reduction of water contaminants. RSC Advances 8 (7), 3599–3610. Available at https://doi.org/10.1039/c7ra12030c.

Narayanan, A., T, P., K, N.K., M, P.B., Padaki, M., 2024. Electrochemical elegance: a review on capacitive deionization for fluoride and arsenic mitigation in brackish water. Electrochimica Acta 503 Available from. Available at https://doi.org/10.1016/j.electacta.2024.144875, https://www.sciencedirect.com/science/journal/00134686.

Neto, G.L.O., Oliveira, N.G.N., Delgado, J.M.P.Q., Nascimento, L.P.C., Gomez, R.S., Cabral, A.S., et al., 2020. A new design of tubular ceramic membrane module for oily water treatment: multiphase flow behavior and performance evaluation. Membranes 10 (12), 1–27. Available from. Available at https://doi.org/10.3390/membranes10120403, https://www.mdpi.com/2077-0375/10/12/403/pdf.

Ng, I.S., Ooi, C.W., Liu, B.L., Peng, C.T., Chiu, C.Y., Chang, Y.K., 2020. Antibacterial efficacy of chitosan- and poly (hexamethylene biguanide)-immobilized nanofiber membrane. International Journal of Biological Macromolecules 154, 844–854. Available from. Available at https://doi.org/10.1016/j.ijbiomac.2020.03.127, www.elsevier.com/locate/ijbiomac.

Nguyen, C.H., Tran, M.L., Tran, T.T.V., Juang, R.S., 2020. Enhanced removal of various dyes from aqueous solutions by UV and simulated solar photocatalysis over TiO2/ZnO/rGO composites. Separation and Purification Technology 232 Available from. Available at https://doi.org/10.1016/j.seppur.2019.115962, http://www.journals.elsevier.com/separation-and-purification-technology/.

Ni, X., Chen, C., Wang, Q., Li, Z., 2019. One-step hydrothermal synthesis of SnO2-MoS2 composite heterostructure for improved visible light photocatalytic performance. Chemical Physics 525, 110398. Available at https://doi.org/10.1016/j.chemphys.2019.110398.

Nie, L., Goh, K., Wang, Y., Lee, J., Huang, Y., Karahan, H.E., et al., Realizing small-flake graphene oxide membranes for ultrafast size-dependent organic solvent nanofiltration. Science Advances 6 (17), eaaz9184.

Ojha, A., Thareja, P., 2018. Electrolyte induced rheological modulation of graphene oxide suspensions and its applications in adsorption. Applied Surface Science 435, 786–798. Available from. Available at https://doi.org/10.1016/j.apsusc.2017.11.157, http://www.journals.elsevier.com/applied-surface-science/.

Oron, G., Gillerman, L., Buriakovsky, N., Bick, A., 2024. Modeling of spiral wound ultrafiltration membrane forward flushing during secondary wastewater upgrading for unrestricted irrigation. Desalination 573, 117129. Available at https://doi.org/10.1016/j.desal.2023.117129.

Ostovar, A., Tarkhani, M., Mousavi, S.A., Asadollahi, M., 2024. Surface modification of GO/PVDF composite membrane by MXene/protonated g-C_3N_4 nanosheets hybrids with improved dye separation and antifouling properties. Polymer Engineering and Science 64 (2), 565–576. Available from. Available at https://doi.org/10.1002/pen.26566, http://onlinelibrary.wiley.com/journal/10.1002/(ISSN)1548-2634.

Othman, N.H., Alias, N.H., Shahruddin, M.Z., Abu Bakar, N.F., Nik Him, N.R., Lau, W.J., 2018. Adsorption kinetics of methylene blue dyes onto magnetic graphene oxide. Journal of Environmental Chemical Engineering 6 (2), 2803–2811. Available from. Available at https://doi.org/10.1016/j.jece.2018.04.024, http://www.journals.elsevier.com/journal-of-environmental-chemical-engineering/.

Pang, Y., Su, C., Jia, G., Xu, L., Shao, Z., 2021. Emerging two-dimensional nanomaterials for electrochemical nitrogen reduction. Chemical Society Reviews 50 (22), 12744–12787. Available from. Available at https://doi.org/10.1039/d1cs00120e, http://pubs.rsc.org/en/journals/journal/cs.

Park, S., Kim, S., Yea, Y., Saravanakumar, K., Lee, E., Yoon, Y., et al., 2023. Adsorptive and photocatalytic performance of cobalt-doped ZnTiO3/Ti3C2Tx MXene nanohybrids towards tetracycline: kinetics and mechanistic insight. Journal of Hazardous Materials 443 (Pt A), 130165. Available at https://doi.org/10.1016/j.jhazmat.2022.130165.

Peng, Q., Guo, J., Zhang, Q., Xiang, J., Liu, B., Zhou, A., et al., 2014. Unique lead adsorption behavior of activated hydroxyl group in two-dimensional titanium carbide. Journal of the American Chemical Society 136 (11), 4113–4116. Available from. Available at https://doi.org/10.1021/ja500506k, http://pubs.acs.org/journal/jacsat.

Peng, D., Jiang, W., Li, F.F., Zhang, L., Liang, R.P., Qiu, J.D., 2018. One-pot synthesis of boron carbon nitride nanosheets for facile and efficient heavy metal ions removal. ACS Sustainable Chemistry and Engineering 6 (9), 11685–11694. Available from. Available at https://doi.org/10.1021/acssuschemeng.8b01951, http://pubs.acs.org/journal/ascecg.

Peng, G., Sharshir, S.W., Wang, Y., An, M., Ma, D., Zang, J., et al., 2021. Potential and challenges of improving solar still by micro/nano-particles and porous materials - a review. Journal of Cleaner Production 311, 127432. Available at https://doi.org/10.1016/j.jclepro.2021.127432.

Peng, C., Zhou, T., Wei, P., Yan, X., Kong, Y., Xu, W., et al., 2022. Steering interfacial charge kinetics: synergizing cocatalyst roles of Ti3C2M (MXene) and NCDs for superior photocatalytic performance over TiO2. Applied Surface Science 599, 154001. Available at https://doi.org/10.1016/j.apsusc.2022.154001.

Qin, Y., Qi, P., Hao, S., Shi, W., Xiao, J., Wang, J., et al., 2025. Methylation of reverse osmosis membrane for superior anti-fouling performance via blocking carboxyl groups in polyamide. Nature Water 3 (1), 110–121. Available from. Available at https://doi.org/10.1038/s44221-024-00371-x, https://www.nature.com/natwater/.

Rajakumaran, R., Boddu, V., Kumar, M., Shalaby, M.S., Abdallah, H., Chetty, R., 2019. Effect of ZnO morphology on GO-ZnO modified polyamide reverse osmosis membranes for desalination. Desalination 467, 245–256. Available from. Available at https://doi.org/10.1016/j.desal.2019.06.018, https://www.journals.elsevier.com/desalination.

Ramesh, K., Gnanavel, B., Shkir, M., 2021. Enhanced visible light photocatalytic degradation of bisphenol A (BPA) by reduced graphene oxide (RGO)–metal oxide (TiO2, ZnO and WO3) based nanocomposites. Diamond and Related Materials 118, 108514. Available at https://doi.org/10.1016/j.diamond.2021.108514.

Rasool, K., Mahmoud, K.A., Johnson, D.J., Helal, M., Berdiyorov, G.R., Gogotsi, Y., 2017. Efficient antibacterial membrane based on two-dimensional Ti3C2Tx (MXene) nanosheets. Scientific Reports 7 (1) Available from. Available at https://doi.org/10.1038/s41598-017-01714-3, www.nature.com/srep/index.html.

Rathod, P.P., Bhosle, V.K., Ubale, A.U., Khan, Z.S., 2025. Ultrasonically synthesized ZnO-GO nanocomposite for photocatalytic and antibacterial applications. Composites Communications 57, 102457. Available at https://doi.org/10.1016/j.coco.2025.102457.

Ray, S.S., Sharma, T.S.K., Myung, S., Choi, W.M., Kwon, Y.N., 2024. Holistic approach for fabrication of superhydrophobic MXene-based membrane for enhanced membrane distillation. Desalination 580 Available from. Available at https://doi.org/10.1016/j.desal.2024.117536, https://www.sciencedirect.com/science/journal/00119164.

Ren, C.E., Hatzell, K.B., Alhabeb, M., Ling, Z., Mahmoud, K.A., Gogotsi, Y., 2015. Charge- and size-selective ion sieving through Ti3C2Tx MXene membranes. Journal of Physical Chemistry Letters 6 (20), 4026–4031. Available from. Available at https://doi.org/10.1021/acs.jpclett.5b01895, http://pubs.acs.org/journal/jpclcd.

Safarpour, M., Khataee, A., Vatanpour, V., 2015. Thin film nanocomposite reverse osmosis membrane modified by reduced graphene oxide/TiO2 with improved desalination performance. Journal of Membrane Science 489, 43–54. Available from. Available at https://doi.org/10.1016/j.memsci.2015.04.010, www.elsevier.com/locate/memsci.

Saravanakumar, K., Yun, K., Maheskumar, V., Yea, Y., Jagan, G., Park, C.M., 2023. Construction of novel In2S3/Ti3C2 MXene quantum dots/SmFeO3 Z-scheme heterojunctions for efficient photocatalytic removal of sulfamethoxazole and 4-chlorophenol: degradation pathways and mechanism insights. Chemical Engineering Journal 451, 138933. Available at https://doi.org/10.1016/j.cej.2022.138933.

Saroyan, H., Kyzas, G.Z., Deliyanni, E.A., 2019. Effective dye degradation by graphene oxide supported manganese oxide. Processes 7 (1) Available from. Available at https://doi.org/10.3390/pr7010040, https://www.mdpi.com/2227-9717/7/1/40/pdf.

Shahzad, A., Nawaz, M., Moztahida, M., Tahir, K., Kim, J., Lim, Y., et al., 2019. Exfoliation of titanium aluminum carbide (211 MAX Phase) to form nanofibers and two-dimensional nanosheets and their application in aqueous-phase cadmium sequestration. ACS Applied Materials and Interfaces 11 (21), 19156–19166. Available from. Available at https://doi.org/10.1021/acsami.9b03899, http://pubs.acs.org/journal/aamick.

Shahzad, A., Rasool, K., Miran, W., Nawaz, M., Jang, J., Mahmoud, K.A., et al., 2017. Two-dimensional Ti3C2Tx MXene nanosheets for efficient copper removal from water. ACS Sustainable Chemistry and Engineering 5 (12), 11481–11488. Available from. Available at https://doi.org/10.1021/acssuschemeng.7b02695, http://pubs.acs.org/journal/ascecg.

Shahzad, A., Rasool, K., Miran, W., Nawaz, M., Jang, J., Mahmoud, K.A., et al., 2018. Mercuric ion capturing by recoverable titanium carbide magnetic nanocomposite. Journal of Hazardous Materials 344, 811–818. Available from. Available at https://doi.org/10.1016/j.jhazmat.2017.11.026, www.elsevier.com/locate/jhazmat.

Shao, B., Wang, J., Liu, Z., Zeng, G., Tang, L., Liang, Q., et al., 2020. Ti3C2T: X MXene decorated black phosphorus nanosheets with improved visible-light photocatalytic activity: experimental and theoretical studies. Journal of Materials Chemistry A 8 (10), 5171–5185. Available from. Available at https://doi.org/10.1039/c9ta13610j, http://pubs.rsc.org/en/journals/journal/ta.

Shen, J.H., Li, M.M., Chu, L.F., Guo, C.X., Guo, Y.J., Guo, Y.P., 2021. Effect mechanism of copper ions on photocatalytic activity of TiO2/graphene oxide composites for phenol-4-sulfonic acid photodegradation. Journal of Colloid and Interface Science 586, 563–575. Available from. Available at https://doi.org/10.1016/j.jcis.2020.10.121, http://www.elsevier.com/inca/publications/store/6/2/2/8/6/1/index.htt.

Shi, H., Xu, M., Wang, L., Ma, Q., Zhao, M., Li, Q., et al., 2025. Application of capacitive deionization in heavy metal ions removal and recovery: a review. Separation and Purification Technology 364, 132027. Available at https://doi.org/10.1016/j.seppur.2025.132027.

Shi, Y.C., Wang, A.J., Wu, X.L., Chen, J.R., Feng, J.J., 2016. Green-assembly of three-dimensional porous graphene hydrogels for efficient removal of organic dyes. Journal of Colloid and Interface Science 484, 254–262. Available from. Available at https://doi.org/10.1016/j.jcis.2016.09.008, http://www.elsevier.com/inca/publications/store/6/2/2/8/6/1/index.htt.

Skuse, C., Alberto, M., Luque-Alled, J.M., Mercadillo, V.O., Asuquo, E., Gallego-Schmid, A., et al., 2024. Spray coating of 2D materials in the production of antifouling membranes for membrane distillation. Journal of Membrane Science 711 Available from. Available at https://doi.org/10.1016/j.memsci.2024.123162, https://www.sciencedirect.com/science/journal/03767388.

Solangi, N.H., Mubarak, N.M., Karri, R.R., Mazari, S.A., Kailasa, S.K., Alfantazi, A., 2023. Applications of advanced MXene-based composite membranes for sustainable water desalination. Chemosphere 314 Available from. Available at https://doi.org/10.1016/j.chemosphere.2022.137643, www.elsevier.com/locate/chemosphere.

Song, J., Wu, X., Zhang, M., Liu, C., Yu, J., Sun, G., et al., 2020. Highly flexible, core-shell heterostructured, and visible-light-driven titania-based nanofibrous membranes for antibiotic removal and E. coil inactivation. Chemical Engineering Journal 379, 122269. Available at https://doi.org/10.1016/j.cej.2019.122269.

Song, T., Yu, C., He, X., Lin, J., Liu, Z., Yang, X., et al., 2018. Synthesis of magnetically separable porous BN microrods@Fe3O4 nanocomposites for Pb(II) adsorption. Colloids and Surfaces A: Physicochemical and Engineering Aspects 537, 508–515. Available from. Available at https://doi.org/10.1016/j.colsurfa.2017.10.060, www.elsevier.com/locate/colsurfa.

Soroush, A., Ma, W., Silvino, Y., Rahaman, M.S., 2015. Surface modification of thin film composite forward osmosis membrane by silver-decorated graphene-oxide nanosheets. Environmental Science: Nano 2 (4), 395–405.

Available from. Available at https://doi.org/10.1039/c5en00086f, http://www.rsc.org/publishing/journals/en/about.asp.

Srivastava, A., Murthy, Z.V.P., 2021. Investigating the effect of PEG200 and two-dimensional h-BN on PVDF membrane performance for membrane distillation–crystallization. Materials Today Chemistry 22, 100545. Available at https://doi.org/10.1016/j.mtchem.2021.100545.

Su, P., Wang, F., Li, Z., Tang, C.Y., Li, W., 2020. Graphene oxide membranes: controlling their transport pathways. Journal of Materials Chemistry A 8 (31), 15319–15340. Available from. Available at https://doi.org/10.1039/d0ta02249g, http://pubs.rsc.org/en/journals/journal/ta.

Su, T., Zhang, Y., Chang, Y., Lin, X., Dong, M., Wu, S., et al., 2025. Nanozyme meets pathogen: A M-MXene@CuS platform for triple-mode detection and synergistic inactivation of MRSA. Chemical Engineering Journal 519, 165639. Available at https://doi.org/10.1016/j.cej.2025.165639.

Sun, X.F., Qin, J., Xia, P.F., Guo, B.B., Yang, C.M., Song, C., et al., 2015. Graphene oxide-silver nanoparticle membrane for biofouling control and water purification. Chemical Engineering Journal 281, 53–59. Available from. Available at https://doi.org/10.1016/j.cej.2015.06.059, www.elsevier.com/inca/publications/store/6/0/1/2/7/3/index.htt.

Sun, B., Tao, F., Huang, Z., Yan, W., Zhang, Y., Dong, X., et al., 2021. Ti3C2 MXene-bridged Ag/Ag3PO4 hybrids toward enhanced visible-light-driven photocatalytic activity. Applied Surface Science 535, 147354. Available at https://doi.org/10.1016/j.apsusc.2020.147354.

Tang, R., Gong, D., Deng, Y., Xiong, S., Deng, J., Li, L., et al., 2022. π-π Stacked step-scheme PDI/g-C3N4/TiO2@Ti3C2 photocatalyst with enhanced visible photocatalytic degradation towards atrazine via peroxymonosulfate activation. Chemical Engineering Journal 427, 131809. Available at https://doi.org/10.1016/j.cej.2021.131809.

Tang, Y., Yang, C., Que, W., 2018. A novel two-dimensional accordion-like titanium carbide (MXene) for adsorption of Cr(VI) from aqueous solution. Journal of Advanced Dielectrics 08 (05), 1850035. Available at https://doi.org/10.1142/s2010135x18500352.

Tao, P., Ni, G., Song, C., Shang, W., Wu, J., Zhu, J., et al., 2018. Solar-driven interfacial evaporation. Nature Energy 3 (12), 1031–1041. Available from. Available at https://doi.org/10.1038/s41560-018-0260-7, www.nature.com/nenergy/.

Taye, A., Yifru, A., Getachew, N., Mehretie, S., Admassie, S., 2023. Adsorption of hexavalent chromium using water hyacinth leaf protein concentrate/graphene oxide hydrogel. Environmental Monitoring and Assessment 195 (11) Available from. Available at https://doi.org/10.1007/s10661-023-11960-w, https://www.springer.com/journal/10661.

Thirumal, V., Yuvakkumar, R., Kumar, P.S., Ravi, G., Keerthana, S.P., Velauthapillai, D., 2022. Facile single-step synthesis of MXene@CNTs hybrid nanocomposite by CVD method to remove hazardous pollutants. Chemosphere 286 (Pt 2), 131733. Available at https://doi.org/10.1016/j.chemosphere.2021.131733.

Velusamy, S., Roy, A., Sundaram, S., Kumar Mallick, T., 2021. A review on heavy metal ions and containing dyes removal through graphene oxide-based adsorption strategies for textile wastewater treatment. The Chemical Record 21 (7), 1570–1610. Available at https://doi.org/10.1002/tcr.202000153.

Viet, T.Q.Q., Khoi, V.H., Thi Huong Giang, N., Thi Van Anh, H., Dat, N.M., Phong, M.T., et al., 2021. Statistical screening and optimization of photocatalytic degradation of methylene blue by ZnO–TiO2/rGO nanocomposite. Colloids and Surfaces A: Physicochemical and Engineering Aspects 629 Available from. Available at https://doi.org/10.1016/j.colsurfa.2021.127464, www.elsevier.com/locate/colsurfa.

Vigneshwaran, S., Karthikeyan, P., Park, C.M., Meenakshi, S., 2020. Boosted insights of novel accordion-like (2D/2D) hybrid photocatalyst for the removal of cationic dyes: mechanistic and degradation pathways. Journal of Environmental Management 273, 111125. Available at https://doi.org/10.1016/j.jenvman.2020.111125.

Vigneshwaran, S., Park, C.M., Meenakshi, S., 2021. Designed fabrication of sulfide-rich bi-metallic-assembled MXene layered sheets with dramatically enhanced photocatalytic performance for Rhodamine B removal. Separation and Purification Technology 258, 118003. Available at https://doi.org/10.1016/j.seppur.2020.118003.

Wan, J., Lacey, S.D., Dai, J., Bao, W., Fuhrer, M.S., Hu, L., 2016. Tuning two-dimensional nanomaterials by intercalation: materials, properties and applications. Chemical Society Reviews 45 (24), 6742–6765. Available from. Available at https://doi.org/10.1039/c5cs00758e, http://pubs.rsc.org/en/journals/journal/cs.

Wang, F., Zhang, L., Wang, Y., Liu, X., Rohani, S., Lu, J., 2017a. Fe 3 O 4 @SiO 2 @CS-TETA functionalized graphene oxide for the adsorption of methylene blue (MB) and Cu(II). Applied Surface Science 420, 970–981. Available from. Available at https://doi.org/10.1016/j.apsusc.2017.05.179, http://www.journals.elsevier.com/applied-surface-science/.

Wang, L., Tao, W., Yuan, L., Liu, Z., Huang, Q., Chai, Z., et al., 2017b. Rational control of the interlayer space inside two-dimensional titanium carbides for highly efficient uranium removal and imprisonment. Chemical Communications 53 (89), 12084–12087. Available from. Available at https://doi.org/10.1039/c7cc06740b, www.rsc.org/chemcomm.

Wang, W., Li, G., Xia, D., An, T., Zhao, H., Wong, P.K., 2017c. Photocatalytic nanomaterials for solar-driven bacterial inactivation: recent progress and challenges. Environmental Science: Nano 4 (4), 782–799. Available from. Available at https://doi.org/10.1039/C7EN00063D, http://www.rsc.org/publishing/journals/en/about.asp.

Wang, W., Cai, K., Wu, X., Shao, X., Yang, X., 2017d. A novel poly(m-phenylenediamine)/reduced graphene oxide/nickel ferrite magnetic adsorbent with excellent removal ability of dyes and Cr(VI). Journal of Alloys and Compounds 722, 532–543. Available at https://doi.org/10.1016/j.jallcom.2017.06.069.

Wang, C., Cheng, R., Hou, P.X., Ma, Y., Majeed, A., Wang, X., et al., 2020. MXene-carbon nanotube hybrid membrane for robust recovery of Au from trace-level solution. ACS Applied Materials and Interfaces 12 (38), 43032–43041. Available from. Available at https://doi.org/10.1021/acsami.0c09310, http://pubs.acs.org/journal/aamick.

Wang, Y., Wu, X., Wu, P., Zhao, J., Yang, X., Owens, G., et al., 2021a. Enhancing solar steam generation using a highly thermally conductive evaporator support. Science Bulletin 66 (24), 2479–2488. Available from. Available at https://doi.org/10.1016/j.scib.2021.09.018, http://link.springer.com/journal/11434.

Wang, Z., Ma, C., Xu, C., Sinquefield, S.A., Shofner, M.L., Nair, S., 2021b. Graphene oxide nanofiltration membranes for desalination under realistic conditions. Nature Sustainability 4 (5), 402–408. Available from. Available at https://doi.org/10.1038/s41893-020-00674-3, www.nature.com/natsustain/.

Wang, K., Chen, L., Zhu, G., Xu, X., Wan, L., Lu, T., et al., 2022a. Ferroferric oxide@titanium carbide MXene heterostructure with enhanced sodium storage ability for efficient hybrid capacitive deionization. Desalination 522, 115420. Available at https://doi.org/10.1016/j.desal.2021.115420.

Wang, L., Liu, D., Jiang, L., Ma, Y., Yang, G., Qian, Y., et al., 2022b. Advanced 2D–2D heterostructures of transition metal dichalcogenides and nitrogen-rich nitrides for solar water generation. Nano Energy 98, 107192. Available at https://doi.org/10.1016/j.nanoen.2022.107192.

Wang, Q., Mei, Y., Zhou, R., Komarneni, S., Ma, J., 2022c. Persulfate activation of CuS@Ti3C2-based MXene with Bi-active centers toward Orange II removal under visible light. Colloids and Surfaces A: Physicochemical and Engineering Aspects 648, 129315. Available at https://doi.org/10.1016/j.colsurfa.2022.129315.

Wang, S., Liang, S., Chen, L., Fang, H., 2022d. Realizing ultrahigh nanofiltration performance based on small flake reduced graphene oxide membranes. Desalination 528 Available from. Available at https://doi.org/10.1016/j.desal.2022.115601, https://www.journals.elsevier.com/desalination.

Wang, Q., Zhu, F., Cheng, H., Komarneni, S., Ma, J., 2023a. Efficient activation of persulfate by Ti3C2 MXene QDs modified ZnFe2O4 for the rapid degradation of tetracycline. Chemosphere 328, 138546. Available at https://doi.org/10.1016/j.chemosphere.2023.138546.

Wang, S., Du, X., Yao, C., Cai, Y., Ma, H., Jiang, B., et al., 2023b. S-scheme heterojunction/Schottky junction tandem synergistic effect promotes visible-light-driven catalytic activity. Nano Research 16 (2), 2152–2162. Available from. Available at https://doi.org/10.1007/s12274-022-4960-8, https://www.springer.com/journal/12274.

Wang, Y., Xie, F., Li, R., Yu, Z., Jian, X., Gao, X., et al., 2023c. Bi3O4Br/Ti3C2 Schottky junction with enhanced active species generation for boosting visible-light photodegradation bisphenol A activity. Separation and Purification Technology 318, 124001. Available at https://doi.org/10.1016/j.seppur.2023.124001.

Wang, M., Bodirsky, B.L., Rijneveld, R., Beier, F., Bak, M.P., Batool, M., et al., 2024a. A triple increase in global river basins with water scarcity due to future pollution. Nature Communications 15 (1) Available from. Available at https://doi.org/10.1038/s41467-024-44947-3, https://www.nature.com/ncomms/.

Wang, Y., Wang, T., Wu, X., Wang, J., Chang, E., Huang, Z., et al., 2024b. High performance Zn–Al LDH modified forward osmosis membrane with antibacterial, anti-membrane fouling, and photocatalytic degradation ability. Journal of Membrane Science 705, 122899. Available at https://doi.org/10.1016/j.memsci.2024.122899.

Wang, M., Chen, X., Zhang, C., Wang, L., Zhuang, S., 2025. Photocatalytic degradation of contaminants in water by Ti3C2 MXene: a review. International Journal of Environmental Science and Technology 22 (9), 8553–8580. Available at https://doi.org/10.1007/s13762-025-06417-1.

Wei, Mp, Chai, H., Cao, Yl, Jia, Dz, 2018. Sulfonated graphene oxide as an adsorbent for removal of Pb2+ and methylene blue. Journal of Colloid and Interface Science 524, 297–305. Available from. Available at https://doi.org/10.1016/j.jcis.2018.03.094, http://www.elsevier.com/inca/publications/store/6/2/2/8/6/1/index.htt.

Wen, C., Li, X., Yan, S., Wen, J., Zheng, R., Wang, X., et al., 2025. Progress in MXene-based photothermal materials for solar-driven water evaporation and desalination. Chemical Engineering Journal 510, 161873. Available at https://doi.org/10.1016/j.cej.2025.161873.

Wojciechowski, T., Rozmysłowska-Wojciechowska, A., Matyszczak, G., Wrzecionek, M., Olszyna, A., Peter, A., et al., 2019. Ti 2 C MXene modified with ceramic oxide and noble metal nanoparticles: synthesis, morphostructural properties, and high photocatalytic activity. Inorganic Chemistry 58 (11), 7602–7614. Available at https://doi.org/10.1021/acs.inorgchem.9b01015.

Wu, Z.L., Liu, F., Li, C.K., Chen, X.Q., Yu, J.G., 2016. A sandwich-structured graphene-based composite: preparation, characterization, and its adsorption behaviors for Congo red. Colloids and Surfaces A: Physicochemical and Engineering Aspects 509, 65–72. Available from. Available at https://doi.org/10.1016/j.colsurfa.2016.08.084, www.elsevier.com/locate/colsurfa.

Wu, Y., Li, X., Yang, Q., Wang, D., Yao, F., Cao, J., et al., 2020a. Mxene-modulated dual-heterojunction generation on a metal-organic framework (MOF) via surface constitution reconstruction for enhanced photocatalytic activity. Chemical Engineering Journal 390 Available from. Available at https://doi.org/10.1016/j.cej.2020.124519, www.elsevier.com/inca/publications/store/6/0/1/2/7/3/index.htt.

Wu, Z., Liang, Y., Yuan, X., Zou, D., Fang, J., Jiang, L., et al., 2020b. MXene Ti3C2 derived Z–scheme photocatalyst of graphene layers anchored TiO2/g–C3N4 for visible light photocatalytic degradation of refractory organic pollutants. Chemical Engineering Journal 394, 124921. Available at https://doi.org/10.1016/j.cej.2020.124921.

Wu, Y., Fu, C.F., Huang, Q., Zhang, P., Cui, P., Ran, J., et al., 2021. 2D heterostructured nanofluidic channels for enhanced desalination performance of graphene oxide membranes. ACS Nano 15 (4), 7586–7595. Available from. Available at https://doi.org/10.1021/acsnano.1c01105, http://pubs.acs.org/journal/ancac3.

Wu, F.D., Chen, J.C., Hu, J.P., 2022. Synthesis of TiO2/Ti3C2Tx/AgI Z-scheme photocatalyst for tetracycline hydrochloride photocatalytic degradation. Journal of Environmental Chemical Engineering 10 (1) Available from. Available at https://doi.org/10.1016/j.jece.2021.107117, http://www.journals.elsevier.com/journal-of-environmental-chemical-engineering/.

Wu, D., Dai, Z., Wang, C., Gao, Y., Chen, B., Zhang, X., 2023a. Scalable and high-efficiency lignocellulose sponge-based evaporators for solar-driven desalination and desiccant regeneration. Separation and Purification Technology 318 Available from. Available at https://doi.org/10.1016/j.seppur.2023.124032, http://www.journals.elsevier.com/separation-and-purification-technology/.

Wu, T., Liu, B., Liu, C., Wan, J., Yang, A., Liu, K., et al., 2023b. Solar-driven efficient heterogeneous subminute water disinfection nanosystem assembled with fingerprint MoS2. Nature Water 1 (5), 462–470. Available from. Available at https://doi.org/10.1038/s44221-023-00079-4, https://www.nature.com/natwater/.

Wu, G., Wang, C., Ma, Y., Zheng, H., Huang, L., Zhu, X., et al., 2025. Enhanced photocatalytic performance and a mechanistic study of novel black phosphorus/graphene/TiO2 composite membrane for o-chlorophenol removal. Environmental Research 267, 120636. Available at https://doi.org/10.1016/j.envres.2024.120636.

Xing, F., Li, T., Li, J., Zhu, H., Wang, N., Cao, X., 2017. Chemically exfoliated MoS2 for capacitive deionization of saline water. Nano Energy 31, 590–595. Available from. Available at https://doi.org/10.1016/j.nanoen.2016.12.012, http://www.journals.elsevier.com/nano-energy/.

Xing, J., Zhang, H., Wei, G., Du, L., Chen, S., Yu, H., et al., 2023. Improving the performance of the lamellar reduced graphene oxide/molybdenum sulfide nanofiltration membrane through accelerated water-transport channels and capacitively enhanced charge density. Environmental Science and Technology 57 (1), 615–625. Available from. Available at https://doi.org/10.1021/acs.est.2c06697, http://pubs.acs.org/journal/esthag.

Xu, J., Xu, D., Zhu, B., Cheng, B., Jiang, C., 2018. Adsorptive removal of an anionic dye Congo red by flower-like hierarchical magnesium oxide (MgO)-graphene oxide composite microspheres. Applied Surface Science 435, 1136–1142. Available from. Available at https://doi.org/10.1016/j.apsusc.2017.11.232, http://www.journals.elsevier.com/applied-surface-science/.

Xu, R., Kang, Y., Zhang, W., Pan, B., Zhang, X., 2023a. Two-dimensional MXene membranes with biomimetic sub-nanochannels for enhanced cation sieving. Nature Communications 14 (1) Available from. Available at https://doi.org/10.1038/s41467-023-40742-8, https://www.nature.com/ncomms/.

Xu, S., Zhao, C., Li, G., Shi, Z., Liu, B., 2023b. In situ oxidized TiO2/MXene ultrafiltration membrane with photocatalytic self-cleaning and antibacterial properties. RSC Advances 13 (23), 15843–15855. Available from. Available at https://doi.org/10.1039/d3ra02230g, http://pubs.rsc.org/en/journals/journal/ra.

Xu, X., Zeng, X., Zhang, C., Huang, R., Ding, W., 2023c. Enhanced electrocatalytic removal of tetracycline using dual carbon material combined particle electrodes in a three-dimensional electrochemical system: degradation pathway and mechanism. Journal of Cleaner Production 419 Available from. Available at https://doi.org/10.1016/j.jclepro.2023.138257, https://www.journals.elsevier.com/journal-of-cleaner-production.

Xu, J., Jiang, R., Qiu, Z., Chen, C., Liu, B., Hu, X., et al., 2025. Conjugated microporous polymer for membrane separation: a review. Separation and Purification Technology 362, 131795. Available at https://doi.org/10.1016/j.seppur.2025.131795.

Yadav, A., Patel, R.V., Awasthi, P., Sinha, N., Labhasetwar, P.K., 2022. Hexagonal boron nitride nanosheets incorporated photocatalytic polyvinylidene fluoride mixed matrix membranes for textile wastewater treatment via vacuum-assisted distillation. Colloids and Surfaces A: Physicochemical and Engineering Aspects 653 Available from. Available at https://doi.org/10.1016/j.colsurfa.2022.129955, www.elsevier.com/locate/colsurfa.

Yang, Y., Xu, L., Wang, H., Wang, W., Zhang, L., 2016. TiO2/graphene porous composite and its photocatalytic degradation of methylene blue. Materials and Design 108, 632–639. Available at https://doi.org/10.1016/j.matdes.2016.06.104.

Yang, L., Hu, J., He, L., Tang, J., Zhou, Y., Li, J., et al., 2017. One-pot synthesis of multifunctional magnetic N-doped graphene composite for SERS detection, adsorption separation and photocatalytic degradation of Rhodamine 6G. Chemical Engineering Journal 327, 694–704. Available from. Available at https://doi.org/10.1016/j.cej.2017.06.162, www.elsevier.com/inca/publications/store/6/0/1/2/7/3/index.htt.

Yang, X., Yang, Y., Fu, L., Zou, M., Li, Z., Cao, A., et al., 2018a. An ultrathin flexible 2D membrane based on single-walled nanotube–MoS2 hybrid film for high-performance solar steam generation. Advanced Functional Materials 28 (3) Available from. Available at https://doi.org/10.1002/adfm.201704505, http://onlinelibrary.wiley.com/journal/10.1002/(ISSN)1616-3028.

Yang, Q., Lu, R., Ren, S.S., Chen, C., Chen, Z., Yang, X., 2018b. Three dimensional reduced graphene oxide/ZIF-67 aerogel: effective removal cationic and anionic dyes from water. Chemical Engineering Journal 348, 202–211. Available from. Available at https://doi.org/10.1016/j.cej.2018.04.176, www.elsevier.com/inca/publications/store/6/0/1/2/7/3/index.htt.

Yang, Y., Yang, X., Liang, L., Gao, Y., Cheng, H., Li, X., et al., 2019. Large-area graphene-nanomesh/ carbon-nanotube hybrid membranes for ionic and molecular nanofiltration. Science (New York, N.Y.) 364 (6445), 1057–1062. Available from. Available at https://doi.org/10.1126/science.aau5321, https://science.sciencemag.org/content/364/6445/1057/tab-pdf.

Yang, R., Fan, Y., Yu, R., Dai, F., Lan, J., Wang, Z., et al., 2021. Robust reduced graphene oxide membranes with high water permeance enhanced by K+ modification. Journal of Membrane Science 635, 119437. Available at https://doi.org/10.1016/j.memsci.2021.119437.

Yang, C., Long, M., Ding, C., Zhang, R., Zhang, S., Yuan, J., et al., 2022a. Antifouling graphene oxide membranes for oil-water separation via hydrophobic chain engineering. Nature Communications 13 (1) Available from. Available at https://doi.org/10.1038/s41467-022-35105-8, https://www.nature.com/ncomms/.

Yang, L., Chen, J., Que, M., Liu, X., Zheng, H., Yang, T., et al., 2022b. Construction of 2D/2D/2D BiOBr/(001)-TiO2/Ti3C2Tx heterojunction for enhancing photocatalytic degradation of dye. Journal of Alloys and Compounds 893 Available from. Available at https://doi.org/10.1016/j.jallcom.2021.162289, https://www.journals.elsevier.com/journal-of-alloys-and-compounds.

Yang, L., Cao, X., Wang, Y., Zhu, Z., Sun, H., Liang, W., et al., 2023. Enhanced visible light catalytic activity of holey Ti3C2 based catalyst for azo dyes. Applied Surface Science 623, 157002. Available at https://doi.org/10.1016/j.apsusc.2023.157002.

Yin, M., Lu, S., Gao, G., Zeng, H., Zheng, W., Xu, J., et al., 2025. PVDF based Janus composite membrane for collaborative solar-driven interface evaporation and photocatalysis with high salt-resistance and anti-biofouling. Desalination 601, 118565. Available at https://doi.org/10.1016/j.desal.2025.118565.

Yin, Z., Wang, H., Jian, M., Li, Y., Xia, K., Zhang, M., et al., 2017. Extremely black vertically aligned carbon nanotube arrays for solar steam generation. ACS Applied Materials and Interfaces 9 (34), 28596–28603. Available from. Available at https://doi.org/10.1021/acsami.7b08619, http://pubs.acs.org/journal/aamick.

Ying, Y., Liu, Y., Wang, X., Mao, Y., Cao, W., Hu, P., et al., 2015. Two-dimensional titanium carbide for efficiently reductive removal of highly toxic chromium(VI) from water. ACS Applied Materials and Interfaces 7 (3), 1795–1803. Available from. Available at https://doi.org/10.1021/am5074722, http://pubs.acs.org/journal/aamick.

Yu, L., Zhang, Y., Zhang, B., Liu, J., Zhang, H., Song, C., 2013. Preparation and characterization of HPEI-GO/PES ultrafiltration membrane with antifouling and antibacterial properties. Journal of Membrane Science 447, 452–462. Available at https://doi.org/10.1016/j.memsci.2013.07.042.

Yu, M., Liang, H., Zhan, R., Xu, L., Niu, J., 2021. Sm-doped g-C3N4/Ti3C2 MXene heterojunction for visible-light photocatalytic degradation of ciprofloxacin. Chinese Chemical Letters 32 (7), 2155–2158. Available from. Available at https://doi.org/10.1016/j.cclet.2020.11.069, http://www.elsevier.com/wps/find/journaldescription.cws_home/997/description#description.

Yu, Y., Zhou, Z., Huang, G., Cheng, H., Han, L., Zhao, S., et al., 2022. Purifying water with silver nanoparticles (AgNPs)-incorporated membranes: recent advancements and critical challenges. Water Research 222, 118901. Available at https://doi.org/10.1016/j.watres.2022.118901.

Yu, Y., Lv, Z., Zhao, S., Han, L., Shen, L., Zhou, Z., 2025. How graphene oxide properties guide transport channel construction and impact desalination performance in composite membranes. Desalination 607, 118810. Available at https://doi.org/10.1016/j.desal.2025.118810.

Yuan, J., Passeport, E., Hofmann, R., 2022. Understanding adsorption and biodegradation in granular activated carbon for drinking water treatment: a critical review. Water Research 210, 118026. Available at https://doi.org/10.1016/j.watres.2021.118026.

Zeng, Q., Zhou, X., Shen, L., Zhao, D.L., Kong, N., Li, Y., et al., 2024. Exceptional self-cleaning MXene-based membrane for highly efficient oil/water separation. Journal of Membrane Science 700, 122691. Available at https://doi.org/10.1016/j.memsci.2024.122691.

Zhang, Z.B., Wu, J.J., Su, Y., Zhou, J., Gao, Y., Yu, H.Y., et al., 2015. Layer-by-layer assembly of graphene oxide on polypropylene macroporous membranes via click chemistry to improve antibacterial and antifouling performance. Applied Surface Science 332, 300–307. Available from. Available at https://doi.org/10.1016/j.apsusc.2015.01.193, http://www.journals.elsevier.com/applied-surface-science/.

Zhang, M., Liu, Z., Gao, Y., Shu, L., 2017. Ag modified g-C3N4 composite entrapped PES UF membrane with visible-light-driven photocatalytic antifouling performance. RSC Advances 7 (68), 42919–42928. Available from. Available at https://doi.org/10.1039/c7ra07775k, http://pubs.rsc.org/en/journals/journalissues.

Zhang, P., Gong, J.L., Zeng, G.M., Song, B., Liu, H.Y., Huan, S.Y., et al., 2018a. Ultrathin reduced graphene oxide/MOF nanofiltration membrane with improved purification performance at low pressure. Chemosphere 204, 378–389. Available from. Available at https://doi.org/10.1016/j.chemosphere.2018.04.064, www.elsevier.com/locate/chemosphere.

Zhang, Y., Yan, X., Yan, Y., Chen, D., Huang, L., Zhang, J., et al., 2018b. The utilization of a three-dimensional reduced graphene oxide and montmorillonite composite aerogel as a multifunctional agent for wastewater treatment. RSC Advances 8 (8), 4239–4248. Available from. Available at https://doi.org/10.1039/c7ra13103h, http://pubs.rsc.org/en/journals/journal/ra.

Zhang, M., Guan, K., Ji, Y., Liu, G., Jin, W., Xu, N., 2019. Controllable ion transport by surface-charged graphene oxide membrane. Nature Communications 10 (1) Available from. Available at https://doi.org/10.1038/s41467-019-09286-8, http://www.nature.com/ncomms/index.html.

Zhang, Y., Xiong, T., Nandakumar, D.K., Tan, S.C., 2020. Structure architecting for salt-rejecting solar interfacial desalination to achieve high-performance evaporation with in situ energy generation. Advanced Science 7 (9) Available from. Available at https://doi.org/10.1002/advs.201903478, http://onlinelibrary.wiley.com/journal/10.1002/(ISSN)2198-3844.

Zhang, C., Li, C., Chen, G., Ji, F., Shen, Y., Peng, J., et al., 2021a. In situsynthesis of a hybrid Fe(Co)/MXene/ZSM-5 catalyst for phenol abatement. New Journal of Chemistry 45 (36), 16862–16871. Available from. Available at https://doi.org/10.1039/d1nj02810c, http://pubs.rsc.org/en/journals/journal/nj.

Zhang, L., Zhang, M., Liu, G., Jin, W., Li, X., 2021b. Fungal cell wall-graphene oxide microcomposite membrane for organic solvent nanofiltration. Advanced Functional Materials 31 (23) Available from. Available at https://doi.org/10.1002/adfm.202100110, http://onlinelibrary.wiley.com/journal/10.1002/(ISSN)1616-3028.

Zhang, W.H., Yin, M.J., Zhao, Q., Jin, C.G., Wang, N., Ji, S., et al., 2021c. Graphene oxide membranes with stable porous structure for ultrafast water transport. Nature Nanotechnology 16 (3), 337–343. Available from. Available at https://doi.org/10.1038/s41565-020-00833-9, http://www.nature.com/nnano/index.html.

Zhang, W., Xu, H., Xie, F., Ma, X., Niu, B., Chen, M., et al., 2022. General synthesis of ultrafine metal oxide/reduced graphene oxide nanocomposites for ultrahigh-flux nanofiltration membrane. Nature Communications 13 (1) Available from. Available at https://doi.org/10.1038/s41467-022-28180-4, http://www.nature.com/ncomms/index.html.

Zhang, J., Li, Z. Zhang, Q., Zhang, L., Ma, T., Ma, X., et al., 2023a. Nanoconfined MXene-MOF nanolaminate film for molecular removal/collection and multiple sieving. ACS Applied Materials and Interfaces 15 (13), 17222–17232 Available from. Available at https://doi.org/10.1021/acsami.3c00909, http://pubs.acs.org/journal/aamick.

Zhang, L., Tan, L., Yuan, Z., Xu, B., Chen, W., Tang, Y., et al., 2023b. Engineering of Bi2O2CO3/Ti3C2Tx heterojunctions co-embedded with surface and interface oxygen vacancies for boosted photocatalytic degradation of levofloxacin. Chemical Engineering Journal 452, 139327. Available at https://doi.org/10.1016/j.cej.2022.139327.

Zhang, X., Qiao, D., Yang, M., 2024. The manipulation of the interfacial evaporation property of graphene porous membranes. Molecular Physics 122 (10) Available from. Available at https://doi.org/10.1080/00268976.2023.2283141, www.tandf.co.uk/journals/titles/00268976.asp.

Zhang, G., Fan, D., Li, Z., Li, Y., Lu, Y., 2025a. 2D Nb2C MXene-enhanced hierarchical hydrogel for efficient solar-driven water evaporation. RSC Advances 15 (21), 16257–16265. Available from. Available at https://doi.org/10.1039/d5ra01735a, http://pubs.rsc.org/en/journals/journal/ra.

Zhang, J., Wang, W., Huang, S., Lv, Y., Li, M., Wu, M., et al., 2025b. Metal-free photocatalyst with reduced graphene oxide-doped graphitic carbon nitride homojunctions for efficient antibacterial applications. RSC Advances 15 (4), 2444–2451. Available from. Available at https://doi.org/10.1039/d4ra07829b, http://pubs.rsc.org/en/journals/journal/ra.

Zhang, T., Zheng, L., Yang, X., Li, Y., Mol, Z., Demeestere, K., et al., 2025c. Combining ozonation and powdered activated carbon adsorption for organic micropollutants removal in municipal wastewater treatment plants: single versus two-reactor systems. Water Research 284 Available from. Available at https://doi.org/10.1016/j.watres.2025.124035, https://www.sciencedirect.com/science/journal/00431354.

Zhao, S., Chen, D., Wei, F., Chen, N., Liang, Z., Luo, Y., 2018. Synthesis of graphene oxide/metal–organic frameworks hybrid materials for enhanced removal of Methylene blue in acidic and alkaline solutions. Journal of Chemical Technology and Biotechnology 93 (3), 698–709. Available from. Available at https://doi.org/10.1002/jctb.5419, http://onlinelibrary.wiley.com/journal/10.1002/(ISSN)1097-4660.

Zhao, X., Wang, L.Y., Tang, C.Y., Zha, X.J., Liu, Y., Su, B.H., et al., 2020. Smart Ti3C2TxMXene fabric with fast humidity response and joule heating for healthcare and medical therapy applications. ACS Nano 14 (7), 8793–8805. Available from. Available at https://doi.org/10.1021/acsnano.0c03391, http://pubs.acs.org/journal/ancac3.

Zheng, S., Tu, Q., Urban, J.J., Li, S., Mi, B., 2017. Swelling of graphene oxide membranes in aqueous solution: characterization of interlayer spacing and insight into water transport mechanisms. ACS Nano 11 (6), 6440–6450. Available from. Available at https://doi.org/10.1021/acsnano.7b02999, http://pubs.acs.org/journal/ancac3.

Zheng, S., Tu, Q., Wang, M., Urban, J.J., Mi, B., 2020. Correlating interlayer spacing and separation capability of graphene oxide membranes in organic solvents. ACS Nano 14 (5), 6013–6023. Available from. Available at https://doi.org/10.1021/acsnano.0c01550, http://pubs.acs.org/journal/ancac3.

Zheng, H., Huang, S., Huang, J., Xu, M., Shao, B., Ma, X., et al., 2025. Trade-offs in microplastic-adsorbed iopamidol degradation by UV-AOPs: molecular-level insights into deiodination pathways versus iodinated disinfection by-products formation. Water Research 286, 124234. Available at https://doi.org/10.1016/j.watres.2025.124234.

Zhou, W., Yu, B., Zhu, J., Li, K., Tian, S., 2022. Hierarchical ZnO/MXene (Nb2C and V2C) heterostructure with efficient electron transfer for enhanced photocatalytic activity. Applied Surface Science 590, 153095. Available at https://doi.org/10.1016/j.apsusc.2022.153095.

Zhou, H., Gong, J., Li, J., Song, B., Fang, S., Wang, Y., et al., 2023. Cross-linked and doped graphene oxide membranes with excellent antifouling capacity for rejection of antibiotics and salts. ACS Applied Materials and Interfaces 15 (6), 8636–8652. Available from. Available at https://doi.org/10.1021/acsami.2c19789, http://pubs.acs.org/journal/aamick.

Zhou, Y., 2022. Energy sharing and trading on a novel spatiotemporal energy network in Guangdong-Hong Kong-Macao Greater Bay Area. Applied Energy 318 Available from. Available at https://doi.org/10.1016/j.apenergy.2022.119131, https://www.journals.elsevier.com/applied-energy.

Zhu, M., Xiang, X., Jiang, X., Yang, Y., Zhang, K., 2025. Engineered molybdenum disulfide-based nanomaterials for capacitive deionization applications. Inorganic Chemistry Frontiers 12 (11), 3739–3790. Available from. Available at https://doi.org/10.1039/d4qi03147d, http://pubs.rsc.org/en/journals/journal/qi.

Zhuang, J., Wang, Y., Lin, L., Xiao, J., Cai, J., Hu, Y., et al., 2025. Water flow-driven high piezo-photocatalytic efficacy in superthin and densely porous GO-TiO2/PVDF-HFP composite film for ROS generation. Chemical Engineering Journal 520, 166270. Available at https://doi.org/10.1016/j.cej.2025.166270.

Zou, G., Guo, J., Peng, Q., Zhou, A., Zhang, Q., Liu, B., 2016. Synthesis of urchin-like rutile titania carbon nanocomposites by iron-facilitated phase transformation of MXene for environmental remediation. Journal of Materials Chemistry A 4 (2), 489–499. Available at https://doi.org/10.1039/c5ta07343j.

Zou, X., Zhao, X., Zhang, J., Lv, W., Qiu, L., Zhang, Z., 2021. Photocatalytic degradation of ranitidine and reduction of nitrosamine dimethylamine formation potential over MXene–Ti3C2/MoS2 under visible light irradiation. Journal of Hazardous Materials 413, 125424. Available at https://doi.org/10.1016/j.jhazmat.2021.125424.

Zu, D., Song, H., Wang, Y., Chao, Z., Li, Z., Wang, G., et al., 2020. One-pot in-situ hydrothermal synthesis of CdS/Nb2O5/Nb2C heterojunction for enhanced visible-light-driven photodegradation. Applied Catalysis B: Environmental 277, 119140. Available at https://doi.org/10.1016/j.apcatb.2020.119140.

CHAPTER

4

The characterization of 2D nanomaterials techniques: morphology and surface

Xiaohui Song[1] *and Xingyu Zhang*[2]

[1]Department of Materials Science and Engineering, Hefei University of Technology, Hefei, P.R. China [2]School of Mathematics, Statistics and Mechanics, Beijing University of Technology, Beijing, P.R. China

Fig. 4.1.

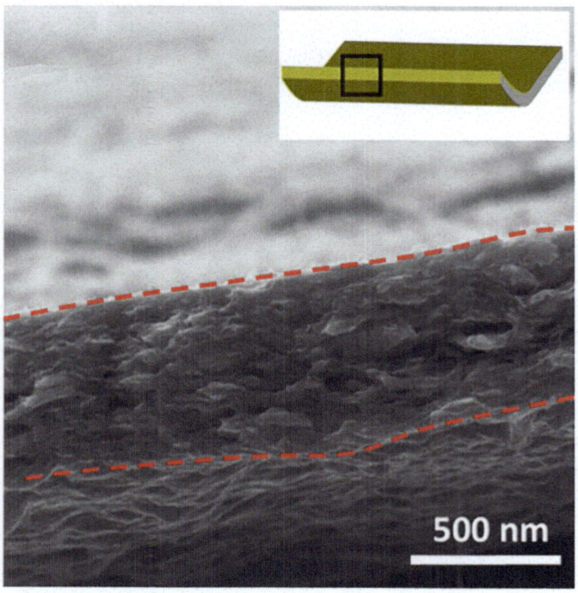

FIGURE 4.1 Field emission scanning electron microscope (FESEM) image of MoS_2 (Achari et al., 2016).

Two-Dimensional Nanomaterials for Water Purification
DOI: https://doi.org/10.1016/B978-0-443-44530-9.00003-1

4.1 Introduction

Two-dimensional (2D) membrane materials represent a class of novel nanomaterials characterized by their infinite lateral dimensions (length and width) and nanoscale thickness, typically confined to a single atom or just a few atomic layers. Key examples include 2D transition metal carbides, nitrides, and carbonitrides (MXenes), transition metal dichalcogenides (TMDs), hexagonal boron nitride (h-BN), and black phosphorus, as illustrated in Fig. 4.2.

Thanks to their layered atomic structures and exceptionally high specific surface area, these materials exhibit a range of unique physical and chemical properties. These include ultrahigh mechanical strength, selective gas permeability, outstanding thermal and electrical conductivity, and rich surface chemistry. Such characteristics make 2D membranes not only an ideal platform for fundamental research in condensed matter physics and materials science, but also highly promising for a variety of engineering applications. These include separation membranes, flexible electronics, energy storage and conversion, sensing, and protective coatings. As such, they are regarded as one of the key materials capable of driving the next generation of technological innovation.

In modern materials science and engineering, the relentless pursuit of superior material performance necessitates a precise understanding and control over their intrinsic properties. The remarkable characteristics of 2D membrane materials—from their exceptional strength and ultrafast ion transport to their unique selectivity—are fundamentally dictated by their atomic-level structure, defects, interlayer spacing, and interfacial interactions. For instance, precisely tuning the pore size and surface properties of 2D membranes can significantly enhance permeation rates and separation efficiency, as shown in Fig. 4.3.

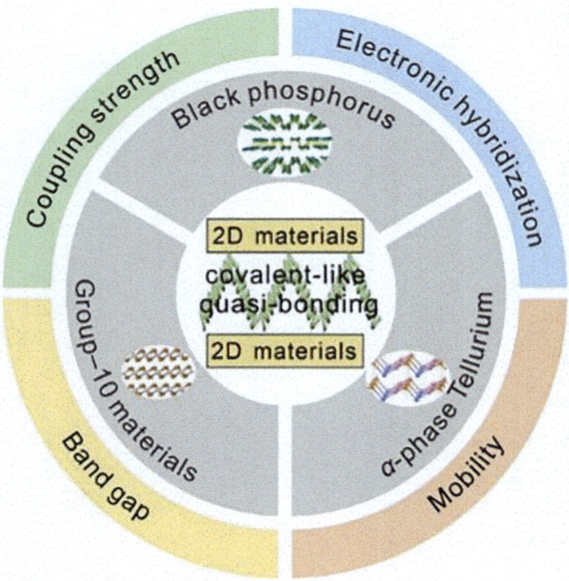

FIGURE 4.2 Classification and characteristics of 2D membrane materials (Bian et al., 2022).

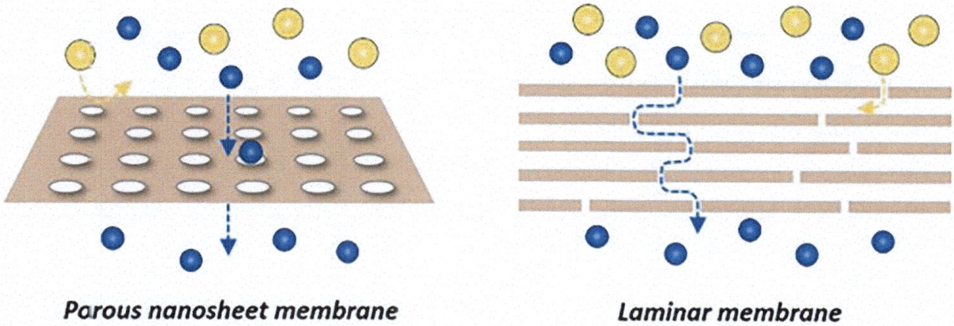

FIGURE 4.3 Schematic of two types of 2D membrane materials: porous nanosheet membranes and laminated membranes (Cheng et al., 2021).

Consequently, the development and application of sophisticated, multidimensional characterization techniques are essential. These methods provide a crucial link between macroscopic performance and microscopic atomic structure, establishing a fundamental understanding of structure–property relationships. This understanding forms the cornerstone of rational material design, process optimization, and ultimately, successful industrial application. Systematically mastering these characterization techniques is therefore indispensable for transitioning 2D membrane materials from laboratory discoveries to real-world technologies.

This chapter provides a comprehensive and systematic overview of the characterization methodologies for 2D membrane materials and their representative applications. Section 4.2 begins by categorizing characterization techniques into three primary groups: structural characterization methods, such as scanning electron microscopy (SEM), transmission electron microscopy (TEM), atomic force microscopy (AFM), X-ray diffraction (XRD), Raman spectroscopy, and computed tomography (CT), which are employed to elucidate morphological features, crystal structure, layer number, and defect information; performance characterization methods, including gas separation efficiency, water flux and permeability, mechanical strength, thermal properties, and electrical performance measurements, which directly assess functional behavior under realistic operating conditions; and supplementary techniques such as molecular dynamics (MD) simulations, density functional theory (DFT) calculations, and Fourier-transform infrared spectroscopy (FTIR), which offer theoretical modeling and auxiliary analysis.

In Section 4.3, we examine how these characterization tools are applied to address key scientific challenges in the study of 2D membranes. Through specific examples, we illustrate how an integrated approach yields a holistic understanding of material properties. The chapter concludes with a critical assessment of the strengths and limitations of each technique, along with perspectives on future developments, aiming to serve as a practical reference for researchers in selecting and applying appropriate characterization strategies.

4.2 Classification of characterization methods for 2D membrane materials

4.2.1 Structural characterization methods

4.2.1.1 Scanning electron microscopy

4.2.1.1.1 Principle of scanning transmission electron microscopy

Scanning electron microscopy operates by generating a high-energy electron beam from an electron gun, which is then focused through an electron optical system into a fine probe. Under the control of scanning coils, this probe systematically raster-scans the sample surface. The interaction between the incident electrons and the specimen produces various signals, including secondary electrons, backscattered electrons, and characteristic X-rays, which are collected by specialized detectors. These signals are subsequently converted into high-resolution images, revealing critical microstructural details of 2D membrane materials such as surface topography, thickness, pore architecture, and localized defects, as illustrated in Fig. 4.4.

4.2.1.1.2 Sample preparation requirements

Step 1. Chemical stability: Samples must maintain structural and compositional integrity under high vacuum conditions and during electron beam exposure.

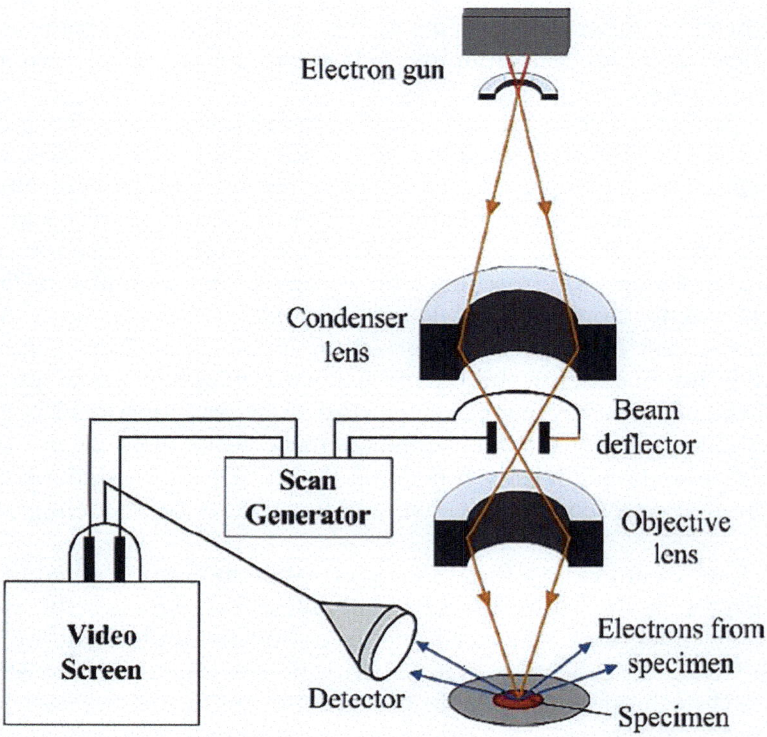

FIGURE 4.4 Schematic diagram of scanning electron microscopy (Zhu et al., 2014).

Step 2. Dryness: Complete removal of moisture and organic volatiles is essential to prevent outgassing or structural damage in vacuum.

Step 3. Surface cleanliness: Surfaces must be free from dust and contaminants to ensure clear imaging without interference.

Step 4. Conductive coating: Nonconductive specimens require surface deposition of thin metal coatings (e.g., gold, carbon) to prevent charge accumulation.

Step 5. Demagnetization: Magnetic materials must be predemagnetized to avoid deflection of the electron beam.

Step 6. Structural integrity: Samples in powder, thin-film, bulk, or fractured forms should preserve their native microstructure without artificial alterations.

Step 7. Dimensional compatibility: Specimen dimensions must fit the microscope stage, typically not exceeding 3 cm in diameter and 1 cm in thickness.

4.2.1.1.3 Application of scanning electron microscopy in 2D membrane materials

Fig. 4.5 presents a series of SEM images comparing the surface morphology of thin-film composite (TFC) membranes with thin-film nanocomposite (TFN) membranes incorporating varying loadings of MoS_2 nanosheets. These images demonstrate the pronounced influence of MoS_2 as nanofiller on the polyamide (PA) active layer formation.

The morphological evolution observed through SEM reveals distinct effects of MoS_2 incorporation: unlike the characteristic 'leaf-like' structures of conventional TFC membranes, TFN surfaces develop increased nodular features, suggesting altered monomer reaction kinetics and modified PA layer formation mechanisms. Furthermore, 2D MoS_2 nanosheets become clearly identifiable within the membrane matrix, transitioning from

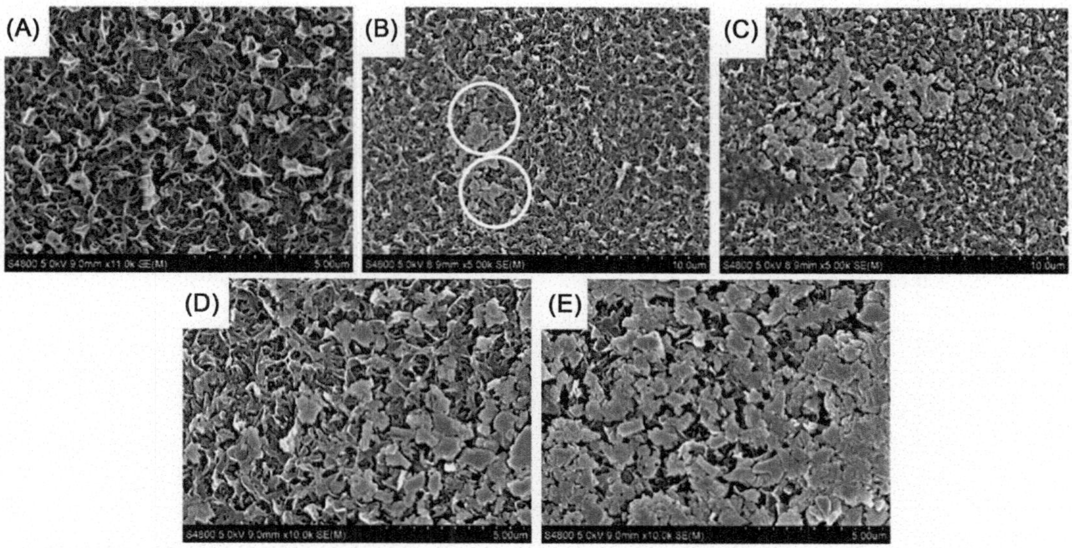

FIGURE 4.5 SEM images of (A) TFC membrane, (B) 0.005 wt% MoS_2-TFN, (C) 0.01 wt% MoS_2-TFN, (D) 0.02 wt% MoS_2-TFN, and (E) 0.05 wt% MoS_2-TFN membranes (Li et al., 2019).

well-dispersed states to aggregated clusters as their loading increases from 0.005 to 0.05 wt%, eventually forming continuous coverage on the PA surface.

These SEM findings provide direct evidence of successful MoS_2 integration and its substantial impact on membrane microstructure, establishing a morphological basis for the enhanced hydrophilicity, water permeability, and antifouling properties discussed in subsequent sections.

4.2.1.2 *Transmission electron microscopy*

4.2.1.2.1 Working principle of transmission electron microscopy

As illustrated in Fig. 4.6, Transmission Electron Microscopy operates by transmitting a high-energy electron beam through an ultrathin specimen. The interaction between these electrons and the sample generates transmitted, scattered, and diffracted signals. These signals are subsequently converted into images by the imaging system, enabling analysis at the nanoscale. This technique reveals details such as lattice fringes, dislocations, crystal defects, and interfacial bonding states. Furthermore, when combined with selected area electron diffraction (SAED), it provides specific information on crystal structure and orientation.

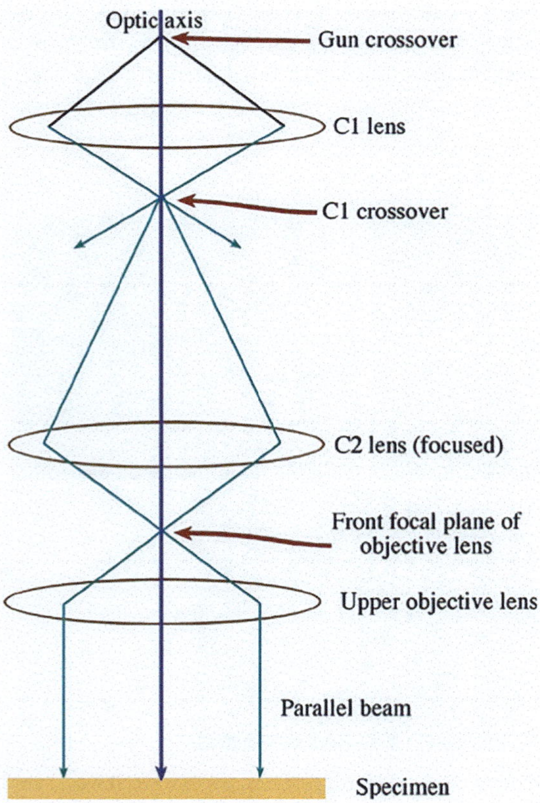

FIGURE 4.6 Schematic diagram of transmission electron microscopy (Pennycook, 2009).

4.2.1.2.2 Sample preparation requirements

1. Ultrathin sectioning: Specimens must be prepared as ultrathin slices with a thickness typically below 100 nm to permit sufficient electron transmission.
2. Vacuum and beam stability: Samples must exhibit stability under high vacuum conditions and resist degradation under electron beam irradiation to maintain integrity during analysis.
3. Radiation-sensitive materials: Beam-sensitive materials, such as polymers or organic films, require specialized techniques like cryo-cooling or low-dose imaging to minimize structural damage.
4. Powder and particulate samples: Powders or nanoparticles should be uniformly dispersed in a solvent and deposited onto a TEM grid coated with a support film (e.g., lacey or continuous carbon).
5. Bulk and thick films: Solid materials require thinning to electron transparency via methods such as ultramicrotomy, ion milling, or focused ion beam (FIB) sectioning.

4.2.1.2.3 Application of transmission electron microscopy in 2D membrane materials

Fig. 4.7 presents TEM micrographs and corresponding SAED patterns of GeS and TiS_2 nanomaterials, comparing their states before and after surface functionalization with polydopamine (PDA). These images clearly reveal the nanosheets' morphology, dimensions,

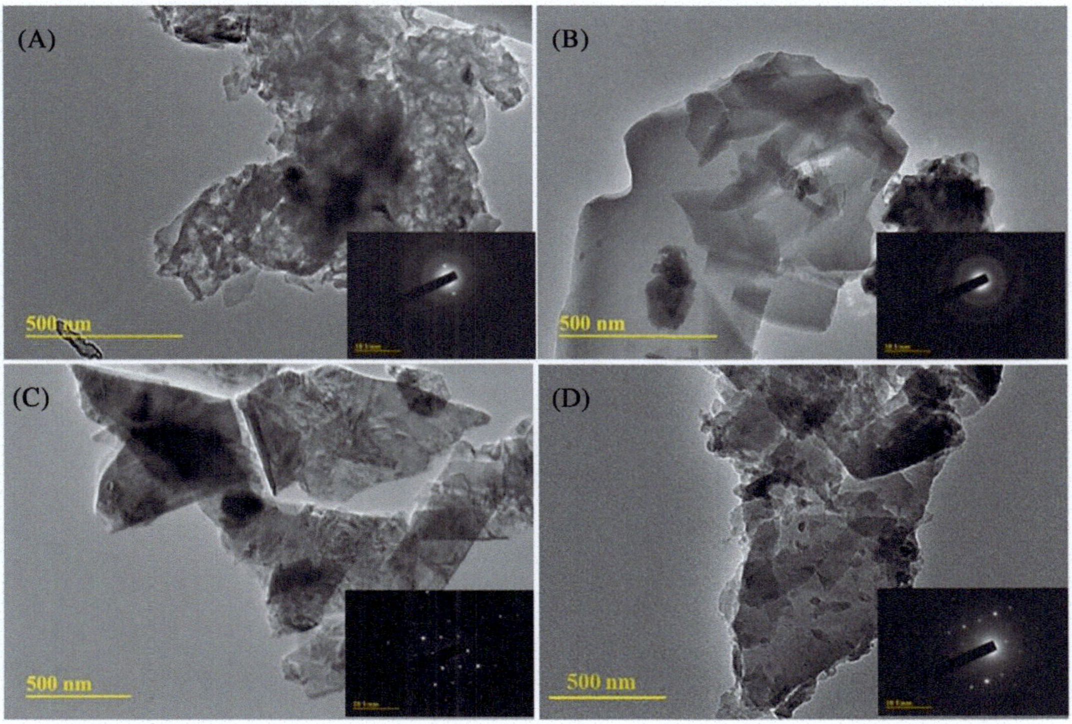

FIGURE 4.7 Transmission electron microscopy images and selected-area electron diffraction patterns of (A) GeS, (B) PDA-GeS, (C) TiS, and (D) PDA-TiS_2 (Jamiu et al., 2025).

and the microstructure of their edge coatings, while the diffraction patterns reflect the materials' crystalline state.

This comparative analysis serves to clarify two key questions: Whether PDA modification was successfully constructed on the nanosheet surfaces, and whether this treatment affected the materials' inherent 2D morphology and crystalline integrity. The evidence shown by the continuous, uniform "halo-like" coating of PDA and the preservation of intact, well-defined lattice fringes beneath it-confirms that PDA forms an ultrathin, conformal, and noninvasive layer on the material surfaces without disrupting their intrinsic crystal structure. The persistence of bright diffraction spots in the SAED patterns further verifies that the functionalized materials retain high crystallinity. Together, these TEM observations provide direct microscale evidence that PDA modification enhances surface properties while preserving the original structure of the nanosheets.

4.2.1.3 *Atomic force microscopy*

4.2.1.3.1 Working principle of atomic force microscopy

Atomic AFM operates by scanning a nanoscale probe across the sample surface. The system measures cantilever deflection caused by interatomic forces between the probe tip and the sample surface. This deflection is detected via laser reflection and converted into topographical data, enabling reconstruction of 3D surface morphology. The technique simultaneously characterizes surface roughness, identifies localized defects, and evaluates nanomechanical properties of thin films, as illustrated in Fig. 4.8.

4.2.1.3.2 Sample preparation requirements

1. The sample surface must be clean and free of particles, dust, or oil contamination that could interfere with scanning.
2. The surface should be reasonably flat to avoid abrupt height variations that may cause probe instability.

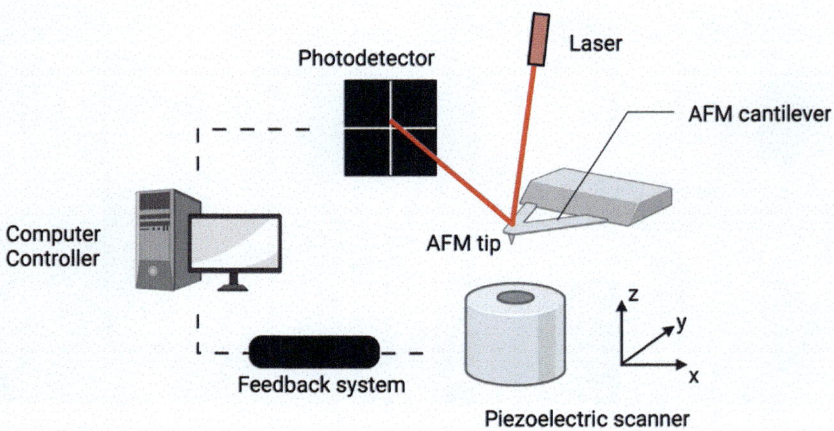

FIGURE 4.8 Schematic diagram of atomic force microscopy (Alsteens, 2024).

3. For soft or easily damaged films, tapping or noncontact mode should be selected to minimize sample damage.
4. The specimen must be securely fixed to prevent movement or deformation during the scanning process.
5. If testing is conducted in a liquid environment, the liquid must be pure and chemically compatible with the film.

4.2.1.3.3 Application of atomic force microscopy in 2D membrane materials

Fig. 4.9 presents 3D AFM surface topography images of an unmodified MXene/CA membrane (M0) and a polydopamine-modified PDA@MXene/CA membrane (M4), characterizing their microscale roughness structures. The images visually demonstrate significant differences in surface morphology: The M4 membrane exhibits pronounced undulations and nodular features, whereas the M0 surface remains relatively smooth.

This comparison aims to elucidate how PDA functionalization modulates the membrane's surface topology and to explore the resulting structure-performance relationship in terms of hydrophilicity and antifouling properties. Quantitative analysis of the AFM data reveals an increased average roughness of the M4 membrane, indicating that the self-polymerized PDA layer effectively enhances surface roughness and expands the effective filtration area, thereby contributing to improved water permeability. Although greater roughness may potentially facilitate contaminant deposition, the significantly enhanced hydrophilicity induced by PDA promotes the formation of a protective hydration layer, which shields the membrane surface from direct foulant contact. As a result, the overall antifouling performance is improved under the dominant influence of hydrophilicity. These AFM observations provide microstructural evidence supporting the potential application of this modified membrane in practical wastewater treatment.

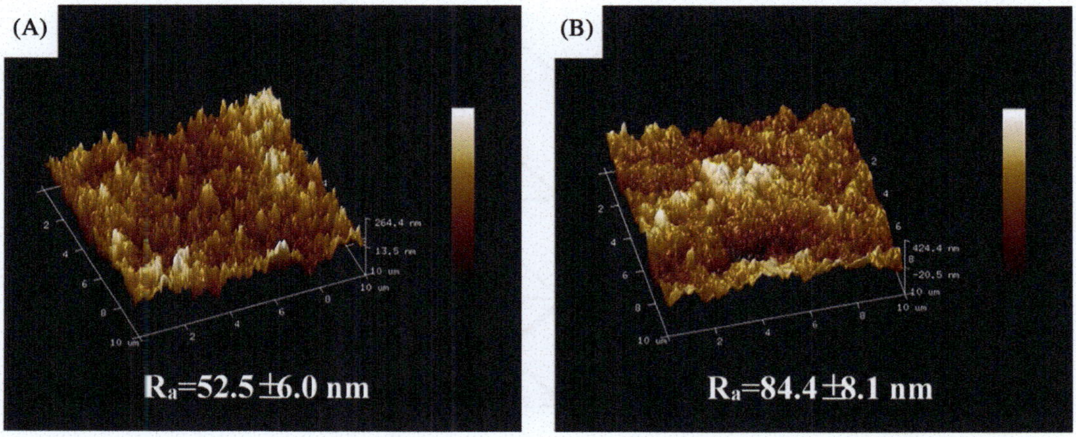

FIGURE 4.9 3D AFM images of (A) M0 membrane and (B) M4 membrane (Lin et al., 2021).

4.2.1.4 X-ray diffraction

4.2.1.4.1 Working principle of X-ray diffraction

As illustrated in Fig. 4.10, X-ray diffraction (XRD) operates based on the Bragg diffraction of incident X-rays by crystal lattices. By analyzing the positions and intensities of the resulting diffraction peaks, this technique enables characterization of a material's crystal structure, preferred orientation, and degree of crystallinity.

When applied to 2D membrane materials, XRD serves to identify crystalline phases, assess crystal quality and stress states, and evaluate the effectiveness of structural modifications in the membranes.

4.2.1.4.2 Sample preparation requirements

1. The sample surface must be flat and continuous, as cracks or uneven areas can adversely affect diffraction intensity.
2. Brittle or flexible membranes should be securely fixed prior to testing to prevent warping or fracture during measurement.
3. The specimen must be clean and free of dust, oil, or other contaminants that could interfere with the diffraction signal.
4. For ultra-thin films with weak signal intensity, grazing-incidence XRD (GI-XRD) or enhanced light sources (such as synchrotron radiation) are recommended to improve the signal-to-noise ratio.
5. Prior to testing, ensure that the diffraction peaks of the membrane can be clearly distinguished from those of the substrate to avoid signal overlap.

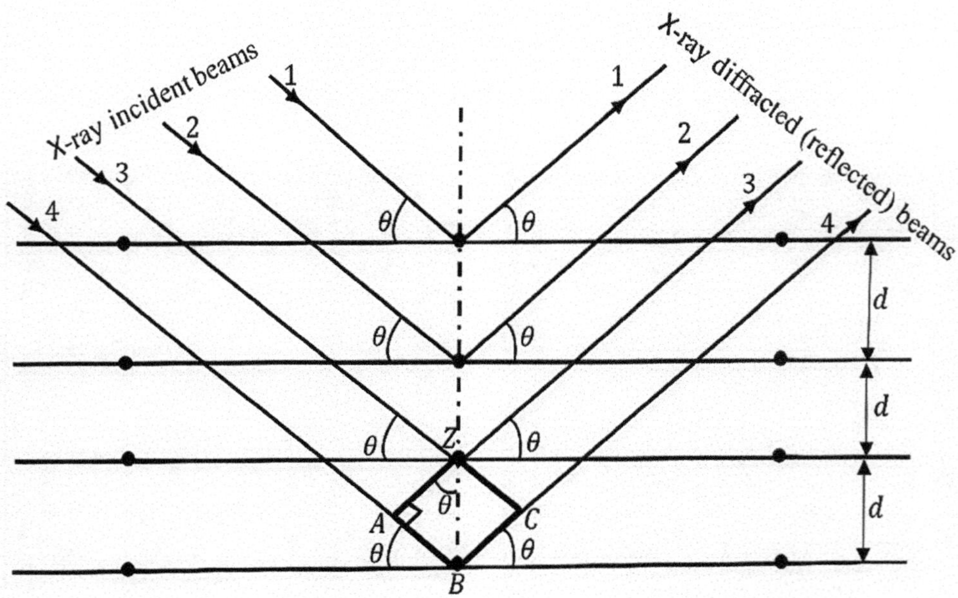

FIGURE 4.10 Schematic diagram of Bragg diffraction (Ameh, 2019).

4.2.1.4.3 Application of X-ray diffraction in 2D membrane materials

Fig. 4.10 XRD pattern of the PEI-based mixed matrix membrane (Yue et al., 2013).

Fig. 4.10 displays XRD patterns of pure polyetherimide (PEI) membranes and PEI-based mixed matrix membranes (MMMs) incorporated with ZIF-8, MoS_2, and h-BN nanofillers, respectively. These patterns are used to analyze the crystalline structure of each membrane and verify the successful incorporation of the nanofillers.

The results show a typical amorphous diffraction halo for the pure PEI membrane, while the functionalized membranes exhibit distinct sharp diffraction peaks corresponding to the characteristic crystal planes of ZIF-8, MoS_2, and h-BN. This confirms that both the metal-organic framework and the 2D nanomaterials have been successfully integrated into the membranes while retaining their original crystal structures.

Through this XRD analysis, the authors effectively demonstrate the successful composite formation and preservation of crystallinity of the nanofillers within the PEI matrix. The sharp diffraction peaks further indicate that the fillers are well-dispersed and present in appropriate concentrations. These findings provide key structural evidence supporting the enhanced mechanical strength, improved hydrophilicity, and superior antibiotic separation performance observed in the subsequent evaluation of these membranes.

4.2.1.5 *Raman spectroscopy*

4.2.1.5.1 Working principle of Raman spectroscopy

As illustrated in Fig. 4.11, Raman spectroscopy relies on the inelastic scattering of light by matter. By analyzing the frequency shifts in the scattered light, this technique provides information about molecular vibrations, lattice vibrations, and other structural characteristics of materials.

When applied to membrane materials, Raman spectroscopy can identify chemical bonds, functional groups, crystalline phases, defect densities, and doping states. This makes it particularly valuable for investigating the structural integrity and chemical stability of 2D membranes.

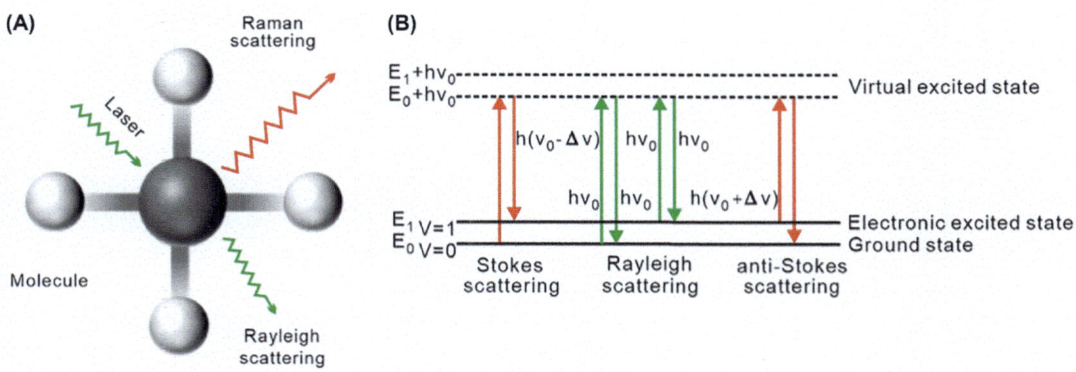

FIGURE 4.11 (A) Schematic of Raman spectroscopy and (B) the corresponding energy level diagram (Li et al., 2024).

4.2.1.5.2 Sample preparation requirements

1. The sample surface should be clean and flat to avoid contamination or fluorescence interference with the Raman signal.
2. The specimen must be securely fixed to prevent displacement during laser irradiation.
3. For thermally sensitive membranes, laser power should be reduced or pulsed lasers employed to prevent localized heating.
4. Sample thickness and transparency should be compatible with the laser penetration depth to ensure acquisition of representative signals.
5. If the sample exhibits strong fluorescence background, an appropriate laser wavelength should be selected to suppress this interference.

4.2.1.5.3 Application of Raman spectroscopy in 2D membrane materials

Fig. 4.12 shows Raman spectra of magnetic iron oxide nanoparticles (MNPs) functionalized with varying concentrations (5%–20% v/v) of oleic acid (OA), obtained using 785 nm laser excitation. These spectra are used to compare the effect of surface modification chemistry on the intrinsic vibrational modes of the nanoparticles.

The spectra reveal high similarity in both characteristic peak positions and intensities across all samples, with no systematic variations observed as a function of OA concentration. This indicates that the OA modification does not alter the core chemical structure or phase composition of the MNPs.

Through this Raman analysis, the authors verify the chemical consistency of MNPs functionalized with different OA concentrations, ruling out significant interference of the modification process with the nanoparticles' intrinsic properties. This establishes a reliable baseline for subsequent investigation of interfacial interactions in composite membranes.

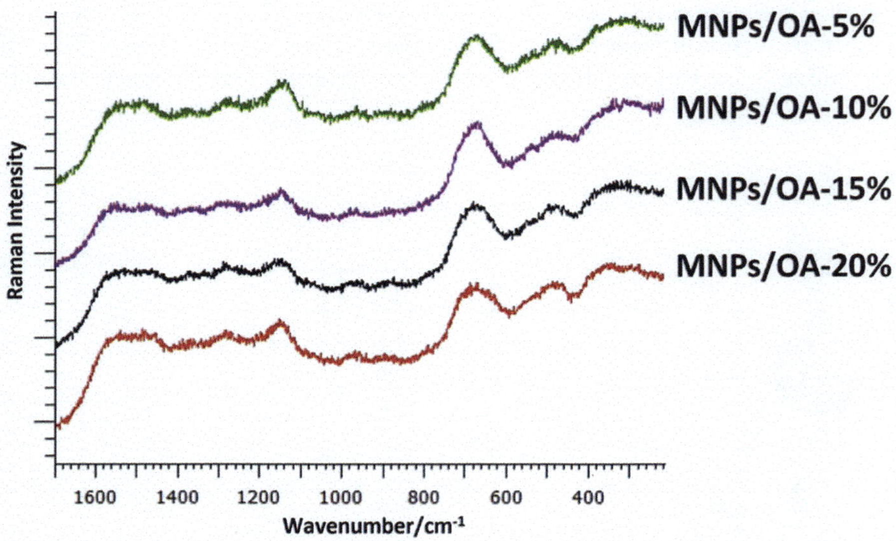

FIGURE 4.12 Raman spectra of MNPs/OA obtained with different oleic acid concentrations (laser line: 785 nm) (Kołodziej et al., 2020).

The consistent iron oxide Raman signatures across all OA-modified samples confirm that varying the surface modifier concentration introduces neither chemical structural differences nor new phase formation. This ensures that any observed performance variations in polymer nanocomposites can be primarily attributed to interfacial interactions rather than variations in filler properties, thereby providing a solid material foundation for elucidating the interfacial coupling mechanism between MNPs and the PCL matrix.

4.2.1.6 *Computed tomography*

4.2.1.6.1 Working principle of X-ray computed tomography

X-ray computed tomography (CT) operates by measuring variations in X-ray absorption as radiation passes through a sample. By acquiring scans from multiple angles and computationally reconstructing cross-sectional images, this technique enables 3D, nondestructive visualization of the internal structure of materials.

When applied to membrane materials, X-ray CT can characterize overall architecture, layer separation, pore networks, cracks, and thickness uniformity.

4.2.1.6.2 Sample preparation requirements

1. The sample must possess adequate mechanical strength to maintain structural integrity during scanning.
2. Specimen dimensions should be compatible with both the rotation stage and the detector's field of view.
3. Sufficient density contrast within the membrane material is required to generate adequate imaging contrast.
4. Flexible or ultrathin membranes may be rolled or stacked, provided deformation does not compromise measurement accuracy.
5. Samples must be thoroughly dried, as residual moisture or volatiles can affect X-ray transmission and reconstruction quality.

4.2.1.6.3 Application of computed tomography in 2D membrane materials

Fig. 4.13 presents localized thickness measurements of a membrane electrode assembly (MEA) at different degradation stages—from beginning-of-life (BOL) to end-of-life

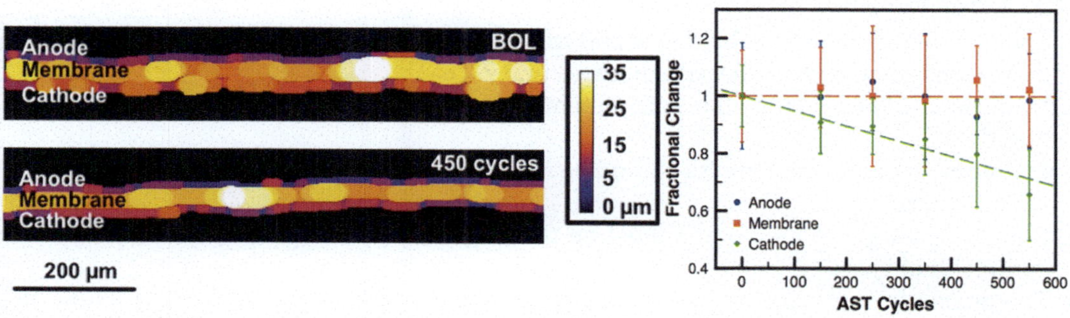

FIGURE 4.13 Local thickness measurement of MEA components via XCT, connecting the gradual thinning of the cathode catalyst layer to corrosion during voltage-cycling AST (White et al., 2017).

(EOL)—obtained through X-ray computed tomography (XCT) during an accelerated stress test (AST). The images visually display thickness variations and their statistical distribution across MEA components.

Quantitative analysis reveals that under voltage-cycling conditions, the cathode catalyst layer (CCL) undergoes a progressive thickness reduction (approximately 40%), attributed to carbon support corrosion and subsequent structural collapse. In contrast, both the membrane and anode catalyst layer maintain relatively stable thicknesses throughout the testing.

This study employs nondestructive, in situ imaging to precisely quantify the spatial degradation of MEA components, particularly the CCL, during AST, establishing carbon corrosion as the dominant degradation mechanism. The correlation between CCL thickness loss and the reduction in electrochemically active surface area, alongside overall cell performance decay, is clearly demonstrated. Furthermore, these results highlight XCT's unique capability for tracking temporal evolution at identical locations, providing direct morphological evidence for understanding the spatial heterogeneity and progressive nature of fuel cell electrode degradation.

4.2.2 Property characterization methods

4.2.2.1 *Gas separation performance testing*

4.2.2.1.1 Testing principle of gas separation performance

Gas separation performance is typically evaluated using either the constant-pressure/variable-volume method or the constant-volume/variable-pressure method. These approaches determine gas permeance and selectivity by monitoring either the flow rate on the low-pressure side or the pressure change over time on the high-pressure side.

The separation process in dense polymer membranes is governed by the solution-diffusion mechanism. This mechanism involves three sequential steps: gas molecules first dissolve into the membrane matrix at the high-pressure side, then diffuse through the membrane material driven by a concentration gradient, and finally desorb from the low-pressure side. The differing solubilities and diffusion rates of gas species in the membrane enable their separation.

4.2.2.1.2 Sample preparation requirements

1. The membrane must be fully dense and free of pinholes or cracks to prevent gas leakage that would compromise measurement accuracy.
2. Uniform thickness across the entire membrane is essential to ensure reliable permeation data.
3. Ultrathin or fragile membranes should be supported by a porous substrate with properly sealed edges.
4. Prior to testing, the membrane must be thoroughly vacuum-dried to remove moisture and residual gases.
5. The assembly must ensure complete gas-tight sealing of all connections to avoid any bypass leakage.

4.2.2.2 *Water flux and permeation performance testing*

4.2.2.2.1 Testing principle

Water flux and solute rejection are evaluated through rejection experiments or cross-flow filtration tests. Under a transmembrane pressure or concentration gradient, the volume of water passing through the membrane per unit time is measured. By comparing the concentrations of permeate and feed solutions, both water flux and solute rejection rates are calculated. The separation mechanisms primarily involve size exclusion, electrostatic repulsion, and interfacial hydrophilic/hydrophobic interactions.

4.2.2.2.2 Sample preparation requirements

Step 1. The membrane must be prewetted to ensure full pore opening and minimize initial flux fluctuations.

Step 2. The sample should be securely fixed to prevent liquid bypass between the membrane and the testing device.

Step 3. The membrane surface must be intact and flat to ensure uniform permeation.

Step 4. For composite membranes, the substrate should not interfere with the test results.

Step 5. Prior to testing, the sample must be cleaned to avoid any influence of residual contaminants on permeation behavior.

4.2.2.3 *Mechanical properties testing*

4.2.2.3.1 Testing principle

The mechanical properties of 2D membranes can be characterized through uniaxial tensile testing, nanoindentation, or blister tests. Tensile testing evaluates strength and ductility by applying controlled stress and recording the corresponding strain response. Nanoindentation determines hardness and elastic modulus by analyzing the interaction curve between an indenter tip and the sample. The blister test measures mechanical behavior by applying a gas pressure differential to induce membrane deformation, with displacement-pressure relationships reflecting key mechanical parameters.

4.2.2.3.2 Sample preparation requirements

Step 1. Membranes should be cut into appropriate dimensions with uniform edges to prevent stress concentration.

Step 2. Fragile membranes generally require a supportive substrate or frame to ensure test feasibility.

Step 3. Samples must be stored under dry or controlled environmental conditions to avoid humidity- or swelling-induced alterations in mechanical properties.

Step 4. For nanoindentation, the membrane surface must be smooth and clean to prevent tip slippage or measurement inaccuracies.

Step 5. Blister tests require airtight sealing to prevent gas leakage and ensure data reliability.

4.2.2.4 *Thermal properties testing*

4.2.2.4.1 Testing principle

The thermal properties of materials are commonly characterized using differential scanning calorimetry (DSC), thermogravimetric analysis (TGA), and laser flash analysis (LFA). DSC measures the heat flow difference between a sample and a reference, revealing phase transition temperatures and specific heat characteristics. TGA monitors mass change as temperature increases, providing information on thermal stability and decomposition behavior. LFA determines thermal diffusivity and conductivity by applying a short laser pulse to one side of the sample and recording the temperature rise curve on the opposite side.

4.2.2.4.2 Sample preparation requirements

Step 1. Samples should be uniformly prepared with thickness appropriate for the specific instrument requirements.

Step 2. The sample surface must be flat to avoid thermal flow distortion caused by local thickness variations.

Step 3. Prior to testing, samples should be preconditioned under inert atmosphere or vacuum to remove adsorbed moisture.

Step 4. Ultrathin film materials often require stacking or pellet compression to meet detection sensitivity requirements.

Step 5. Ensure optimal contact between the sample and the sample holder to guarantee accurate heat transfer measurements.

4.2.2.5 Electrical properties testing

4.2.2.5.1 Testing principle

The electrical properties of 2D membranes can be characterized using four-point probe measurement, electrochemical impedance spectroscopy (EIS), and current–voltage (I-V) characteristic analysis. The four-point probe method separates current injection and voltage sensing paths to eliminate contact resistance, enabling accurate conductivity measurement. EIS applies an AC signal across a frequency range to obtain impedance spectra, which are analyzed via equivalent circuit modeling to elucidate charge transport mechanisms and interfacial processes. I-V characterization records current response under applied DC voltage to evaluate conductive behavior and charge transfer efficiency.

4.2.2.5.2 Sample preparation requirements

Step 1. Specimens must provide sufficient area for electrode or probe placement.

Step 2. Reliable electrical contact should be ensured through deposited metal electrodes or conductive adhesives.

Step 3. Surfaces must be clean and free from contaminants to minimize contact resistance.

Step 4. Humidity-sensitive samples require testing under vacuum or inert atmosphere.

Step 5. Flexible membranes should be mounted on rigid substrates to prevent measurement artifacts induced by bending.

4.2.3 Other characterization methods

4.2.3.1 Molecular dynamics simulation

4.2.3.1.1 Physical and mathematical foundations of MD simulation

Molecular dynamics (MD) simulations derive their theoretical framework from statistical mechanics and classical or quantum mechanics, enabling the prediction of macroscopic physicochemical properties through the analysis of microscopic particle dynamics. Central to this approach is the ergodic hypothesis, which posits that over sufficiently long time scales, a single system will explore all accessible microscopic states of its statistical ensemble. This principle allows the extraction of ensemble averages, such as temperature and pressure, from particle trajectory data. Temperature is computed from the average kinetic energy of particles via the equipartition theorem, while pressure is evaluated using the virial theorem, incorporating contributions from both kinetic energy and interatomic forces acting on the simulation boundary.

Two primary methodological paradigms exist based on the level of mechanical description. Classical molecular dynamics neglects explicit quantum electronic effects, representing atoms as classical point particles that evolve according to Newton's equations of motion. This approximation offers computational efficiency, enabling simulations of systems comprising hundreds of thousands to millions of atoms, and is widely applied in studies of protein folding, polymer dynamics, and metallic diffusion. In contrast, quantum molecular dynamics incorporates electronic structure effects by solving the Schrödinger equation, often within the Born-Oppenheimer approximation, using density functional theory or Hartree–Fock methods. While essential for modeling chemical reactions, charge transfer, and materials under extreme conditions, its high computational cost restricts application to systems of several hundred atoms. A hybrid approach, quantum-mechanical/molecular-mechanical (QM/MM) simulation, strategically combines both treatments: a small, chemically active region is described quantum-mechanically, while the surrounding environment is modeled classically, thus balancing accuracy and feasibility in systems such as enzymatic catalysts.

The mathematical description of interatomic interactions is encapsulated in the potential energy function, which forms the computational core of classical MD. The total potential energy is typically partitioned into bonded terms, including bond stretching modeled by harmonic oscillators, angle bending described by quadratic potentials, and dihedral torsion represented with Fourier series, and nonbonded interactions composed of van der Waals forces modeled by Lennard-Jones potentials and electrostatic interactions governed by Coulomb's law. Specialized force fields such as AMBER and CHARMM have been developed for biological macromolecules, while packages like LAMMPS provide parameter sets for materials science applications. To extend the accessible spatial and temporal scales, coarse-grained potentials such as MARTINI reduce computational demand by grouping multiple atoms into single interaction sites, enabling the simulation of supramolecular assemblies and biomembranes.

4.2.3.1.2 Molecular dynamics simulation workflow and control logic

MD simulation serves as a fundamental methodology for investigating material behavior at the atomic scale, establishing a critical connection between microscopic interactions and macroscopic properties through the framework of classical Newtonian mechanics. The core of MD lies in constructing physically realistic system models that accurately represent crystal structures, molecular compositions, and appropriate molecular representations. This is achieved through potential functions that describe electrostatic, van der Waals, and bonded interactions, complemented by numerical algorithms that solve Newton's equations of motion at femtosecond-level time steps while maintaining system stability through thermodynamic ensembles.

A comprehensive MD investigation follows a systematic workflow encompassing system construction, initialization and execution, dynamic monitoring, and result analysis, with each phase clearly illustrated in the accompanying schematic. The construction phase involves preparing solid structures by creating supercells and modifying surfaces to eliminate dangling bonds, followed by configuring realistic solvent environments and chemical components. During initialization, energy minimization eliminates unphysical configurations before establishing ensemble conditions, temperature parameters, and simulation duration according to the procedural guidelines. Simulations are typically implemented using packages such as GROMACS with compatible force fields. The monitoring phase focuses on tracking atomic trajectories and capturing essential dynamic behaviors, while subsequent analysis employs quantitative methods, including radial distribution functions, to establish correlations between microscopic mechanisms and experimental observations.

This visualized procedural framework (Fig. 4.14) not only provides an operational roadmap but also ensures reproducibility and reliability throughout the simulation process. The documented parameters at each stage enable exact replication, while standardized procedures guarantee result accuracy. Ultimately, this methodology offers atomic-level insights into the origins of macroscopic phenomena, establishing a critical foundation for related scientific investigations.

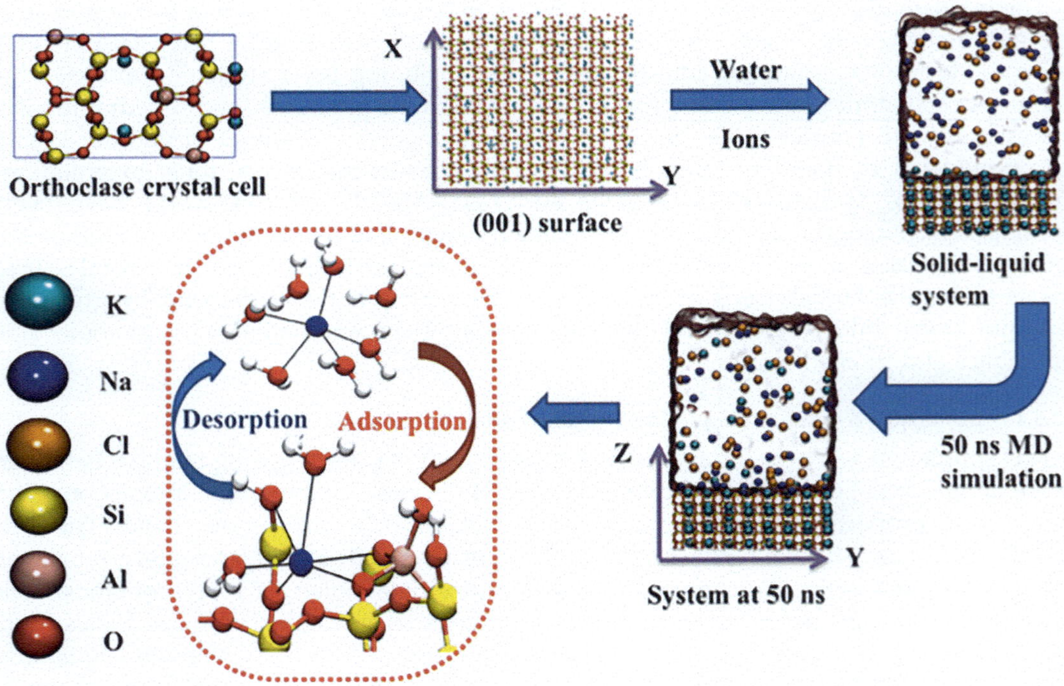

FIGURE 4.14 Workflow diagram of molecular dynamics simulation (Liu et al., 2021).

4.2.3.1.3 Construction criteria for molecular dynamics virtual samples

The integrity of the system constitutes the fundamental prerequisite in molecular simulation, requiring comprehensive inclusion of all components relevant to the research objectives while maintaining consistency with experimental conditions. For solute modeling, biomolecular systems demand complete sequences, with any missing segments properly reconstructed, and accurate incorporation of posttranslational modifications. In materials systems, such as lithium-ion battery simulations, all functional components, including electrodes, electrolytes, and interface layers, must be explicitly represented. The solvent environment should fully encapsulate the solute, with simulation box dimensions ensuring a minimum distance of 1.0 nm between solute atoms and periodic boundaries. Ion concentrations must match experimental conditions, and initial ion placements require energy minimization to avoid artificial salt bridges or aggregation.

Structural rationality must adhere to established chemical and physical principles. Bond parameters, including bond lengths and angles, should remain within physically plausible ranges (e.g., C-C single bond equilibrium length ≈ 0.154 nm, with deviations ≤ 0.01 nm), achievable through molecular mechanics optimization. Dihedral angles must reflect appropriate conformational states, with aberrant torsions systematically corrected. Spatial clashes can be resolved through a hierarchical relaxation protocol: initially fixing the solute backbone while optimizing solvent and ion positions, followed by full-system energy minimization without constraints.

Computational feasibility necessitates balancing accuracy with resource constraints. System size should align with research objectives and available resources: classical MD typically handles 10 (Li et al., 2019)-10 (Jamiu et al., 2025) atoms (with coarse-grained models or GPU clusters enabling larger systems), while quantum MD remains limited to ≤10 (Zhu et al., 2014) atoms (employing cutoff radii to reduce computational load). Parameter consistency is critical for instance, AMBER potential functions must pair with corresponding atomic charges and force field parameters, and NPT ensemble simulations require time steps compatible with pressure control response times.

Reproducibility forms the cornerstone of reliable results. Detailed documentation of construction parameters-including PDB identifiers, solvent models, and initial structure files-should accompany publicly available input files specifying all simulation parameters (duration, time step, etc.). Validation may involve comparison with experimental data (e.g., RMSD analysis against NMR protein structures) or extended simulations to verify the stability of physical observables.

4.2.3.1.4 Molecular dynamics output formats and analytical interpretation

MD simulations generate data files and graphical outputs that require integrated quantitative and qualitative analysis to verify their reliability.

Output files generally fall into four categories. Log files (e.g., GROMACS.log) record real-time simulation data, enabling runtime monitoring and efficiency evaluation. Trajectory files (such as CHARMM.dcd formats) store atomic coordinates and velocities, supporting both parameter calculation and visualization through tools like VMD or PyMOL. Energy files (e.g., GROMACS.edr) contain total energy values and detailed energy components, allowing analysis of system stability and interaction contributions. Structure files (including

GROMACS.gro formats) preserve system configurations, useful for parameter verification or as starting points for subsequent simulations.

Key metrics are interpreted across four dimensions. Energy indicators determine equilibrium states-total energy fluctuations within 5% typically signify equilibrium-while subdivided energy terms reveal underlying interaction mechanisms. Structural metrics include RMSD (root mean square deviation), where values stabilized within 0.2 nm suggest conformational convergence in protein simulations, and RMSF (root mean square fluctuation), which reflects local atomic flexibility. Dynamic behavior indicators encompass hydrogen bond lifetime analysis for interaction stability, diffusion coefficients for molecular mobility quantification, and binding free energy calculations for ligand-receptor affinity. Macroscopic properties such as elastic modulus and viscosity derived from microscopic behaviors provide opportunities for experimental validation.

Result validation employs three primary approaches: direct comparison with experimental data (where deviations ≤10 % generally indicate reasonable parameterization); convergence testing through extended simulations to verify result stability; and parameter sensitivity analysis to assess dependency on key inputs. When experimental references are unavailable, cross-validation with comparable studies or alternative computational methods provides additional verification pathways.

4.2.3.1.5 Application of molecular dynamics simulation in 2D membrane materials

Following the explanation of MD testing principles, we now turn to Fig. 4.15 to further illustrate the practical outcomes of MD simulations.

Fig. 4.15 provides an atomic-scale visualization of the adsorption states of four cations (Na^+, K^+, Ca^{2+}, and Mg^{2+}) on the orthoclase (001) surface. The simulation clearly reveals distinct adsorption behaviors: Na^+, K^+, and Ca^{2+} preferentially adsorb at specific sites, forming direct coordination bonds with surface oxygen atoms, characteristic of inner-sphere adsorption. For instance, Ca^{2+} establishes strong coordination with surface oxygen atoms, forming a stable inner-sphere configuration consistent with the principles of ion-surface interactions discussed in MD simulations, where electrostatic attraction governs adsorption behavior and ionic properties such as charge and radius determine specific adsorption sites and modes.

In contrast, Mg^{2+} exhibits a markedly different adsorption pattern, maintaining a greater distance from the surface and remaining surrounded by a hydration shell-a typical outer-sphere adsorption configuration. This behavior arises from Mg^{2+}'s high hydration energy, which prevents direct contact with the surface by preserving its water coordination environment.

This atomic-level analysis demonstrates MD simulations' capability to reveal ion adsorption configurations that are inaccessible through conventional experimental techniques. By precisely identifying adsorption sites and distinguishing adsorption types, the simulations provide direct evidence for understanding cation selectivity on mineral surfaces. These results highlight MD's powerful role in elucidating microscopic solid-liquid interface mechanisms, offering fundamental insights into complex processes such as ion adsorption and ligand exchange. Such theoretical guidance proves invaluable for practical applications, including ionic separation and wastewater treatment, ultimately supporting experimental design and advancing related technological developments.

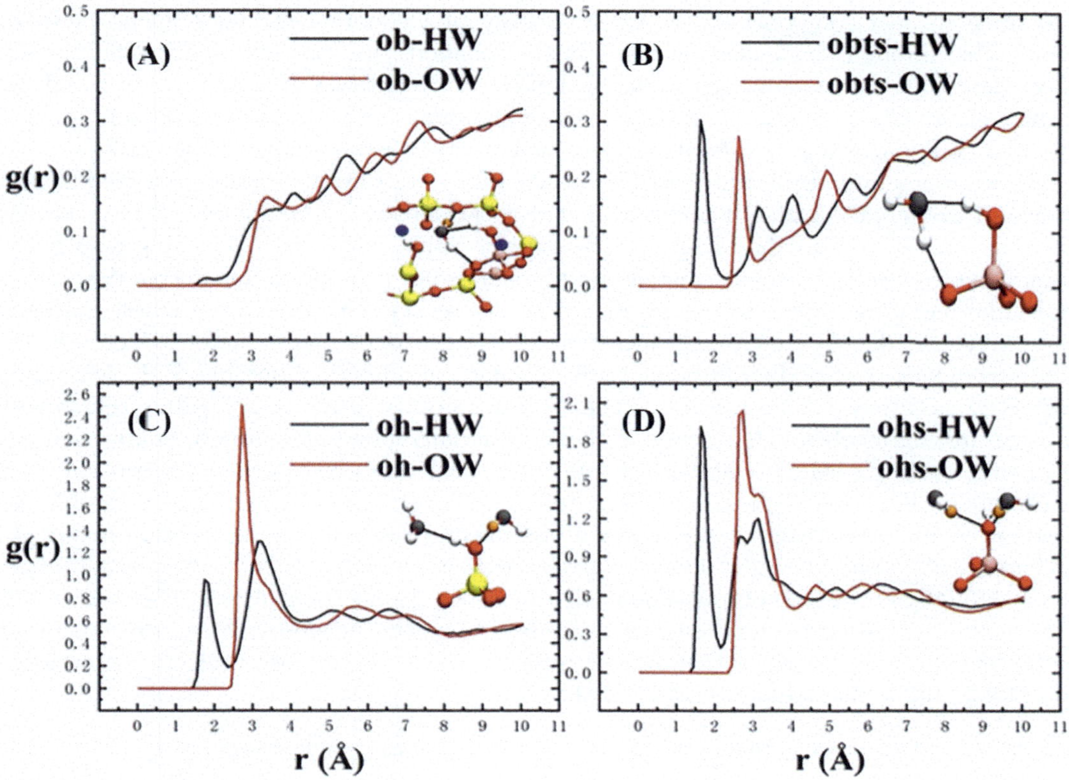

FIGURE 4.15 Radial distribution function (RDF) and atomic visualization derived from molecular dynamics simulation, showing interactions between lattice oxygen and water molecules in a NaCl system (Liu et al., 2021).

4.2.3.2 *Density functional theory calculations*

4.2.3.2.1 Theoretical foundation: quantum-mechanical and mathematical basis of density functional theory

DFT stands as a cornerstone computational method in quantum mechanics for investigating electronic structure, fundamentally relying on electron density distribution to determine ground-state properties of many-electron systems while circumventing the formidable complexity inherent in direct many-electron wavefunction approaches. The theoretical framework rests upon the Hohenberg-Kohn theorems, which establish that the ground-state energy of an interacting electron system can be uniquely expressed as a functional of its electron density. This profound insight enables the complete determination of both the external potential and all ground-state properties, including energy and charge distribution, solely from the electron density $\rho(r)$, thereby reducing the computational problem from the intractable 3N-dimensional wavefunction domain to the tractable 3D density domain.

The practical implementation of DFT is achieved through the Kohn-Sham scheme, which constructs a fictitious system of noninteracting electrons that exactly reproduces the electron density of the original interacting system. Within this framework, the total energy is partitioned into three distinct components: the kinetic energy of the noninteracting reference system, computable through single-electron Schrödinger equations; the classical electrostatic interactions comprising electron–nucleus attraction and electron–electron Coulomb repulsion; and the exchange-correlation energy that encapsulates all quantum mechanical many-body effects. It is this final component that necessitates approximation, as no exact analytic form exists.

The accuracy and applicability of DFT calculations are predominantly governed by the choice of exchange-correlation functional. The local density approximation (LDA), which depends exclusively on the local electron density value, provides reasonable results for homogeneous electron systems such as simple metals but proves inadequate for describing molecular bonding and band gaps in solids. More sophisticated generalized gradient approximation (GGA) functionals incorporate electron density gradients to account for inhomogeneous charge distributions, with popular parameterizations like PBE delivering improved accuracy for molecular structures and surface processes. Further refinement emerges through meta-GGA functionals that additionally incorporate kinetic energy density gradients, enabling better treatment of transition metal complexes and hydrogen-bonding interactions. For the highest accuracy in challenging electronic structure problems, particularly molecular excitation energies and semiconductor band gaps, hybrid functionals such as B3LYP and HSE06 blend exact Hartree-Fock exchange with DFT correlation, though at substantially increased computational expense. The selection of an appropriate functional thus requires careful consideration of the specific system and properties under investigation, with PBE commonly employed for metallic systems, B3LYP for organic molecular processes, and HSE06 for semiconductor electronic structures.

4.2.3.2.2 Density functional theory calculation workflow and core control parameters

DFT serves as a fundamental computational approach in condensed matter physics and quantum chemistry for determining electronic structures. Its core innovation lies in using electron density rather than many-electron wavefunctions to characterize ground-state properties, thereby significantly simplifying the quantum many-body problem. Established on the Hohenberg–Kohn theorems, DFT posits that the ground-state electron density uniquely determines the external potential (up to an additive constant) and that a universal energy functional exists whose minimum corresponds to the ground-state energy. The practical implementation relies on the Kohn–Sham equations, which map the interacting electron system onto a fictitious system of noninteracting electrons moving within an effective potential comprising external, Hartree, and exchange-correlation components. Solving these single-electron equations yields the electron density, from which the total energy and structural properties can be derived.

The computational procedure involves two interconnected cycles: self-consistent field iteration and structural optimization. The self-consistent cycle begins with an initial guess for the electron density, typically constructed from atomic density superposition. This initial density is used to construct the Kohn–Sham effective potential, followed by solving the Kohn–Sham equations to obtain single-electron wavefunctions and eigenvalues. A new electron density is then computed from these wavefunctions, and its deviation from the previous density is evaluated. If this deviation exceeds the convergence threshold, the process repeats-rebuilding the potential, resolving the equations, and recalculating the density-until self-consistency is achieved.

Once electronic self-consistency is attained, structural optimization commences based on the Hellmann–Feynman theorem, which enables efficient calculation of atomic forces as the energy gradient with respect to nuclear positions. These forces, pointing toward energy minimization, guide atomic position adjustments through algorithms such as the conjugate gradient method. Each new atomic configuration requires renewed electronic self-consistency to compute updated energies and forces. The optimization cycle continues until all atomic forces fall below a specified threshold, typically 0.01 eV/Å, indicating a stable equilibrium geometry.

The complete DFT mechanism forms a closed-loop workflow, visually represented in Fig. 4.16. Beginning with the initial atomic configuration and electron density guess, the process proceeds through KS potential construction, KS equation solution, electron density and energy calculation, and force computation via the Hellmann–Feynman theorem. The force convergence check determines whether to continue atomic position adjustment or to output the final stable structure and electronic properties, effectively illustrating the coupled nature of electronic structure calculation and geometric optimization.

The accompanying DFT mechanistic diagram offers an intuitive understanding of this computational logic. Presented as a flowchart for structural optimization, the diagram illustrates an iterative closed-loop beginning with "initial atomic configuration and electron density guess." progressing sequentially through "construct KS effective potential," "solve

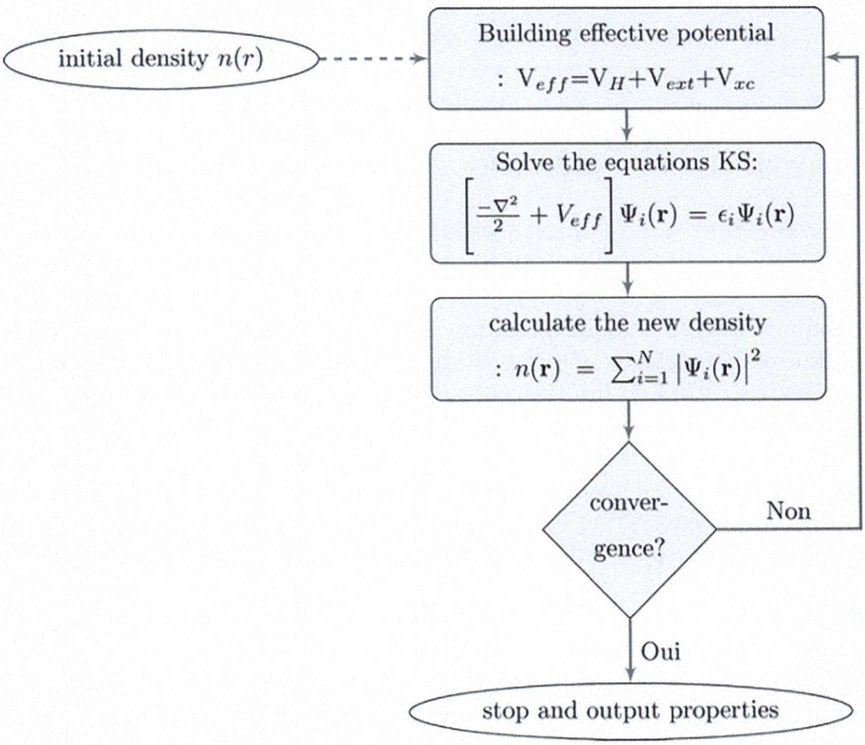

FIGURE 4.16 DFT calculation workflow diagram (Barhoumi, 2022).

KS equations," "calculate electron density and system energy," and "compute atomic forces via the Hellmann–Feynman theorem," ultimately reaching the decision node "whether forces meet convergence criterion."

When the decision yields "no," the workflow proceeds to "adjust atomic positions based on forces' and returns to 'construct KS effective potential" to initiate another iteration. A "yes" decision leads to termination at the final node: "system energy minimized, obtaining stable structure and electronic properties."

This diagram clearly demonstrates the coupling between electronic structure calculation and geometric optimization in DFT simulations. Each atomic position adjustment requires self-consistent electronic structure calculation to verify energy and force changes, with force convergence serving as the fundamental criterion for system stability. This closed-loop process ensures DFT results faithfully represent the system's ground-state properties.

4.2.3.2.3 Construction criteria for density functional theory computational models

In molecular simulation system construction, system integrity serves as the primary foundation, requiring comprehensive coverage of all components relevant to the research objectives while maintaining consistency with the actual system. Molecular systems must include reactants, products, and catalyst molecules to prevent missing atoms that could lead to incorrect reaction pathways. For solid-state systems calculating bulk properties, unit cells or supercells (such as 2×2×2 supercells) should be used, while surface property calculations require slab models with 3–5 atomic layers, reserving 10–15 Å vacuum layers perpendicular to the surface to avoid interactions between adjacent slabs. Interface systems (e.g., metal-oxide interfaces) must ensure reasonable atomic arrangements on both sides, with metal layer numbers and oxide crystal orientations matching experimental conditions, while initial molecular adsorption positions (top, bridge, etc.) should be selected according to chemical bond types.

Structural rationality must conform to chemical bonding rules and crystal structure characteristics. Initial molecular structures can be obtained from experimental data (XRD, NMR) or molecular mechanics optimization (GROMACS, Amber), ensuring bond lengths and angles fall within normal ranges (e.g., C-C single bond ~1.54 Å, O-H bond angle ~104.5°). Unknown structures should undergo preoptimization using semi-empirical methods like AM1 or PM3. Solid structures should acquire cell parameters from crystallographic databases such as ICDD or COD (e.g., Cu face-centered cubic lattice constant ~3.61 Å). Defect systems should introduce defects based on perfect crystal cells: vacancy defects by removing specific atoms, doping defects by replacing target atoms and adjusting surrounding coordinates to reduce lattice distortion, and dislocation defects by constructing supercells containing dislocation cores to eliminate boundary effects.

Parameter consistency requires that input parameters (functionals, basis sets, pseudopotentials, k-point grids) align with research objectives. Functionals should be selected according to the system: PBE for metals, B3LYP for organic optical properties, HSE06 for semiconductor band gaps. Basis sets and pseudopotentials must match, with molecular systems using Gaussian basis sets (6–31+G(d,p) with diffuse functions for hydrogen-bonded systems) and solid-state systems employing plane-wave basis sets (including relativistic pseudopotentials for heavy elements), with cutoff energies 20%–30 % higher than the kinetic energy of the highest occupied orbital. K-points should coordinate with supercells, pairing 2×2×2 supercells with 3×3×3 k-points and unit cells with 8×8×8 k-points.

Convergence control requires systematic testing to determine optimal parameters. Energy convergence should be achieved by adjusting cutoff energy, k-points, and self-consistent thresholds, such as increasing cutoff energy until total energy changes < 10^(-5) eV/atom and densifying k-points until fluctuations < 10^(-4) eV/atom. Structural convergence should be controlled by adjusting force thresholds, such as reducing them until coordinate changes < 0.001 Å. All systems calculating differential properties (reaction energy barriers, adsorption energies) should employ identical parameters.

4.2.3.2.4 Density functional theory calculation output and analysis

DFT results are typically presented through data files, graphical representations, and numerical reports, combining quantitative and qualitative analysis to reveal electronic structure characteristics and physicochemical properties while verifying reliability. Output files generally fall into four categories: log files (such as VASP's OUTCAR and Gaussian's.log files) record computational process data including iteration energies, convergence status, total energies, and atomic coordinates, enabling calculation monitoring and key parameter extraction; wavefunction files (like VASP's WAVECAR) store Kohn–Sham single-electron wavefunctions and energy levels, supporting band structure calculations, density of states analysis, and visualization through software like VESTA and XCrySDen; charge density files (such as VASP's CHGCAR) contain electron density distributions, enabling 3D visualization of chemical bonding and facilitating Bader charge and differential charge density analysis; structure files (including VASP's POSCAR) preserve optimized atomic coordinates and unit cell parameters for subsequent calculations or comparison with experimental techniques like XRD and TEM.

Key metrics are interpreted across four dimensions. The electronic structure dimension focuses on band structure characteristics (where metal valence and conduction bands overlap at the Fermi level without band gaps, semiconductors exhibit band gaps that determine optoelectronic properties), density of states (with projected density of states revealing atomic orbital contributions, such as transition metal d-band centers near the Fermi level indicating high catalytic activity), and charge distribution (where Bader charge quantifies electron transfer and differential charge density maps visualize electron accumulation/depletion regions, elucidating chemical bonding nature). The geometric structure dimension examines bond lengths and angles (e.g., water molecule O-H~0.96 Å and H-O-H~104.3°, matching experimental values when parameters are appropriate), bond energies (higher values indicating greater bond stability, such as C-C single bond at 347 kJ/mol), lattice parameters (e.g., calculated Cu lattice ~3.60 Å versus experimental 3.61 Å with <0.3 % deviation), and defect-induced lattice distortions (where smaller atomic displacements indicate greater system stability). The energy properties dimension concentrates on total energies (lower values indicating greater stability among structures with identical composition), reaction energy barriers (lower barriers suggesting more favorable reactions), adsorption energies (larger negative absolute values indicating stronger adsorption interactions), and formation energies (lower values implying easier defect generation)—all serving as essential criteria for assessing stability and reaction feasibility.

4.2.3.2.5 Application of density functional theory in 2D membrane materials

Following the introduction of DFT testing principles, we can further understand its applications through practical examples. In a study published in a renowned materials science journal, researchers employed DFT to conduct an in-depth investigation of various doping scenarios in zinc anodes, with the corresponding results presented in Fig. 4.17. The study aimed to optimize the performance of zinc anodes by calculating the work functions of tin (Sn)-, bismuth (Bi)-, silver (Ag)-, and cobalt (Co)-doped systems using DFT. This approach theoretically predicted the influence of different doping elements on the electronic structure of zinc anodes, thereby inferring their mechanisms of action on practical performance.

In these calculations, based on density functional theory, the complex problem of the many-electron system was simplified by treating electron density as the fundamental

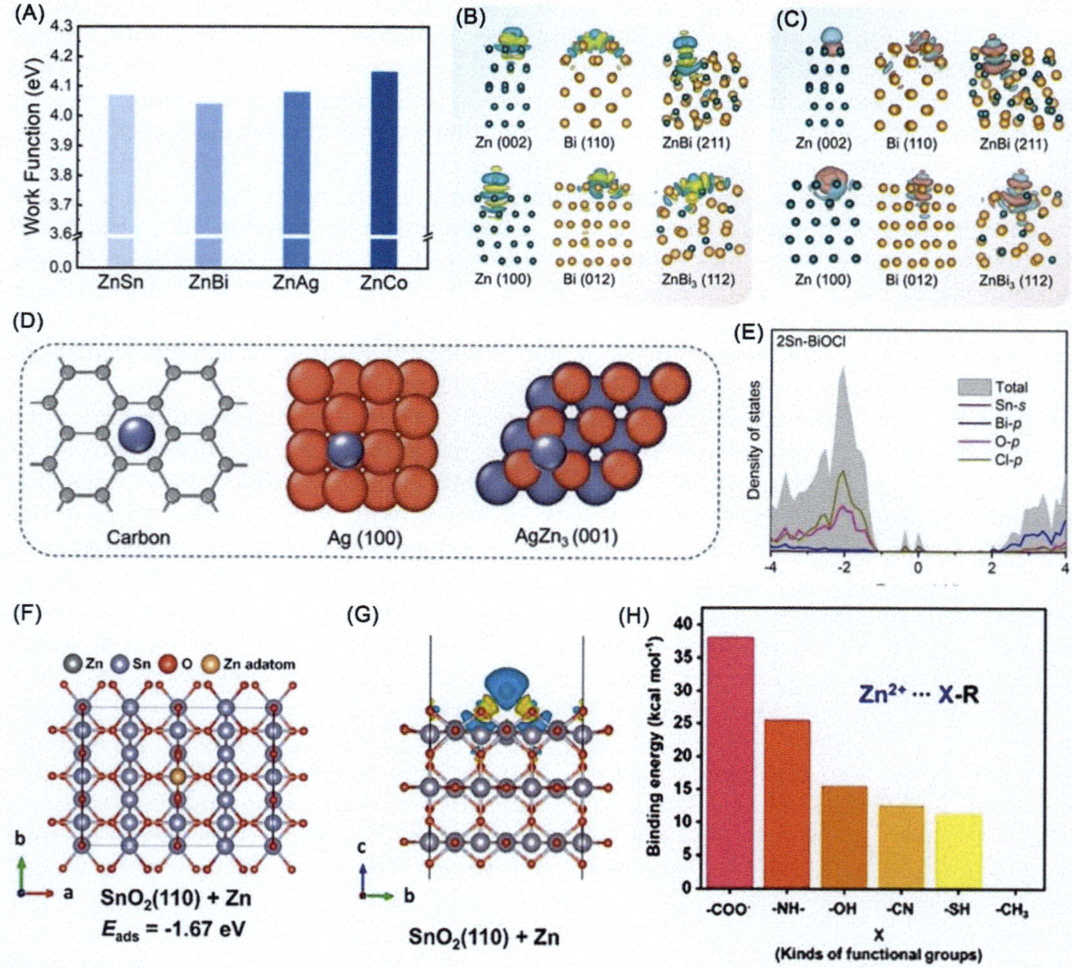

FIGURE 4.17 Example of work function calculation obtained from DFT (Cao et al., 2024).

variable. Through the construction of appropriate models and the selection of suitable exchange-correlation functionals, the energy and electronic structure of the systems were accurately solved. As shown in the results, different doping elements exhibited distinct work function values, providing strong theoretical guidance for the experimental screening of optimal dopants.

This example fully demonstrates the critical role of DFT in materials research. It enables theoretical prediction of material properties at atomic and electronic scales prior to experimental work, helping researchers gain deep insights into intrinsic material mechanisms. This allows for targeted design and optimization of materials, significantly reducing experimental costs and time, and accelerating the development of novel materials.

4.2.3.3 *Fourier-transform infrared spectroscopy*

4.2.3.3.1 Fundamental theory: molecular vibrations and spectroscopic principles of fourier-transform infrared

Fourier-transform infrared (FTIR) spectroscopy relies on the selective absorption of infrared light by molecules, with the resulting absorption spectrum providing insights into molecular structure and chemical composition. This technique is fundamentally based on the matching of energy between molecular vibrational transitions and infrared photons. Molecules, consisting of atoms connected by chemical bonds, undergo vibrations around their equilibrium positions at room temperature. Different chemical bonds (e.g., C-H, O-H, C=O) exhibit characteristic vibrational modes with specific natural frequencies, corresponding to energies in the infrared region (wavelength 2.5–25 μm, wavenumber 400–4000 cm^{-1}). When the energy of an infrared photon matches the energy required for a vibrational transition, the molecule absorbs the photon and transitions to a higher vibrational energy level, producing a characteristic absorption peak. The position, intensity, and shape of these peaks reflect the type of chemical bond, functional group, and molecular environment.

Molecular vibrations are categorized into two main types: stretching and bending vibrations. Stretching vibrations involve the rhythmic movement of atoms along the bond axis and include symmetric stretching (e.g., simultaneous extension of two C-H bonds in a -CH_2- group) and asymmetric stretching (alternating extension of the bonds). These vibrations require higher energy and typically produce absorption peaks at higher wavenumbers (e.g., O-H stretching at 3200–3600 cm^{-1}, C-H stretching at 2800–3000 cm^{-1}). Bending vibrations involve movements perpendicular to the bond axis and are classified as in-plane (scissoring, rocking) or out-of-plane (wagging, twisting) vibrations. These require less energy and appear at lower wavenumbers (e.g., C-H bending at 1300–1470 cm^{-1}, C-O-C bending at 1000–1150 cm^{-1}). It is important to note that only vibrations causing a change in the molecular dipole moment, termed 'infrared-active vibrations', produce absorption peaks. Vibrations with no dipole moment change (e.g., symmetric stretching of O=C=O) are "infrared-inactive" and do not yield observable peaks.

A key distinction between FTIR and traditional dispersive infrared spectroscopy lies in the use of Fourier-transform technology. In FTIR, broadband infrared light passes through a Michelson interferometer, generating an interferogram that encodes absorption information. This interferogram is then mathematically transformed via the Fourier transform into a conventional infrared spectrum, with wavenumber as the horizontal axis and absorbance or transmittance as the vertical axis. This approach eliminates the need for sequential wavelength scanning, allowing simultaneous acquisition of data across the entire infrared range and reducing measurement time to seconds or tens of seconds. Additionally, multiple scans can be accumulated to minimize noise, enhancing resolution and signal-to-noise ratio, and improving sensitivity for detecting weak absorption signals.

4.2.3.3.2 Testing principles: instrument configuration and workflow of fourier-transform infrared

FTIR spectroscopy operates on the principle of selective absorption of infrared radiation by molecules. Different molecular structures or functional groups, characterized by distinct vibrational and rotational energy levels, absorb specific energies of infrared light, producing characteristic absorption peaks in the spectrum. By analyzing the position, intensity, and shape of these peaks, qualitative and quantitative analysis of material composition and structure can be achieved.

In geological sample analysis, FTIR can identify mineral types by detecting functional groups such as O-H and C-O in rocks and minerals. When infrared radiation passes through a sample, molecules absorb radiation at specific wavenumbers according to their structure. The unabsorbed radiation is captured by the detector and converted into a spectrum with wavenumber (cm^{-1}) as the horizontal axis and transmittance (% T) or absorbance as the vertical axis. Regions with transmittance close to 100 % indicate no absorption, while troughs (absorption peaks) correspond to characteristic wavenumbers of functional groups. Compositional identification is accomplished by comparing sample spectra with those of reference materials.

Dispersive infrared spectrometers (Fig. 4.18) represent a classical implementation of this principle, with a workflow that aligns closely with the underlying theory. The instrument begins by generating continuous infrared radiation from a heated filament, which is directed and split into two parallel beams of equal intensity: one serving as the sample beam passing through the specimen (e.g., a pellet composed of mineral powder mixed with potassium bromide), and the other as the reference beam bypassing the sample. Both beams enter the monochromator, where a rapidly rotating beam chopper alternately directs them to a diffraction grating (or a prism in older instruments), dispersing the light into a continuous infrared spectrum by wavenumber. As the diffraction grating rotates slowly, the radiation reaching the thermocouple detector varies in wavenumber. The detector compares the intensities of the two beams and calculates the transmittance using the formula $\% T = I_s/I_r \times 100$ (where I_s is the sample beam intensity and I_r is the reference beam intensity). The signal is amplified and recorded by a recorder, which plots the final spectrum. This process clearly illustrates the complete path from radiation generation to detection, providing a concrete reference for understanding infrared analysis of rocks and minerals at the instrumental level.

4.2.3.3.3 Sample preparation requirements: core standards for fourier-transform infrared sample preparation

The reliability of FTIR analysis fundamentally depends on proper sample preparation, which must adhere to specific criteria tailored to the sample's physical state and analytical objectives.

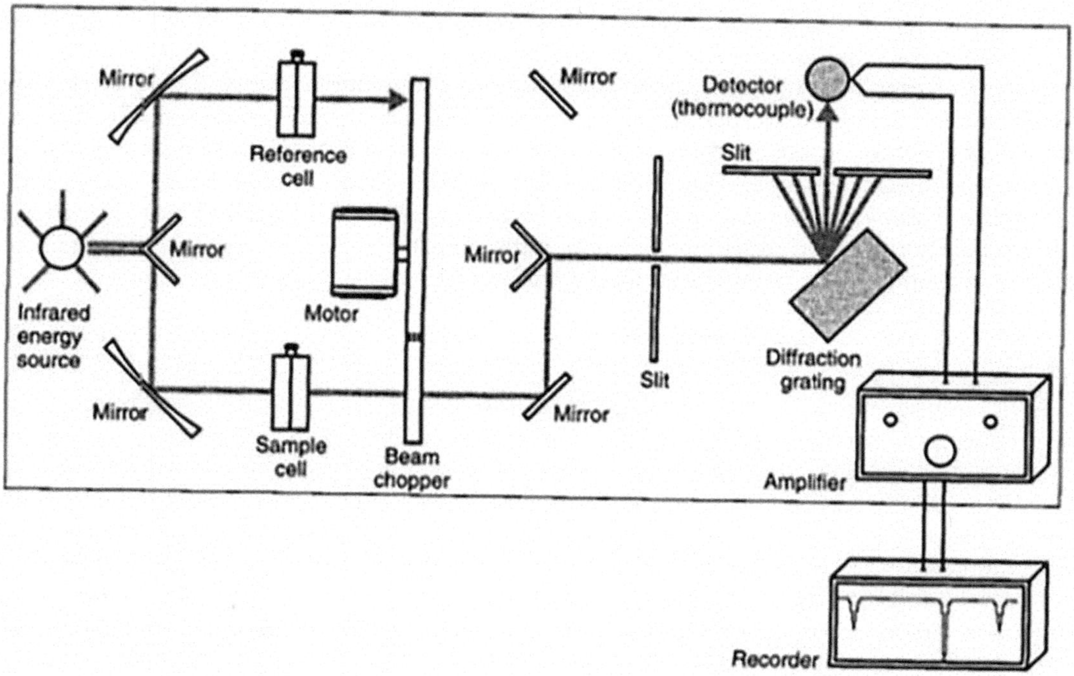

FIGURE 4.18 Schematic diagram of a dispersive infrared spectrometer (Risakotta and Andayany, 2020).

For solid samples, the most common preparation method involves creating potassium bromide (KBr) pellets. This requires the sample to be finely ground to a particle size below 5 micrometers to minimize scattering losses, then homogeneously mixed with dry, spectroscopic-grade KBr at a typical concentration of 0.2%–2% sample by weight. The mixture is compressed under high vacuum (approximately 8–10 tons of pressure) to form a transparent pellet. This method is unsuitable for moisture-sensitive samples or those that may react with halides. Alternatively, the attenuated total reflectance (ATR) technique allows for direct analysis of solid samples with minimal preparation, requiring only that the sample make firm, uniform contact with the ATR crystal surface. For polymer films, a thickness of 10–50 μm is optimal to avoid saturation of absorption bands.

Liquid samples are typically analyzed as thin films between two salt plates (using NaCl or KBr windows) or, for volatile compounds, in sealed liquid cells with precisely controlled path lengths. The path length must be carefully selected based on the sample's absorption characteristics, generally 0.01–0.1 mm for strongly absorbing functional groups like C=O, and 0.5–1.0 mm for weaker absorbers. ATR-FTIR has largely replaced traditional liquid cell methods for most applications, requiring only a small droplet placed directly on the crystal.

Gas phase analysis employs specialized gas cells with significantly longer path lengths (typically 5–20 cm, extendable to multiple meters with mirrored systems) to compensate for lower molecular density. The sample must be free of particulate matter and moisture to prevent interference with the spectrum. The system should be thoroughly purged and evacuated before the introduction of the analyte gas.

Universal preparation requirements include ensuring sample purity to avoid spectral contamination, complete dryness to prevent broad O-H absorption bands that can obscure important spectral regions, and homogeneity to guarantee representative sampling. For all preparation methods, the sample must be appropriately concentrated, neither too dilute (resulting in weak signals) nor too concentrated (causing peak saturation and distortion). The preparation method must also be compatible with the sample's chemical properties to prevent reactions or decomposition during analysis.

4.2.3.3.4 Test results: output formats and analytical interpretation of fourier-transform infrared spectra

FTIR testing results are presented as infrared spectra (typically transmittance or absorbance spectra) accompanied by data reports. By analyzing the wavenumber position, intensity, peak shape, and splitting patterns of absorption peaks in the spectrum, the functional group composition, molecular structure, and chemical environment of the sample can be inferred. Reliability must be ensured through background subtraction and spectral validation.

Analysis follows the principle of "functional group region first (4000–1300 cm^{-1}), fingerprint region second; strong peaks first, weak peaks later": initial identification of major functional groups through strong peaks in the functional group region (e.g., a strong peak at 1715 cm^{-1} combined with peaks at 2800–3000 cm^{-1} suggests a ketone), followed by structural confirmation using peaks in the fingerprint region (e.g., substitution positions of aromatic rings). Intensity enables semi-quantitative analysis, with concentration calculated according to the Lambert-Beer law ($A=\varepsilon bc$); peak shape reflects the molecular environment (e.g., free O-H appears as a sharp peak at 3600 cm^{-1}, while hydrogen bonding causes a redshift to 3200–3400 cm^{-1}).

Reliability verification requires: no significant H_2O or CO_2 interference peaks after background subtraction, with a flat baseline and no drift; peak position deviation ≤2 cm^{-1} and intensity deviation ≤5% for repeated tests of the same sample; comparison of unknown samples with standard spectral libraries (e.g., Sadtler, NIST), with a match ≥90%, followed by confirmation using NMR or mass spectrometry.

4.2.3.3.5 Case study

To further understand the application of infrared spectroscopy in practical testing, we refer to the following mineral identification study conducted using FTIR spectroscopy, as shown in Fig. 4.19.

The spectrum reveals a series of distinct absorption peaks, each corresponding to specific minerals or functional groups. A strong absorption peak observed at approximately 3618 cm^{-1} is identified as kaolinite, resulting from the vibrational modes of specific chemical bonds in the kaolinite structure matching the energy of infrared radiation at this wavenumber. The peak at 1612 cm^{-1} indicates the presence of montmorillonite, while the absorption at 2924 cm^{-1} suggests organic carbon components. Additional characteristic peaks correspond to quartz at 1874 cm^{-1}, calcite at 1797 cm^{-1}, albite at 1026 cm^{-1}, illite at 756 cm^{-1}, and hematite at 532 cm^{-1}. Through precise identification and analysis of these

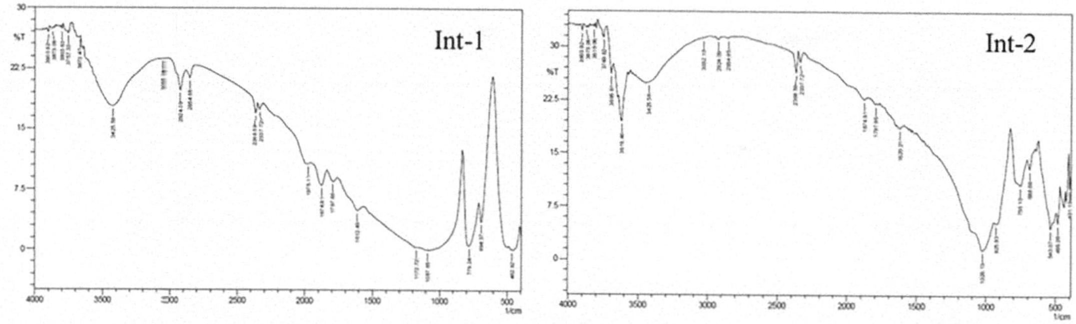

FIGURE 4.19 FTIR spectra of rock samples from the Pohon Batu hot spring area (Risakotta and Andayany, 2020).

absorption features, researchers can accurately determine the mineral composition of rock samples from the investigated region.

4.3 Characterization methods and applications of 2D membrane materials

4.3.1 Application of structural characterization methods in 2D membrane materials

In the microscopic structural characterization of amino acid cross-linked graphene oxide membranes (GOM-A), scanning electron microscopy (SEM) serves as a fundamental technique for observing their layered architecture. Figure 4.20 clearly demonstrates the value of SEM applications in this study.

Figure 4.20 comprises eight SEM images, presenting the surface and cross-sectional characteristics of an uncrosslinked graphene oxide membrane (GOM) and three amino acid-cross-linked membranes (GO-Ala, GO-Ser, GO-PHE). Images (A), (C), (E), and (G) display the surface morphology of the membranes, all showing the characteristic wrinkled topography typical of graphene-based membranes. Images (B), (D), (F), and (H) depict the cross-sectional views, which are critical for observing the layered structure.

The cross-sectional images reveal that all graphene-based membranes exhibit a distinct layered stacking structure, consistent with the membrane formation principle wherein graphene oxide (GO) nanosheets self-assemble via vacuum filtration. This validates the fundamental structural integrity of the membranes.

A detailed comparison shows that the uncrosslinked GOM in (B), while layered, displays relatively poor ordering and compactness. In contrast, the cross-linked membranes, GO-Ala (D), GO-Ser (F), and GO-PHE (H), show more regular and tightly stacked layered structures. This structural difference visually demonstrates the role of amino acid crosslinking: amino acid molecules act as "bridges," connecting GO nanosheets via interactions between their carboxyl/amino groups and the hydroxyl/carboxyl groups on GO, thereby reducing interlayer spacing and optimizing the layered architecture.

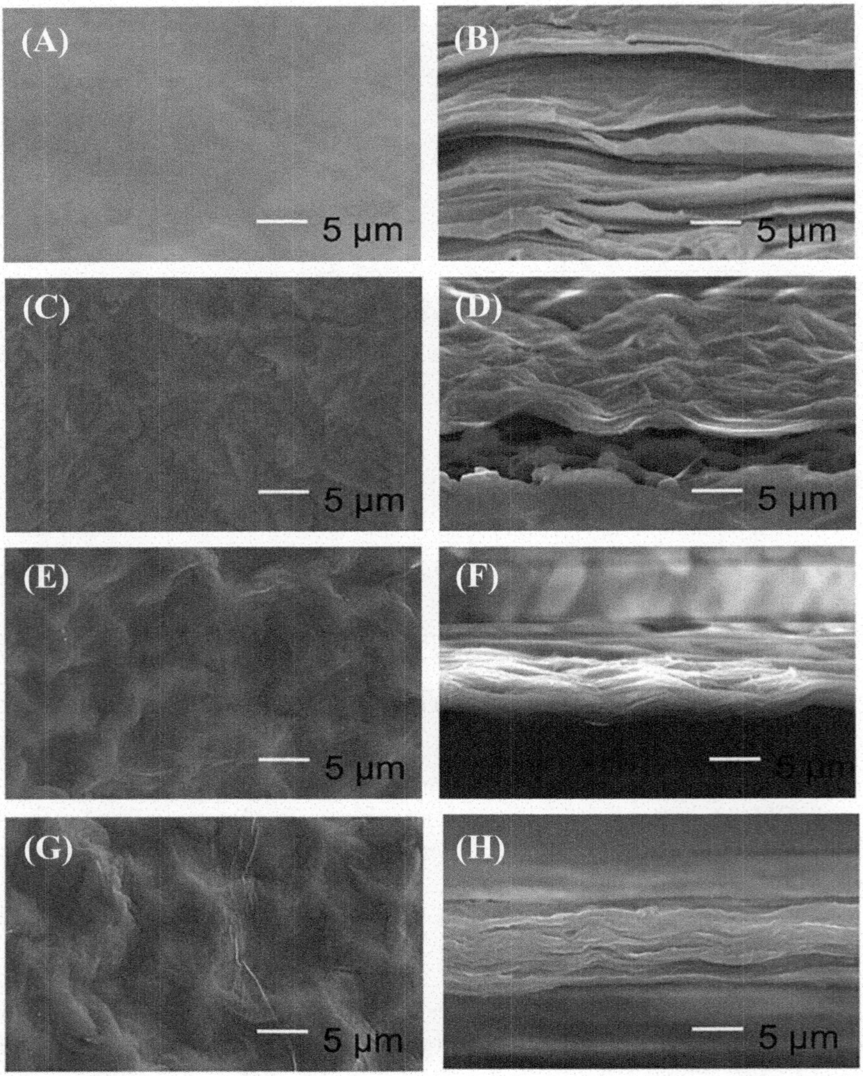

FIGURE 4.20 Application of SEM in observing the layered structure of graphene-based membranes (Qian et al., 2020).

Furthermore, the layered structure observed by SEM correlates with membrane performance. For instance, the more compact layered architecture provides a structural basis for the high ion rejection rates reported subsequently (e.g., exceeding 94% for alkali metal ions), while the wrinkled surface morphology contributes to enhanced hydrophilicity and water flux. In summary, Fig. 4.20 not only confirms SEM's capability to clearly capture the detailed layered structure of graphene-based membranes but also provides direct microscopic evidence for analyzing the impact of crosslinking modification on membrane structure and performance.

4.3.1.1 *Example: application of transmission electron microscopy for analyzing the nanopore structure of MXene membranes*

Figure 4.21 presents a bright-field TEM image, which serves as essential characterization data for analyzing the fundamental structure of the MXene membrane. This image provides critical structural reference for subsequent nanopore fabrication and performance investigations.

Figure 4.21(C) and (D) focus on Ti_2CT_x nanosheets: (C) displays Ti_2CT_x nanosheets drop-cast on a porous carbon film TEM grid, clearly revealing their 2D morphology and dispersion state. This allows assessment of potential agglomeration, where uniformly dispersed

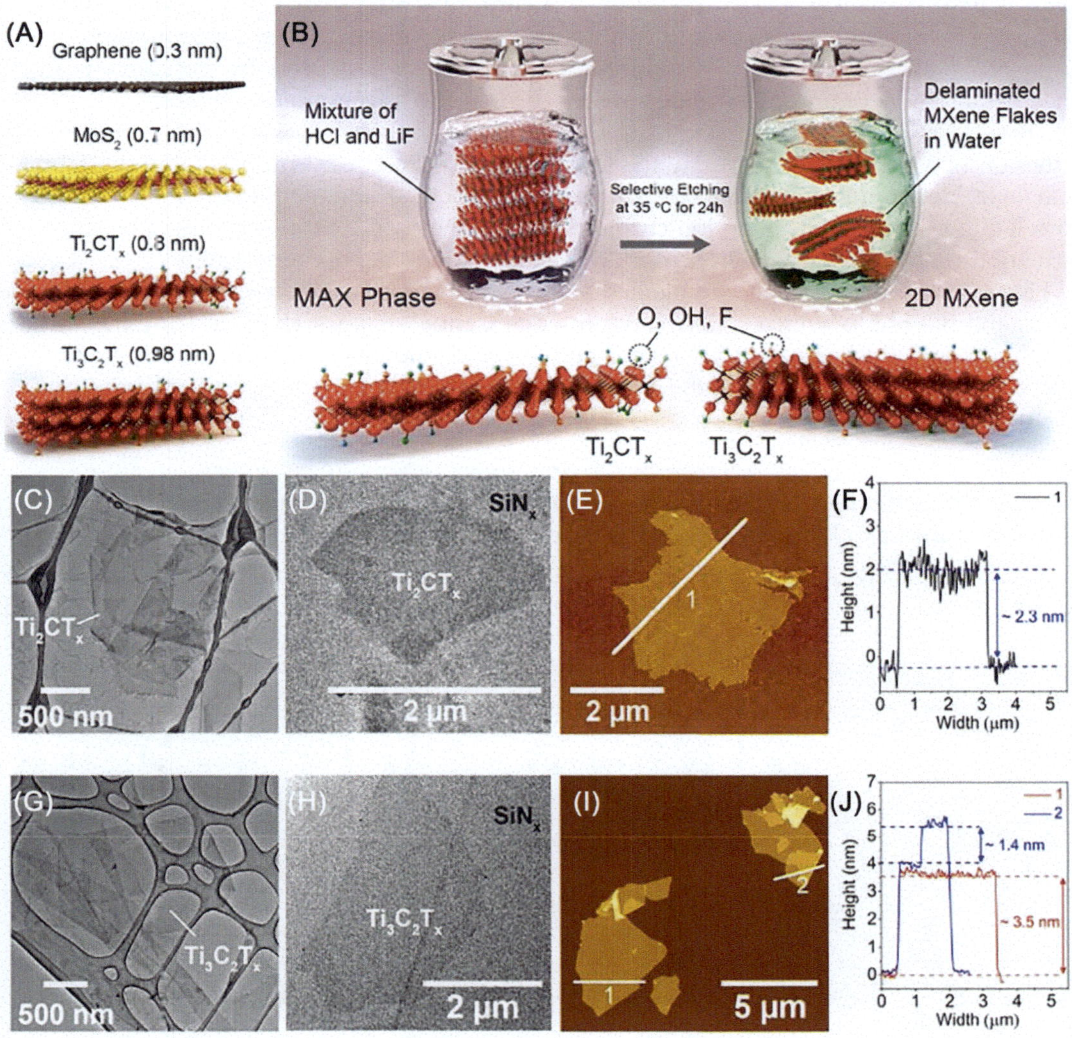

FIGURE 4.21 Example of transmission electron microscopy analysis for nanopore structure in MXene membranes (Mojtabavi et al., 2019).

nanosheets are essential for subsequent nanopore fabrication and ensuring structural homogeneity. (D) Presents Ti_2CT_x nanosheets drop-cast on a Si chip, directly illustrating their spreading behavior on a solid substrate and enabling evaluation of the nanosheet–substrate adhesion. This is critical for the subsequent transfer of MXene membranes to nanopore support substrates (e.g., SiN_x membranes), as it helps anticipate potential membrane damage during transfer due to poor adhesion, which could compromise nanopore fabrication.

Figure 4.21(G) and (H) focus on $Ti_3C_2T_x$ nanosheets: (G) shows exfoliated monolayer $Ti_3C_2T_x$ nanosheets drop-cast on a porous carbon film TEM grid, allowing clear identification of nanosheet integrity. The presence of edge defects or voids could directly impact the precision of subsequent nanopore fabrication (e.g., leading to irregular pore edges). (H) Depicts $Ti_3C_2T_x$ nanosheets drop-cast on a Si chip, revealing their lateral size distribution. Nanosheet dimensions influence the support stability of nanopores. Appropriately sized nanosheets prevent structural fragility around nanopores caused by excessively small sheets, thereby avoiding membrane rupture during ion current measurements.

These TEM images, by characterizing the morphology, dispersion, integrity, and substrate adhesion of MXene nanosheets, lay the foundation for subsequent precise nanopore fabrication (e.g., determining electron beam drilling parameters) and performance analysis (e.g., correlating pore structure with ion current signals). They serve as indispensable preliminary structural characterization references in MXene membrane nanopore research.

Fig. 4.22 comprises three complementary AFM images: (A) topography, (B) amplitude error, and (C) phase, which collectively provide essential data for characterizing the surface roughness and thickness of a 2D membrane material (monolayer graphene, 1 L graphene). These images fully demonstrate the core value of AFM in the microscopic characterization of 2D materials.

4.3.1.2 Example: application of atomic force microscopy for characterizing surface roughness and thickness of 2D membrane materials

The topography image in Figure 4.22(A–C) serves as the primary basis for surface roughness analysis, directly revealing distinct microscopic undulations and dispersed

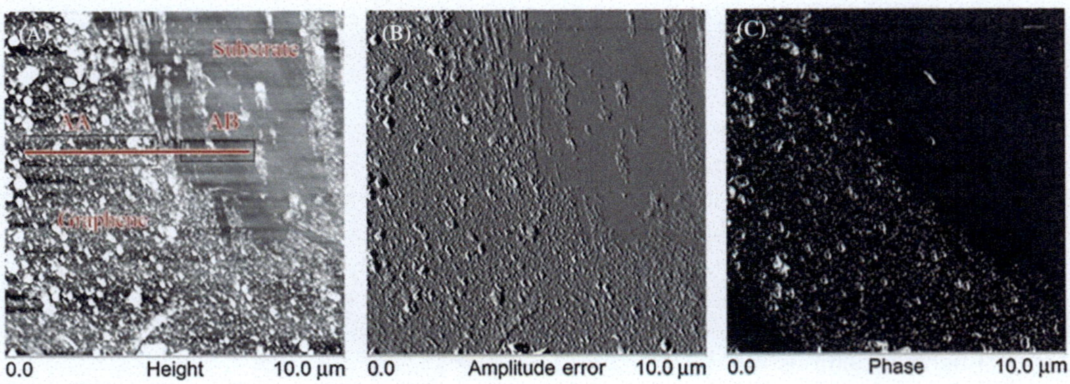

FIGURE 4.22 Example of atomic force microscopy analysis for surface roughness and thickness characterization of 2D membrane materials (Yao et al., 2017).

adsorbates on the monolayer graphene surface. These features originate from graphene's intrinsic tendency to form wrinkles through localized C-C bond elongation and to adsorb molecules for thermodynamic stabilization. The image visually captures the roughness distribution across the material surface, providing direct morphological evidence for correlating surface topography with material properties.

The amplitude error image (B) and phase image (C) play essential roles in thickness analysis. The amplitude error image, by detecting variations in the vibration amplitude of the AFM probe, clearly outlines the boundary between the graphene and substrate. This reduces boundary identification errors that may arise from surface contaminants when relying solely on topography. The phase image utilizes phase differences in tip-sample interactions to further distinguish the graphene region (AA area) from the substrate (AB area), defining precise target zones for thickness measurement. Together, these complementary images establish a reliable foundation for accurate thickness calculation using height profiles and histogram analysis, demonstrating the systematic and robust capability of multimode AFM imaging in characterizing 2D membrane materials.

4.3.2 Application of property characterization methods in 2D membrane materials

XRD serves as an essential and nondestructive technique for characterizing the crystal structure and phase composition of 2D membrane materials. The method is based on Bragg's law ($n\lambda=2d \sin\theta$), where a monochromatic X-ray beam incident on a material interacts with its periodically arranged atomic layers, acting as diffraction gratings. When the geometric conditions satisfy the Bragg equation, constructive interference produces distinct diffraction peaks. By analyzing the position, intensity, and shape of these peaks, critical structural information such as interlayer spacing, crystallinity, preferred orientation, and layer number can be determined.

XRD proves particularly valuable for the characterization of 2D membrane materials. The characteristic basal plane diffraction peak directly corresponds to the interlayer spacing of the material, where peak position shifts sensitively indicate the intercalation of ions or molecules between layers. Furthermore, the peak width at half maximum correlates closely with crystallite size, and crystallinity-peak broadening typically signifies smaller crystalline domains or the presence of lattice defects.

Using specialized grazing-incidence configurations, XRD can significantly enhance weak signals originating from thin film surfaces and interfaces, enabling highly sensitive analysis of ultrathin membranes. Consequently, XRD serves not only to identify phases and crystal structures in 2D materials like graphene and molybdenum disulfide, but also plays an indispensable role in evaluating their stacking mode, layer number, and structural evolution during applications such as energy storage, catalysis, and sensing.

The X-ray diffraction patterns clearly reveal the phase transition characteristics of MoS_2 materials (Fig. 4.23). The bulk MoS_2 with typical 2H-phase structure (black curve) exhibits a strong (002) diffraction peak at $2\theta \approx 14.4°$, corresponding to a tight interlayer spacing of 0.62 nm, characteristic of the semiconducting trigonal prismatic coordination.

After lithiation treatment, the sample (red curve) shows a significant phase transformation: the main diffraction peak shifts to $2\theta \approx 7.8°$, attributed to the (001) crystal plane, with an expanded interlayer spacing of 1.1 nm. This 77 % interlayer expansion directly confirms the

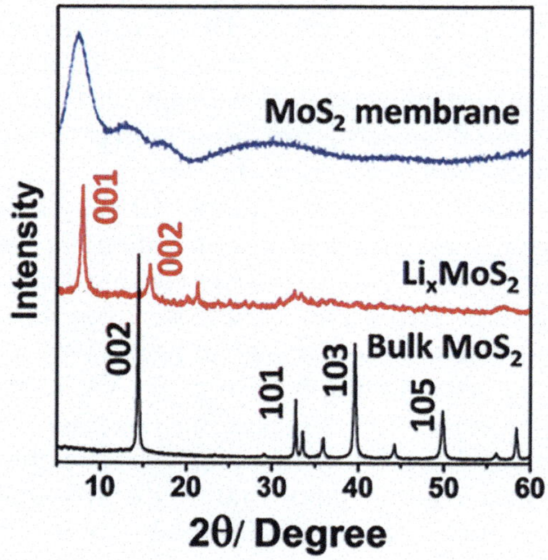

FIGURE 4.23 XRD pattern of MoS_2 (Achari et al., 2016).

lithium-ion intercalation-induced transition from 2 H to 1 T phase, forming a metallic octahedral coordination structure.

Notably, the MoS_2 membrane deposited on an AAO substrate (blue curve), while maintaining the characteristic peak position of the 1 T phase, shows reduced intensity and increased full width at half maximum, along with a weak residual peak at 14.4°.

This transformation in crystal structure critically determines gas separation performance. The expanded interlayer spacing (1.1 nm) of the 1 T phase creates more capacious transport channels, enabling highly efficient hydrogen permeation (with a permeability of 1175 Barrer) while significantly restricting CO_2 molecules (kinetic diameter: 3.30 Å), which exhibit a permeability of only 142 Barrer. This results in an H_2/CO_2 selectivity of 8.29.

As illustrated in Fig. 4.24, gas permeability demonstrates an inverse correlation with molecular size. The nanoconfined interlayer channels unique to the 1 T phase achieve molecular-scale precision sieving. This structure–function relationship, where performance is governed by architecture, confirms that precisely tuning the interlayer spacing of 2D materials through phase engineering can effectively optimize their molecular selectivity in separation processes.

4.3.2.1 *Example: Raman spectroscopy for analysis of chemical bond distribution in rGO membranes*

Raman spectroscopy serves as a powerful nondestructive analytical technique that probes the inelastic scattering of light resulting from interactions between laser photons and molecular vibrational modes in materials. This process yields unique spectral fingerprints highly sensitive to chemical bonding, crystal symmetry, electronic structure, and external perturbations, establishing Raman spectroscopy as an indispensable tool for characterizing 2D membrane materials. The fundamental principle lies in the direct correspondence

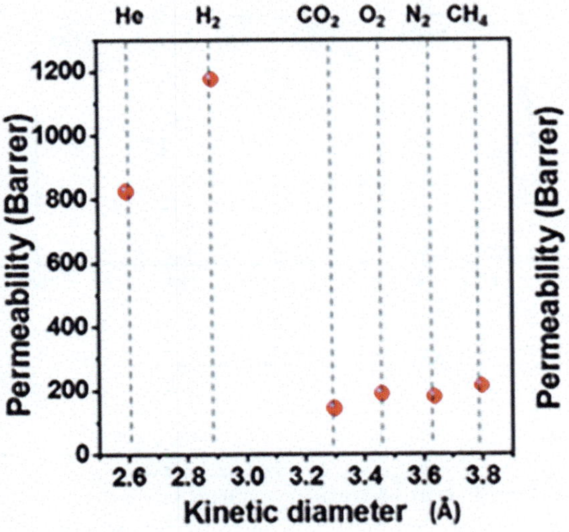

FIGURE 4.24 Permeability of different gases through MoS_2 membrane as a function of kinetic diameter (Achari et al., 2016).

between characteristic spectral features—peak positions, intensities, linewidths, and peak numbers—and specific lattice vibrational modes, where any microstructural alterations manifest as detectable spectral responses.

For 2D membrane analysis, Raman spectroscopy provides critical information regarding layer number, stacking configuration, defect density, doping levels, and strain state. The phonon confinement effect, which varies with layer thickness, induces systematic frequency shifts and line shape modifications in characteristic vibrational modes, enabling rapid and nondestructive layer counting. Additionally, defect- or disorder-related vibrational modes produce distinct scattering peaks whose intensities permit quantitative assessment of crystal quality. Furthermore, given the exceptional strain sensitivity of 2D materials, applied tensile or compressive stresses directly cause characteristic peak shifts, offering valuable insights into their mechanical properties.

Raman spectroscopy exhibits perfect complementarity with X-ray diffraction (XRD): While XRD reveals long-range periodic crystal structures, Raman spectroscopy sensitively captures local chemical bonding and vibrational information. Together, these techniques form the analytical foundation for comprehensively understanding the structure-property relationships in 2D membrane materials.

The molecular vibrational modes of GO, rGOv, and rGOL films were investigated using a 532 nm laser Raman spectroscopy system. The spectra were deconvoluted using a five-peak fitting model, as illustrated in Fig. 4.25. This model identifies five characteristic vibrational peaks: Peak I (~1120–1180 cm^{-1}, corresponding to C-O-C stretching and edge defects), the D peak (~1348–1349 cm^{-1}, attributed to sp (Zhu et al., 2014) carbon disorder or defect-induced vibrations), the D" peak (~1490–1521 cm^{-1}, reflecting amorphous sp (Cheng et al., 2021) bonds and cleavage of epoxy C-O-C groups), the G peak (~1582–1591 cm^{-1}, representing in-plane vibrations of the sp (Cheng et al., 2021) carbon network), and the D' peak (~1605–1616 cm^{-1}, associated with oxidized sp (Cheng et al., 2021) carbon or edge functional groups).

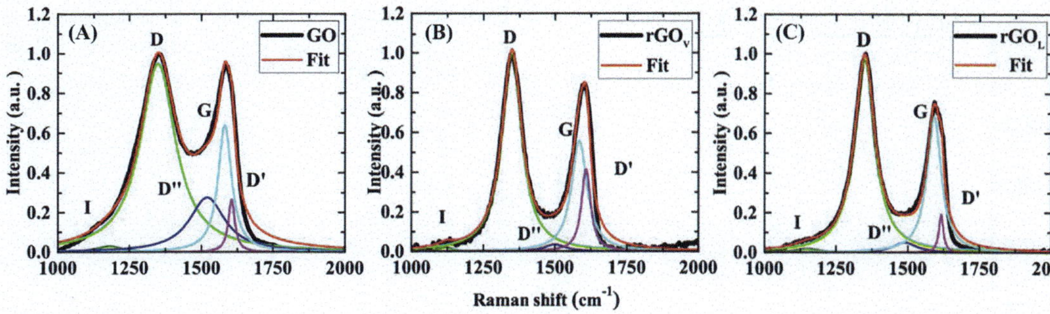

FIGURE 4.25 Raman spectra peak deconvolution of (A) GO, (B) rGOV, and (C) rGOL films spray-pyrolyzed at 150°C (Bargaoui et al., 2022).

This fitting approach accurately captures bond configuration changes during the reduction process. The shifts in the D'' and G peaks are particularly significant, revealing the extent of oxygen-containing functional group removal and the restoration of the carbon backbone. As a nondestructive characterization tool, Raman spectroscopy not only provides fingerprint information on chemical bonding but also complements XRD results, jointly elucidating the microstructural evolution of the thin films.

4.3.2.2 *Example: gas separation performance testing for evaluating pore selectivity of 2D membrane materials*

Gas separation performance testing serves as a critical step in evaluating the practical potential of 2D membrane materials, aiming to quantitatively measure their separation capability and permeation efficiency for components in gas mixtures. The testing is typically conducted in specially designed membrane modules or gas permeation cells, operating on the fundamental principle of utilizing a driving force across the membrane to enable different gas components to permeate at distinct rates, thereby achieving separation.

During testing, the prepared 2D membrane is mounted in a test chamber where a mixed gas with a specific composition is introduced on the upstream side. Under precisely controlled temperature and pressure conditions, the permeated gas is collected downstream and analyzed using instruments such as gas chromatography to determine its composition and flow rate. This process yields two key performance parameters: gas permeability and selectivity. Permeability represents the rate at which gas molecules traverse the membrane, reflecting the membrane's throughput capacity, while selectivity indicates the membrane's ability to discriminate between target gas molecules and others, directly determining separation precision.

For 2D membranes, their unique interlayer nanochannels and surface chemical properties enable preferential and rapid permeation of small gas molecules, resulting in exceptional molecular-sieving selectivity. Consequently, this testing not only provides an objective, quantitative standard for comparing different membrane materials but also serves as a crucial bridge connecting the material's microstructure with macroscopic separation applications. It represents a decisive step in predicting material performance in practical industrial separation environments.

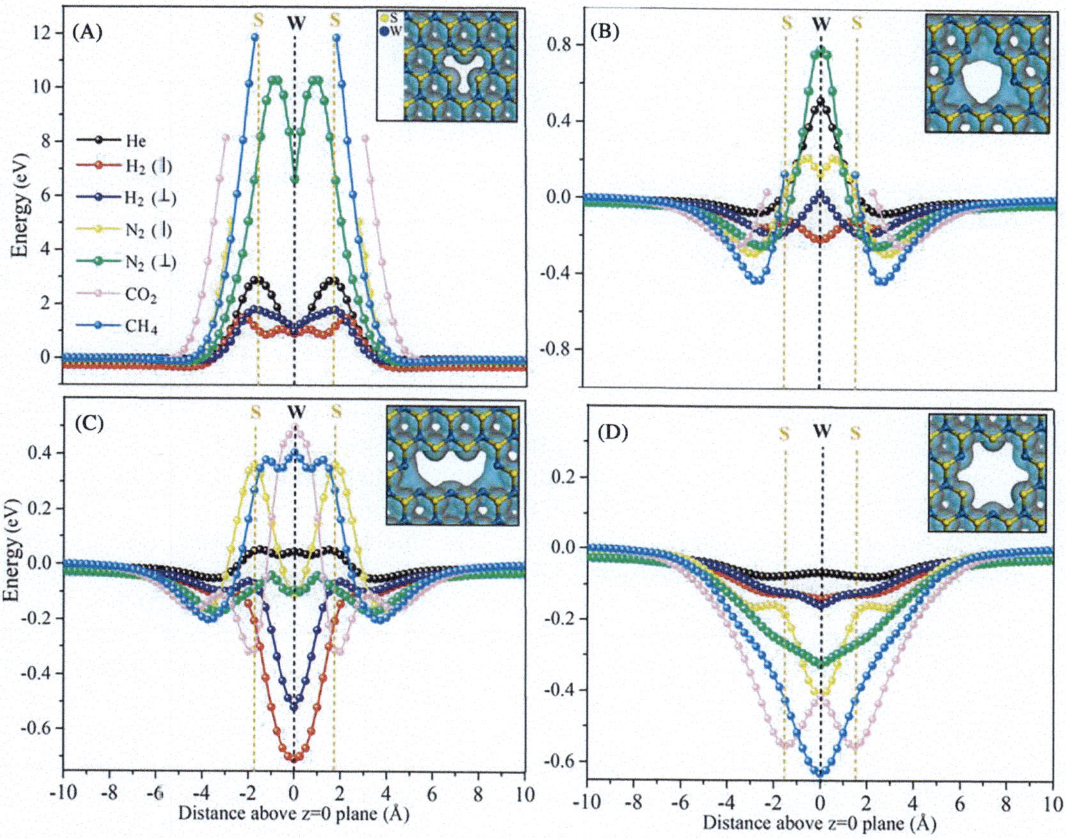

FIGURE 4.26 Interaction energy curves for gas molecules (He, H_2 [⊥,∥], N_2 [⊥,∥], CO_2, CH_4) across porous WS_2 monolayers: (A) 1 W, (B) 1W6S, (C) 2W6S, and (D) 3W6S. Electron density isosurfaces (0.2 e/Å (Zhu et al., 2014) are shown as insets (Hassani et al., 2021).

Combining first-principles calculations and molecular dynamics simulations, we systematically investigated the gas permeation performance of four vacancy-type pores (1 W, 1W6S, 2W6S, 3W6S) in WS_2 monolayer membranes (Fig. 4.26). The 2W6S pore (effective diameter: 2.10 Å) demonstrated exceptional H_2 selectivity, achieving H_2/CO_2 and H_2/CH_4 selectivities of 10 (Yisheng et al., 2023) and 10 (Ameh, 2019), respectively, at 300 K, surpassing traditional MoS_2 membranes (selectivity: 3.4–4.4) by nine orders of magnitude.

This outstanding performance arises from a synergistic mechanism of size sieving and chemical adsorption. When H_2 molecules pass through the pore in a parallel orientation (∥), they encounter a negligible energy barrier of 0.00 eV, whereas CO_2, with its larger kinetic diameter (3.30 Å), faces a significant barrier of 0.82 eV.

The energy barrier distribution curve clearly illustrates the selectivity mechanism: At the center of the 2W6S pore (z=0 Å), the energy barrier for H_2 disappears entirely, while CO_2 exhibits a prominent barrier peak (0.82 eV). This distinction stems from the overlap between the electron clouds of H_2 and the W atomic orbitals, which induces charge transfer from WS_2

to H_2 and enhances adsorption. In contrast, the linear structure of CO_2 leads to strong repulsion with the S atomic orbitals, creating a penetration barrier.

4.3.3 Application of other characterization techniques in 2D membrane materials

4.3.3.1 Example: molecular dynamics simulation for probing molecular transport behavior in 2D membrane materials

MD simulation is a key computational technique for investigating material structure and dynamic behavior at the atomic scale. It numerically solves Newton's equations of motion for each atom in a system under the influence of interatomic potentials, thereby simulating the evolutionary trajectory of a material under specific conditions. The core of this method lies in using empirical or first-principles force fields to describe interatomic interactions, enabling the calculation of a system's dynamic evolution over time. MD provides instantaneous, atomic-level details-such as precise atomic positions, movement pathways, and energy fluctuations-that are often difficult to capture through experimental means.

In the study of 2D membrane materials, MD simulations play an indispensable role. They can construct idealized models of single or multilayer membrane structures and simulate the entire process of adsorption, diffusion, and permeation of gas molecules, ions, or liquids within their nanochannels, interlayer spaces, or defect sites. By analyzing particle mean squared displacement, concentration profiles, and free energy barriers, MD simulations quantitatively reveal the microscopic mechanisms underlying ultrafast transport or highly selective sieving, for instance, explaining why certain small gas molecules permeate preferentially while others are effectively blocked. Furthermore, these simulations provide deep insights into the structural response, mechanical properties, and phase transition behavior of 2D membranes under external forces, electric fields, or thermal perturbations, offering fundamental theoretical explanations for experimentally observed phenomena.

Through all-atom molecular dynamics simulations, we systematically analyzed the transport process of water molecules through functionalized nanoporous graphene membranes (Fig. 4.27). By statistically analyzing the spatial distribution of water molecules near

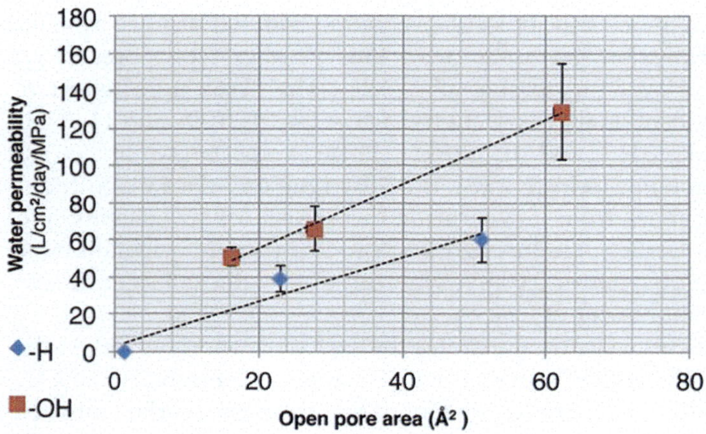

FIGURE 4.27 Computed water permeability of nanoporous graphene with different pore sizes and hydrogen/hydroxyl functionalization (Cohen-Tanugi and Grossman, 2012). (Water permeability shows an approximately linear relationship with the area of hydroxylated or hydrogenated pores.)

the pore channels and capturing their instantaneous permeation events, we clearly revealed the regulatory effect of chemical functionalization on molecular transport.

The simulation results demonstrate that the high-density region of water molecules formed around the hydroxylated (hydrophilic) pore edges significantly reduces the energy barrier for water permeation, thereby substantially enhancing water flux. This atomic-scale mechanistic understanding not only reasonably explains the permeability differences observed experimentally but, more importantly, establishes an intrinsic connection between "pore chemical structure," "molecular transport behavior," and "macroscopic separation performance." This provides a solid theoretical basis for directionally enhancing transport properties through precise control of pore-edge chemistry.

4.3.3.2 *Example: density functional theory for probing the electronic structure of 2D membrane materials and its impact on catalytic performance*

DFT stands as a fundamental computational method rooted in quantum mechanics, dedicated to unraveling the microscopic physical and chemical mechanisms of materials at the electronic level by solving for the ground-state properties of electron systems. Its core concept involves reformulating the complex interactions in many-electron systems into a problem of functionals of the electron density, enabling highly accurate predictions of key material properties, such as electronic structure, adsorption energy, band gap, and density of states, at a relatively low computational cost.

In the study of 2D membrane materials, DFT serves as a theoretical cornerstone. It enables the construction of atomically precise models of 2D materials based on first principles and provides deep insights into their intrinsic electronic properties, the adsorption behavior of gas molecules on material surfaces, and the diffusion barriers for ion migration between layers. For instance, by calculating the adsorption energies of different gas molecules on membrane surfaces, DFT can theoretically predict a material's selective adsorption capacity for specific gases. Through analysis of changes in band structure and electronic density of states, it elucidates how doping or defects modulate electrical conductivity and catalytic activity.

Moreover, DFT possesses strong predictive capabilities, allowing for the screening of potential 2D materials with desirable interlayer spacing, high selectivity, or specific catalytic sites prior to experimental synthesis. This offers precise theoretical guidance for the design and development of new materials.

In the study of catalytic mechanisms in 2D membrane materials, the key advantage of DFT calculations lies in their ability to visualize and quantify electron redistribution processes on material surfaces. As shown in Fig. 4.28, by comparing the band structure changes of intrinsic 2D MoS_2 membranes before and after NO_2 molecule adsorption, the emergence of new electronic states near the Fermi level can be clearly observed. These DFT results indicate that the adsorption process induces unoccupied states approximately 0.31 eV below the conduction band. These newly formed energy levels serve as catalytic active centers, significantly enhancing the material's electron exchange capability.

DFT further elucidates the physical origin of catalytic enhancement through density of states calculations and charge density difference ($\Delta\rho$) analysis. Charge density difference maps visually demonstrate the pathways and direction of charge transfer between gas

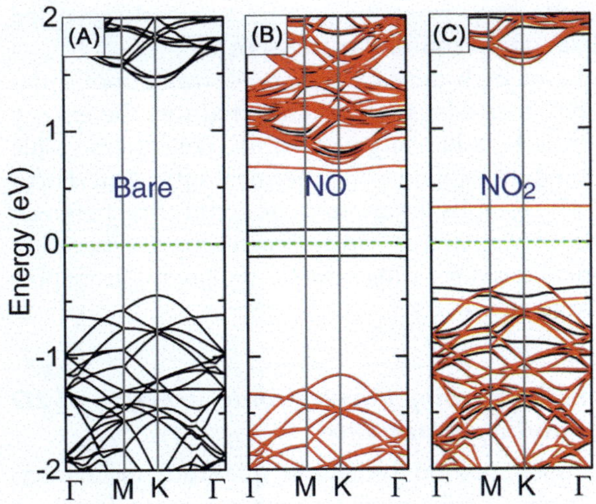

FIGURE 4.28 Band structures of (A) pristine, (B) NO-adsorbed, and (C) NO_2-adsorbed monolayer MoS_2 (Yue et al., 2013). *Black (red)* lines correspond to spin-down (spin-up) bands, with the *dashed line* indicating the Fermi level.

molecules and the membrane surface, while projected density of states calculations precisely identify the contribution of specific atomic orbitals to the catalytic process. These computational results provide electronic-level insights into the origin of catalytic activity, enabling researchers to move beyond traditional trial-and-error approaches and achieve rational design of 2D catalytic membrane materials.

4.4 Conclusion and perspectives

This chapter provides a systematic overview of the characterization methodology for 2D membrane materials and their applications. The field has evolved a multidimensional, interdisciplinary framework of techniques, broadly categorized into structural characterization, performance evaluation, and theoretical modeling with auxiliary analysis. Methods such as scanning electron microscopy (SEM), transmission electron microscopy (TEM), and atomic force microscopy (AFM) reveal surface and cross-sectional morphology from the micro- to nanoscale. X-ray diffraction (XRD) and Raman spectroscopy offer rapid, nondestructive means to diagnose crystal structure, layer number, and stacking configuration. Meanwhile, performance tests including gas permeation, water flux, and mechanical strength directly reflect the material's practical applicability.

Furthermore, theoretical approaches such as molecular dynamics (MD) and density functional theory (DFT) simulations have become indispensable tools for bridging atomic-scale structure with macroscopic properties, enabling prediction and validation of experimental phenomena.

In practice, no single technique can fully capture the complex characteristics of 2D membrane materials. These methods are highly complementary: for instance, TEM provides atomic-resolution images but only of localized regions and with demanding sample preparation; XRD offers statistically averaged structural information but lacks direct defect visualization; Raman spectroscopy is sensitive to layer number and strain but has limited spatial resolution. Therefore, integrating multiple techniques for cross-validation and comprehensive analysis is essential to draw reliable conclusions. Correlating AFM-measured thickness, XRD-derived interlayer spacing, SEM-observed macroscopic defects, and gas separation performance data allows a complete understanding of the material's structure–performance relationship.

These advanced characterization methods collectively drive the application of 2D membranes in areas such as gas separation (e.g., CO_2 capture, hydrogen purification), water treatment (high-flux desalination, nanofiltration), and energy conversion (ion-sieving membranes, fuel cell proton exchange membranes). They establish a solid scientific foundation for translating laboratory research into real-world applications, directly revealing the significant potential and unique advantages of these materials.

Nevertheless, current characterization techniques still face multiple challenges. A primary issue is the accurate resolution of complex real-world structures. Practical 2D membrane materials often contain defects, wrinkles, interfacial disorder, and nonuniform pore structures, placing high demands on spatial resolution, sensitivity, and adaptability of characterization tools under complex conditions. In addition, the need for multiscale analysis is increasingly critical: from atomic-scale transport mechanisms and nanoscale pores/defects to microscale grain boundaries and macroscopic membrane integrity, a methodology spanning multiple scales must be established.

Future developments will focus on advancing in situ and operando characterization, such as real-time structural observation under applied pressure, temperature, electric fields, or in liquid environments, as well as integrated techniques combining spectroscopy and microscopy, and artificial intelligence-assisted data analysis. These approaches will enable deeper understanding of the dynamic behavior and performance limits of 2D membrane materials, ultimately supporting the designed and precise fabrication of high-performance membranes.

AI disclosure

During the preparation of this work, the authors used Deepseek in order to polish the language. After using this tool/service, the authors reviewed and edited the content as needed and take full responsibility for the content of the publication.

References

Achari, A., Sahana, S., Eswaramoorthy, M., 2016. High performance MoS 2 membranes: effects of thermally driven phase transition on CO 2 separation efficiency. Energy & Environmental Science 9 (4), 1224–1228. Available at https://doi.org/10.1039/C5EE03856A.

Alsteens, D., 2024. Probing living cell dynamics and molecular interactions using atomic force microscopy. Biophysical Reviews 16 (6), 663–677. Available at https://doi.org/10.1007/s12551-024-01258-7.

Ameh, E.S., 2019. A review of basic crystallography and x-ray diffraction applications. The International Journal of Advanced Manufacturing Technology 105 (7-8), 3289–3302. Available at https://doi.org/10.1007/s00170-019-04508-1.

Bargaoui, I., Bitri, N., Ménard, J.-M., 2022. A comparative investigation of chemically reduced graphene oxide thin films deposited via spray pyrolysis. ACS Omega 7 (14), 11973–11979. Available at https://doi.org/10.1021/acsomega.2c00108.

Barhoumi, M., 2022. The Density Functional Theory and Beyond: Example and Applications. IntechOpen. Available at 10.5772/intechopen.100618.

Bian, Z., Miao, J., Zhao, Y., Chai, Y., 2022. Strong interlayer interaction for engineering two-dimensional materials. Accounts of Materials Research 3 (12), 1220–1231. Available at https://doi.org/10.1021/accountsmr.2c00146.

Cao, J., Zhao, F., Yang, X., Yan, L., Zhao, Q., Gao, L., et al., 2024. Material design and mechanism study for zinc ion batteries: applications of density functional theory calculations and molecular dynamic simulations. Chemical Engineering Journal 494, 153239. Available at https://doi.org/10.1016/j.cej.2024.153239.

Cheng, L., Liu, G., Zhao, J., Jin, W., 2021. Two-dimensional-material membranes: manipulating the transport pathway for molecular separation. Accounts of Materials Research 2 (2), 114–128. Available at https://doi.org/10.1021/accountsmr.0c00092.

Cohen-Tanugi, D., Grossman, J.C., 2012. Water desalination across nanoporous graphene. Nano Letters 12 (7), 3602–3608. Available at https://doi.org/10.1021/nl3012853.

Hassani, N., Ghorbani-Asl, M., Radha, B., Drndić, M., Krasheninnikov, A.V., Neek-Amal, M., 2021. Gas permeability and selectivity of a porous WS2monolayer. Journal of Physical Chemistry C 125 (45), 25055–25066. Available at https://doi.org/10.1021/acs.jpcc.1c06894, http://pubs.acs.org/journal/jpccck.

Jamiu, Z.A., Jagaba, A.H., Abdulazeez, I., Salhi, B., Lawal, D.U., Baig, N., et al., 2025. 2D transition metal dichalcogenide membranes for efficient oil-water separation. Colloids and Surfaces A: Physicochemical and Engineering Aspects 726. Available at https://doi.org/10.1016/j.colsurfa.2025.137933, https://www.sciencedirect.com/science/journal/09277757.

Kołodziej, A., Wesełucha-Birczyńska, A., Świętek, M., Skalniak, Ł., Błażewicz, M., 2020. A 2D-Raman correlation spectroscopy study of the interaction of the polymer nanocomposites with carbon nanotubes and human osteoblast-like cells interface. Journal of Molecular Structure 1212, 128135. Available at https://doi.org/10.1016/j.molstruc.2020.128135.

Li, Y., Yang, S., Zhang, K., Van der Bruggen, B., 2019. Thin film nanocomposite reverse osmosis membrane modified by two dimensional laminar MoS2 with improved desalination performance and fouling-resistant characteristics. Desalination 454, 48–58. Available at https://doi.org/10.1016/j.desal.2018.12.016.

Li, N., Hussain, N., Ding, Z., Qu, C., Li, Y., Chu, L., et al., 2024. Guidelines for Raman spectroscopy and imaging techniques in food safety analysis. Food Safety and Health 2 (2), 221–237. Available at https://doi.org/10.1002/fsh3.12040.

Lin, Q., Liu, Y., Zeng, G., Li, X., Wang, B., Cheng, X., et al., 2021. Bionics inspired modified two-dimensional MXene composite membrane for high-throughput dye separation. Journal of Environmental Chemical Engineering 9 (4), 105711. Available at https://doi.org/10.1016/j.jece.2021.105711.

Liu, Q., Zhang, X., Jiang, B., Li, J., Li, T., Shao, X., et al., 2021. Molecular dynamics simulation of ion adsorption and ligand exchange on an orthoclase surface. ACS Omega 6 (23), 14952–14962. Available at https://doi.org/10.1021/acsomega.1c00826.

Mojtabavi, M., VahidMohammadi, A., Liang, W., Beidaghi, M., Wanunu, M., 2019. Single-molecule sensing using nanopores in two-dimensional transition metal carbide (MXene) membranes. ACS Nano 13 (3), 3042–3053. Available at https://doi.org/10.1021/acsnano.8b08017.

Pennycook, S.J., 2009. Transmission Electron Microscopy: A Textbook for Materials Science, 932.

Qian, L., Wang, H., Yang, J., Chen, X., Chang, X., Nan, Y., et al., 2020. Amino acid cross-linked graphene oxide membranes for metal ions permeation, insertion and antibacterial properties. Membranes 10 (10), 296. Available at https://doi.org/10.3390/membranes10100296.

Risakotta, M.Y.S., Andayany, H., 2020. Mineral identification of rocks from Pohon Batu hot springs in West Seram using FTIR spectroscopy. Journal of Physics: Conference Series 1572 (1), 012045. Available at https://doi.org/10.1088/1742-6596/1572/1/012045.

White, R.T., Wu, A., Najm, M., Orfino, F.P., Dutta, M., Kjeang, E., 2017. 4D in situ visualization of electrode morphology changes during accelerated degradation in fuel cells by X-ray computed tomography. Journal of Power Sources 350, 94–102. Available at https://doi.org/10.1016/j.jpowsour.2017.03.058, https://www.journals.elsevier.com/journal-of-power-sources.

Yao, Y., Ren, L., Gao, S., Li, S., 2017. Histogram method for reliable thickness measurements of graphene films using atomic force microscopy (AFM). Journal of Materials Science & Technology 33 (8), 815–820. Available at https://doi.org/10.1016/j.jmst.2016.07.020.

Yisheng, M., Chengpeng, X., Xuelong, W., Zhongyao, L., Junsheng, W., 2023. Seeing through materials: X-Ray imaging using computed tomography. Journal of Beijing Institute of Technology (English Edition) 32 (4), 494–516. Available at https://doi.org/10.15918/j.jbit1004-0579.2023.069, http://journal.bit.edu.cn/jbit/en/article/doi/10.15918/j.jbit1004-0579.2023.069.

Yue, Q., Shao, Z., Chang, S., Li, J., 2013. Adsorption of gas molecules on monolayer MoS2 and effect of applied electric field Nanoscale Research Letters 8 (1). Available at https://doi.org/10.1186/1556-276x-8-425.

Zhu, F.Y., Wang, Q.Q., Zhang, X.S., Hu, W., Zhao, X., Zhang, H.X., 2014. 3D nanostructure reconstruction based on the SEM imaging principle, and applications. Nanotechnology 25 (18). Available at https://doi.org/10.1088/0957-4484/25/18/185705, http://iopscience.iop.org/0957-4484/25/18/185705/pdf/0957-4484_25_18_185705.pdf.

CHAPTER

5

Application of computational methodologies in 2D nanomaterials

İdil Bali[1,], Tuğba Baysal[1,*], Melih Doğancı[1,*], Muhammed Sarıkaya[1,*], Mütesir Temel[2,*], Hasan Topaçlı[1,*], Eyüp Yazkan[1,*], and Sadiye Velioğlu[1,3]*

[1]Institute of Nanotechnology, Institute of Nanotechnology, Gebze Technical University, Kocaeli, Gebze, Türkiye [2]Chemical Engineering Division, Chalmers University of Technology, Gothenburg, Chalmers, Sweden [3]Department of Chemical Engineering, Istanbul Technical University, Istanbul, Maslak, Türkiye

5.1 Introduction

The advent of graphene in 2004 (Novoselov et al., 2004) marked a turning point in materials science, opening the door to the exploration of two-dimensional (2D) materials, atomically thin, crystalline structures with extraordinary physical and chemical properties. While graphene initially captured widespread attention for its unique characteristics, its success quickly inspired the scientific community to investigate other 2D nanomaterials with application-specific functionalities. Among various fields, water treatment has emerged as a particularly promising area where 2D nanomaterials show significant potential. The discovery that graphene oxide (GO) nanosheets can create interlayer channels enabling ultrafast water molecule transport catalyzed extensive research into 2D nanomaterials for separation and purification technologies. This has led to the development of a broad range of candidates, including transition metal dichalcogenide (TMD) (Cao et al., 2020), MXene (Liu et al., 2024), borophene (Alomar et al., 2025), hexagonal boron nitride (h-BN) (Liu et al., 2020; Song et al., 2020; Vafa et al., 2023), 2D metal-organic framework (MOF) (Baysal et al., 2025), and covalent organic framework (COF) (Wei et al., 2020). The intrinsic characteristics of 2D nanomaterials, such as high surface area, tunable surface chemistry, adjustable interlayer spacing, and remarkable mechanical (strong in-plane stiffness with high flexibility

*These authors contributed equally to this work.

Two-Dimensional Nanomaterials for Water Purification
DOI: https://doi.org/10.1016/B978-0-443-44530-9.00007-9

out-of-plane) and thermal (excellent thermal conductivity) properties, make them ideal candidates for designing next-generation water purification materials in various water treatment applications.

Although experimental studies have advanced understanding of 2D nanomaterials, they often fall short in capturing molecular-level details, especially under nanoscale confinement or during dynamic water flow. Molecular simulations, performed at both quantum and atomistic scales, offer powerful tools to explore nanostructure behavior in ways that experiments cannot. The rapid progression of computational power over the past decade, coupled with increasingly sophisticated and realistic simulation algorithms, has brought simulations closer to replicating experimental conditions with high fidelity. These advancements have paved the way for the widespread application of molecular simulations in water treatment, where they now serve as indispensable tools. Specifically, simulations can evaluate interaction energies between water or solutes and 2D nanomaterials, predict separation efficiency, analyze fouling behavior, reveal transport pathways and water distribution patterns, and predict material characteristics like hydrophilicity or surface charge (Cheng et al., 2021). Despite their advantages, simulations alone are limited in the investigation of the vast number of materials, and the recent integration of machine learning (ML) has further elevated material design strategies. ML excels at identifying complex patterns from large datasets, enabling the rapid prediction of key material properties and accelerating the discovery of high-performance materials for water purification (Dansawad et al., 2023; Frontistis et al., 2023). Together, molecular simulation at quantum- and atomistic-scales and ML form a synergistic framework that drives the development of next-generation 2D nanomaterials for efficient and sustainable water treatment solutions.

Collectively, as the demand for clean water intensifies globally, the exploration of 2D nanomaterials continues to play a vital role in developing advanced, energy-efficient solutions for sustainable water treatment. Therefore, it is important to benefit from all these theoretical methods to gain a deeper understanding of 2D nanomaterial behavior, aid in interpreting experimental results, and apply this knowledge to forecast their effectiveness in water treatment processes. This chapter examines the fundamental principles of these computational approaches, emphasizes their relevance to 2D nanomaterials, and highlights their applications in specific water purification scenarios.

5.2 Quantum-scale simulations

5.2.1 Concepts and methods

Quantum-level computational studies provide an opportunity to investigate the interactions between 2D nanomaterials and various pollutants at the atomic scale. In particular, density functional theory (DFT) simulations allow for a more detailed evaluation of the physical and chemical events occurring on the surface during water treatment processes.

Quantum chemical/mechanical (QM) and other electronic structure approaches use the Born–Oppenheimer approximation to calculate the interaction mechanism between the electron and the atomic core (total energy of a system), which is covered by the time-independent Schrödinger equation (Eqs. 5.1 and 5.2) (Kohn and Sham, 1965).

$$\hat{H}|\Psi\rangle = E|\Psi\rangle \tag{5.1}$$

$$\left[-\frac{1}{2}\nabla^2 - \frac{Z}{r}\right]\psi(r) = E\psi(r) \tag{5.2}$$

Both the electron's kinetic and potential energy and the quantity enclosed in square brackets imply a charge-carrying nucleus **Z**. Atomic units are used to quantify electronic energy (***E***), and the wavefunction ψ(r) describes the probability amplitude of finding the electron at position r. The basic demonstration of those two equations, known as the many-body Hamiltonian $\hat{H}$, can be defined as in Eq. 5.3.

$$\hat{H} = \hat{H}_{eZ} + \hat{H}_{ee} + \hat{H}_{ZZ} \tag{5.3}$$

The contribution of electron–electron interaction (ee), nuclei–nuclei interaction (ZZ), and electron–nuclei interaction (eZ) is calculated via this expression. However, in the case of more than one electron and nuclei, the time-independent Schrödinger equation does not have an analytical solution due to electron–electron interaction. Therefore, the approximations that were developed must be utilized to perform an analytical solution.

5.2.1.1 *Exchange correlation functionals*

Wavefunction and DFT methods are the two categories of QM techniques used for approximations (Becke, 2014). Rather than using wavefunctions, DFT rewrites the problem in terms of electron density. The application of DFT to many-electron systems that make computations manageable is known as Kohn–Sham (KS) theory. The core theorem states that the total electron density $\rho(r)$ of a nondegenerate ground state system uniquely determines the groud state energy through a universal functional, expressed as follows:

$$E[\rho] = T_s[\rho] + \int V_{ext}(r)\rho(r)dr + J[\rho] + E_{xc}[\rho] \tag{5.4}$$

By picking a single number from it and defining it, like the definite integral, the system's energy can be expressed as the functional E of ρ, or E[(ρ)]. This provides a major benefit over wavefunction-based methods since the ρ(r) for a system of N electrons in DFT only depends on three spatial coordinates, as opposed to the corresponding wavefunction's dependency on 3 N spatial and N spin variables. As a result, depending on the function E[ρ], the DFT method can be even less expensive than Hartree–Fock (HF), which is the simplest wavefunction based method, while still incorporating a significant portion of the missing instantaneous electron correlation. Among all the terms in the energy expression, only the exchange-correlation energy E_{xc} is not known exactly. Therefore the accuracy of the calculated properties depends strongly on the approximations used for E_{xc}. As a result, this has led to the introduction of several approximate exchange-correlation functionals in the literature. The most commonly used ones are the local density approximation (LDA), semilocal approximations such as generalized gradient approximation (GGA), and meta-GGA (Mardirossian and Head-Gordon, 2017).

LDA is basically used for a homogeneous electron gas, and the electron density is constant throughout space. GGA remains widely used due to its favorable balance between accuracy and efficiency and is supported in virtually all electronic structure codes. Some common GGA approaches are PBE, OPBE, PW91, and BLYP. Meta-GGA generally improves the accuracy of GGA by incorporating the local kinetic energy density. This allows meta-GGAs to treat different types of chemical bonds (such as covalent, metallic, and weak bonds) more accurately.

5.2.1.2 *Hybrid functionals in density functional theory*

As DFT has advanced, exchange-correlation functionals have become more complex and accurate. A significant improvement among these is represented by hybrid functionals. A relatively simple method for enhancing the computation of numerous molecular characteristics, including bond lengths and angles, single point energy (SPE), and vibrational frequencies, which are frequently inadequately characterized by basic local density functionals, is hybridization with HF exchange (Becke, 1993). These functionals provide a useful trade-off between accuracy and computational cost by combining the advantages of traditional DFT approximations and HF theory (Janesko, 2021). Since LDA and GGA functionals frequently have self-interaction errors (SIE) and inadequate handling of nonlocal exchange effects (Perdew et al., 1996), to overcome this intrinsic electronic SIE and restore the piecewise-linear behavior of the exact exchange-correlation functional between integer occupations, hybrid functionals have become a standard technique for molecular calculations (Ivády et al., 2017). Thus, the crucial aspects of the system, such as bandgap prediction, charge transport, or electron localization, are enhanced significantly (Cramer, 2002). Common hybrid functionals are B3LYP, BHLYP, PBE0, and wB97X.

5.2.2 Applications in water treatment

5.2.2.1 *Applications of catalytic process in water treatment*

Photocatalysis has emerged as one of the most important methods in advanced oxidation processes (AOPs), followed by catalytic degradation, to treat contaminants in wastewater (Lee et al., 2023; Iyyappan et al., 2024). A more effective photocatalysis process for treating pollutants in wastewater requires a multidisciplinary approach that identifies and develops new materials, as well as understands the underlying mechanisms at the atomic level. To understand the catalytic water treatment ability of 2D nanomaterials via DFT, metal reduction, pharmaceutical degradation, and oil-water separation processes were identified to guide the experiments (Lin et al., 2022; Wang et al., 2025; Fu et al., 2020). DFT enables the performance investigation across various metal-free photocatalysts like C_3N_5 and g-C_3N_4 (Zhang et al., 2021), and heterojunctions like g-C_3N_4/BiOI S-scheme and Z-scheme Fe-g-C_3N_4/Bi_2WO_6 (Dou et al., 2023; Liu et al., 2022), TiO_2/α-Fe_2O_3/Fe-g-C_3N_4 (Zhao et al., 2024). For instance, for the piezo-catalytic reduction of Cr^{6+} and antibiotics degradation, W^{6+}-doped 1T-MoS_2 was examined, and the potential 19 intermediate compounds (see Fig. 5.1) were produced throughout the reaction process, which helps to understand the breakdown pathway of tetracycline (Chen et al., 2025). As these aspects are impossible to detect with real-time experiments, DFT revealed that doping of metal narrowed the bandgap, enhanced charge carrier separation, and supported the stabilization of intermediate states, as illustrated in Fig. 5.1, which are necessary for efficient redox reactions (Chen et al., 2025).

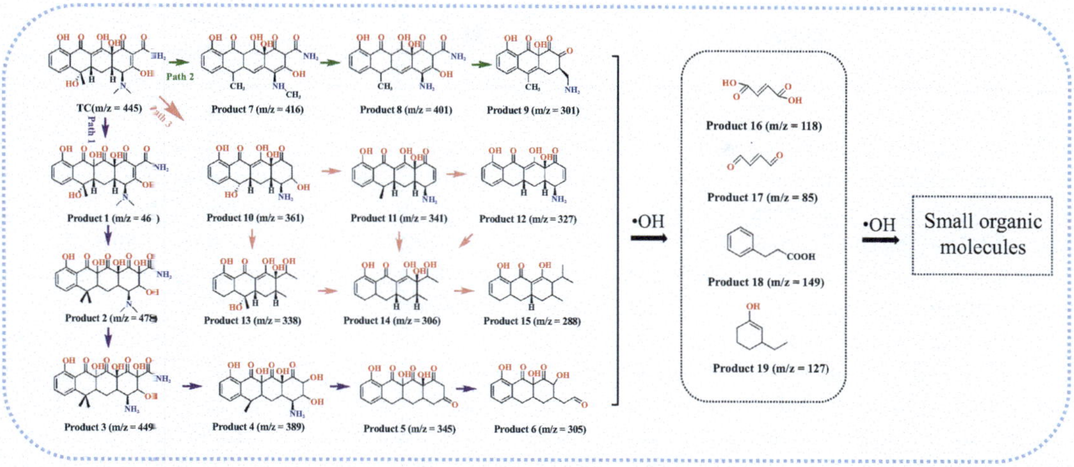

FIGURE 5.1 Degradation pathways of tetracycline by W^{6+}-doped 1T-MoS_2 (Chen et al., 2025). *Reprinted with permission from Chen X, Pan R, Xu H, Wu S, Liu C, Zheng Z, Wnag Y, Chen S, Meng F. High-yield preparation of 1T-MoS_2 with excellent piezo-catalytic activity via W^{6+} doping for antibiotics degradation and Cr^{6+} reduction: Mechanistic insights from characterizations to DFT calculations. Appl Catal B: Envir and Energy, 2025;361:124688. https://doi.org/10.1016/j.apcatb.2024.124688. Copyright 2025 Elsevier.*

To improve the catalytic activity of materials, the incorporation of transition metals such as Fe, Ni, and Zn into the metal centers within the nanomaterial and defects has attracted attention. To investigate the effect of this integration, DFT provides to be a valuable tool, either revealing intrinsic catalytic performance or the mutual contribution of catalytic and adsorption behaviour (Hassan et al., 2021). For instance, for MOF with metal center (2D π–d conjugated MOFs), the extended conjugation pathways and metal centers that facilitate electron transitions and pollutant oxidation in photo-Fenton-like processes were captured by DFT simulations (Xing et al., 2021). On the other hand, DFT revealed electronic structure differences and the electron transfer mechanism after the integration of sulfur/oxygen dual defects into MoS_2@TiO_2 composites for tetracycline degradation (Wu et al., 2022). Moreover, DFT was used to reveal charge transfer from peroxymonosulfate (PMS) to Co–TiO_x nanosheets via electron density difference maps, which help to find active contact positions for tetracycline and ofloxacin degradation (Meng and Ding, 2022). These various parameters cannot be detected or computed via experimental approaches at such small scales. To sum up, these studies and related explanations strongly support the idea that DFT is not only a simple and mere computational tool, but a cornerstone of the design and revealing features of photocatalysts. It fills the gaps between experimental performance and atomistic behavior by simulating reaction processes, charge transfer kinetics, adsorption energies, and band structures.

5.2.2.2 Applications in desalination

DFT simulations have also played a key role in understanding how 2D nanomaterials affect both salt ions and water behavior in desalination applications by identifying adsorption energies, charge transfer mechanisms, and electron distributions in detail. For instance, for Na^+ and Cl^- ions, the binding efficiency of a hydroxyl-functionalized $Mo_{1.33}C$ MXene surface was examined by Guerrero-Sanchez et al. (2022). They reported an adsorption energy of −0.99 eV for Na^+ (physisorption) and −2.04 eV for Cl^- (chemisorption) (Guerrero-Sanchez et al., 2022). Bader charge analysis showed that Na^+ donated about 0.87 electrons, while Cl^- gained around 0.62 electrons from the surface, as illustrated in Fig. 5.2. DFT-based charge transfer analysis visualizes that the Cl^- ion attracts electrons from the substrate, whereas Na^+ loses electrons to it, supporting selective interactions. These findings highlight that both electrostatic forces and charge transfer effects on nanomaterials are important in ion selectivity. Additionally, they clarify the importance of the balance between surface charge and nanochannel structure on desalination.

Water interaction with 2D nanomaterial surfaces is also crucial in gaining information about material performance for desalination. Abbaspour and Morsali (2024) investigated how water molecules behave when confined between flat or bent graphene oxide (GO) surfaces, by successfully using quantum-scale (DFT) and atomic scale (molecular dynamics [MD]) simulation approaches together. They revealed that bent GO surfaces slowed down water diffusion by about 30% compared to flat ones. The study also found that water molecules tended to align with epoxy groups at angles of 0° or 180°, which helped stabilize hydrogen-bond networks. Collectively, DFT simulations help to understand how molecular orientation, surface geometry, and chemical terminations can strongly affect water behavior in desalination membranes (Ai et al., 2022).

5.2.2.3 *Applications in dye adsorption and degradation*

DFT simulations provide valuable insights into surface interactions, charge transfer, and bond formation during the adsorption and degradation of dye molecules. In addition to pristine 2D nanomaterials (Saadh et al., 2024; Bangari et al., 2022), the

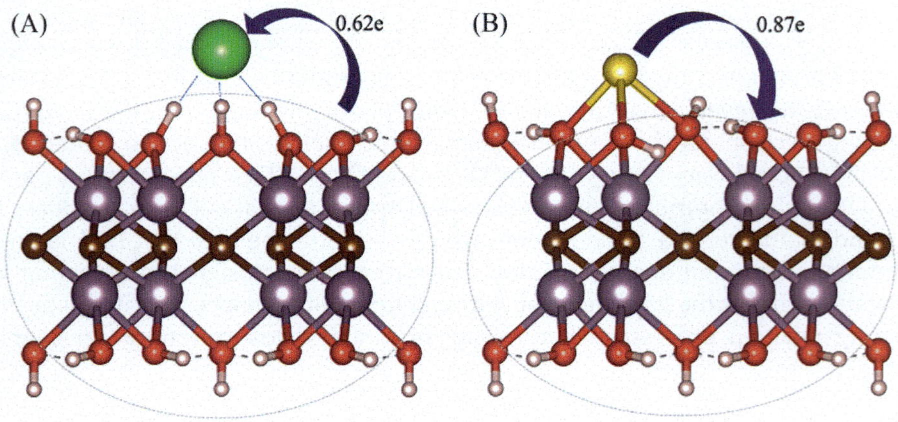

FIGURE 5.2 (A) Schematic of the charge transfer from the $Mo_{1.33}C$ MXene surface to the Cl^-, and (B) atomic model for the charge transfer from the Na^+ to the $Mo_{1.33}C$ MXene surface. Arrows denote the charge transfer direction (Guerrero-Sanchez et al., 2022). *Reprinted with permission from Guerrero-Sanchez J, Muñoz-Pizza DM, Moreno-Armenta MG, Takeuchi N. Atomic-scale understanding of the Na and Cl trapping on the $Mo_{1.33}C(OH)_2$-MXene, Sci Rep 2022; 12: 8340. https://doi.org/10.1038/s41598-022-12177-6. Copyright 2022 Nature.*

impact of several chemical modifications on materials, such as metal doping, functional groups, and heterojunctions, can be easily elucidated via DFT approaches. For example, Liu et al. (2024) reported that pristine Ti_3C_2 MXene shows a remarkably strong chemisorption tendency toward methylene blue (MB) with the adsorption energy of −13.02 eV. However, when the surface was terminated with oxygen or hydroxyl groups, the adsorption strength weakened, with energies of −1.47 and −2.82 eV, indicating that the interactions were mainly limited to hydrogen bonding and electrostatic forces. Similarly, Alomar et al. (2025) studied cobalt- and nickel-decorated borophene and found that metal decoration clearly increased the adsorption energies compared to the bare surface, reaching −2.99 and −2.27 eV, respectively.

More recent studies have expanded this understanding to a wider range of 2D nanomaterials and heterostructures. Wang et al. (2022) designed CdS/WS_2 heterojunctions and revealed strong orbital hybridization between Cd and W atoms, significantly reducing the bandgap. This structural tuning enhanced the photocatalytic degradation of dyes such as Rhodamine B and MB under visible light irradiation. Similarly, Yang et al. (2023) studied $BiOBr/Bi_2MoO_6$@MXene ternary heterojunction and revealed the promoted effective electron–hole separation and improved dye removal efficiencies exceeding 98%.

Identification of which types of interactions are governed during dye separation is crucial in defining the potential 2D nanomaterials. While boron nitride nanosheets (BNNs) revealed higher electronic conductivity due to a narrowed band gap (1.54 eV), even though the interactions were mainly governed by van der Waals (vdW) forces for dye adsorption (Bangari et al., 2022), metal doping to BNNs had led to a decrease in bandgap, creating new electronic states near the Fermi level (Hassan et al., 2021), which demonstrates high performance in degrading organic dyes. On the other hand, electrophilicity plays a crucial role in driving adsorption, as evidenced by the charge distributions of both the materials involved and the dyes used (Yadav and Dindorkar, 2022; AbdulAmeer et al., 2023). Moreover, DFT simulations display via the calculation of binding energy that these interactions also play a role in the stability of 2D nanomaterials in the dye adsorption or degradation processes (Junejo et al., 2022; Liu et al., 2022). Collectively, pristine or minimally functionalized 2D surfaces often show the highest physisorption energies, while doping, heterojunction construction, and ionic environment significantly affect electronic structure, charge transfer pathways, and stability, resulting in chemisorption.

5.2.2.4 *Applications in pharmaceutical removal*

Pharmaceutical pollutants, particularly antibiotics, have emerged as one of the major pollutants owing to their excessive use in the pharmaceutical industry and human medicine (Salahshoori et al., 2024). Mostly 2D nanostructured materials for the efficient removal of antibiotics have been used in various water treatment methods, including adsorption and membrane filtration (Chen et al., 2023). To gain an insight into the efficiency of 2D nanomaterials at the atomic level, quantum-based simulation approaches have been introduced to calculate adsorption energy and find active sites (Firmansyah et al., 2025). DFT helps measure the strength of interaction between antibiotics and adsorbent surfaces, displays the particular surface sites that are involved, and shows the electronic interactions, H-bonding, π–π stacking, and metal coordination (Mutahir et al., 2024). For instance, the interaction of tetracycline with nitrogen-doped BNNs by stacking π–π on the basal plane and generating charge-assisted hydrogen bonds with nitrogen-rich edge sites at the same time was revealed by DFT simulations (Bangari et al., 2022). In another case, DFT and MD were utilized to investigate 2D covalent organic frameworks (COFs) such as DAAQ-TFP, COF-78, and COF-OEt for the adsorption of drugs (Ghasemi et al., 2022). Ciprofloxacin had a stronger adsorption affinity to COF-78 because of the combined effects of π–π and hydrogen bonding (Ghasemi et al., 2022).

The significance of chemically functionalizing the surface of 2D nanomaterials is crucial, as it influences the strength of adsorption. Additionally, identifying an optimal configuration for pharmaceuticals on the surface of these 2D nanomaterials is vital for determining the most favorable drug–material complexes. DFT simulations, particularly those involving graphene or graphene oxide (GO) as 2D nanomaterials, have highlighted these aspects effectively. For instance, it was revealed that functionalizing rGO with hydroxyl (–OH), carboxyl (–COOH), and sulfoxide (–OS) groups enhances paclitaxel (PTX) adsorption (Huwaimel and Alqarni, 2025). The introduction of –OH and –COOH groups effectively increases the hydrophilicity of the rGO surface, which fosters stronger interactions with the hydrophilic regions of PTX. In addition, these functional groups facilitate hydrogen bonding and ionic interactions, contributing to greater stability of the drug-substrate complex and improving adsorption efficiency (Huwaimel and Alqarni, 2025). On the other hand, another study evaluated the adsorption efficiency of pristine graphene (GN) for favipiravir (FPV) using the DFT method, specifically analyzing FPV···GN complexes in both perpendicular and parallel configurations (Huwaimel and Alqarni, 2025). The results indicated that the parallel configuration was significantly more advantageous than the perpendicular arrangement, achieving an impressive maximum adsorption energy of $-15.95\ kcal\cdot mol^{-1}$ (Huwaimel and Alqarni, 2025). To sum up, quantum-scale simulations not only predict adsorption energies and pollutant–adsorbent conformations that are important for material design, but they also measure molecular interactions, including electrostatic effects, hydrogen bonding, and π–π stacking. With this level of accuracy, it is possible to quickly and cheaply design next-generation adsorbents that are particularly customized for certain types of antimicrobial pollutants.

5.2.2.5 *Applications in heavy metal removal*

Heavy metal pollution caused by common heavy metals known as lead (Pb), cadmium (Cd), mercury (Hg), chromium (Cr), arsenic (As), nickel (Ni), and zinc (Zn) originates from wastewater streams of various industrial processes. After discharge of heavy metals, they have the ability to bioaccumulate in living organisms easily, enter the food chain, and harm humans and animals by causing neurotoxicity, mutagenesis, and cancer (Qasem et al., 2021). In addition to chemical interaction-based separation approaches, physical interaction-based separation approaches such as membrane filtration, ion exchange, and adsorption are extensively exploited to separate heavy metal ions from wastewater via various 2D nanomaterials (Ihsanullah, 2020).

DFT has improved the ability to predict where and how heavy metal ions will bind to adsorbent surfaces. Obtaining this information experimentally is challenging, especially for nanomaterials with complex morphologies or functional groups. For instance, it was revealed that boron nanoparticles could be utilized in various ways to treat wastewater, including heavy metal ions, either adsorbing spontaneously with effective charge redistribution (Milon et al., 2021) or forming stable complexes and increasing electrical conductivity (Kaviani et al., 2022). Moreover, considering graphene and graphdiyne, a greater adsorption was observed for graphdiyne, particularly toward Ni^{2+}, due to its enhanced structural porosity and π-electron delocalization (Mashhadzadeh et al., 2016). For another type of 2D nanomaterial, MXene (Ca_2C and Cr_2C), DFT with a broad functional range (PBE, B3LYP, CAM-B3LYP, and so on) offered a chance to choose the best MXene for removing ions based on local reactivity characteristics (Ajaj et al., 2024). Through electronic density and bond length calculations, DFT reveals subtle performance changes that are difficult to measure using conventional spectroscopic or surface analysis methods (urRehman et al., 2023; Oyetade et al., 2017).

An essential aspect of effective heavy metal remediation is the selective capture of toxic ions from diverse water matrices with multiple contaminants. Several studies utilized DFT to understand why some 2D nanomaterials have a higher tendency to adsorb certain heavy metal ions more than others (Ajaj et al., 2024; urRehman et al., 2023; Abu Tarboush et al., 2018). For instance, it was reported that B30 nanosheets had the highest adsorption energy for As^{3+} over seven different metal ions (Kaviani et al., 2022); functionalized BN nanosheets revealed high Cu^{2+}/Hg^{2+} selectivity, supported by binding energy and coordination (Azamat and Khataee, 2016); and nitrogen-functionalized carbon nanotubes demonstrated a stronger attraction to Pb^{2+} than Zn^{2+} as a result of more stable inner-sphere complexes with nitrogen donor atoms (Oyetade et al., 2017).

DFT provides a great understanding of the electronic structure of 2D nanomaterials, which affects their efficiency in heavy metal adsorption. For instance, for MOFs and porphyrin-based adsorbents, it was investigated how specific functional groups ($-NH_2$ vs. $-OH$) alter frontier orbital positions (Kyung and Choi, 2023). Even though adsorption data obtained from experimental tests may suggest a similar trend, DFT validated and quantified the molecular orbital-level cause (Kyung and Choi, 2023). Moreover, for MoS_2 monolayers, DFT analyzed band edge shifts and adsorption-induced electronic perturbations for several heavy metal ions (Zhang et al., 2021). The results ranked binding affinities based on energy alignment and orbital hybridization, providing insights that would not emerge from experimental adsorption data alone (Zhang et al., 2021).

5.2.2.6 *Other applications in water treatment*

DFT has become an essential computational approach for predicting and elucidating the interactions of emerging pollutants with 2D nanomaterials. While conventional studies have mostly focused on dyes, heavy metals, or pharmaceutical compounds, recent DFT research has expanded to encompass a much wider spectrum of contaminants, including volatile disinfection by-products, endocrine-disrupting chemicals, persistent organics, and radioactive ions. More specific examples are given as follows. Gomaa et al. (2024) systematically investigated the adsorption of trihalomethanes (THMs) on $Ti_3C_2T_x$ MXene with different surface terminations ($-O_2$, $-OH_2$, $-Cl_2$, and $-F_2$) and showed that $-Cl_2$ significantly enhanced the binding affinity, with adsorption energies reaching −1.12 eV, highlighting the critical role of surface functionalization. Martínez-Jódar et al. (2023) focused on the catalytic reduction of nitroarenes via $MoSe_2$ nanosheets and revealed that 3-nitroaniline in the vertical configuration achieved the most stable adsorption energy, confirming that isomer selectivity arose from the specific adsorption geometry at active sites. Wang et al. (2016) presented one of the first combined experimental and DFT studies on uranyl ion capture using V_2CT_x MXene and showed that uranyl ions were predominantly adsorbed via bidentate inner-sphere complexation with $-OH_2$ groups on MXene. Kong et al. (2024) comprehensively investigated the role of structural modification in MOF-based catalysts for the removal of ozone-resistant organic pollutants. Wei et al. (2020) elucidated the adsorption mechanism of bisphenol-A (BPA) on COF decorated with Fe_3O_4 and revealed that adsorption of BPA was governed by the mutual effects of possible active sites on the COF surface, the optimized BPA adsorption configuration where strong hydrogen bonding occurs at O1 and O2 atoms, and the differential charge density induced by BPA interaction, confirming charge redistribution and noncovalent interaction regions, as each of them illustrated in Fig. 5.3. Despite all this progress, it is unfortunate that the number of comprehensive studies addressing a broader range of pollutant types remains limited. Additional and diverse classes of environmental components and pollutants should be systematically investigated using DFT to provide valuable guidance for the rational design of next-generation adsorbents and photocatalysts with enhanced water treatment efficiency.

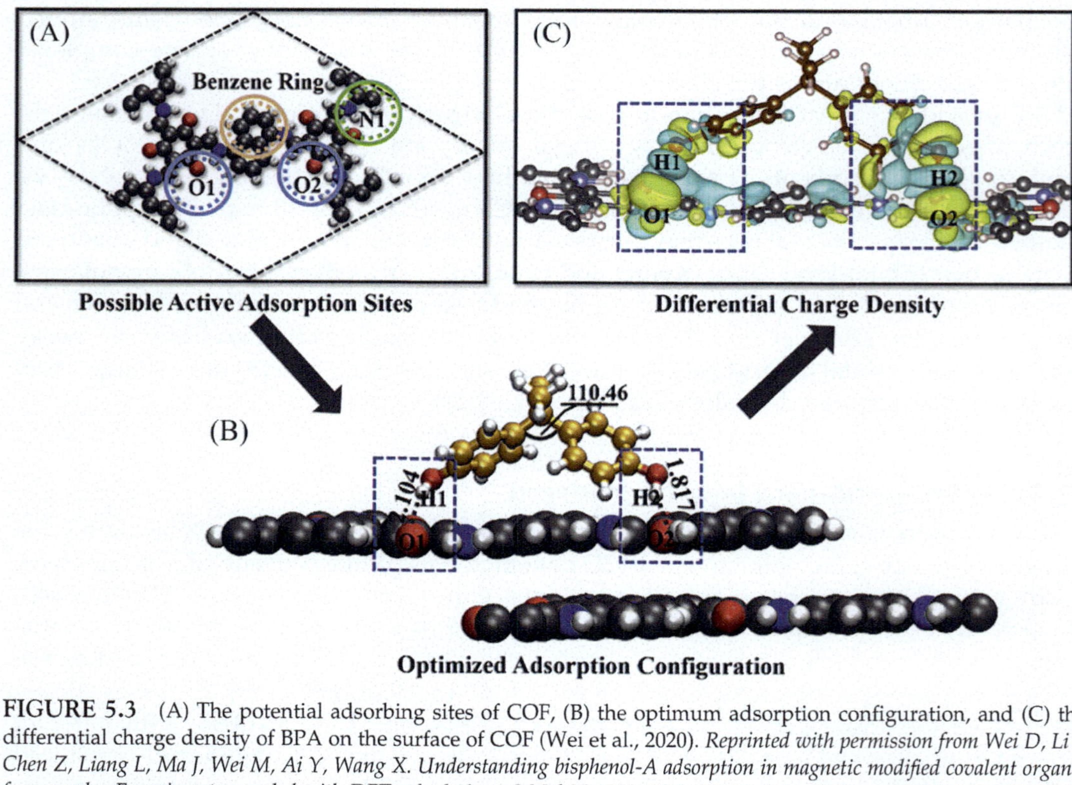

FIGURE 5.3 (A) The potential adsorbing sites of COF, (B) the optimum adsorption configuration, and (C) the differential charge density of BPA on the surface of COF (Wei et al., 2020). *Reprinted with permission from Wei D, Li J, Chen Z, Liang L, Ma J, Wei M, Ai Y, Wang X. Understanding bisphenol-A adsorption in magnetic modified covalent organic frameworks: Experiments coupled with DFT calculations. J Mol Liq 2020;301:112431. https://doi.org/10.1016/j.molliq.2019.112431. Copyright 2020 Elsevier.*

5.3 Atomic scale simulations

5.3.1 Concepts and methods

5.3.1.1 *Concepts of molecular dynamics*

Molecular dynamics (MD) simulations are vital tools for predicting the physical properties of systems by analyzing the motion and behavior of atoms within a system, thereby solving Newton's equations of motion (Frenkel and Smit, 2023). As MD methods investigate system behavior under steady-state conditions, they are also classified as equilibrium molecular dynamics (EMD) methods. Using molecular mechanics force fields at every timestep, EMD splits time into discrete stages and calculates forces on each particle. Thus, it updates the positions and velocities of all particles. The characteristics, as well as the duration and timelines that need to be explored, determine the nature and complexity of EMD simulation. Therefore the potential energy, namely the forces influencing the particles, must be understood to resolve the equation. These forces, or in other words, interactions among particles, are generally represented using two types of empirical force fields: reactive and nonreactive force fields (Frenkel and Smit, 2023). The fundamental distinction between the two lies in the nature of nonreactive force fields, where atomic or molecular bonds may be twisted or stretched but cannot be broken or formed (i.e., no chemical reactions occur), whereas reactive force fields permit the breaking and formation of bonds, thereby enabling the modeling of chemical reactions (Van Duin et al., 2001).

The nonreactive force fields comprise diverse combinations of bonded and nonbonded interactions. The bonded component generally encompasses bonds, angles, and dihedral strains depicted by harmonic potentials, whereas the nonbonded interactions comprise dispersion, illustrated by vdW potentials and Coulombic interactions (Frenkel and Smit, 2023). Combining all the components in the force fields, the total interaction energy owing to the nonreactive forces can be expressed as follows (Frenkel and Smit, 2023):

$$E_{Nonreactive} = E_{vdW} + E_{Coulomb} + E_{Bond} + E_{Angle} + E_{Torsion} \tag{5.5}$$

All interactions in nonreactive force fields are defined by several forces contingent upon the specific atoms/molecules or material types employed in the simulation, which are documented in well-known force fields such as UFF, Dreiding, OPLS, AMBER, CHARMM, etc. (Frenkel and Smit, 2023). On the other hand, in reactive force fields, several approaches have been developed to simulate instantaneous bond cleavage, including various reactive force fields such as ReaxFF, AIREBO, Stillinger–Weber (SW), and force field–based methodologies such as adiabatic reactive molecular dynamics (ARMD) and multisurface adiabatic reactive molecular dynamics (MSARMD) (Van Duin et al., 2001).

5.3.1.2 *Concepts of nonequilibrium molecular dynamics*

Nonequilibrium molecular dynamics (NEMD) is a simulation method in which an external gradient or driving force (such as pressure, temperature, density, or chemical potential) is intentionally applied to create a nonequilibrium state within the system. The system's response, in terms of mass, momentum, and energy fluxes, is then monitored (Bedeaux et al., 2015). By modifying classical molecular dynamics to introduce these perturbations, NEMD enables the investigation of time-dependent macroscopic behaviors at the microscopic scale using statistical averages on many nonequilibrium trajectories starting from well-defined initial conditions (Ciccotti and Ferrario, 2016). Therefore, NEMD has gained prominence as a powerful tool for studying transport behavior in confined environments such as nanopores and nanochannels of advanced 2D nanomaterials, including applications in desalination (Banan Baghbani et al., 2020; He et al., 2023), heavy metal removal (Binazir et al., 2025; Dehhaghi et al., 2020), and solvent separation (Alizadeh et al., 2024). To address the varying demands of different filtration scenarios, several NEMD variants have been developed. Among these (Fig. 5.4), the most commonly employed approaches in water treatment applications are movable-wall NEMD (MW-NEMD), boundary-driven NEMD (BD-NEMD), and concentration-driven NEMD (CD-NEMD) (Güvensoy-Morkoyun et al., 2024; Ang et al., 2019; Ozcan et al., 2017; Karmakar et al., 2023).

The MW-NEMD is a widely used and computationally efficient technique. By applying different levels of force to these movable walls, a pressure difference is generated across the membrane. A vacuum region is also introduced on the permeate side to prevent interaction with the periodic image of the feed side (Fig. 5.4A; Baysal et al., 2025). However, because of the limited system size, the total simulation time is often short. This makes it difficult to gather statistically meaningful data for fast-permeating solvents, where transport events occur quickly and require longer observation periods for accurate analysis (Baysal et al., 2025).

In BD-NEMD, an external force was applied to water molecules within a predefined slab in the feed region (Fig. 5.4B) (Güvensoy-Morkoyun et al., 2024; Ang et al., 2019). Due to

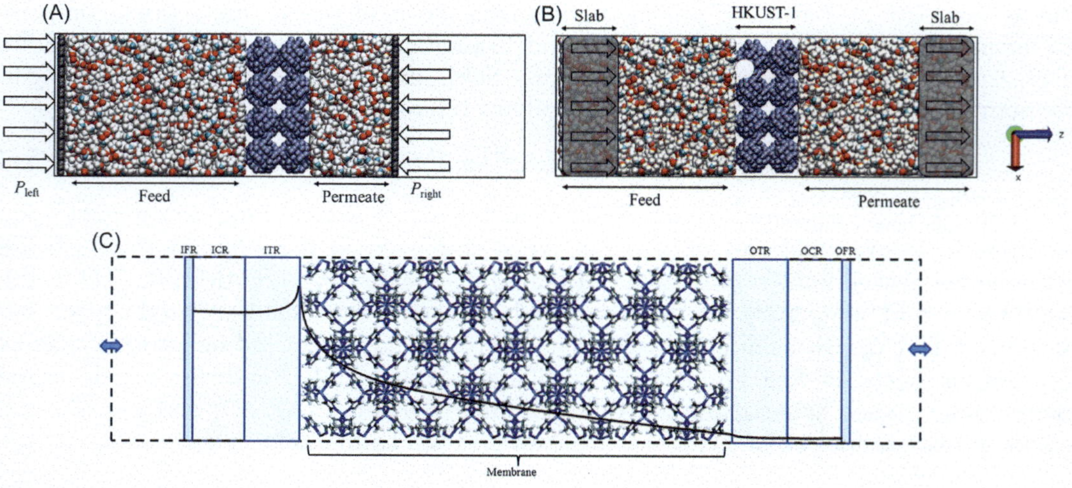

FIGURE 5.4 A representative HKUST-1 (MOF)-methanol system for (A) MW-NEMD and (B) BD-NEMD simulations (Baysal et al., 2025). (C) Conceptual representation of the CD-NEMD simulation (Ozcan and Perego, 2017). *(A) and (B) Reprinted with permission from Baysal T, Dahanayaka M, Velioğlu S, Chew J W. Organic solvent permeation through HKUST-1 MOF membranes: Molecular-scale insights. J Membr Sci 2025;733:124328. https://doi.org/10.1016/j.memsci.2025.124328. Copyright 2025 Elsevier.) (C) Reprinted with permission from Ozcan A, Perego C, Salvalaglio M, Parrinello M, Yazaydin O. Concentration gradient driven molecular dynamics: A new method for simulations of membrane permeation and separation. Chem Sci 2017;8(5):3858–3865. https://doi.org/10.1039/C6SC04978H. Copyright 2017 Royal Society of Chemistry.*

periodic boundary conditions, permeated molecules reenter the feed region, allowing steady-state flux calculations over extended simulation times. However, it requires more careful parameter optimization, particularly regarding slab location, slab thickness, and applied force magnitude, to ensure accuracy and reproducibility (Güvensoy-Morkoyun et al., 2024).

Unlike MW-NEMD and BD-NEMD methods, CD-NEMD is specifically designed for simulating separation processes involving solvent–solute mixtures. It enables accurate modeling by maintaining fixed solute concentrations at both the feed and permeate sides of the membrane, thereby capturing the effects of osmotic pressure more realistically (Fig. 5.4C). This hybrid technique combines Monte Carlo (MC) and EMD approaches. However, it becomes inefficient for dense fluids due to the extremely low acceptance rates for particle insertion and deletion. Additionally, reliable implementation requires careful optimization of the ratio between MC and MD steps, which complicates the simulation protocol (Ozcan et al., 2017; Karmakar et al., 2023).

5.3.1.3 *Concepts of Monte Carlo*

The Monte Carlo (MC) simulation method is a statistical analysis technique based on random sampling, widely used to model uncertainty in complex systems (Metropolis and Ulam, 1949). The fundamental principle of this approach involves generating random values for certain model parameters and repeatedly simulating the system to obtain statistically meaningful results. Unlike EMD simulations, which provide time-dependent trajectories, MC simulations are time-independent but offer significant advantages in sampling efficiency, particularly at low temperatures and high densities (Frenkel and Smit, 2023). Among various MC approaches, the Grand Canonical Monte Carlo (GCMC) method is widely used in adsorption and phase equilibrium studies. GCMC allows fluctuations in the number of particles while maintaining constant temperature, volume, and chemical potential, making it well-suited for modeling open systems.

5.3.1.4 *Water models*

Simulation studies focusing on water molecules have demonstrated that water exhibits different behaviors depending on temperature, pressure, and interaction type, often leading to deviations from experimental values. For this reason, various water models have been developed to ensure more accurate representations under diverse scenarios and interaction types. Although not the first, one of the most widely used water models is the simple point charge (SPC) model developed by Berendsen et al. (1981). Subsequently, the SPC/E (Extended SPC) model, an improved version with energy correction, was proposed by the same group (Berendsen et al., 1987). These models are categorized as three-point water models. Another example of a three-point water model is the TIP3P (Transferable Intermolecular Potential 3-Point) model developed by Jorgensen et al. (1983). Unlike SPC and SPC/E, TIP3P assigns the partial charge solely to the oxygen atom (Jorgensen et al., 1983). To date, more than 30 different water models have been developed, focusing on their accurate representation of various experimental properties such as density, coordination number, surface tension, dielectric constant, and self-diffusion coefficient. It was revealed that there is no universally "best" water model; rather, the choice of model should depend on

TABLE 5.1 The atom's charge, bond, and angle values of different water models (Kadaoluwa Pathirannahalage et al., 2021).

Water models	O-σ (Å)	O-ε (kcal/mol)	O (charge)	H (charge)	M (charge)	L (charge)	H-O-H angle (°)
SPC	3.166	0.155	−0.82	0.41	–	–	109.47
SPC/E	3.166	0.155	−0.84	0.42	–	–	109.47
CVFF	3.165	0.155	−0.82	0.41	–	–	104.50
TIP3P	3.150	0.152	−0.83	0.17	–	–	104.52
TIP3P-Ew	3.188	0.102	−0.83	0.415	–	–	104.52
TIP4P	3.153	0.155	0	0.520	−1.040	–	104.52
OPC	3.166	0.128	0	0.679	−1.358	–	103.60
TIP4P-2005	3.158	0.852	0	0.556	−1.112	–	104.52
TIP5P	3.120	0.160	0	0.241	–	−0.241	104.52
TIP5P-2018	3.145	0.188	−0.64	0.394	–	−0.073	104.52

the specific aim of the simulation (Kadaoluwa Pathirannahalage et al., 2021). Some of these water models, along with their specific parameters, are presented in Table 5.1. Depending on the specific research objective, as well as the temperature and pressure conditions of interest, different water models can be selected.

5.3.2 Applications in water treatment

5.3.2.1 Applications in desalination

A wide variety of 2D nanomaterials have been investigated for desalination applications via atomistic-scale simulation approaches. Among the most studied are graphene and its derivatives (GO, graphdiyne, graphenylene) (Jahangirzadeh et al., 2022; Xue et al., 2013; Banan Baghbani et al., 2020; Cohen-Tanugi and Grossman, 2012; Cohen-Tanugi et al., 2016; Dahanayaka et al., 2020; Chen et al., 2017; Gao et al., 2022), MoS_2 (Abal et al., 2021; Oviroh et al., 2021; Song et al., 2020), MXene (Meidani et al., 2021; Ma et al., 2024; Si et al., 2023; Ding et al., 2020; Osti et al., 2016), h-BN (Liu et al., 2020; Song et al., 2020; Vafa-1 et al., 2023), C_3N (Liu et al., 2021; Liu et al., 2018), COF (Lv and Zhao, 2024; Liu et al., 2022), and MOF (Cao et al., 2019; Yuan et al., 2023). These materials have been evaluated under different structural and environmental conditions to achieve high desalination performance and to understand the effects of various influencing parameters, such as nanochannel size (Vafa-1 et al., 2023; Vafa-2 et al., 2023), pore geometry (Jahangirzadeh et al., 2022), applied pressure (Ma et al., 2024) and temperature (Oviroh et al., 2021), salt type and concentration (Vafa-1 et al., 2023; Vafa-2 et al., 2023), hydration radius (Risplendi et al., 2020), surface charge (Ma et al., 2024), and chemical functionalization (Hosseini et al., 2018), collectively govern the performance of membrane-based desalination.

MC and EMD simulations are employed as a preliminary step, aiming to determine the most stable and energetically favorable initial arrangement of either water molecules or ions within nanochannels under equilibrium conditions before proceeding to investigate the transport behavior. MC and EMD simulations provide detailed insights into water, ion, and membrane interactions that are difficult to access experimentally. Techniques such as radial distribution function (RDF) analysis and potential of mean force (PMF) calculations are employed to evaluate intermolecular binding tendencies and energy barriers, respectively. Density distribution analysis obtained from the output of MC or EMD simulations provides the distribution of water molecules or ions confined within the nanochannels of 2D nanomaterials. All these analysis tools are dominantly used for water-based analysis, rarely for ion-based, in EMD simulations to understand the effects of the above-mentioned influencing parameters. For instance, through visualization of EMD simulation results, the formation of well-ordered layers of water molecules adjacent to the surface of the $MoSe_2$ nanochannel was demonstrated (Wang et al., 2024). These structured layers provided a low-energy pathway for water transport. Furthermore, for the nanopores in the 2D nanomaterial, the density distribution analysis revealed a pronounced accumulation of Na^+ ions near the pore wall, while a substantial number of Cl^- ions gathered in the central region of the pore, despite electrostatic repulsion from the COO^- functional groups (Fang et al., 2017). Moreover, RDF analysis revealed that the preference of water confined within h-BN and graphene nanochannels for Na^+, Cl^-, and Mg^{2+} ions changes with the interlayer distance (Vafa-1 et al., 2023). Monovalent ions exhibited a single RDF peak, whereas Mg^{2+} displayed two distinct peaks, corresponding to the presence of two hydration shells, leading to the enhancement in ion rejection (Vafa-1 et al., 2023). In boron nitride nanochannels, a greater number of confined water molecules resulted in higher ion hydration, which may restrict the movement of water molecules, thereby contributing to restricted water transport across boron nitride nanochannels. Finally, PMF provides information about the energy barriers that water molecules or ions encounter when entering or residing within nanochannels or nanopores. PMF analysis revealed that water molecules passing through the C_3N nanoslit experienced the lowest free energy barrier (6.2 $kJ \cdot mol^{-1}$), significantly lower than that of Na^+ (27.3 $kJ \cdot mol^{-1}$) and Cl^- (14.2 $kJ \cdot mol^{-1}$), indicating efficient water transport and effective salt ion rejection (Gu et al., 2022). These examples from the literature illustrate the application of such tools, which can be adapted to diverse conditions. Insights gained from these tools, gained from equilibrium simulations, also inform the understanding of transport phenomena under nonequilibrium conditions.

MC simulations have been utilized in the determination of water adsorption capacities of 2D nanomaterial surfaces as a result of various modifications (Takamatsu and Ohba, 2021). It was revealed that either oxygen-containing functional groups promote water adsorption, followed by dominant water-water interactions (Takamatsu and Ohba, 2021) or deprotonated hydroxyl groups bind water molecules via hydrogen bonds (Shaharudin et al., 2022). For a specific example, Zaman et al. (2021) investigated the interaction behavior of water molecules on $Ti_3C_2T_x$ (MXene) surfaces modified with Li^+ and K^+ ions using GCMC simulations and demonstrated that the presence and type of surface ions directly influence the interaction with water molecules (see Fig. 5.5) and alter the water adsorption capacity.

Water–membrane and ion–membrane interactions have been widely studied through EMD and MC simulations, providing insight into the underlying mechanisms. In particular,

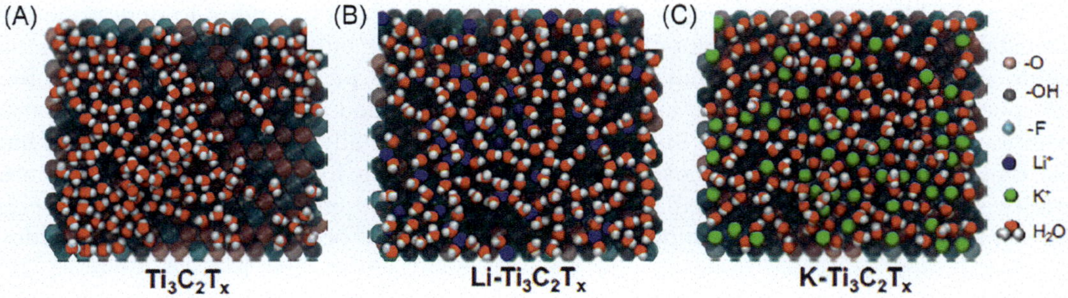

FIGURE 5.5 Snapshot of top views of water-surface interactions at (A) $Ti_3C_2T_x$, (B) Li-$Ti_3C_2T_x$, and (C) K-$Ti_3C_2T_x$ (Zaman et al., 2021). *Reprinted with permission from Zaman W, Matsumoto R A, Thompson M W, Liu Y H, Bootwala Y, Dixit M B, Nemsak S, Crumlin E, Hatzell M C, Cumming P T, Hatzell K B. In situ investigation of water on MXene interfaces. PNAS 2021;118(49):e2108325118. https://doi.org/10.1073/pnas.2108325118. Copyright 2020 PNAS.*

NEMD simulations allow for direct observation of permeation behavior under these conditions, enabling a quantitative evaluation of water flux and salt rejection (Cohen-Tanugi and Grossman, 2012; Si et al., 2023). Therefore, the impact of several parameters on the desalination performance of 2D nanomaterials is discussed based on NEMD simulation results. For instance, Ma et al. (2024) performed a systematic NEMD study on a single-layer MXene nanomaterial with nanopores to display water flux dependence of applied pressure (50–300 MPa). For $Ti_3C_2(OH)_2$ with small pores, flux values ranged from 12 to 110 $nm \cdot ns^{-1}$, reaching up to 212 $nm \cdot ns^{-1}$ in membranes with larger pores. Overall, with the help of applied pressure in NEMD simulations, the critical role of surface properties (Ma et al., 2024; Lv and Zhao, 2024) during transport, especially hydrophilicity and hydrophobicity, in determining membrane performance, was identified more realistically, and the following information was proposed by NEMD simulations.

Hydrophobic nanomaterials like graphene have weak interactions with water molecules, leading to high water flux (Vafa-1 et al., 2023). In contrast, hydrophilic materials like h-BN, MoS_2, and functionalized GO or MXene interact more strongly with water, forming extensive hydrogen bonds at the membrane-water interface (Song et al., 2020; Vafa-1 et al., 2023). This interaction can hinder water mobility but often improves ion exclusion. Hydrophobic membranes rely on size-exclusion mechanisms, while hydrophilic nanochannels/nanopores enhance ion rejection through strong hydration interactions or ion adsorption. The surface of 2D nanomaterials can be improved by incorporating functional groups like hydroxyl (Cohen-Tanugi and Grossman, 2012), carboxyl (Dahanayaka et al., 2020), or sulfonic acid (Lv and Zhao, 2024). These modifications can lower the energy barrier for water transport and enhance permeability but may also restrict flow by promoting hydrogen-bond formation.

Since NEMD simulations provide an effective way to model water transport under realistic desalination conditions, it is more applicable than experimental measurements to screen various materials, concentrating on the fixed structural properties like pore size. For instance, single-layer MoS_2, $MoSe_2$, graphene, boron nitride, and phosphorene nanomaterials with identical pore sizes were compared using NEMD simulations (Cao et al., 2020). MoS_2 exhibited the highest water permeability, exceeding other materials by 20%–38%, while maintaining ion rejection rates above 99%. This performance was linked to its lower energy barrier and favorable water–pore interactions. Such comparisons are difficult to achieve experimentally due to the challenge of fabricating identical pores in different materials. In light of this, intrinsically nanoporous materials, such as graphyne, graphenylene, and 2D MOF membranes, have recently gained significant attention for desalination (Jahangirzadeh et al., 2022; Xue et al., 2013; Banan Baghbani et al., 2020; Cao et al., 2019). Owing to their well-defined native pore structures, these materials eliminate the need for additional pore engineering processes.

Notably, while the NEMD simulation results suggest promising performance for single-layer 2D nanomaterials, their experimental realization remains a challenge. Therefore, concentrating simulation efforts on multilayered 2D nanomaterials is crucial to provide more practical insights for experimental studies. Recent NEMD simulations involving multilayered nanoporous materials such as MoS_2 (Abal et al., 2021), g-C_3N_4 (Liu et al., 2018), and GO (Cohen-Tanugi et al., 2016) have demonstrated that water exhibits far more complex behavior in these systems. In such environments, localized electrostatic fields can attract water molecules toward the channel walls or pores (Abal et al., 2021), while geometric confinement and potential energy barriers may restrict their motion (Abal et al., 2021; Cohen-Tanugi et al., 2016 Liu et al., 2018).

Beyond surface chemistry and pore structure, interlayer spacing is a critical parameter that strongly influences the balance between water permeability and ion rejection in layered 2D nanomaterial membranes (Vafa-1 et al., 2023; Ma et al., 2022; Maleki et al., 2022). Experimental techniques such as solvent swelling or intercalant insertion can tune the interlayer spacing (Ding et al., 2020; Yi et al., 2021; Lu et al., 2019), yet they fall short in identifying the ideal channel height. Fortunately, NEMD simulations effectively bridge this gap. PMF analysis shows that the increase in interlayer spacing lowers the energy barriers for both water and salt, which enhances water transport but decreases ion selectivity (Wang et al., 2024).

Similarly, in experimental tests, isolating the effects of ion hydration, ion–nanomaterial and ion–ion interactions are challenging, especially in systems containing mixed salts such as seawater. When membranes encounter competing ions (e.g., Na^+ and Ca^{2+}), overlapping behaviors obscure the contributions of individual ions (Ding et al., 2020). Both EMD and NEMD simulations overcome this limitation by systematically isolating these factors, enabling researchers to observe ion transport mechanisms at atomic resolution, test arbitrary salt combinations, and independently examine hydration effects (Liu et al., 2018). To conclude, simulations reflect the usual salt rejection hierarchy in experiments ($NaCl > MgCl_2 > Na_2SO_4$) and show that membrane-specific ion affinities may change these outcomes.

5.3.2.2 *Applications in heavy metal ion removal*

Industrial wastewater contains toxic heavy metals such as Cu^{2+}, Pb^{2+}, Cd^{2+}, Zn^{2+}, Hg^{2+}, Fe^{3+}, and Co^{2+}, making their removal essential before environmental discharge (Kommu et al., 2016). While many studies focus on adsorption-based removal methods, others employ membrane-based separation techniques akin to those used in desalination, where salts are removed instead of heavy metals. Rejecting heavy metals is generally easier than rejecting salts since most heavy metal ions are divalent or multivalent. Atomistic-scale simulations significantly advance our understanding of heavy metal removal efficiency in adsorption and membrane separation by providing detailed atomistic and dynamic insights, often difficult to obtain experimentally. Adsorption characteristics are investigated via MC, and membrane characteristics are evaluated via EMD or NEMD simulations. As we have already discussed the effects of pore size, functional groups, interlayer spacing, and nanomaterial modifications on water flux in the desalination section, this section focuses exclusively on parameters influencing heavy metal adsorption- and membrane-specific rejection.

MC simulations allow for the identification of the most stable adsorption configurations and geometries of heavy metal ions on a given nanomaterial surface. By scanning the potential energy landscape, these simulations enable the analysis of adsorbate–adsorbent interactions and preferred adsorption sites. For instance, Khnifira et al. (2022) investigated the adsorption of several heavy metals onto a graphite-based carbon surface and suggested that adsorption processes for all studied metals were spontaneous and exothermic. Most importantly, a key insight from MC simulations is the critical role of exploration of acidic conditions, which are difficult or impractical to realize in laboratory settings, in enhancing adsorption performance. As a result, it was theoretically demonstrated that acidic conditions enhance adsorption efficiency, whereas basic environments diminish it (Khnifira et al., 2022; Khalaf et al., 2022).

EMD simulations conducted under equilibrium conditions provide valuable data for analyses that offer insights into membrane behavior. For example, PMF analyses of various heavy metals interacting with 2D nanomaterials have shown that, regardless of the type of nanomaterial, there is a clear selectivity for heavy metal ions in the order of $Hg^{2+} > Cd^{2+} > Cu^{2+}$, with only slight variations in PMF values (Majidi et al., 2022; Azamat et al., 2014; Dehhaghi et al., 2020). Notably, Hg^{2+} exhibits a higher energy barrier for transport compared to other metals. It is also important to highlight that the PMF for Cl^- is significantly higher than for these metal ions, suggesting that salt ions can be more easily separated from metal ions. In addition, coordination numbers derived from RDF analysis provide insight into the hydration shells of metal ions confined within 2D nanochannels. This information reflects the strength of ion–water interactions and allows comparison of ion hydration before and after confinement. For instance, Giri and Cordeiro (2021) studied the coordination numbers of various metal ions within graphene nanochannels and found that the first coordination number for all metal ions, except Pb^{2+}, was approximately 5 both in bulk solution and inside the membrane nanochannels. Pb^{2+}, on the other hand, showed a coordination number of 6, indicating that the inner hydration shell of these metal ions remains largely intact upon confinement (Giri and Cordeiro, 2021). This contrasts with salt ions, whose coordination numbers change after confinement.

Since NEMD simulations effectively capture the real-time transport of heavy metal ions through membranes under applied pressure or concentration gradients, the following section evaluates the heavy metal ion separation performance of 2D nanomaterials based on various parameters observed from NEMD studies. For instance, the type of 2D nanomaterial significantly influences the removal efficiency of heavy metal ions. Köhler et al. (2018) examined the effect of cation valence on metal ion rejection in graphene and MoS_2 membranes. MoS_2 membranes consistently outperform graphene across all cation valences, achieving 100% rejection of Zn^{2+} and Fe^{3+} ions at all pore sizes and pressures. Moreover,

NEMD simulations provide an effective means to evaluate how different chemical modifications on the surface or in the pore of 2D nanomaterials influence heavy metal rejection. Wang et al. (2024) and Tabasi et al. (2023) investigated functionalized GO membranes for heavy metal rejection. Wang's protonated GO achieved 100% rejection in small pores and over 90% in larger pores at 150 MPa, while Tabasi's amide-, thiourea-, sulfonic acid-, and carbamate-functionalized GO exhibited similar rejection rates. However, Tabasi's membranes demonstrated significantly higher water permeability due to optimized functional groups, highlighting how chemical modifications can improve performance beyond pore size effects. Alongside excellent consistency of NEMD simulations with experimental observations, they also deliver pinpoint data that guides the targeted development and optimization of membrane materials. For instance, it was observed that for graphene membranes, an interlayer spacing of 0.8 nm, a three–layer structure, a gap width of 1.23 nm, and the use of small flakes were found to be optimal for achieving complete metal ion rejection alongside considerable water permeation (Zheng et al., 2023; Giri and Cordeiro, 2021).

Complementary NEMD simulation studies by Ou et al. (2024) and Binazir et al. (2025) demonstrated that the metal ion rejection mechanism strongly depends on interlayer spacing in GO membranes. Narrow spacings (~9–12.4 Å) exclude ions via steric hindrance, effectively blocking hydrated heavy metal ions such as As^{3+} and Hg^{2+}. As the spacing widens (10–16.1 Å), electrostatic interactions between the negatively charged membrane and cationic metals become dominant, enhancing ion adsorption, particularly for As^{3+} due to its higher charge density. At even larger spacings (>17.3 Å), rejection rates for As^{3+} decline and approach those of Na^+, as weakened electrostatic forces combined with partial preservation of hydration shells facilitate ion permeation. These findings highlight a critical balance between steric exclusion at narrow spacings and electrostatic adsorption at intermediate spacings in tuning membrane rejection and permeability.

Unlike pressure-driven NEMD studies, recent research has applied external electric fields as the driving force to facilitate heavy metal ions across modified nanomaterials, including γ-graphyne-1 (Majidi et al., 2021), C_2N (Majidi et al., 2022), graphene (Azamat et al., 2014), and h-BN (Azamat et al., 2015) to reduce the heavy metal ion concentration in wastewater. At high electric fields, these metal cations permeated rapidly through the membrane. Notably, Cu^{2+} ions exhibited the strongest electrostatic interaction with the $-NH_2$ functionalized pores, leading to high ion permeation. In contrast, –COOH functionalized pores imposed a higher energy barrier to Hg^{2+} ions, resulting in low Hg permeation. Interestingly, Karimzadeh et al. (2021) applied both pressure and an electrical voltage in opposite directions to enhance the rejection of positively charged heavy metal ions through silicon carbide (SiC) membranes with functionalized pores (–F, –N, –OH). It is also worth noting that simulations offer a feasible and efficient means to investigate such combined effects, enabling a deeper understanding of their synergistic contributions to separation performance. Increasing the voltage enhances ion rejection, especially in pristine SiC with larger pores and lower energy barriers. In functionalized SiC, negatively charged pore edges attract ions, reducing the electric field's effect; rejection remains high, though less sensitive to voltage variation. High rejection stems from strong interactions between ions with hydration shells and functional groups at pore edges, where hydration shells add steric and energetic barriers to ion transport.

5.3.2.3 Applications in dye removal

Adsorption has emerged as one of the most promising technologies for dye removal, owing to its simplicity, high efficiency, and economic viability. The efficiency of the adsorption process is largely determined by the surface properties of the adsorbent material and the nature of its interaction with the adsorbate (dye) molecules. While experimental studies provide macroscopic insights into adsorption capacity and kinetics, they often fall short of fully elucidating the molecular-level mechanisms of these interactions. At this point, MC and EMD simulations stand out as a common approach, unraveling the molecular-level mechanism of adsorption. Through these simulations, it is understood that adsorption processes are generally spontaneous and governed by various forces such as vdW interactions, electrostatic forces, or π–π stacking (Grouli et al., 2024; Wakjira et al., 2025; Narayanaswamy et al., 2020; Tanveer et al., 2024).

MC simulations investigating the interaction behavior of Methylene Blue (MB) and Rhodamine B (RB) dyes on graphene and GO surfaces showed how two dyes with different structures bind to the same surface via different mechanisms (Narayanaswamy et al., 2020). The planar MB molecule maximized π–π stacking interactions by orienting parallel to the graphene surface. In contrast, the bulkier RB molecule prefers to bind through electrostatic attraction with the oxygenated functional groups on GO. EMD simulations provide energy component analysis and interspatial distance analysis for the identification of dye adsorption performance of 2D nanomaterials. For instance, the total interaction energies of the aromatic and cationic dyes, crystal violet (CV) and brilliant blue (BB), with V_2CT_x were determined as $-149.95\ kJ{\cdot}mol^{-1}$ and $-104.85\ kJ{\cdot}mol^{-1}$, respectively (Bina et al., 2025). However, MB adsorbed onto bio-based carbon surfaces at a rate of around $-85\ kcal{\cdot}mol^{-1}$ (Haghbin et al., 2024). This energy component analysis indicated that π-π interactions and electrostatic attractions with the carbon surface were paramount. On the other hand, RDF analysis revealed that CV and BB molecules were clustered at an average distance of 0.9 nm on the V_2CT_x surface (Bina et al., 2025). This configuration, coupled with poor mobility, signifies the formation of stable adsorption complexes. The research by Haghbin et al. (2024) and Hou et al. (2020) indicated that the distance between dyes and carbon surface was primarily found in a narrower range. This indicates robust π–π stacking interactions between the dye and the aromatic carbon framework. These MC and EMD simulation results clearly demonstrate how decisive the surface chemistry and the structure of the dye molecule are in adsorbent design. However, in the context of membrane processes rather than adsorption, NEMD simulations have not been applied to dye separation studies. This is primarily due to the bulky and complex structures of dye molecules, which make the simulation systems computationally demanding and challenging to model. As an alternative, sampling-based approaches combined with EMD simulations can be employed to obtain membrane-related insights, such as dye rejection performance, without the full computational cost of NEMD simulations.

5.3.2.4 Applications in pharmaceutical separation

Due to similar challenges encountered in dye removal, such as the bulky and structurally complex nature of the molecules, which increases the computational cost and complexity of simulations, pharmaceutical separation studies are also largely limited to either MC or EMD simulations. These approaches primarily assess adsorption performance through RDF analysis or structural visualization. Since experimental data alone may not be sufficient to comprehensively evaluate the pharmaceutical separation performance of adsorbents, MC or EMD simulation-derived information revealing the interactions between pharmaceutical molecules and material surfaces at the atomic level is highly valuable. Such insights help define the strength of drug adsorption on various surfaces, determine the selectivity of these surfaces, and elucidate the competitive behavior among different molecules. For instance, Bahamon and Vega (2017) used GCMC simulation to identify how effectively various activated carbon (AC) models represent the adsorption of pharmaceutical molecules in aqueous environments. AC surfaces exhibited much higher selectivity toward pharmaceutical molecules compared to water, with calculated drug/water selectivity ratios ranging from 3000 to 7000 (Bahamon and Vega, 2017). The adsorption preference among the investigated pharmaceuticals followed the order of paracetamol (PRM) > diclofenac (DCF), naproxen (NPX) > ibuprofen (IBP) > amoxicillin (AMX). IBP was adsorbed at approximately 130 $mg \cdot g^{-1}$, while NPX reached 241 $mg \cdot g^{-1}$ (Bahamon and Vega, 2017). This difference is related to the molecular size and the chemical affinity of the molecules to the surface. EMD simulations performed by Ghasemi et al. (2022) have shown that specific COFs can achieve complete removal of certain pharmaceuticals, such as atenolol (ATN) and carbamazepine (CBZ), with favorable Gibbs free energy values (−24.48 and −21.15 $kJ \cdot mol^{-1}$, respectively). Other COFs, like COF-OEt, exhibit the highest adsorption affinities and interaction energies, effectively capturing compounds such as ciprofloxacin (CIF), DCF, IBP, and gemfibrozil (GEM) (Ghasemi et al., 2022). These studies also indicate that hydrogen bonding and vdW forces are the dominant adsorption mechanisms, highlighting the combined role of specific and nonspecific interactions in pharmaceutical separation. Although simulation studies in the field of pharmaceutical separation remain limited, the findings from the aforementioned works demonstrate that simulations can be reliably applied in this area and can serve as a valuable tool for obtaining preliminary insights, screening drugs, or evaluating materials prior to experimental studies.

5.3.2.5 *Applications in pervaporation*

Pervaporation is a separation process that selectively transports specific components from a liquid mixture through a membrane, followed by evaporation on the permeate side. It is particularly advantageous for separating heat-sensitive materials, azeotropic mixtures, and recovering solvents. One of the most widely studied applications of pervaporation is the separation of water–alcohol mixtures. Although EMD and NEMD simulations are more feasible for the investigation of pervaporation, dual ensemble MC simulations, such as grand canonical (μVT) in combination with a canonical (NVT) ensemble (Takaba et al., 2000) and united isobaric–isothermal and Gibbs ensemble (Zou and Lin, 2020), have been applied to various zeolites for ethanol-water separation, and a separation factor exceeding 400 for 2D nanomaterials was reported. On the other hand, to simulate a pervaporation system, in NEMD simulation, vacuum is applied on the permeate side, while the feed chamber is filled with a water–alcohol mixture. Therefore, the primary driving force in pervaporation is the difference in chemical potential across the membrane (Liu et al., 2023).

Liu et al. (2023) and Shi et al. (2017) investigated graphene membranes having the pores functionalized with hydrophobic hydrogen (G-H) and hydrophilic hydroxyl groups (G–OH) to separate a water–ethanol mixture via NEMD simulations. G-H membranes exhibited reverse selectivity, favoring ethanol due to strong adsorption on the hydrophobic sp^2 surface (Liu et al., 2023), while G–OH favored water via interactions with oxygen-containing groups (Liu et al., 2023; Shi et al., 2017). RDF and MSD analyses observed from EMD simulations showed ethanol accumulation and higher mobility in larger G-H (≥3.6 Å), promoting ethanol-selective transport (Liu et al., 2023). PMF results further revealed lower energy barriers near hydrophilic pores, enhancing water flux in G–OH membranes (Shi et al., 2017). Moreover, novel graphyne membranes have recently been explored as reverse-selective materials, exhibiting exceptionally high ethanol permeability (850 $kg \cdot m^{-2} \cdot h^{-1} \cdot bar^{-1}$) with a moderate selectivity of 5 (Zhang et al., 2019). This unusual separation behavior is attributed to the combination of a hydrophobic surface and a unique polyporous, single-atom-thick structure, which together promote ethanol adsorption and pore trapping. Interestingly, although pure water permeates faster than pure ethanol, in mixtures, ethanol sorption dominates and hinders water transport, resulting in overall ethanol-selective performance. This finding is consistent with the study by Liu et al. (2019); however, their objective was to achieve complete ethanol rejection and high water permeance. They designed GO membranes with varying pore sizes and O/C ratios, showing that higher O/C ratios enhance water selectivity (Liu et al., 2019; Yu et al., 2018) and that water flux increases with both oxidation level and pore size. Based on RDF and MSD analyses, they concluded that at high oxidation levels, water permeation is governed primarily by sorption, whereas at lower oxidation degrees, diffusion becomes the dominant transport mechanism. Moreover, Yu et al. (2018) demonstrated that membrane affinity toward water or ethanol varies depending on both pore size and surface functionality. Ethanol permeation increased at low oxidation with small pores and high oxidation with large pores, while water permeation dominated in other cases. In mixtures, water flux decreased compared to pure water, likely due to hydrogen bonding with slower ethanol molecules hindering water transport. These results highlight the complexity of experimentally optimizing membrane design and emphasize the essential role of molecular simulations in guiding such efforts.

Apart from graphene-based membranes, COFs, such as TyPa-1, have also been utilized in pervaporation due to their ease of functionalization (Gupta et al., 2023). By introducing various functional groups, a range of pore sizes and surface affinities was achieved. Overall, all COF membranes exhibited a higher permeability toward ethanol. Notably, when using the hydrophilic TyPa-1-OC_4H_8OH membrane, ethanol permeation was completely suppressed due to its narrow 2.7 Å pore size, allowing only water to pass through and resulting in a high separation factor. The other nanomaterial that was investigated in pervaporation processes is zeolite. Tsai and Lin (2024) investigated 14 different zeolite nanosheets for pervaporative separation of isopropanol (IPA)/water mixtures. The MRE nanosheet demonstrated optimal performance, achieving an IPA flux of 4.15×10^4 $kg \cdot m^{-2} \cdot h^{-1}$ and an IPA/water separation factor of ~432, enriching the permeate to 99.7 wt.% IPA in a single stage. The separation was adsorption-driven, with IPA exhibiting a stronger affinity to zeolite and a lower energy barrier compared to water. Hydrophobic pore entrances enhanced IPA adsorption, while an IPA-rich permeate interface facilitated evaporation.

These findings underscore the critical roles of pore geometry, adsorption energetics, and interfacial design in high-performance IPA-selective membranes.

5.3.2.6 *Other applications in water treatment*

Beyond these applications, simulation studies involving 2D nanomaterials have also been conducted to explore their potential in separating disinfection by-products, organic micropollutants, oil-water and solvent-water systems, as well as to gain a deeper understanding of the underlying transport and separation mechanisms, even in the absence of specific target compounds. However, such studies remain scarce, despite the need for their expansion.

Yang et al. (2021) investigated the separation of small organics (e.g., alcohols, urea, NDMA, phenol) using nanoporous graphene (NPG) membranes functionalized with –OH or –H terminated pores via NEMD simulations. Membranes with small, OH-edged pores showed higher water/organic selectivity by facilitating water transport and impeding organic contaminant passage. In line with the role of functionalization in modulating separation performance, Azamat et al. (2015) conducted EMD simulations to investigate the separation of trihalomethanes chlorination disinfection by-products using NPG membranes functionalized with –F or –H terminated pores. Their results indicated that the solute concentration must remain below its solubility limit in water to ensure effective separation. Furthermore, NPG membranes functionalized with –F-terminated pores exhibited higher permeance for both water and solutes compared to those with –H terminations. These findings reinforce the significance of surface functionalization in tailoring membrane performance for specific contaminant removal. Focusing purely on steric exclusion, Risplendi et al. (2020) conducted a NEMD simulation study using graphene membranes to separate boric acid from water. By analyzing the RDF between water and boric acid, they identified the first hydration shell radius of boric acid as approximately 3.6 Å. Based on this, they concluded that the optimal pore diameter for high boric acid rejection should be around 7.2 Å. Graphene membranes with pore sizes of 6.28 and 8.37 Å achieved boric acid rejection rates of 95% and 80%, respectively. Since boric acid is nonpolar and the pores were neutral, the separation mechanism was found to rely entirely on size exclusion.

To provide better insight into water transport through 2D nanomaterials, Willcox and Kim (2017) investigated water transport across different regions of GO membranes (namely pristine, oxidized, and mixed) through NEMD simulations. The results showed that water flows most rapidly through pristine regions, while transport is significantly hindered in oxidized regions due to strong interactions with hydroxyl groups. On the other hand, Zhang et al. (2025) identified the impact of tortuous MXene nanochannels, compared with regular ones, for water transport. Water permeation through regular nanochannels was much faster than through tortuous nanochannels, demonstrating the rapid movement of water molecules under capillary pressure without external interference. In this way, the effects of different simulation designs on water transport have been demonstrated.

Recent studies have also focused on improving the accuracy of classical NEMD simulations applied to the determination of solvent transport (Liu et al., 2023). Güvensoy-Morkoyun et al. (2024) optimized simulation parameters for $Ti_3C_2O_2$–solvent systems, showing that applying force to entire molecules containing at least one atom within the slab

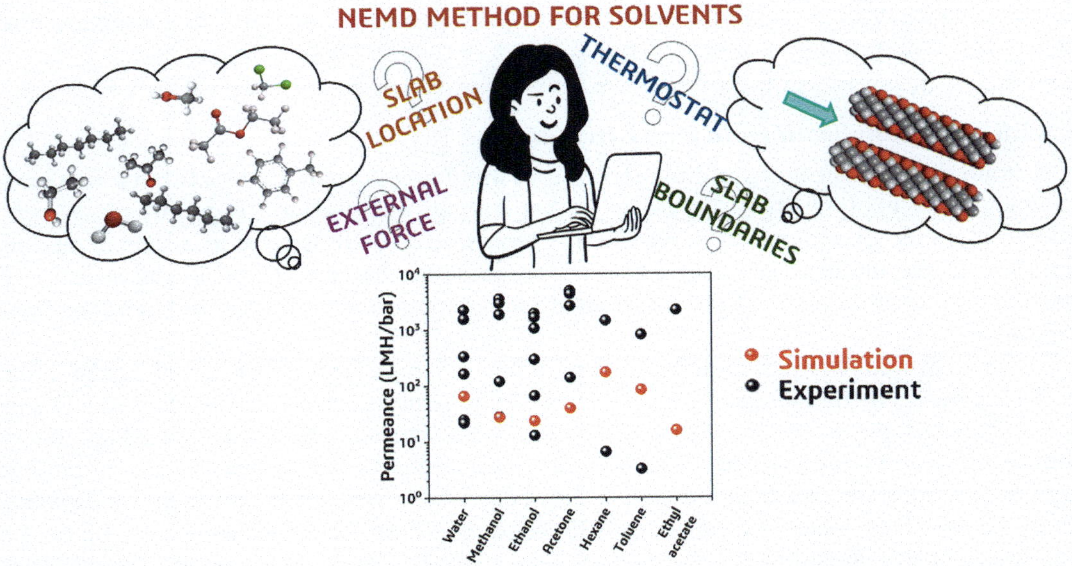

FIGURE 5.6 Illustration of the evaluated NEMD simulation parameters for organic solvent transport through MXene nanochannel as a case study (Güvensoy-Morkoyun and Baysal, 2024). *Reprinted with permission from Guvensoy-Morkoyun A, Baysal T, Tantekin-Ersolmaz S B, Velioglu S. Guide for nonequilibrium molecular dynamics simulations of organic solvent transport in nanopores: The case of 2D MXene membranes. J Chem Theory Comput, 2024;20(21):9642–9654. https://doi.org/10.1021/acs.jctc.4c00693. Copyright 2024 ACS.*

region, rather than to each atom individually, ensures more accurate force distribution. In addition, thermostating the MXene nanosheets instead of the solvent and tethering membrane atoms with a spring constant of 100 kcal·mol^{-1}·Å^{-2} improved structural stability and reproducibility. These are representative examples of simulation parameters (see Fig. 5.6) that were evaluated by Güvensoy-Morkoyun et al. (2024) for solvent flow. More broadly, factors such as force field selection and validation, thermostat type, and the location where force is applied all have a significant impact and must be carefully considered in NEMD simulations.

5.4 Machine learning approaches

5.4.1 Concepts and methods

ML is an area of computer science and statistics that enables algorithms to improve their performance without being explicitly programmed, by learning patterns in data (Jordan and Mitchell, 2015). Traditionally, physics-based or mechanistic models rely on predefined equations and expert knowledge-based linkages, whereas ML algorithms directly extract complicated and frequently nonlinear patterns from experimental or simulation data (Butler et al., 2018; Ramprasad et al., 2017). In recent years, breakthroughs in high-throughput experiments, material databases, and computer capacity have led to the rapid use of ML across scientific disciplines. Particularly in the domains of materials science and environmental engineering, it gives considerable advantages in processes such as the discovery of complex systems, design optimization, and real-time control (Agrawal and Choudhary, 2016; Cortes and Vapnik, 1995).

Supervised learning enables models to be trained using labeled datasets, where input features (X) are matched with known outputs (Y). The main purpose of this strategy is to train functions that can generalize to previously unseen data and produce reliable outcomes. This technique is applicable to both regression (continuous outputs) and classification (categorical outputs) problems. Among supervised learning methods, random forest (RF) is frequently utilized in materials science due to its ability to efficiently handle complex, nonlinear connections (Chen and Guestrin, 2016). Support vector machines (SVMs) are also widely used, especially in small or high-dimensional datasets; this method can describe nonlinear interactions through kernel functions (Ramprasad et al., 2017; Cortes and Vapnik, 1995). More recently, gradient boosting (GB) algorithms such as XGBoost have demonstrated excellent predictive performance on various structured scientific datasets, such as adsorption and membrane performance predictions (Chen and Guestrin, 2016; Jablonka et al., 2020).

Unsupervised learning, on the other hand, seeks to reveal hidden structures, patterns, or clusters by working with unlabeled data (Butler et al., 2018). Clustering methods such as K-means and hierarchical clustering minimize variance within clusters by grouping comparable cases together, which is important for tasks such as class identification and material categorization (Jablonka et al., 2020). Additionally, dimensionality reduction techniques such as principal component analysis (PCA) and t-distributed stochastic neighbor embedding (t-SNE) simplify visualization and feature extraction operations in high-dimensional datasets (Ward et al., 2016; van der Maaten and Hinton, 2008). In materials research, unsupervised learning is utilized for compound classification, exposing structure–property correlations, and directing feature engineering for prediction models (Jablonka et al., 2020; Ward et al., 2016).

Reinforcement learning (RL) enables an algorithm to learn the effective decision-making techniques by interacting with an environment and obtaining reward-based feedback (Sutton and Barto, 1999). Unlike supervised learning, RL explores through trial and error, making it particularly suitable for the complex and sequential optimization issues faced in materials science (Butler et al., 2018). Recent research has shown that RL has achieved successful outcomes in sectors such as molecular design, synthesis optimization, and nanostructure engineering by efficiently navigating enormous chemical spaces (Jablonka et al., 2020; Zhou et al., 2019). However, identifying scientifically valid reward functions and ensuring sufficient exploration remain key difficulties to be solved.

Deep learning (DL) automatically extracts hierarchical features from raw data using multilayer artificial neural network (ANN), hence enabling higher prediction performance in complex systems (Lecun et al., 2015). In materials research, domain-adapted DL architectures, particularly graph neural network (GNN), can predict material properties without the requirement for predefined descriptors by directly encoding atomic structures (Butler et al., 2018; Ramprasad et al., 2017). Leading examples in this field, such as the Crystal Graph Convolutional Neural Network (CGCNN) (Xie and Grossman, 2018) and SchNet (Schütt et al., 2018), have shown success in precisely forecasting formation energies, band gaps, and adsorption behaviors. While these models can effectively capture structure–property interactions, the interpretability of their decision mechanisms remains an important topic under investigation (Jablonka et al., 2020; Murdoch et al., 2019).

The ML model-building process generally follows a sequential order consisting of data preparation, feature engineering, model training, and evaluation processes. In the preprocessing stage, tasks such as filling in missing data, scaling data, and partitioning data into training, validation, and test sets are performed (Butler et al., 2018). Feature engineering integrates domain knowledge (e.g., material composition descriptors) with methodologies such as DL to accomplish automatic feature learning (Ward et al., 2016). During the model optimization phase, loss functions such as mean squared error (MSE) for regression or cross-entropy for classification are reduced using algorithms such as stochastic gradient descent (SGD) (Lecun et al., 2015). Overfitting, which refers to the model memorizing the training data and losing its generalization capacity, can be controlled using regularization, cross-validation, and dropout methods (Murdoch et al., 2019).

Despite all these achievements, several fundamental issues persist. Explainability is of tremendous importance in scientific ML; consequently, research on explainable artificial intelligence (XAI) methodologies, such as feature importance analysis and interpretable model architectures, is increasingly rising (Roscher et al., 2020; Barredo Arrieta et al., 2020). The model's generalization ability may deteriorate in cases of distribution shifts; physics-based ML techniques are presented as a solution to such challenges (Karpatne et al., 2017). Additionally, data scarcity is a major concern in materials research. This issue is commonly solved by complementing experimental data with simulations and applying transfer learning approaches (Butler et al., 2018; Jablonka et al., 2020). Fortunately, open-source software ecosystems such as Scikit-learn (Pedregosa et al., 2011), TensorFlow (Abadi et al., 2016), PyTorch, and Matminer (Ward et al., 2018) have substantially advanced the use of ML in several areas, such as materials and water treatment research.

5.4.2 Applications in water treatment

ML technologies are rapidly being employed to solve crucial problems by delivering both predictive modeling and optimization capabilities in various water treatment processes. Among water treatment applications via 2D nanomaterials, up to now, ML approaches, albeit limited in number, have been applied for membrane, adsorption, and photocatalytic technologies. **(1)** In membrane-based desalination processes, supervised ML algorithms such as ANN, SVM, and GB were applied to estimate basic membrane performance parameters such as flux, salt rejection, and fouling tendency based on structural and operational variables (Dansawad et al., 2023; Frontistis et al., 2023). **(2)** In adsorption-based water treatment, ML models such as RF and GB successfully predicted adsorption capacities for various contaminants by considering variables such as surface area, pore characteristics, solution pH, and functional groups (Liu et al., 2022; Zhang et al., 2023). **(3)** ML techniques were utilized in the optimization of advanced oxidation processes (AOPs); in this context, the selection of parameters such as oxidant dosage, reaction time, and catalyst properties is guided to maximize contaminant degradation efficiency (Huo et al., 2021; Lu et al., 2024). On the other hand, beyond material-level optimization, DL models were utilized to assess complex sensor data for real-time monitoring of water treatment plant operation parameters and for predicting effluent water quality (Lowe et al., 2022; Huang et al., 2021). Additionally, RF algorithms were being researched for dynamic process management, enabling the adaptive optimization of parameters such as chemical dosing, membrane cleaning cycles, and energy usage under varying influent water quality conditions (Yang et al., 2022).

ML has become a crucial tool in the field of membrane-based desalination processes, especially when combined with the structural flexibility of 2D nanomaterials. RF, Gaussian Process Regression (GPR), Kernel Support Vector Regression (k-SVR), ANNs, deep neural networks (DNN), and deep reinforcement learning (DRL) have been successfully applied for purposes such as structure optimization of membrane material, separation performance prediction, and the design of new nanoporous membranes (Ma et al., 2024; Priya et al., 2022; Guo et al., 2023; Wang et al., 2024). These models examine several structural aspects of membranes, such as pore size, surface charge, functionalization state, and hydrophilicity, in conjunction with operating conditions, providing high-accuracy forecasts of performance parameters such as salt rejection, flux, and fouling propensity. For example, ML models applied in 2D nanomaterials such as MXene and graphene have exhibited substantial success in achieving high ion rejection performance even in challenging multi-component systems (Ma et al., 2024; Wang et al., 2021). Wang et al. (2021) integrated a DRL framework with a convolutional neural network (CNN) to surpass the limitations of conventional MD simulations, enabling the automated design and optimization of graphene nanopores. While screening of 1000 nanopore configurations via simulations would require over four years of computational effort, the DRL-based approach facilitated the generation and evaluation of 7999 nanopores in under a week, representing a gain of several orders of magnitude in computational efficiency (Wang et al., 2021). This AI-driven methodology provided novel insights into ion rejection mechanisms in nanopores characterized by irregular geometries, rough edges, and high surface-to-area ratios. By combining the predictive accuracy of the CNN model (R^2 = 0.996) with the exploratory power of DRL, the study demonstrated a robust framework not only for performance prediction but also for the autonomous discovery of high-performing nanomaterial candidates (Wang et al., 2021). On the other hand, Ma et al. (2024) developed high-accuracy ($R^2 > 0.92$) XGBoost-based ML models based on 288 data points obtained from NEMD simulations of MXene nanomaterials. Similarly, the applied ML model reduced the computational time of traditional MD simulations, which took approximately 10000 CPU hours, to a matter of seconds, achieving approximately 1000-fold computational efficiency (Ma et al., 2024). Via feature importance map, it was revealed that in order of pressure, charge distribution on MXene surface, size, material, concentration, and shape of pore have an importance in water flux and ion rejection, as illustrated in Fig. 5.7. In addition to focusing on specific features, ML-based virtual screening algorithms have permitted the systematic examination of vast material design spaces, supporting the quick identification of high-performance membrane candidates (Priya et al., 2022). In the study by Priya et al. (2022), an ML model trained with approximately 260 EMD simulation data compiled from the literature was used to quickly scan 3814 different 2D nanomaterials. Initially, linear models such as LASSO, Ridge, and Elastic Net were tested but proved insufficient; the highest prediction accuracy was achieved using ensemble algorithms such as XGBoost and RF, which reached an R^2 value of 0.99 (Priya et al., 2022). Among the top candidates identified by the ML model, InH_2O_2 exhibited the highest water flow rate, while $Au_2Cl_2N_2$ demonstrated the highest salt rejection (Priya et al., 2022). It was also revealed that

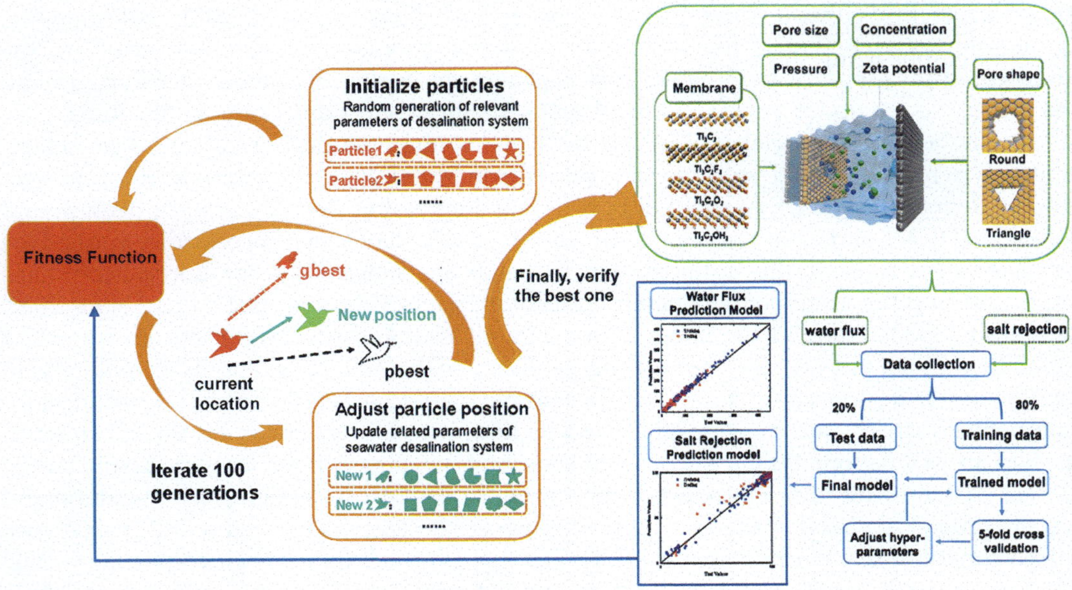

FIGURE 5.7 Illustration of an optimal system design for the MXene membrane based on ML and optimization algorithm (Ma et al., 2024). *Reprinted with permission from Ma X, Lan C, Lin H, Peng Y, Li T, Wang J, Azamat J, Liang L. Designing desalination MXene membranes by machine learning and global optimization algorithm. J Membr Sci, 2024;702:122803. https://doi.org/10.1016/j.memsci.2024.122803. Copyright 2024 Elsevier.*

the water permeability of these new materials is approximately four times higher than that of common materials such as graphene or MoS_2. All these findings emphasize the great potential of synergistic usage of ML models in the prediction of next-generation 2D nanomaterial membranes for desalination processes.

ML techniques have made major contributions in recent years to the optimization and reverse engineering of AOPs using 2D nanomaterials as photocatalysts. ANN, Adaptive Neuro-Fuzzy Inference Systems (ANFIS), SVM, and XGBoost have been successfully applied in modeling complex degradation mechanisms, optimizing operating conditions, and predicting pollutant removal efficiencies (Wang et al., 2024; Dashti et al., 2024; Moslehi et al., 2024; Sun et al., 2023). ZnO-based 2D photocatalysts and $ZnFe_{12}O_{19}$/BiOI magnetic nanocomposites have garnered great interest in the removal of pollutants such as metronidazole and dyes (Dashti et al., 2024; Moslehi et al., 2024). Dashti et al. (2024) have proposed a pioneering approach that systematically uses ML for the first time to model the degradation of pesticides with ZnO-based photocatalysts. Among the four different ML models developed based on experimental data in the literature, the RBF model stood out by accurately predicting pesticide degradation with an $R^2 = 0.978$ and a 4.80% error rate (Dashti et al., 2024). These models not only made highly accurate predictions but also precisely analyzed the effects of process parameters such as irradiation time, dopant ratio, and pesticide concentration on degradation efficiency. Another study (Moslehi et al., 2024) developed an ANN model to predict and optimize the photocatalytic degradation efficiency of a pharmaceutical contaminant called metronidazole (MNZ). The model accurately modeled the complex relationships between process variables and degradation efficiency. The model demonstrated very high fit values of R^2 = 0.992 for the training data and R^2 = 0.986 for the test data (Moslehi et al., 2024). Optimization outputs demonstrated the potential to significantly accelerate the experimental process and reduce costs, showing that 91.7% MNZ degradation could be achieved in just 25 minutes (Moslehi et al., 2024). On the other hand, thanks to the optimization of ML-supported UV/H_2O_2 and ozonation systems, more effective control has been established over oxidant dosages, irradiation periods, and pollutant concentrations (Sun et al., 2023; Seyedin et al., 2024). Seyedin et al. (2024) used an ANN model in a photoreactor design for UV water disinfection, significantly reducing the computational cost of traditional CFD simulations and achieving a low MSE of 8.9×10^{-4} with an average error of only 1.2% in the test set. At the same time, robust optimization reduced performance variability from 11.93% to 5.06%, thereby increasing system reliability (Seyedin et al., 2024). Sun et al. (2023) used the XGBoost model in the reverse design of electrochemical oxidation processes, incorporating reaction conditions into the model and successfully increasing the R^2 score from 0.63 to 0.84 and reducing the RMSE from 0.96 to 0.79. All these studies suggest that ML approaches are playing an increasingly crucial role in strengthening AOP performance through material design and optimization of operating conditions.

ML is also rapidly being employed in adsorption-based water treatment, particularly in the use of highly tunable adsorbents for removing diverse contaminants. ANN was frequently utilized for predicting the adsorption capacity of various materials and optimizing several process parameters such as adsorbent dosage, contact duration, pH, and temperature (Rao et al., 2023; Syah et al., 2021; Syah et al., 2022). For instance, ML-based modeling of nanocomposites composed of layered double hydroxide (LDH) and MOF has exhibited good accuracy rates in predicting adsorption performance using a limited dataset (Syah et al., 2021; Syah et al., 2022). Additionally, ANN-based optimization frameworks based on advanced learning algorithms have been effectively implemented to optimize water recovery processes involving nanoparticle-modified systems (Rao et al., 2023). This study demonstrates the power of ML to simplify and interpret information from data-intensive and analysis-heavy systems such as transmission electron microscopy (TEM). The proposed ML method outperforms classical ANN and CNN approaches in both BF-TEM and ADF-STEM image datasets, achieving impressive results such as 85% training accuracy, 79% specificity, and 66% average precision for ADF-STEM (Rao et al., 2023). Thanks to its ability to learn complex and nonlinear structures, the ML algorithm has made nanoparticle size and shape classification more efficient and automated, thereby reducing data processing requirements.

Despite these successes, several critical challenges remain. The scarcity of high-quality, standardized datasets continues to limit the full potential of ML-driven material discovery and process optimization. The generalizability of trained models across different operational conditions, material systems, and pollutant profiles is often restricted, requiring the development of hybrid modeling approaches that incorporate both physical knowledge and data-driven learning. Furthermore, model explainability remains a major concern, particularly in scientific domains where a mechanistic understanding is essential. Continued advances in XAI, transfer learning, and physics-informed ML will be pivotal to overcoming these challenges. Looking forward, future research should emphasize the integration of ML models with high-throughput experimental platforms and advanced simulation tools to create closed-loop autonomous systems for material screening and process design. Additionally, more importantly, the development of large, open-access material databases specific to water treatment applications will be critical for training more generalizable and interpretable models. Moreover, interdisciplinary collaboration among materials scientists, chemists, environmental engineers, and data scientists will be essential to fully realize the transformative potential of ML in sustainable water treatment using 2D nanomaterials. As a final note, the use of ML marks a turning point in material discovery, where data-driven modeling and optimization techniques transcend traditional methods and pave the way to the "fourth paradigm".

5.5 Conclusion and perspectives

By providing distinct structural and physicochemical characteristics that allow for exact control over transport and separation processes like membrane, adsorption, and photocatalysis, the investigation of 2D nanomaterials has transformed water treatment research. Molecular simulations have been crucial in determining nanoscale mechanisms and directing customized material design, even though experimental research has yielded insightful information about the performance of 2D nanomaterials. The expansion of the material pool and the remarkably accurate prediction of performance metrics have been further accelerated by the recent incorporation of machine learning into simulation-driven research. In order to overcome current obstacles in the water treatment process, such as scalability, fouling resistance, and stability under realistic operating conditions, it will be crucial to jointly employ experiments, simulations, and data-driven methodologies.

Clarifying the atomic level mechanisms underlying water treatment processes with 2D nanomaterials has been made possible by quantum-scale simulations, especially DFT, which use electronic configurations to provide incredibly accurate insights but are frequently restricted to relatively small atomic systems. DFT has given precise insights into adsorption energies, charge transfer pathways, electronic structure modifications, and intermediate reaction states that are otherwise unavailable to experiments in a variety of fields, including dye adsorption, pharmaceutical removal, heavy metal remediation, catalytic degradation, and desalination. By directing techniques like doping, defect engineering, heterojunction formation, and surface functionalization to maximize the performance of photocatalysts, membranes, and adsorbents, these findings allow for the predictive design of 2D nanomaterials. Moving forward, to create a thorough, multiscale, and data-driven framework for 2D nanomaterial discovery in water treatment, it will be crucial to integrate DFT with atomistic simulations, machine learning, and experimental validation. Despite impressive advancements, there are still issues, especially with the small number of pollutant types studied and the oversimplification of model systems in comparison to actual aquatic environments. Systematic investigation of various contaminants and the inclusion of practical operating conditions (e.g., competing ions, variable pH, fouling) are necessary to address these problems.

The water treatment performance of 2D nanomaterials has been clarified with atomic resolution by also atomistic-scale simulations, specifically EMD, NEMD, and MC, which have revealed many details that are not possible to observe experimentally. Through RDF/PMF and density distribution analyses, parameters like nanochannel size/geometry, interlayer spacing, surface charge and functionalization, salt type/concentration, hydration radius, temperature, and pressure have been quantitatively linked to the water flux–salt rejection trade-off in desalination, which are defined with the mutual contribution of EMD and NEMD simulations. While EMD/NEMD have measured selectivity and the impact of functional groups (such as protonation/anion edges) via hydration shells and energy barriers, MC has mapped the most stable adsorption configurations and the role of pH in heavy metal removal. For bulky/complex molecules such as dyes and pharmaceuticals, MC/EMD has distinguished the dominance of π–π, electrostatic, and H-bonding interactions, as well as competitive adsorption, thereby enabling material–target matching for screening studies. Finally, the overall consistency between simulation predictions and experimental trends has been confirmed, highlighting the impact of methodological choices in NEMD, such as driving force application schemes, thermostatting protocols, force field selection, and mechanical constraints on membranes, on result reliability. Moving from idealized single-layer and defect-free pores to realistic multilayer membranes that incorporate defects, oxidation/aging, swelling, interlayer heterogeneity, and tortuosity should be the next step. Multiscenario (EMD/NEMD/MC) simulations that can break down hydration, ion–ion, and ion–surface interactions are required to represent and support realistic feed water chemistry, which includes mixed salts, multivalent ions, organics/NOM, variable pH, and competitive separation. However, atomistic-scale simulations ought to be incorporated into multiscale frameworks, combining DFT to gain insights into reactivity and charge transfer and to investigate the combined effects of electric fields, pressure, and concentration gradients that are pertinent to industrial operations. On the application side, manufacturability, mechanical/chemical durability, and fouling resistance must be cooptimized, while expanding focus toward underexplored contaminant classes such as disinfection by-products, per-/polyfluorinated substances, radionuclides, and solvent separations.

In addition to simulation-based methods, machine learning has quickly become a potent enabler for water treatment applications involving 2D nanomaterials. ML models, such as RF, XGBoost, SVM, ANN/DNN, and DRL, have continuously shown excellent accuracy in forecasting separation efficiency, adsorption capacity, degradation kinetics, and fouling propensity across membrane desalination, adsorption, and photocatalysis. In comparison to traditional simulations, ML has significantly reduced computational costs while speeding up performance prediction and design optimization by capturing intricate relationships between material features (pore size, charge distribution, hydrophilicity, and functionalization) and operating conditions (pH, pressure, concentration, and oxidant dosage). Despite its promise, ML applications in 2D nanomaterial-based water treatment face several challenges. Model transferability and generalizability across various nanomaterial classes and operating conditions are severely limited by the lack and fragmentation of high-quality, standardized datasets. To facilitate autonomous material discovery and process optimization, ML should be combined with high-throughput experiments and molecular simulations (MC/EMD/NEMD) in hybrid closed-loop systems. Additionally, to convert algorithmic advancements into deployable water treatment solutions, interdisciplinary cooperation between data scientists, materials scientists, chemists, and environmental engineers will be crucial. Long-term, the combination of ML, multiscale simulations, and experimental automation could propel the field toward self-driving labs, allowing for adaptive water treatment system operation and inverse nanomaterial design. Collectively, all these observations highlight a key advantage of theoretical water treatment studies and provide the current challenges associated with them. They offer a faster and more cost-effective pathway through precise, atomistic-level optimization without the trial-and-error limitations of experimental methods.

Acknowledgments

Financial support by the Scientific and Technological Research Council of Turkey (TUBITAK, Grant No. 223M298) is acknowledged.

AI disclosure

During the preparation of this work, the authors used Grammarly to correct the grammar, improve the readability, and polish the language of the work. After using this tool/service, the authors reviewed and edited the content as needed and take full responsibility for the content of the publication.

References

Abadi, M., Agarwal, A., Barham, P., Brevdo, E., Chen, Z., Citro, C., et al., 2016. TensorFlow: Large-scale machine learning on heterogeneous distributed systems. Software Ecosystem. Available at https://doi.org/10.48550/arXiv.1603.04467.

Abal, J.P.K., Dillenburg, R.F., Köhler, M.H., Barbosa, M.C., 2021. Molecular dynamics simulations of water anchored in multilayered nanoporous MoS_2 membranes: implications for desalination. ACS Applied Nano Materials 4 (10), 10467–10476. Available at https://doi.org/10.1021/acsanm.1c01982.

Abbaspour, M., Morsali, A., 2024. Density functional theory and molecular dynamics simulation of water molecules confined between two-dimensional graphene oxide surfacesz. Journal of Molecular Graphics and Modelling 133. Available at https://doi.org/10.1016/j.jmgm.2024.108862.

AbdulAmeer, S.A., Madani, A.H., Mohammed, I.M., Jalil, A.T., Cao, Y., Altalbawy, F.M.A., et al., 2023. Utility of hexagonal boron carbide nanosheets for removing harmful dyes: Electronic study via DFT. Inorganic Chemistry Communications 148, 110322. Available at https://doi.org/10.1016/J.INOCHE.2022.110322.

Abu Tarboush, B.J., Chouman, A., Jonderian, A., Ahmad, M., Hmadeh, M., Al-Ghoul, M., 2018. Metal-organic framework-74 for ultratrace arsenic removal from water: Experimental and density functional theory studies. ACS Applied Nano Materials 1 (7), 3283–3292. Available at https://doi.org/10.1021/acsanm.8b00501.

Agrawal, A., Choudhary, A., 2016. Perspective: Materials informatics and big data: Realization of the 'fourth paradigm' of science in materials science. APL Materials 4 (5). Available at https://doi.org/10.1063/1.4946894.

Ai, Z., Zhao, Y., Chen, L., Wen, T., Song, S., Zhang, T., 2022. Floating MMT/MXene janus membrane for solar steam generation and mechanism of improving water transportation by DFT calculation. Separation and Purification Technology 300, 121918. Available at https://doi.org/10.1016/j.jmgm.2024.108774.

Ajaj, Y., Basem, Khaddour, M.H., Yadav, A., Kaur, Sharma, et al., 2024. Removal of heavy metal ions from wastewater using two-dimensional transition metal carbides. Journal of Molecular Graphics and Modelling 130, 108774. Available at https://doi.org/10.1016/j.jmgm.2024.108774.

Alizadeh, M., Hasanzadeh, A., Ajalli, N., Azamat, J., 2024. A computational investigation of DMSO/water separation through functionalized GO multilayer nanosheet membrane using molecular dynamics simulation and deep neural network model for membrane performance prediction. Journal of Azamat, Chemosphere 349, 140802. Available at https://doi.org/10.1016/j.chemosphere.2023.140802.

Alomar, M., Khan, A.A., Yar, M., 2025. DFT investigation of Co/Ni decorated borophene for efficient removal of methylene blue from wastewater effluent. Journal of Molecular Liquids 417. Available at https://doi.org/10.1016/j.molliq.2024.126627.

Ang, E.H., Velioğlu, S., Chew, J.W., 2019. Tunable affinity separation enables ultrafast solvent permeation through layered double hydroxide membranes. Journal of Membrane Science 591, 117318. Available at https://doi.org/10.1016/j.memsci.2019.117318.

Azamat, J., Khataee, A., Joo, S.W., 2014. Functionalized graphene as a nanostructured membrane for removal of copper and mercury from aqueous solution: A molecular dynamics simulation study. Journal of Molecular Graphics and Modelling 53, 112–117. Available at https://doi.org/10.1016/j.jmgm.2014.07.013.

Azamat, J., Khataee, A., Joo, S.W., 2015. Molecular dynamics simulation of trihalomethanes separation from water by functionalized nanoporous graphene under induced pressure. Chemical Engineering Science 127, 285–292. Available at https://doi.org/10.1016/j.ces.2015.01.048, http://www.journals.elsevier.com/chemical-engineering-science/.

Azamat, J., Khataee, A., Joo, S.W., 2016. Separation of copper and mercury as heavy metals from aqueous solution using functionalized boron nitride nanosheets: A theoretical study. Journal of Molecular Structure 108, 144–149. Available at https://doi.org/10.1016/j.molstruc.2015.11.058.

Bahamon, D., Vega, L.F., 2017. Pharmaceutical removal from water effluents by adsorption on activated carbons: A Monte Carlo simulation study. Langmuir 33 (42), 11146–11155. Available at https://doi.org/10.1021/acs.langmuir.7b01967.

Banan Baghbani, N., Azamat, J., Erfan-Niya, H., Majidi, S., Khazini, L., 2020. Molecular insights into water desalination performance of pristine graphdiyne nanosheet membrane. Journal of Molecular Graphics and Modelling 101, 107729. Available at https://doi.org/10.1016/j.jmgm.2020.107729.

Bangari, R.S., Yadav, A., Awasthi, Sinha, N., 2022. Experimental and theoretical analysis of simultaneous removal of methylene blue and tetracycline using boron nitride nanosheets as adsorben. Colloids and Surfaces A: Physicochemical and Engineering Aspects 634, 127943. Available at https://doi.org/10.1016/j.colsurfa.2021.127943.

Barredo Arrieta, A., Díaz-Rodríguez, N., Del Ser, J., Bennetot, A., Tabik, S., Barbado, A., et al., 2020. Explainable Artificial Intelligence (XAI): Concepts, taxonomies, opportunities and challenges toward responsible AI. Information Fusion 58, 82–115. Available at https://doi.org/10.1016/j.inffus.2019.12.012.

Baysal, T., Dahanayaka, M., Velioğlu, S., Chew, J.W., 2025. Organic solvent permeation through HKUST-1 MOF membranes: Molecular-scale insights. Journal of Membrane Science 733, 124328. Available at https://doi.org/10.1016/j.memsci.2025.124328.

Becke, A.D., 2014. Perspective: Fifty years of density-functional theory in chemical physic. Journal of Chemical Physics 140 (18), 331. Available at https://doi.org/10.1063/1.4869598.

Becke, A.D., 1993. A new mixing of Hartree-Fock and local density-functional theories. The Journal of Chemical Physics 98 (2), 1372–1377. Available at https://doi.org/10.1063/1.464304.

Bedeaux, D., Kjelstrup, S., Sengers, J., 2015. Experimental thermodynamics volume x: non-equilibrium thermodynamics with applications. Royal Society of Chemistry. Available at https://doi.org/10.1039/9781782622543.

Berendsen, H.J.C., Postma, J.P.M., van Gunsteren, W.F., Hermans, J., 1981. Interaction models for water in relation to protein hydration. Intermolecular forces: proceedings of the fourteenth Jerusalem symposium on quantum chemistry and biochemistry held in jerusalem, Israel, April 13–16, 1981, Springer, 1981: pp. 331–342.

Berendsen, H.J.C., Grigera, J.R., Straatsma, T.P., 1987. The missing term in effective pair potentials. The Journal of Physical Chemistry 91 (24), 6269–6271. Available at https://doi.org/10.1021/j100308a038.

Bina, A., Raissi, H., Crystal, C.V., 2025. A strategy to improve the adsorption capacity of OPs dye pollutants from the aqueous environment using adsorbents based on 2D transition metal carbides (V_2CT_x). Applied Water Science 15 (1). 1–16. Available at https://doi.org/10.1007/s13201-024-02331-y.

Binazir, I.A., Vafa, N., Firoozabadi, B., 2025. Layered graphene oxide membrane for heavy metal separation via molecular dynamics simulation. Journal of Molecular Liquids 417, 126639. Available at https://doi.org/10.1016/j.molliq.2024.126639.

Butler, K.T., Davies, D.W., Cartwright, H., Isayev, O., Walsh, A., 2018. Machine learning for molecular and materials science. Nature 559 (7715), 547–555. Available at https://doi.org/10.1038/s41586-018-0337-2.

Cao, Z., Liu, V., Barati Farimani, A., 2020. Why is single-layer MoS_2 a more energy efficient membrane for water desalination? ACS Energy Letters 5 (7), 2217–2222. Available at https://doi.org/10.1021/acsenergylett.0c00923.

Cao, Z., Liu, V., Barati Farimani, A., 2019. Water desalination with two-dimensional metal–organic framework membranes. Nano Letters 19 (12), 8638–8643. Available at https://doi.org/10.1021/acs.nanolett.9b03225.

Chen, L., Shi, G., Shen, J., Peng, B., Zhang, B., Wang, Y., et al., 2017. Ion sieving in graphene oxide membranes via cationic control of interlayer spacing. Nature 550 (7676), 380–383. Available at https://doi.org/10.1038/nature24044.

Chen, T., Guestrin, C. 2016, XGBoost: A Scalable Tree Boosting System. Proceedings of the 22nd ACM SIGKDD, 785–794. Available at https://doi.org/10.1145/2939672.2939785.

Chen, W., Li, G., Zhang, X., Wang, S., Li, L., Huang, Q., 2023. Weakly supervised text-based actor-action video segmentation by clip-level multi-instance learning. ACM Transactions on Multimedia Computing, Communications, and Applications 19 (1), 1–22. Available at https://doi.org/10.1145/3514250.

Chen, X., Pan, R., Xu, H., Wu, S., Liu, C., Zheng, Z., et al., 2025. High-yield preparation of 1T-MoS_2 with excellent piezo-catalytic activity via W^6 doping for antibiotics degradation and Cr(VI) reduction: Mechanistic insights from characterizations to DFT calculations. Applied Catalysis B: Environmental 361, 124688. Available at https://doi.org/10.1016/j.apcatb.2024.124688.

Cheng, L., Liu, G., Zhao, J., Jin, W., 2021. Two-dimensional-material membranes: manipulating the transport pathway for molecular separation. Accounts of Materials Research 2 (2), 114–128. Available at https://doi.org/10.1021/accountsmr.0c00092.

Ciccotti, G., Ferrario, M., 2016. Non-equilibrium by molecular dynamics: a dynamical approach. Molecular Simulation 42 (16), 1385–1400. Available at https://doi.org/10.1080/08927022.2015.1121543.

Cohen-Tanugi, D., Grossman, J.C., 2012. Water desalination across nanoporous graphene. Nano Letters 12 (7), 3602–3608. Available at https://doi.org/10.1021/nl3012853.

Cohen-Tanugi, D., Lin, L.C., Grossman, J.C., 2016. Multilayer nanoporous graphene membranes for water desalination. Nano Letters 16 (2), 1027–1033. Available at https://doi.org/10.1021/acs.nanolett.5b04089.

Cortes, C., Vapnik, V., 1995. Support-vector networks. Machine Learning 20 (3), 273–297. Available at https://doi.org/10.1023/A:1022627411411.

Cramer, C.J., 2002. Essentials of computational chemistry: theories and models. J. Wiley, West Sussex, England; New York.

Dahanayaka, M., Liu, B., Srikanth, N., Zhou, K., 2020. Ionised graphene oxide membranes for seawater desalination. Desalination 496, 114637. Available at https://doi.org/10.1016/j.desal.2020.114637.

Dansawad, P., Li, Y., Li, Y., Zhang, J., You, S., Li, W., et al., 2023. Machine learning toward improving the performance of membrane-based wastewater treatment: A review. Advanced Membranes 3, 100072. Available at https://doi.org/10.1016/j.advmem.2023.100072.

Dashti, A., Navidpour, A.H., Amirkhani, F., Zhou, J.L., Altaee, A., 2024. Application of machine learning models to improve the prediction of pesticide photodegradation in water by ZnO-based photocatalysts. Chemosphere 362, 142792. Available at https://doi.org/10.1016/j.chemosphere.2024.142792.

Dehhaghi Y. Boroumandpour F. Frank I. Nadimi E. 2020. Iran Water treatment with cation selective porous graphene and boron nitride membrane: A molecular dynamics investigation, In: 2020 28th Iranian Conference on Electrical Engineering, ICEE. Institute of Electrical and Electronics Engineers Inc. Available at https://doi.org/10.1109/ICEE50131.2020.9260741.

Ding, Li, L., Liu, Y., Wu, Y., Lu, Z., Deng, J., et al., 2020. Effective ion sieving with $Ti_3C_2T_x$ MXene membranes for production of drinking water from seawater. Nature Sustainability 3 (4), 296–302. Available at https://doi.org/10.1038/s41893-020-0474-0.

Ding, M., Xu, H., Chen, W., Yang, G., Kong, Q., Ng, D., et al., 2020. 2D laminar maleic acid-crosslinked MXene membrane with tunable nanochannels for efficient and stable pervaporation desalination. Journal of Membrane Science 600, 117871. Available at https://doi.org/10.1016/j.memsci.2020.117871.

Dou, K., Peng, C., Wang, Cao, H., Yao, C., Qiu, J., et al., 2023. S-scheme tubular g-C_3N_4/BiOI heterojunctions for boosting photodegradation of tetracycline and Cr(VI): Mechanism insight, degradation pathway and DFT calculation. Chemical Engineering Journal 455. Available at https://doi.org/10.1016/j.cej.2022.140813.

Fang, C., Yu, Z., Qiao, R., 2017. Impact of surface ionization on water transport and salt leakage through graphene oxide membranes. Journal of Physical Chemistry C 121 (24), 13412–13420. Available at https://doi.org/10.1021/acs.jpcc.7b04283.

Firmansyah, M.L., Alwan, Y., Ullah, N., 2025. Talanta Open. 12, 100491.

Frenkel, D., Smit, B., 2023. Understanding molecular simulation: from algorithms to applications. Third edition understanding molecular simulation: from algorithms to applications, third Ed. Elsevier, United Kingdom, pp. 1–728. Available at http://doi.org/10.1016/C2009-0-63921-0.

Frontistis, Z., Lykogiannis, G., Sarmpanis, A., 2023. Machine learning implementation in membrane bioreactor systems: Progress, challenges, and future perspectives: A review. Environmental Science and Pollution Research 10 (7), 127. Available at https://doi.org/10.3390/environments10070127.

Fu, K., Liu, X., Yu, D., Luo, J., Wang, Z., Crittenden, J.C., 2020. Highly efficient and selective Hg(II) removal from water using multilayered Ti3C2O xMXene via adsorption coupled with catalytic reduction mechanism. Environmental Science and Technology 54 (24), 16212–16220. Available at https://doi.org/10.1021/acs.est.0c05532.

Gao, H., Wang, J., Zhang, X., Hu, M., Xu, Q., Xie, Y., et al., 2022. Confined lamellar channels structured by multilayer graphene for high-efficiency desalination. Desalination 530, 115681. Available at https://doi.org/10.1016/j.desal.2022.115681.

Ghasemi, M., Khedri, M., Didandeh, M., Taheri, M., Ghasemy, E., Maleki, R., et al., 2022. Removal of pharmaceutical pollutants from wastewater using 2D Covalent Organic Frameworks (COFs): An in silico engineering Study. Industrial and Engineering Chemistry Research 61 (25), 8809–8820. Available at https://doi.org/10.1021/acs.iecr.2c00924.

Giri, A.K., Cordeiro, M.N.D.S., 2021. CordeiroHeavy metal ion separation from industrial wastewater using stacked graphene Membranes: A molecular dynamics simulation study. Journal of Molecular Liquids 338, 116688. Available at https://doi.org/10.1016/J.MOLLIQ.2021.116688.

Gomaa, I., Hosny, N.M., Elhaes, H., Ezzat, H.A., Elmahgary, M.G., Ibrahim, M.A., 2024. Two-dimensional MXene as a promising adsorbent for trihalomethanes removal: A density-functional theory study. Nanomaterials 14 (5). Available at https://doi.org/10.3390/nano14050454.

Grouli, A., Chraka, A., Bachra, Y., Elkouali, M., Chtita, S., Berrada, M., 2024. An investigation of the adsorption of Congo red dye on two naturally occurring adsorbents Hydroxyapatite and Bentonite: An experimental analysis, DFT calculations, and Monte Carlo simulation. Heliyon 10 (21), e39884. Available at https://doi.org/10.1016/j.heliyon.2024.e39884.

Gu, Z., Lin, G., Xie, G., 2022. Carbon nitride (C_3N) nanoslits guided membrane for efficient seawater desalination. Applied Surface Science 582. Available at https://doi.org/10.1016/j.apsusc.2022.152460.

Guerrero-Sanchez, J., Muñoz-Pizza, D.M., Moreno-Armenta, M.G., Takeuchi, N., 2022. Atomic-scale understanding of the Na and Cl trapping on the $Mo_{1.33}C(OH)_2$-MXene. Scientific Reports 12 (1). Available at https://doi.org/10.1038/s41598-022-12177-6.

Guo, Y., Sun, H., An, M., Mabuchi, T., Zhao, Y., Li, G., 2023. Prediction of water transport properties on an anisotropic wetting surface via deep learning. Nanoscale 15 (30), 12737–12747. Available at https://doi.org/10.1039/d3nr02709k.

Gupta, K.M., Aitipamula, S., Chow, P.S., 2023. In-silico design of covalent organic framework membranes for efficient water/ethanol separation. Journal of Membrane Science 687, 122083. Available at https://doi.org/10.1016/j.memsci.2023.122083.

Güvensoy-Morkoyun, A., Baysal, T., Tantekin-Ersolmaz, S.B., Velioğlu, S., 2024. Guide for Nonequilibrium Molecular Dynamics Simulations of Organic Solvent Transport in Nanopores: The Case of 2D MXene Membranes. Journal of Chemical Theory and Computation 20 (21), 9642–9654. Available at https://doi.org/10.1021/acs.jctc.4c00693.

Haghbin, M.R., Niknam, M., Lotfikatooli, L., 2024. Potential of green-based microporous carbon for advanced water decontamination from azo dyes: Experiment and molecular dynamic simulation studies. Journal of Environmental Chemical Engineering 12 (1), 111875. Available at https://doi.org/10.1016/j.jece.2024.111875.

Hassan, J., Naz, S., Haider, Raza, A., Ul-Hamid, A., Qumar, U., et al., 2021. h-BN nanosheets doped with transition metals for environmental remediation; a DFT approach and molecular docking analysis. Materials Science and Engineering: B 272. Available at https://doi.org/10.1016/j.mseb.2021.115365.

He, S., Zhang, W., Gu, Z., 2023. Ultrafast seawater desalination with phosphorene oxide nanochannel based membrane: Nonmonotonic relationship between oxidation degree and water permeability. Desalination 563, 116751. Available at https://doi.org/10.1016/j.desal.2023.116751.

Hosseini, M., Azamat, J., Erfan-Niya, H., 2018. Improving the performance of water desalination through ultra-permeable functionalized nanoporous graphene oxide membrane. Applied Surface Science 427, 1000–1008. Available at https://doi.org/10.1016/j.apsusc.2017.09.071.

Hou, Y., Chen, W., Shi, C., Bao, X., Hu, 2020. Graphene oxide/methylene blue composite membrane for dyes separation: Formation mechanism and separation performance. Applied Surface Science 505, 144145. Available at https://doi.org/10.1016/j.apsusc.2019.144145.

Huang, R., Ma, C., Ma, J., Huangfu, X., He, Q., 2021. Machine learning in natural and engineered water systems. Water Research 205, 117666. Available at https://doi.org/10.1016/j.watres.2021.117666.

Huo, X., Yi, H., Fu, Y., An, Z., Qin, L., Liu, X., et al., 2021. Porous graphitic carbon nitride nanomaterials for water treatment. Environmental Science: Nano 8 (7), 1835–1862. Available at https://doi.org/10.1039/d1en00171j.

Huwaimel, B., Alqarni, S., 2025. A DFT investigation of the adsorption mechanism of paclitaxel on functionalized graphene oxide for enhanced drug delivery. Scientific Reports 15 (1), 13889. Available at https://doi.org/10.1038/s41598-025-99156-9.

Ihsanullah, I., 2020. MXenes (two-dimensional metal carbides) as emerging nanomaterials for water purification: Progress, challenges and prospects. Chemical Engineering Journal 388, 124340. Available at https://doi.org/10.1016/j.cej.2020.124340.

Ivády, V., Gali, A., Abrikosov, I.A., 2017. Hybrid-DFT Vw method for band structure calculation of semiconducting transition metal compounds: The case of cerium dioxide. Journal of Physics: Condensed Matter 29 (45). Available at https://doi.org/10.1088/1361-648X/aa8b93.

Iyyappan, J., Gaddala, B., Gnanasekaran, R., Gopinath, M., Yuvaraj, D., et al., 2024. Critical review on wastewater treatment using photo catalytic advanced oxidation process: Role of photocatalytic materials, reactor design and kinetic. Case Studies in Chemical and Environmental Engineering 9, 100599. Available at https://doi.org/10.1016/J.CSCEE.2023.100599.

Jablonka, K.M., Ongari, D., Moosavi, S.M., Smit, B., 2020. Big-Data Science in Porous Materials: Materials Genomics and Machine Learning. Chemical Reviews 120 (16), 8066–8129. Available at https://doi.org/10.1021/acs.chemrev.0c00004, http://pubs.acs.org/journal/chreay.

Jahangirzadeh, M., Azamat, J., Erfan-Niya, H., 2022. Molecular insight into water desalination through functionalized graphenylene nanosheet membranes. Computational Materials Science 203, 111126. Available at https://doi.org/10.1016/j.commatsci.2021.111126.

Janesko, B.G., 2021. Replacing hybrid density functional theory: Motivation and recent advances. Chemical Society Reviews 50 (15), 8470–8495. Available at https://doi.org/10.1039/d0cs01074j.

Jordan, M.I., Mitchell, T.M., 2015. Machine learning trends, perspectives, and prospects. Science 349 (6245), 255–260. Available at https://doi.org/10.1126/science.aaa8415.

Jorgensen, W.L., Chandrasekhar, J., Madura, J.D., Impey, R.W., Klein, M.L., 1983. Comparison of simple potential functions for simulating liquid water. The Journal of Chemical Physics 79 (2), 926–935. Available at https://doi.org/10.1063/1.445869.

Junejo, R., Shams Jalbani, N., Kaya, S., Serdaroglu, G., Şimşek, S., Memon, S., 2022. Experimental and DFT modeling studies for the adsorptive removal of reactive dyes from wastewater. Separation Science and Technology 57 (3), 339–353. Available at https://doi.org/10.1080/01496395.2021.1900252.

Kadaoluwa Pathirannahalage, S.P., Meftahi, N., Elbourne, A., Weiss, A.C.G., McConville, C.F., Padua, A., et al., 2021. Systematic comparison of the structural and dynamic properties of commonly used water models for molecular dynamics simulations. Journal of Chemical Information and Modeling 61 (9), 4521–4536. Available at https://doi.org/10.1021/acs.jcim.1c00794.

Karimzadeh, S., Safaei, B., Jen, T.-C., Oviroh, P.O., 2021. Enhanced removal efficiency of heavy metal ions from wastewater through functionalized silicon carbide membrane: A theoretical study. Journal of Water Process Engineering 44, 102413. Available at https://doi.org/10.1016/j.jwpe.2021.102413.

Karmakar, T., Finney, A.R., Salvalaglio, M., Yazaydin, A.O., Perego, C., 2023. Non-equilibrium modeling of concentration-driven processes with constant chemical potential molecular dynamics simulations. Accounts of Chemical Research 56 (10), 1156–1167. Available at https://doi.org/10.1021/acs.accounts.2c00811.

Karpatne, A., Atluri, G., Faghmous, J.H., Steinbach, M., Banerjee, A., Ganguly, A., et al., 2017. Theory-guided data science: A new paradigm for scientific discovery from data. IEEE Transactions on Knowledge and Data Engineering 29 (10), 2318–2331. Available at https://doi.org/10.1109/TKDE.2017.2720168.

Kaviani, S., Tayurskii, D.A., Nedopekin, O.V., Piyanzina, I., 2022. DFT insight into Cd^2, Hg^2, Pb^2, Sn^2, As^3, Sb^3, and Cr^3 heavy metal ions adsorption onto surface of bowl-like B30 nanosheet. Journal of Molecular Liquids 365, 120131. Available at https://doi.org/10.1016/j.molliq.2022.120131.

Khalaf, M.M., Gouda, M., Shalabi, K., Shaaban, S., Abd El-Lateef, H.M., 2022. Structural and adsorptive characteristics of 2D multilayer nanoflakes of NiCo phosphates for chromium(VI) removal: experimental and Monte Carlo simulations. ACS Omega 7 (12), 10738–10750. Available at https://doi.org/10.1021/acsomega.2c00529.

Khnifira, M., Boumya, W., Attarki, J., Mahsoune, A., Sadiq, M., Abdennouri, M., et al., 2022. A combined DFT, Monte Carlo, and MD simulations of adsorption study of heavy metals on the carbon graphite (111) surface. Chemical Physics Impact 5, 100121. Available at https://doi.org/10.1016/j.chphi.2022.100121.

Köhler, M.H., Bordin, J.R., Barbosa, M.C., 2018. 2D nanoporous membrane for cation removal from water: Effects of ionic valence, membrane hydrophobicity, and pore size. Journal of Chemical Physics 148 (22). Available at https://doi.org/10.1063/1.5013926.

Kohn, W., Sham, L.J., 1965. Self-consistent equations including exchange and correlation effects. Physical Review 140 (4A), A1133. Available at https://doi.org/10.1103/physrev.140.a1133.

Kommu, A., Namsani, S., Singh, J.K., 2016. Removal of heavy metal ions using functionalized graphene membranes: A molecular dynamics study. RSC Advances 6 (68), 63190–63199. Available at https://doi.org/10.1039/c6ra05817k.

Kong, X., Ma, J., Garg, S., Waite, T.D., 2024. Tailored metal-organic frameworks for water purification: perfluorinated Fe-MOFs for enhanced heterogeneous catalytic ozonation. Environmental Science and Technology 58 (20), 8988–8999. Available at https://doi.org/10.1021/acs.est.4c01133.

Kyung, R., Choi, I.H., 2023. United States Metal-Organic Frameworks and Porphyrins for Water Filtration Using Density Functional Theory. In: IEEE 13th Annual Computing and Communication Workshop and Conference, CCWC, 268–273. Institute of Electrical and Electronics Engineers Inc. https://doi.org/10.1109/CCWC57344.2023.10099165.

Lecun, Y., Bengio, Y., Hinton, G., 2015. Deep learning. Nature 521 (7553), 436–444. Available at https://doi.org/10.1038/nature14539.

Lee, D.E., Kim, M.K., Danish, M., Jo, W.K., 2023. State-of-the-art review on photocatalysis for efficient wastewater treatment: Attractive approach in photocatalyst design and parameters affecting the photocatalytic degradation. Catalysis Communications 183, 106764. Available at https://doi.org/10.1016/j.catcom.2023.106764.

Lin, Q., Zeng, G., Yan, G., Luo, J., Cheng, X., Zhao, Z., et al., 2022. Self-cleaning photocatalytic MXene composite membrane for synergistically enhanced water treatment: Oil/water separation and dyes removal. Chemical Engineering Journal 427, 131668. Available at https://doi.org/10.1016/j.cej.2021.131668.

Liu, C., Dai, H., Tan, C., Pan, Q., Hu, F., Peng, X., 2022. Photo-Fenton degradation of tetracycline over Z-scheme Fe-g-C_3N_4/Bi_2WO_6 heterojunctions: Mechanism insight, degradation pathways and DFT calculation. Applied Catalysis B: Environmental 310, 121326. Available at https://doi.org/10.1016/J.APCATB.2022.121326.

Liu, H., Xia, J., Zou, L., Huo, R., 2022. Comprehensive quantitative evaluation of the water resource carrying capacity in Wuhan City based on the "human–water–city" framework: Past, present and future. Journal of Cleaner Production 366, 132847. Available at https://doi.org/10.1016/j.jclepro.2022.132847.

Liu, H., Wang, X., Wang, R., Xu, H., Yan, W., An, Q., 2024. Computational investigations on the performance of transition metal carbide (MXene) for methylene blue dye adsorption and remediation from wastewater. Materials Today Communications 339. Available at https://doi.org/10.1016/j.mtcomm.2024.108735.

Liu, J., Zhao, Z., Li, L., Wu, Y., He, H., 2023. Molecular simulation study of 2D MXene membranes for organic solvent nanofiltration. Journal of Membrane Science 677, 121623. Available at https://doi.org/10.1016/J.MEMSCI.2023.121623.

Liu, L., Liu, Y., Qi, Y., Song, M., Jiang, L., Fu, G., et al., 2020. Hexagonal boron nitride with nanoslits as a membrane for water desalination: A molecular dynamics investigation. Separation and Purification Technology 251, 117409. Available at https://doi.org/10.1016/j.seppur.2020.117409.

Liu, Q., Wang, X., Guo, Y., Liu, G., Zhou, K.G., 2023. Mechanism of ethanol/water reverse separation through a functional graphene membrane: a molecular simulation investigation. Frontiers of Chemical Science and Engineering 17 (3), 347–357. Available at https://doi.org/10.1007/s11705-022-2246-z.

Liu, Q., Wu, Y., Wang, X., Liu, G., Zhu, Y., Tu, Y., et al., 2019. Molecular dynamics simulation of water-ethanol separation through monolayer graphene oxide membranes: Significant role of O/C ratio and pore size. Separation and Purification Technology 224, 219–226. Available at https://doi.org/10.1016/j.seppur.2019.05.030.

Liu, X., Graham, N., Yu, W., Shi, Y., Sun, K., Liu, T., 2022. Preparation and evaluation of a high performance $Ti_3C_2T_x$-MXene membrane for drinking water treatment. Journal of Membrane Science 654, 120469. Available at https://doi.org/10.1016/J.MEMSCI.2022.120469.

Liu, Y., Cheng, Z., Song, M., Jiang, L., Fu, G., Liu, L., et al., 2021. Molecular dynamics simulation-directed rational design of nanoporous graphitic carbon nitride membranes for water desalination. Journal of Membrane Science 620, 118869. Available at https://doi.org/10.1016/j.memsci.2020.118869.

Liu, Y., Xie, D., Song, M., Jiang, L., Fu, G., Liu, L., et al., 2018. Water desalination across multilayer graphitic carbon nitride membrane: Insights from non-equilibrium molecular dynamics simulations. Carbon 140, 131–138. Available at https://doi.org/10.1016/j.carbon.2018.08.043.

Lowe, M., Qin, R., Mao, X., 2022. Review on machine learning, artificial intelligence, and smart technology in water treatment and monitoring. Water 14 (9), 1384. Available at https://doi.org/10.3390/w14091384.

Lu, B., Xia, Y., Ren, Y., Xie, M., Zhou, L., Vinai, G., et al., 2024. When machine learning meets 2D materials: A review. Advanced Science 11 (13). Available at https://doi.org/10.1002/advs.202305277.

Lu, Z., Wei, Y., Deng, J., Ding, L., Li, Z.-K., Wang, H., 2019. Self-crosslinked MXene ($Ti_3C_2T_x$) membranes with good antiswelling property for monovalent metal ion exclusion. ACS Nano 13 (9), 10535–10544. Available at https://doi.org/10.1021/acsnano.9b04612.

Lv, Z., Zhao, Z., 2024. Research on the application of hydrophilic modified covalent organic framework membranes in seawater desalination. Surfaces and Interfaces 51, 104638. Available at https://doi.org/10.1016/j.surfin.2024.104638.

Ma, X., Lan, C., Lin, H., Peng, Y., Li, T., Wang, J., et al., 2024. Designing desalination MXene membranes by machine learning and global optimization algorithm. Journal of Membrane Science 702, 122803. Available at https://doi.org/10.1016/j.memsci.2024.122803.

Ma, X., Zhu, X., Huang, C., Fan, J., 2022. Revealing the effects of terminal groups of MXene on the water desalination performance. Journal of Membrane Science 647, 120334. Available at https://doi.org/10.1016/j.memsci.2022.120334.

van der Maaten, L., Hinton, G., 2008. Visualizing data using t-SNE, Journal of Machine Learning Research.

Majidi, S., Erfan-Niya, H., Azamat, J., Cruz-Chú, E.R., Walther, J.H., 2021. Efficient removal of heavy metals from aqueous solutions through functionalized γ-graphyne-1 membranes under external uniform electric fields: Insights from molecular dynamics simulations. Journal of Physical Chemistry B 125 (44), 12254–12263. Available at https://doi.org/10.1021/acs.jpcb.1c06617.

Majidi, S., Erfan-Niya, H., Azamat, J., Cruz-Chú, E.R., Honoré Walther, J., 2022. The performance of a C_2N membrane for heavy metal ions removal from water under external electric field. Separation and Purification Technology 289, 120770. Available at https://doi.org/10.1016/j.seppur.2022.120770.

Maleki, R., Shams, S.M., Chellehbari, Y.M., Rezvantalab, S., Jahromi, A.M., Asadnia, M., et al., 2022. Materials discovery of ion-selective membranes using artificial intelligence. Communications Chemistry 5 (1), 132. Available at https://doi.org/10.1038/s42004-022-00744-x.

Mardirossian, N., Head-Gordon, M., 2017. Thirty years of density functional theory in computational chemistry: An overview and extensive assessment of 200 density functionals. Molecular Physics 115 (19), 2315–2372. Available at https://doi.org/10.1080/00268976.2017.1333644.

Martínez-Jódar, A., Villar-Rodil, S., Salvadó, M.A., Carrasco, D.F., Pertierra, P., Recio, J.M., et al., 2023. Two-dimensional transition metal dichalcogenides beyond MoS_2 for the catalytic reduction of nitroarenes: $MoSe_2$ exhibits enhanced performance. Applied Catalysis B: Environmental 339. Available at https://doi.org/10.1016/j.apcatb.2023.123174.

Mashhadzadeh, A.H., Vahedi, A.M., Ardjmand, M., Ahangari, M.G., 2016. Investigation of heavy metal atoms adsorption onto graphene and graphdiyne surface: A density functional theory study. Superlattices and Microstructures 100, 1094–1102. Available at https://doi.org/10.1016/j.spmi.2016.10.079.

Meidani, K., Cao, Z., Barati Farimani, A., 2021. Titanium carbide MXene for water desalination: a molecular dynamics study. ACS Applied Nano Materials 4 (6), 6145–6151. Available at https://doi.org/10.1021/acsanm.1c00944.

Meng, Y., Ding, B., 2022. Design of automatic motion capture algorithm for Yao's long drum dance based on multieye machine vision. Scientific Programming 2022, 1–10. Available at https://doi.org/10.1155/2022/1126020.

Metropolis, N., Ulam, S., 1949. The Monte Carlo method. Journal of the American Statistical Association 44 (247), 335–341. Available at https://doi.org/10.1080/01621459.1949.10483310.

Milon, Hossain, M.K., Roy, D., Ahmed, F., 2021. Boron nanocluster as a heavy metal adsorbent in aqueous environment A DFT Study. Journal of Molecular Structure 1237. Available at https://doi.org/10.1016/j.molstruc.2021.130302.

Moslehi, M.H., Eslami, M., Ghadirian, M., Nateq, K., Ramavandi, B., Nasseh, N., 2024. Photocatalytic decomposition of metronidazole by zinc hexaferrite coated with bismuth oxyiodide magnetic nanocomposite: Advanced modelling and optimization with artificial neural network. Chemosphere 356, 141770. Available at https://doi.org/10.1016/j.chemosphere.2024.141770.

Murdoch, W.J., Singh, C., Kumbier, K., Abbasi-Asl, R., Yu, B., 2019. Definitions, methods, and applications in interpretable machine learning. Proceedings of the National Academy of Sciences 116 (44), 22071–22080. Available at https://doi.org/10.1073/pnas.1900654116.

Mutahir, S., Akram, S., Asim Khan, M., Deng, H., Naglah, A.M., Almehizia, A.A., et al., 2024. Facile synthesis of Zn/Co LDH for the removal of oxytetracycline from wastewater: Experimental and DFT-Based analysis. Chemical Engineering Science 283, 119399. Available at https://doi.org/10.1016/j.ces.2023.119399.

Narayanaswamy, V., Alaabed, S., AL-Akhras, M.-A., Obaidat, I.M., 2020. Molecular simulation of adsorption of methylene blue and rhodamine B on graphene and graphene oxide for water purification. Materials Today: Proceedings 28, 1078–1083. Available at https://doi.org/10.1016/j.matpr.2020.01.086.

Novoselov, K.S., Geim, A.K., Morozov, S.V., Jiang, D., Zhang, Y., Dubonos, S.V., et al., 2004. Electric field effect in atomically thin carbon films. Science 306 (5696), 666–669. Available at https://doi.org/10.1126/science.1102896.

Osti, N.C., Naguib, M., Ostadhossein, A., Xie, Y., Kent, P.R.C., Dyatkin, B., et al., 2016. Effect of metal ion intercalation on the structure of MXene and water dynamics on its internal surfaces. ACS Applied Materials and Interfaces 8 (14), 8859–8863. Available at https://doi.org/10.1021/acsami.6b01490.

Ou, Y., Gu, Z., Luo, Y., 2024. Efficient heavy metal ion removal by fluorographene nanochannel templated molecular sieve: A molecular dynamics simulation study. Scientific Reports 14 (1), 6298. Available at https://doi.org/10.1038/s41598-024-56908-3.

Oviroh, P.O., Jen, T.C., Ren, J., Mohlala, L.M., Warmbier, R., Karimzadeh, S., 2021. Nanoporous MoS_2 membrane for water desalination: A molecular dynamics study. Langmuir the ACS Journal of Surfaces and Colloids 37 (23), 7127–7137. Available at https://doi.org/10.1021/acs.langmuir.1c00708.

Oyetade, O.A., Skelton, A.A., Nyamori, V.O., Jonnalagadda, S.B., Martincigh, B.S., 2017. Experimental and DFT studies on the selective adsorption of Pb2 and Zn2 from aqueous solution by nitrogen-functionalized multiwalled carbon nanotubes. Separation and Purification Technology 188, 174–187. Available at https://doi.org/10.1016/j.seppur.2017.07.022.

Ozcan, A., Perego, C., Salvalaglio, M., Parrinello, M., Yazaydin, O., 2017. Concentration gradient driven molecular dynamics: A new method for simulations of membrane permeation and separation. Chemical Science 8 (5), 3858–3865. Available at https://doi.org/10.1039/c6sc04978h.

Pedregosa, F., Varoquaux, G., Gramfort, A., Michel, V., Thirion, B., Grisel, O., et al., 2011. Scikit-learn: machine learning in Python. Journal of Machine Learning Research 12, 2825–2830.

Perdew, J.P., Ernzerhof, M., Burke, K., 1996. Rationale for mixing exact exchange with density functional approximations. Journal of Chemical Physics 105 (22), 9982–9985. Available at https://doi.org/10.1063/1.472933.

Priya, P., Nguyen, T.C., Saxena, A., Aluru, N.R., 2022. Machine learning assisted screening of two-dimensional materials for water desalination. ACS Nano 16 (2), 1929–1939. Available at https://doi.org/10.1021/acsnano.1c05345, http://pubs.acs.org/journal/ancac3.

Qasem, N.A.A., Mohammed, R.H., Lawal, D.U., 2021. Removal of heavy metal ions from wastewater: A comprehensive and critical review. NPJ Clean Water 4 (1), 36. Available at https://doi.org/10.1038/s41545-021-00127-0.

Ramprasad, R., Batra, R., Pilania, G., Mannodi-Kanakkithodi, A., Kim, C., 2017. Machine learning in materials informatics: Recent applications and prospects. NPJ Computational Materials 3 (1), 54. Available at https://doi.org/10.1038/s41524-017-0056-5.

Rao, K.S., Tirth, V., Almujibah, H., Alshahri, A.H., Hariprasad, V., Senthilkumar, N., 2023. Optimization of water reuse and modelling by saline composition with nanoparticles based on machine learning architectures. Water Science & Technology 87 (11), 2793–2805. Available at https://doi.org/10.2166/wst.2023.161.

Risplendi, F., Raffone, F., Lin, L.C., Grossman, J.C., Cicero, G., 2020. Fundamental insights on hydration environment of boric acid and its role in separation from saline water. Journal of Physical Chemistry C 124 (2), 1438–1445. Available at https://doi.org/10.1021/acs.jpcc.9b10065.

Roscher, R., Bohn, B., Duarte, M.F., Garcke, J., 2020. Explainable machine learning for scientific insights and discoveries. IEEE Access 8, 42200–42216. Available at https://doi.org/10.1109/ACCESS.2020.2976199.

Saadh, M.J., Hassan, W.H., Nemah, A.K., Khaleel Jameel, M., Sharma, P., Kumar, A., et al., 2024. Removal of erythrosine B dye from wastewater using Ca_2C and Ti_2C MXenes: A theoretical study. Journal of Molecular Liquids 411.

Salahshoori, I., Mahdavi, S., Moradi, Z., Otadi, M., Zare Kazemabadi, F., Nobre, M.A.L., et al., 2024. Advancements in molecular simulation for understanding pharmaceutical pollutant Adsorption: A State-of-the-Art review. Journal of Molecular Liquids 410, 125513. Available at https://doi.org/10.1016/j.molliq.2024.125513.

Schütt, K.T., Sauceda, H.E., Kindermans, P.-J., Tkatchenko, A., Müller, K.-R., 2018. A deep learning architecture for molecules and materials. Journal of Chemical Physics 148 (24). Available at https://doi.org/10.1063/1.5019779.

Seyedin, M., Hassanpour, A., Jalali, A., Raisee, M., 2024. Robust optimization of a novel ultraviolet (UV) photoreactor for water disinfection: A neural network approach. Chemosphere 362, 142788. Available at https://doi.org/10.1016/j.chemosphere.2024.142788.

Shaharudin, M.R.B., Williams, C.D., Carbone, P., 2022. The role of surface ionisation in the hydration-induced swelling of graphene oxide membranes. Journal of Membrane Science 653, 120489. Available at https://doi.org/10.1016/j.memsci.2022.120489.

Shi, Q., He, Z., Gupta, K.M., Wang, Y., Lu, R., 2017. Efficient ethanol/water separation via functionalized nanoporous graphene membranes: insights from molecular dynamics study. Journal of Materials Science 52 (1), 173–184. Available at https://doi.org/10.1007/s10853-016-0319-4.

Si, T., Ma, X., Wang, T., Tak Chu, S., Fan, J., 2023. Improvement of desalination performance by adjusting the arrangement of lamellar MXene membrane. Separation and Purification Technology 322, 124265. Available at https://doi.org/10.1016/j.seppur.2023.124265.

Song, X., Lu, L., Wei, M., Dai, Z., Wang, S., 2020. Molecular dynamics simulations on the water flux in different two-dimension materials. Molecular Simulation 46 (9), 689–698. Available at https://doi.org/10.1080/08927022.2018.1510179.

Sun, Y., Zhao, Z., Tong, H., Sun, B., Liu, Y., Ren, N., et al., 2023. Machine learning models for inverse design of the electrochemical oxidation process for water purification. Environmental Science and Technology 57 (46), 17990–18000. Available at https://doi.org/10.1021/acs.est.2c08771.

Sutton, R.S., Barto, A., 1999. Reinforcement learning: An introduction. MIT Press, pp. 229–235. Available at https://web.stanford.edu/class/psych209/Readings/SuttonBartoIPRLBook2ndEd.pdf.

Syah, R., Al-Khowarizmi, A., Elveny, M., Khan, A., 2021. Machine learning based simulation of water treatment using LDH/MOF nanocomposites. Environmental Technology & Innovation 23, 101805. Available at https://doi.org/10.1016/j.eti.2021.101805.

Syah, R., Piri, F., Elveny, M., Khan, A., 2022. Artificial Intelligence simulation of water treatment using nanostructure composite ordered materials. Journal of Molecular Liquids 345, 117046. Available at https://doi.org/10.1016/j.molliq.2021.117046.

Tabasi, E., Vafa, N., Firoozabadi, B., Salmankhani, a, Nouranian, S., Habibzadeh, S., et al., 2023. Ion rejection performances of functionalized porous graphene nanomembranes for wastewater purification: A molecular dynamics simulation study. Colloids and Surfaces A: Physicochemical and Engineering Aspects 656, 130492. Available at https://doi.org/10.1016/j.colsurfa.2022.130492.

Takaba, H., Koyama, A., Nakao, S.I., 2000. Dual ensemble Monte Carlo simulation of pervaporation of an ethanol/water binary mixture in silicalite membrane based on a Lennard-Jones interaction model. Journal of Physical Chemistry B 104 (27), 6353–6359. Available at https://doi.org/10.1021/jp994384v.

Takamatsu, H., Ohba, T., 2021. Water adsorption control by surface nanostructures on graphene-related materials by grand canonical Monte Carlo simulations. Langmuir the ACS Journal of Surfaces and Colloids 37 (50), 14646–14656. Available at https://doi.org/10.1021/acs.langmuir.1c02372.

Tanveer, H.B., Perveen, F., Azam, S., Arshad, n, Rafique, H., Irfan, A., et al., 2024. Efficient removal of toxic azo dyes from contaminated water by adsorption on the GO surface. PLoS One 19 (3), e0299364.

Tsai, M.Y., Lin, L.C., 2024. Pervaporation separation of isopropanol/water using zeolite nanosheets: A molecular simulation study. Journal of Physical Chemistry B 128 (35), 8546–8556. Available at https://doi.org/10.1021/acs.jpcb.4c04237.

urRehman, S., Sajjad, A., Bibi, S., Tabassum, H., Jamil, S., Khan, S.R., et al., 2023. A DFT study on removal of Mn and V from wastewater using chitosan/graphene oxide composite as adsorbent. Chemistry of Inorganic Materials 1, 100030. Available at https://doi.org/10.1016/J.CINORG.2023.100030.

Vafa, N., Firoozabadi, B., Pishkenari, H.N., Mashhadzadeh, A.H., 2023. Comparative analysis of graphene and h-BN nanochannel membranes for confined water purification: Insights into desalination performance through molecular dynamics simulation. Journal of Molecular Liquids 389. Available at https://doi.org/10.1016/j.molliq.2023.122920.

Vafa, N., Firoozabadi, B., Pishkenari, H.N., 2023. Confined water transport and desalination mechanism through lamellar Boron-Nitride Channels: A molecular dynamics simulation study. Computational Materials Science 226, 112208. Available at https://doi.org/10.1016/j.commatsci.2023.112208.

Van Duin, A.C.T., Dasgupta, S., Lorant, F., Goddard, W.A., 2001. ReaxFF: A reactive force field for hydrocarbons. Journal of Physical Chemistry A 105 (41), 9396–9409. Available at https://doi.org/10.1021/jp004368u.

Wakjira, T.L., Gemta, A.B., Tadele, K., Kassahun, G.B., sherefedin, U., Gurumurthi, T., et al., 2025. Molecular structures and adsorption of dyes on bismuth oxychloride surfaces using density functional theory and Monte Carlo dynamic simulation. Results in Physics 68, 108106. Available at https://doi.org/10.1016/j.rinp.2024.108106.

Wang, G., Dou, K., Cao, H., Du, R., Liu, J., Tsidaeva, N.I., et al., 2022. Designing Z-scheme CdS/WS_2 heterojunctions with enhanced photocatalytic degradation of organic dyes and photoreduction of Cr (VI): Experiments, DFT calculations and mechanism. Separation and Purification Technology 291. Available at https://doi.org/10.1016/j.seppur.2022.120976.

Wang, L., Hou, C., Luo, L., Xiang, B., Zhu, Z., Wang, C., 2025. Confining graphite-like carbon into ultrathin layered iron silicate for notably enhanced photo-Fenton catalysis: Mechanism insight and DFT calculation. Applied Catalysis B: Environmental and Energy 371, 125281. Available at https://doi.org/10.1016/J.APCATB.2025.12528.

Wang, L., Yuan, L., Chen, K., Zhang, Y., Deng, Q., Du, S., et al., 2016. Loading actinides in multilayered structures for nuclear waste treatment: the first case study of uranium capture with vanadium carbide MXene. ACS Applied Materials and Interfaces 8 (25), 16396–16403. Available at https://doi.org/10.1021/acsami.6b02989.

Wang, S., Maganga, I., Zeng, L., Gu, Z., 2024. Graphene crown pore for efficient heavy metal ion Removal: Protonated vs. Non-protonated. Journal of Molecular Liquids 395, 123819. Available at https://doi.org/10.1016/j.molliq.2023.123819.

Wang, Y., Su, Q., Li, L., Zhao, D., 2024. Molecular simulation study of $MoSe_2$ nanochannel for seawater desalination. Journal of Membrane Science 702. Available at https://doi.org/10.1016/j.memsci.2024.122811.

Wang, Y., Sorkun, M.C., Brocks, G., Er, S., 2024. ML-aided computational screening of 2D materials for photocatalytic water splitting. Journal of Physical Chemistry Letters 15 (18), 4983–4991. Available at https://doi.org/10.1021/acs.jpclett.4c00425.

Wang, Y., Cao, Z., Barati Farimani, A., 2021. Efficient water desalination with graphene nanopores obtained using artificial intelligence. NPJ 2D Materials and Applications 5 (1), 66. Available at https://doi.org/10.1038/s41699-021-00246-9.

Ward, L., Dunn, A., Faghaninia, A., Zimmermann, N.E.R., Bajaj, S., Wang, Q., et al., 2018. Matminer: An open source toolkit for materials data mining. Computational Materials Science 152, 60–69. Available at https://doi.org/10.1016/j.commatsci.2018.05.018.

Ward, L., Agrawal, A., Choudhary, A., Wolverton, C., 2016. A general-purpose machine learning framework for predicting properties of inorganic materials. NPJ Computational Materials 2 (1), 16028. Available at https://doi.org/10.1038/npjcompumats.2016.28.

Wei, D., Li, J., Chen, Z., Liang, L., Ma, J., Wei, M., et al., 2020. Understanding bisphenol-A adsorption in magnetic modified covalent organic frameworks: Experiments coupled with DFT calculations. Journal of Molecular Liquids. pp. 301. Available at https://doi.org/10.1021/acsnano.6b08538.

Willcox, J.A.L., Kim, H.J., 2017. Molecular dynamics study of water flow across multiple layers of pristine, oxidized, and mixed regions of graphene oxide. ACS Nano 11 (2), 2187–2193. Available at https://doi.org/10.1021/acsnano.6b08538.

Wu, Y., Chen, X., Cao, J., Zhu, Y., Yuan, W., Hu, Z., et al., 2022. Photocatalytically recovering hydrogen energy from wastewater treatment using $MoS_2@TiO_2$ with sulfur/oxygen dual-defect. Applied Catalysis B: Environmental 303, 120878. Available at https://doi.org/10.1016/j.apcatb.2021.120878.

Xie, T., Grossman, J.C., 2018. Crystal graph convolutional neural networks for an accurate and interpretable prediction of material properties. Physical Review Letters 120 (14), 145301. Available at https://doi.org/10.1103/PhysRevLett.120.145301.

Xing, D., Cui, Z., Liu, Y., Wang, Z., Wang, P., Zheng, Z., et al., 2021. Two-dimensional π–d conjugated metal–organic framework Fe3(hexaiminotriphenylene)2 as a photo-Fenton like catalyst for highly efficient degradation of antibiotics. Applied Catalysis B: Environmental 290. Available at https://doi.org/10.1016/j.apcatb.2021.120029.

Xue, M., Qiu, H., Guo, W., 2013. Exceptionally fast water desalination at complete salt rejection by pristine graphyne monolayers. Nanotechnology 24 (50), 505720. Available at https://doi.org/10.1088/0957-4484/24/50/505720.

Yadav, A., Dindorkar, S.S., 2022. Adsorption behaviour of hexagonal boron nitride nanosheets towards cationic, anionic and neutral dyes: Insights from first principle studies. Colloids and Surfaces A: Physicochemical and Engineering Aspects 640. Available at https://doi.org/10.1016/j.colsurfa.2022.128509.

Yang, J., Shen, Z., He, J., Li, Y., 2021. Efficient separation of small organic contaminants in water using functionalized nanoporous graphene membranes: Insights from molecular dynamics simulations. Journal of Membrane Science 630, 119331. Available at https://doi.org/10.1016/j.memsci.2021.119331.

Yang, Q., Cao, W., Meng, W., Si, J., 2022. Reinforcement-learning-based tracking control of waste water treatment process under realistic system conditions and control performance requirements. IEEE Transactions on Systems, Man, and Cybernetics: Systems 52 (8), 5284–5294. Available at https://doi.org/10.1109/TSMC.2021.3122802.

Yang, Z., Lin, Q., Zeng, G., Zhao, S., Yan, G., Ang, M.B.M.Y., et al., 2023. Ternary hetero-structured $BiOBr/Bi_2MoO_6@MXene$ composite membrane: Construction and enhanced removal of antibiotics and dyes from water. Journal of Membrane Science 669. Available at https://doi.org/10.1016/j.memsci.2022.121329.

Yi, M., Héraly, F., Chang, J., Khorsand Kheirabad, A., Yuan, J., Wang, Y., et al., 2021. A transport channel-regulated MXene membrane: Via organic phosphonic acids for efficient water permeation. Chemical Communications 57 (51), 6245–6248. Available at https://doi.org/10.1039/d1cc01464a.

Yu, T., Bai, L., Xu, Z., Yang, X., 2018. Molecular simulation of permeation behaviour of ethanol/water molecules with single-layer graphene oxide membranes. Molecular Simulation 44 (10), 840–849. Available at https://doi.org/10.1080/08927022.2018.1464161.

Yuan, M., Wang, J., Li, Y., Zhao, M., Li, Y.-Q., Li, W., et al., 2023. Metal-organic frameworks for high performance desalination through thickness control and structural fine-tuning. Water Research 230, 119576. Available at https://doi.org/10.1016/j.watres.2023.119576.

Zaman, W., Matsumoto, R.A., Thompson, M.W., Liu, Y.-H., Bootwala, Y., Dixit, M.B., et al., 2021. In situ investigation of water on MXene interfaces. PNAS 118 (49), e2108325118. Available at https://doi.org/10.1073/pnas.2108325118.

Zhang, J., Jing, B., Tang, Z., Ao, Z., Xia, D., Zhu, M., et al., 2021. Experimental and DFT insights into the visible-light driving metal-free C_3N_5 activated persulfate system for efficient water purification. Applied Catalysis B: Environmental 289, 120023. Available at https://doi.org/10.1016/j.apcatb.2021.120023.

Zhang, P., Liu, C., Lao, D., Nguyen, X.C., Paramasivan, B., Qian, X., et al., 2023. Unveiling the drives behind tetracycline adsorption capacity with biochar through machine learning. Scientific Reports 13 (1), 11512. Available at https://doi.org/10.1038/s41598-023-38579-8.

Zhang, S., Gao, K., Zhao, K., Li, L., Ju, S., Zhang, Q., et al., 2025. Hydrogen-bonding-engineered 2D MXene nanochannels with reduced mass transfer resistance for high-permeability water purification. Journal of Membrane Science 730, 124195. Available at https://doi.org/10.1016/J.MEMSCI.2025.124195.

Zhang, W., Xu, Z., Yang, X., 2019. Molecular simulation of penetration separation for ethanol/water mixtures using two-dimensional nanoweb graphynes. Chinese Journal of Chemical Engineering 27 (2), 286–292. Available at https://doi.org/10.1016/j.cjche.2018.02.028.

Zhang, Z., Chen, K., Zhao, Q., Huang, M., Ouyang, X., 2021. Comparative adsorption of heavy metal ions in wastewater on monolayer molybdenum disulfide. Green Energy and Environment 6 (5), 751–758. Available at https://doi.org/10.1016/j.gee.2020.06.019.

Zhao, Z., Guo, P., Zhang, Y., Li, Y., Zhang, H., He, T., et al., 2024. Fabrication of 2D TiO_2/α-Fe_2O_3/Fe-g-C_3N_4 heterogeneous from metallurgical solid waste for Photo-Fenton catalysis of levofloxacin/Cr(VI): Electron transfer pathway and DFT calculation. Journal of Catalysis 432. Available at https://doi.org/10.1016/j.jcat.2024.115447.

Zheng, B., Chu, X., Peng, Z., Tian, Y., 2023. Improving the separation performance for heavy metals by optimizing the structure of multilayered GO membrane. Journal of Molecular Liquids 370, 121071. Available at https://doi.org/10.1016/j.molliq.2022.121071.

Zhou, Z., Kearnes, S., Li, L., Zare, R.N., Riley, P., 2019. Optimization of molecules via deep reinforcement learning. Scientific Reports 9 (1), 10752. Available at https://doi.org/10.1038/s41598-019-47148-x.

Zou, C., Lin, L.C., 2020. Potential and design of zeolite nanosheets as pervaporation membranes for ethanol extraction. Industrial and Engineering Chemistry Research 59 (28), 12845–12854. Available at https://doi.org/10.1021/acs.iecr.0c01171.

CHAPTER

6

Advances and insights: 2D nanomaterials in membrane filtration for water purification

Meilan Pan[1], Jiaming Hu[1,2], Yuxuan Wei[2], Yueping Bao[1], Qiancheng Xia[3], and Guandao Gao[3]

[1]College of Environmental Science and Engineering, Ministry of Education Key Laboratory of Pollution Processes and Environmental Criteria, Tianjin Key Laboratory of Environmental Remediation and Pollution Control, Nankai University, Tianjin, P.R. China [2]College of Environment, Zhejiang University of Technology, Hangzhou, P.R. China [3]School of Environment, State Key Laboratory of Water Pollution Control and Green Resource Recycling, Nanjing University, Nanjing, P.R. China

6.1 Introduction: overview of 2D nanomaterials in membrane filtration for water purification

Over the past decades, membrane-based filtration has emerged as a key technology in addressing global challenges related to water scarcity and contamination. Owing to their high separation efficiency, operational simplicity, low energy demand, and small footprint, membrane processes have been widely implemented across various water treatment scenarios. The properties of membrane materials fundamentally govern the performance and longevity of these filtration systems. Polymeric membranes remain the most widely used due to their cost-effectiveness and ease of fabrication; however, they are inherently limited by several drawbacks, including trade-offs between permeability and selectivity, low fouling resistance, and limited chemical and mechanical stability. These limitations have motivated the development of novel membrane materials and architectures to address the challenges of water purification.

Two-dimensional (2D) nanomaterials have attracted growing attention as promising candidates for revolutionizing membrane engineering (Chen et al., 2017). 2D nanomaterials are defined by their atomic or molecular-scale thickness and lateral dimensions spanning from nanometers (nm) to micrometers (μm), possessing unique physicochemical properties

Two-Dimensional Nanomaterials for Water Purification
DOI: https://doi.org/10.1016/B978-0-443-44530-9.00006-7

(Xue et al., 2021). Their high aspect ratios, tunable surface chemistries, mechanical flexibility, and facile functionalization render them ideal materials for separation membranes with highly controlled porosity, functionality, and transport channels.

To date, a wide range of 2D nanomaterials has been successfully explored for membrane applications. The 2D nanomaterials can be classified into the nonporous and porous nanosheets (Fig. 6.1) (Kang et al., 2019). (1) Nonporous 2D nanosheets are impermeable to all atoms and molecules under ambient conditions (Loske et al., 2020). The assembled 2D nonporous membranes rely on the formed interlayer spacing between stacked sheets for separation, where the formed tortuous paths are the transport channels for molecules/ions. For instance, graphene-based nanomaterials (GBN) are the most popular group of nanosheets, such as graphene oxide (GO), which features abundant oxygen-containing functional groups and high mechanical strength (Chen et al., 2017). MXenes, a family of transition metal carbides and nitrides, possess the hydrophilic surfaces and rich surface terminations (e.g., Ti–O, –OH), facilitating water permeability and selective rejection of multivalent ions and organic pollutants (Karahan et al., 2020). Molybdenum disulfide (MoS_2) exhibits layered channels with low frictional resistance. Boron nitride (BN) combines chemical inertness with thermal stability for nanofluidic applications. Layered double hydroxides (LDHs) offer structural tunability and well-defined porous architectures. Graphitic carbon nitride (g-C_3N_4) nanosheets have garnered attention for their photocatalytic activity, etc. In contrast, porous 2D nanosheets possess the intrinsic nanoscale pores for molecular sieving (Lu et al., 2025). Examples of 2D nanosheets are as follows: covalent organic frameworks (COFs) are crystalline porous materials, offering high stability, tunable pore structures, and excellent molecular sieving properties (Zhang et al., 2019b). Metal–organic frameworks (MOFs) are hybrid crystalline materials constructed from metal ions or clusters coordinated to organic ligands, forming highly porous frameworks. Zeolites are known for their high thermal stability and selective ion exchange. These 2D nanomaterials enrich the transport channels and reduce the mass-transport resistance (Zhu et al., 2024b).

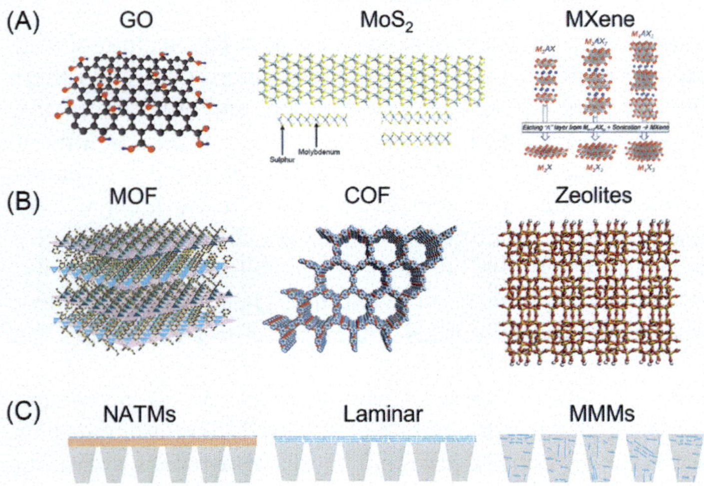

FIGURE 6.1 Schematic diagram of representative (A) nonporous 2D nanosheets, (B) porous 2D nanosheets, and (C) three types of nanosheet membranes (Loske et al., 2020). *Source: Reproduced with permission from Loske, L., Nakagawa, K., Yoshioka, T., Matsuyama, H., 2020. 2D Nanocomposite Membranes: Water Purification and Fouling Mitigation. Membranes, 10(10), 295. https://doi.org/10.3390/membranes10100295.*

Based on structural characteristics and intended applications, 2D nanomaterials are typically assembled in three forms (C) (Yang et al., 2019): nanoporous atomically thin membranes (NATMs), laminar stacked membranes, and mixed-matrix membranes (MMMs). (1) NATMs are composed of monolayer nanosheets (e.g., graphene) with in-plane nanopores that act as molecular sieving channels (Jang et al., 2022). This configuration minimizes transport resistance while enabling precise molecular sieving. However, their broader application is currently constrained by the lack of scalable fabrication methods, difficulty in achieving a uniform pore distribution, and poor mechanical robustness. (2) Laminar membranes, assembled by layer-by-layer stacking of 2D nanosheets, such as graphene oxide, MXenes, MoS_2, BN, or layered double hydroxides, represent a more scalable and structurally tunable platform (Han et al., 2025). In these membranes, mass transport occurs predominantly through interlayer nanochannels. Precise control over the interlayer spacing of nanosheets (Ding et al., 2020), via chemical crosslinking, intercalation, or surface functionalization, enables size- and charge-selective separations at subnanometer resolution. Moreover, the functional groups on the nanosheet surfaces can introduce electrostatic interactions, hydrogen bonding, or affinity-based recognition, thereby enhancing selectivity toward specific solutes (Zhang et al., 2019a). (3) MMMs incorporate 2D nanomaterials as fillers or additives within polymeric matrices to improve properties such as hydrophilicity, antifouling behavior, or selective adsorption. These enhancements arise from synergistic effects, including the formation of rapid transport pathways, improved polymer–filler interfacial compatibility, increased active surface area, and resistance to biofouling. The hybrid architecture of MMMs enables flexible design strategies that leverage both the intrinsic advantages of polymers and the functionality of 2D nanofillers.

Notably, 2D nanomaterial-based membranes have demonstrated significant potential in mitigating the longstanding permeability-selectivity trade-off that constrains conventional membrane performance (Xue et al., 2021; Shen et al., 2025). This capacity primarily arises from the formation of the ultrathin, directionally aligned nanochannels and tunable physicochemical interactions at the nanoscale, which collectively reduce mass transfer resistance while maintaining the high separation performance (Tian et al., 2024). For instance, GO-based membranes exhibit the ultrafast water transport driven by the capillary flow between atomically smooth graphite nanosheets (Wang et al., 2025), while simultaneously achieving solute rejection through size exclusion and electrostatic repulsion mechanisms within functionalized interlayer channels (Pang et al., 2025). Similarly, MXene-based membranes combine the high water permeability with selective pollutant removal, benefiting from their intrinsic hydrophilicity and abundant surface terminations.

The applications of 2D nanomaterials in membrane filtration span a wide range of water treatment scenarios, including seawater desalination, brackish water softening, industrial wastewater remediation, oil/water separation, and the removal of emerging contaminants such as per- and polyfluoroalkyl substances (PFASs), pharmaceuticals, endocrine disruptors, and heavy metal ions (details are shown in the following section). Their integration into existing membrane processes or development into novel modules offers new avenues for energy-efficient, modular, and highly selective separation technologies. In summary, the advent of 2D nanomaterials has unlocked new opportunities and paradigms in membrane-based water purification. Their unique capacity to combine the molecular-level selectivity with impressive permeability presents a transformative pathway to address pressing global water challenges. Furthermore, the interdisciplinary collaboration between academia and industry will be crucial to bridge the gap between laboratory innovations and practical applications, ensuring that 2D membranes contribute meaningfully to sustainable water management and security.

6.2 Mechanisms of water and ion transport in 2D nanomaterial nanofiltration membranes

2D nanomaterial-based membranes have emerged as a promising class of separation membranes for water and ion transport due to their atomic thickness, tunable interlayer spacing, and unique physicochemical properties (Fig. 6.2) (Chen et al., 2017; Yu and Ren, 2024). Understanding the fundamental mechanisms governing water and ion transport within 2D nanomaterial-based nanofiltration membranes is critical to optimizing their separation performance and rationally designing next-generation materials (Kang et al., 2019). This section elucidates the transport behaviors in 2D nanofiltration membranes by examining the interplay of structural, energetic, and physicochemical factors at the nanoscale.

6.2.1 Water transport mechanisms

6.2.1.1 Confined water structure in 2D nanochannel

Within 2D nanochannels, water molecules are subjected to spatial confinement that fundamentally alters their hydrogen-bonding network and structural configuration

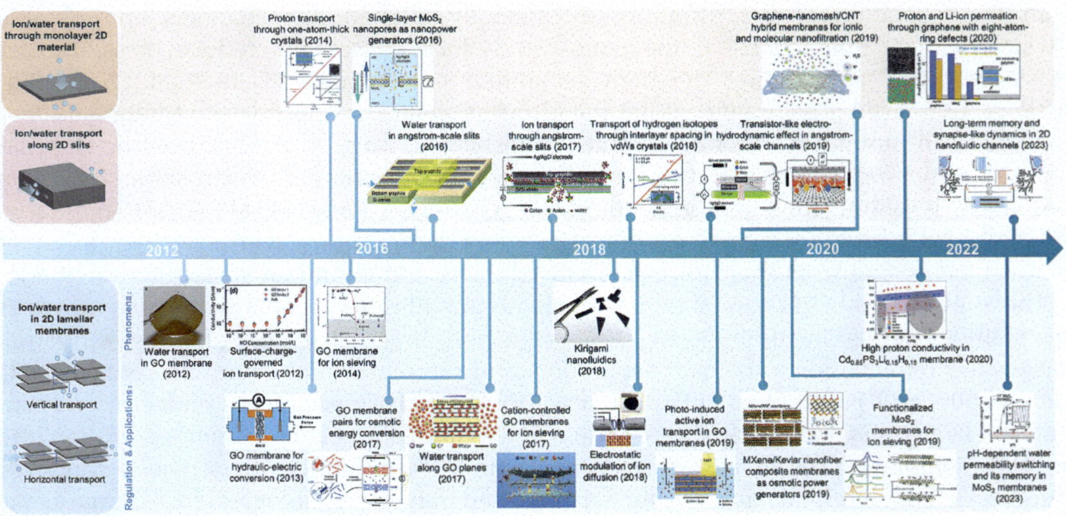

FIGURE 6.2 Progress schedule of water and ion transport research based on two-dimensional materials (Yu and Ren, 2024). Source: *Reproduced with permission from Yu, X., Ren, W., 2024. Ion and Water Transport in 2D Nanofluidic Channels. Advanced Functional Materials, 34(30). https://doi.org/10.1002/adfm.202313968.*

compared to bulk water. In bulk, each water molecule typically forms on average ~3.4 hydrogen bonds. However, when confined within the subnanometer interlayer spaces, particularly below 2 nm, the hydrogen bonds of confined water can decrease to ~2.0–2.5 due to steric hindrance and strong interfacial interactions with the channel walls. And water molecules tend to organize into the stratified configurations, forming monolayers or bilayers aligned parallel to the nanosheet wall (Kim et al., 2023b). This structural reorganization disrupts the tetrahedral hydrogen-bonding network characteristic of bulk water and leads to a more ordered arrangement. As a result, the confined water exhibits a higher overall mobility rate and lower viscosity (Sun et al., 2022). These effects collectively reduce the frictional resistance and enable ultrafast water transport (Wang et al., 2025), which is a key factor contributing to the remarkable permeability of 2D nanofiltration membranes.

6.2.1.2 *Low-friction transport and diffusivity*

Under nanoconfinement in 2D channels, especially within atomically smooth and hydrophobic channels such as graphene (Yang et al., 2017), water transport exhibits the remarkable low-friction characteristics, which are primarily attributed to the presence of the large hydrodynamic slip (Kim et al., 2023b). Both experimental and simulation studies have reported the slip lengths ranging from several tens to even thousands of nanometers, with values exceeding 100 nm on pristine graphene surfaces. This large slip length effectively reduces viscous drag at the channel interface, allowing water molecules to traverse the channel with minimal resistance (Castelletto and Boretti, 2021). Moreover, the unique geometry of 2D membranes, characterized by high aspect ratios and short out-of-plane thicknesses, further promotes efficient transport by shortening the diffusion path (Han et al., 2025). Water transport through the 2D membrane follows a tortuous path from the inlet to the outlet, giving the qualitative estimate of L as described by the following relation:

$$L = (h/d) \times l \quad (1) \tag{6.1}$$

where h is the membrane thickness and d is the interlayer spacing. The ratio h/d reflects both the number of turns the water must take along the transport path and the number of individual nanosheets it traverses. In combination with weak van der Waals interactions at the channel walls and the partial preservation of the hydrogen-bond network along the flow direction, these features contribute to the ultrahigh permeability observed in 2D nanofiltration membranes (Wang et al., 2025; Kim et al., 2023a). This transport behavior highlights the fundamental role of the interfacial nanofluidic phenomena in governing transport efficiency.

6.2.1.3 *Surface energy and water adsorption*

The surface energy of the 2D nanochannels plays a crucial role in governing water transport by modulating the adsorption behavior and interfacial dynamics of confined water. Hydrophilic 2D nanomaterial surfaces, functionalized with groups such as the hydroxyl, carboxyl, or amine moieties, exhibit strong water affinity, forming stable interfacial hydration layers through hydrogen bonding and electrostatic interactions. While such interactions can enhance water uptake, they may simultaneously hinder lateral diffusion by increasing interfacial friction. In contrast, 2D nanomaterials with low surface energy, such as pristine graphene or fluorinated carbon, interact weakly with water, promoting rapid desorption and higher slip lengths. This reduced interaction enhances water flux by minimizing resistance at the solid–liquid interface but may compromise hydration layer stability. Therefore, carefully tuning the surface chemistry of 2D nanomaterials to achieve moderate interactions is essential for balancing adequate water uptake and fast in-plane diffusion. This balance is particularly critical in subnanometer channels, where even minor variations in surface energy can exert disproportionately large effects on transport behavior.

6.2.2 Ion transport mechanisms

6.2.2.1 Ion states and mobility in 2D nanoconfined channels

In 2D nanoconfined environments, ion transport does not follow classical bulk behavior but instead exhibits the distinct migration modes influenced by extreme spatial confinement and strong interfacial interactions. Depending on the interlayer spacing of 2D nanosheets (typically <1 nm) (Ding et al., 2020), ions may (1) diffuse freely in partially hydrated states, (2) migrate via surface adsorption, or (3) move sterically confined along the channel axis. For example, in graphene oxide or MXene 2D membranes with slit widths around 7–10 Å, monovalent ions such as Na^+ or K^+ can maintain partial hydration shells and diffuse through the channel midplane (Chu et al., 2020). In contrast, divalent ions like Mg^{2+} often migrate closer to the channel wall due to their larger effective hydrated diameters (~8.6 Å for Mg^{2+} and ~6.6 Å for K^+) and stronger hydration shells, where they may experience the enhanced electrostatic interactions and steric hindrance (Pang et al., 2025; Yu and Ren, 2024). Additionally, the channel-wall interaction potential also plays a non-negligible role: surfaces terminated with oxygenated groups (–OH, –COOH) can selectively favor or hinder cation adsorption depending on ionic radius and charge density (Kim et al., 2023b). Thus, ion mobility in 2D nanoconfined channels is not solely governed by the concentration gradients or diffusion coefficients, but also by ion-channel compatibility including a complex interplay of steric, electrostatic, and hydration effects that defines the dominant transport mechanism under given physicochemical conditions.

6.2.2.2 Hydration shell and dehydration barrier

The hydration shell of an ion plays a pivotal role in determining its ability to permeate through subnanometer 2D channels. In aqueous environments, ions are stabilized by structured hydration layers, which are quantified by the hydration enthalpy. For transport through angstrom-scale 2D channels, the partial or complete dehydration is often required, imposing a substantial energy barrier to ion permeation (Yu and Ren, 2024; Yang et al., 2022). This barrier is especially significant for strongly hydrated ions, such as Mg^{2+} ($\Delta H_{hyd} \approx$ –1,920 kJ mol^{-1}) and Li^+ ($\Delta H_{hyd} \approx$ –520 kJ mol^{-1}), compared to more weakly hydrated ions like K^+ ($\Delta H_{hyd} \approx$ –330 kJ mol^{-1}). As a result, the permeation rate of multivalent or strongly hydrated ions is substantially reduced in channels narrower than ~7 Å (Chen et al., 2024). Interestingly, this dehydration process can be selectively facilitated by specific surface functional groups (Mozafari et al., 2021). For instance, carbonyl groups can transiently coordinate with dehydrated ions, acting as Lewis bases, to stabilize them during transport. Such interactions lower the dehydration energy barrier and provide preferential pathways for certain ions, such as Na^+ or Ca^{2+}, while suppressing others (Kim et al., 2023b). Notably, in atomically confined graphene channels, an applied potential induces partial dehydration and high-density ion accumulation, triggering a Coulomb-coupled "Newton's cradle"–like mechanism that enables ultrafast ion transport with a diffusion coefficient two orders of magnitude higher than that in bulk water (Xue et al., 2021). These mechanisms govern the energy- and size-selective transport behavior of 2D nanosheet membranes, resembling the selective filtering action of biological ion channels such as KcsA. Therefore, fine-tuning pore chemistry and structural size provides a powerful strategy for designing ultrafast ion transport 2D membranes.

6.2.2.3 Electrostatic interactions and charge exclusion

Electrostatic forces further regulate ion selectivity by introducing long-range interactions that favor or hinder ion entry into the 2D confined channels. In a typical 2D nanofiltration membrane, such as GO, h-BN, and MXene (Liu et al., 2021b), the negatively charged surfaces with functional groups (e.g., $-O^-$, –OH) generate an electrostatic potential that repels anions and attracts cations (Zhang et al., 2019; Zhu et al., 2024a). Once these surface charges are established, they interact electrostatically with surrounding ions in solution, leading to the formation of an electric double layer (EDL) near the channel walls. This EDL is characterized by attraction of counterions and repulsion of co-ions. The spatial extent of this electrostatic screening is described by the Debye length (λ_D), a key parameter derived from the Poisson–Boltzmann equation, which quantifies the thickness of the EDL and reflects how rapidly electrostatic potential decays in an electrolyte solution. The Debye length is given by the following equation (Yu and Ren, 2024):

$$\lambda D = \sqrt{((\varepsilon_0 \varepsilon_r K_B T)/(e^2 \sum_i n_i^\infty z_i^2))} \quad (2) \tag{6.2}$$

where ε_0, ε_r, K_B, T, and e represent the vacuum permittivity, relative permittivity of the medium, Boltzmann constant, absolute temperature, and elementary charge, respectively. The bulk number density of ion i is given by $n_i^\infty = 1000\, N_A c_i$, where N_A is Avogadro's number and c_i is the molar concentration of the ion. z_i denotes the ionic valence. The magnitude and spatial influence of this electrostatic potential depend on the surface charge density and the ionic strength of solution. In low-salinity environments (e.g., 1 mM NaCl), the Debye length can exceed 10 nm, allowing the surface charges of 2D nanomaterials to influence ion transport across the entire channel cross-section (Xie et al., 2023). Under such conditions, cations exhibit enhanced transport rate and selectivity. However, at high ionic strengths (e.g., seawater level, ~0.6 M), electrostatic screening shortens the effective range of charge repulsion to below 1 nm and weakens the surface charge's influence on ion transport. Even so, multivalent ions (e.g., SO_4^{2-}, Al^{3+}) remain largely excluded due to their higher charge density and corresponding Born energy penalties. Furthermore, charge inversion, a phenomenon where specific ion adsorption reverses the effective surface polarity, can occur in membranes containing high-affinity functional groups (e.g., carboxylates), thereby altering ion transport/selectivity by reversing the electrostatic preference for cations over anions. This complex electrostatic environment enables control over the ion transport and rejection, emphasizing the importance of coupling charge effects with pore geometry and hydration considerations in membrane design.

In conclusion, the transport behaviors of water and ions through 2D nanomaterial-based nanofiltration membranes are governed by a synergy of nanoconfinement effects, interfacial molecular interactions, and electrostatic forces, etc. Within angstrom- to nanometer-scale channels, water molecules exhibit near-frictionless flow due to the disruption of hydrogen-bond networks and alignment along channel walls, especially in atomically smooth and hydrophobic channels. This structural ordering reduces energy dissipation, enabling high water permeance that surpasses that of conventional polymeric membranes by orders of magnitude. In contrast, ion transport is far more constrained, governed by a combination of steric exclusion, hydration shell deformation, and electrostatic interactions with charged channel surfaces. Ions with high hydration energies or multivalent charges encounter significant energy barriers in subnanometer pores, resulting in ultrafast transport/strong selectivity in favor of smaller, less hydrated species. Additionally, the interfacial chemistry, such as surface functionalization or oxidation states, modulates local electric fields and adsorption dynamics, fine-tuning ion transport behavior. These mechanisms endow 2D membranes with exceptional capabilities in water/ion transport, offering a promising platform for energy-efficient membrane separation processes.

6.3 Fabrication techniques and structural designs of 2D nanomaterial nanofiltration membranes

The exceptional properties of 2D nanomaterials, including atomic thickness, tunable interlayer spacing, and rich surface chemistry, have revolutionized the design of nanofiltration (NF) membranes (Liu et al., 2021a). These advanced materials enable precise control over water and ion transport while addressing the persistent permeability-selectivity trade-off that limits conventional polymeric membranes (Radha et al., 2016). This section systematically reviews state-of-the-art fabrication techniques and innovative structural designs of 2D nanomaterial-based NF membranes (Fig. 6.3), with particular emphasis on their scalability, separation performance, and practical applications in water purification.

6.3.1 Fabrication techniques for 2D nanomaterial membranes

6.3.1.1 Vacuum-assisted filtration

Vacuum-assisted filtration has emerged as the most widely adopted method for assembling 2D nanosheets (e.g., GO, MXenes, MoS_2) into laminar membranes (Hegab and Zou, 2015). The process involves three key steps: (1) preparation of a homogeneous nanosheet dispersion, (2) filtration through a porous substrate under vacuum conditions, and (3) controlled drying to form densely packed lamellar structures. Recent advances have demonstrated that GO membranes fabricated by this method achieve exceptional water permeance (up to 1000 $L\ m^{-2}\ h^{-1}\ bar^{-1}$) coupled with high solute rejection (>90%) (Kang et al., 2017). The outstanding performance originates from the combination of atomically

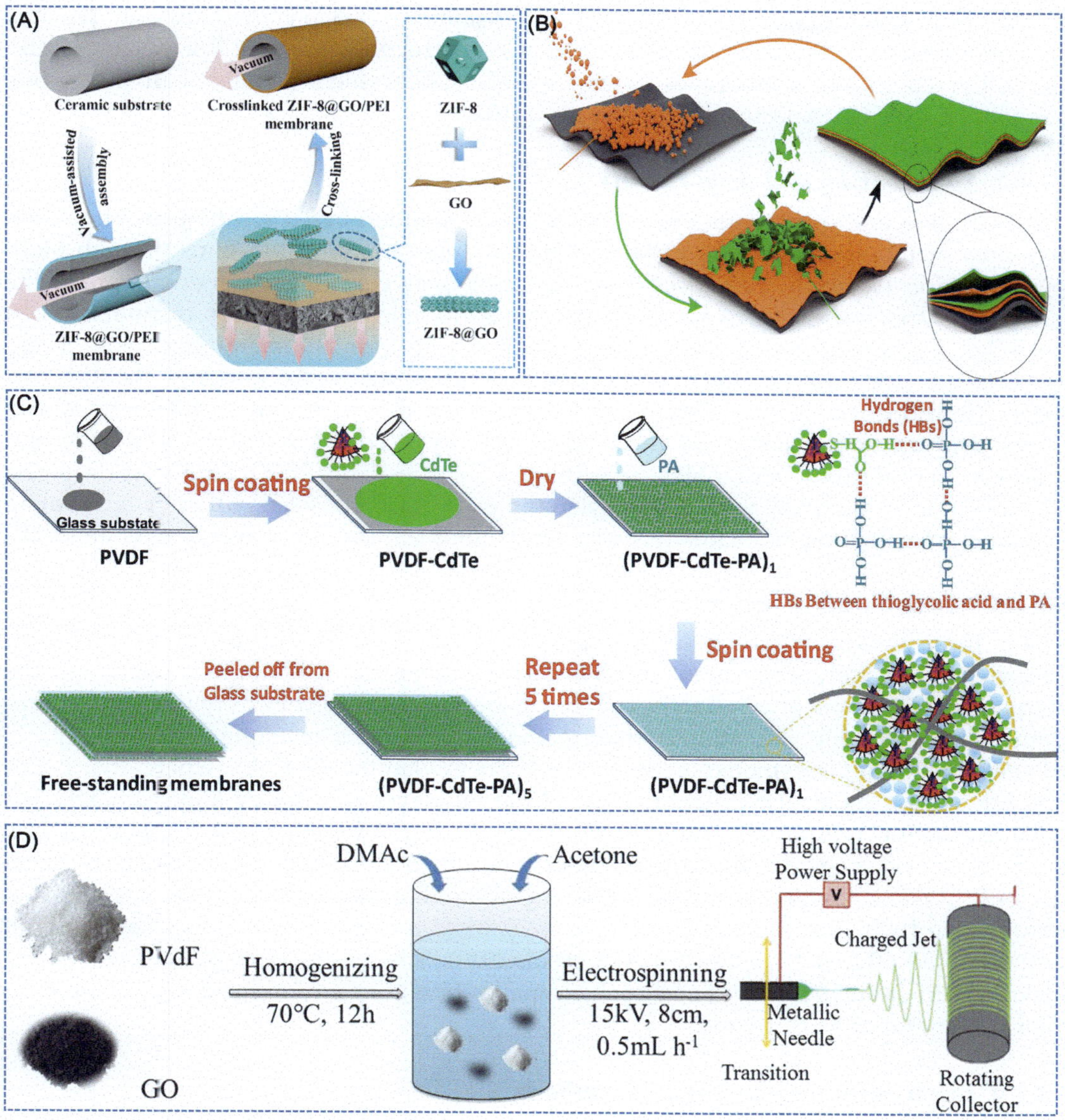

FIGURE 6.3 (A) Schematic illustration of fabrication of crosslinked ZIF-8@GO/PEI composite membrane on tubular ceramic substrate via a vacuum-assisted assembly method (Yang et al., 2018). (B) Schematic illustration of the LbL self-assembly of (MXene/TAEA)n multilayer films onto planar substrates (Tian et al., 2019). (C) Diagrammatic sketch of the (PVDF-CdTe-PA)5 membrane preparation by spin coating (Che, 2019). (D) Preparation of PVdF-GO NFMs through electrospinning. Source: *Reproduced with permission from Che, Q., Li, Z., Pan, B., Duan, X., Jia, T., Liu, L., 2019. Fabrication of layered membrane electrolytes with spin coating technique as anhydrous proton exchange membranes. Journal of Colloid and Interface Science, 555, 722–730. https://doi.org/10.1016/j.jcis.2019.08.034.*

smooth hydrophobic nanochannels and electrostatic repulsion mechanisms. Notable innovations include the incorporation of MOFs as structural spacers. For instance, ZIF-8@GO/PEI composite membranes exhibit a 99% enhancement in methanol flux while maintaining excellent selectivity (Yang et al., 2018).

6.3.1.2 *Spin coating and drop casting*

Spin coating and drop casting represent versatile solution-based techniques suitable for large-area membrane fabrication. Spin coating produces uniform thin films through high-speed rotation (typically 1000–5000 rpm) that generates centrifugal forces to spread the nanosheet dispersion evenly across the substrate surface. In contrast, drop casting relies on controlled solvent evaporation at ambient conditions to form the membrane layer. These methods are particularly effective for creating mixed-matrix membranes (MMMs) or surface-functionalized layers. Recent work has shown that spin-coated MXene/polyelectrolyte bilayers exhibit enhanced mechanical stability (Young's modulus > 10 GPa) and improved ion selectivity (Na^+/Mg^{2+} selectivity > 50) (Lee et al., 2021). Similarly, drop-cast GO membranes modified with polyethylenimine (PEI) demonstrate superior antifouling properties, maintaining >85% flux recovery after multiple filtration cycles (Zhao et al., 2021).

6.3.1.3 *Layer-by-layer assembly*

The layer-by-layer (LbL) assembly technique has gained significant attention for constructing precisely engineered multilayered membranes. This approach involves the alternating deposition of oppositely charged nanosheets or polyelectrolytes through dip-coating, spray-coating, or spin-assisted methods. A notable example includes $Ti_3C_2T_x$ MXene membranes assembled with tris(2-aminoethyl)amine (TAEA), where the TAEA molecules act as molecular pillars to expand the interlayer spacing by approximately 1 Å while enhancing mechanical integrity (Tian et al., 2019). Recent breakthroughs in LbL assembly have focused on creating surface-charged membranes through strategic combinations of GO and polyelectrolytes. These membranes achieve remarkable ion rejection performance (>90% for multivalent ions such as Mg^{2+} and SO_4^{2-}) while maintaining high water flux (50–60 $L\,m^{-2}\,h^{-1}\,bar^{-1}$) (Choi et al., 2013). The precise control over membrane thickness (down to 10 nm) and composition makes LbL assembly particularly valuable for tailoring separation performance.

6.3.1.4 *Electrospinning and hybrid methods*

Electrospinning has emerged as a powerful technique for fabricating nanofibrous membranes embedded with 2D nanomaterials. The process involves applying a high voltage (10–30 kV) to a polymer solution containing dispersed nanosheets, resulting in the formation of continuous fibers with diameters ranging from 50 to 500 nm. Recent studies have demonstrated that MoS_2/C composite nanofibers combine exceptional mechanical flexibility (strain > 20%) with superior electrochemical properties, making them ideal candidates for advanced filtration applications (Xiong et al., 2015). Hybrid fabrication methods, particularly sacrificial templating, have shown great promise in creating novel membrane architectures. For instance, ZIF-8 nanoparticles coated on single-walled carbon nanotube (SWCNT)/polyethersulfone (PES) substrates produce crumpled polyamide layers with significantly enhanced surface area (up to 300 m/g) (Xue et al., 2021) and improved water permeability (Wan et al., 2017).

6.3.2 Structural designs and functionalization strategies

6.3.2.1 *Laminar stacked membranes*

Laminar membranes formed by the ordered stacking of 2D nanosheets represent the most extensively studied architecture for NF applications. The separation performance of these membranes primarily depends on the interlayer nanochannels, which can be precisely engineered through various strategies: (1) Interlayer spacing control: Chemical crosslinking with agents such as glutaraldehyde or calcium ions has proven effective in fine-tuning the interlayer spacing to 0.3–1 nm resolution (Ding et al., 2021). For example, MoS_2 membranes intercalated with Ca^{2+} ions demonstrate Ångstrom-level precision in ion separation, achieving Na^+/K^+ selectivity ratios exceeding 5:1 (Chu et al., 2020). Surface functionalization: The introduction of oxygen-containing groups (e.g., carboxyl, hydroxyl) or polyelectrolytes (e.g., poly(allylamine hydrochloride), polystyrene sulfonate) enables precise control over surface charge and hydrophilicity (Wieszczycka et al., 2021). GO membranes functionalized with polydiallyldimethylammonium chloride (PDDA) exhibit exceptional charge-selective ion rejection (>93% for $MgCl_2$) while maintaining water fluxes above 50 L m^{-2} h^{-1} bar^{-1} (Guo et al., 2021).

6.3.2.2 *Mixed-matrix membranes*

The integration of 2D nanomaterials into polymeric matrices has led to the development of high-performance mixed-matrix membranes (MMMs) with synergistic properties (Huang et al., 2021). Two notable design approaches have emerged: (1) Janus nanosheets: These asymmetrically functionalized nanosheets combine hydrophilic and hydrophobic domains within a single structure. When incorporated into polysulfone (PSf) membranes, Janus GO nanosheets significantly improve both dispersion stability (reducing aggregation by >70%) and antifouling performance (flux decline <15% after 24 h operation) (Akbari et al., 2018). (2) MOF hybrids: The combination of zeolitic imidazolate frameworks (ZIFs) with GO nanosheets creates composite membranes with enhanced mechanical stability (tensile strength > 80 MPa) and tailored microporosity. ZIF-8/GO membranes demonstrate particularly promising results for desalination applications, showing a flux of 6.1 L m^{-2} h^{-1} bar^{-1} with the retention up to 99% (Yang et al., 2018).

6.3.2.3 *Nanoporous atomically thin membranes*

Nanoporous atomically thin membranes (NATMs) represent the cutting edge of membrane technology, offering minimal transport resistance but presenting significant fabrication challenges (Wang et al., 2017). Recent breakthroughs include: (1) graphene nanomeshes (GNMs): the development of GNM/single-walled carbon nanotube (SWNT) hybrid membranes has achieved remarkable performance metrics, including 98.1% NaCl rejection and water flux of 37.2 L m^{-2} h^{-1} bar^{-1} in forward osmosis applications (Zhu et al., 2025; Wang et al., 2013). These membranes also exhibit exceptional mechanical strength (Young's modulus > 1 TPa) due to strong π-π interactions between the components.

(2) Plasma-etched MoS_2: advanced plasma etching techniques have enabled the creation of sub-nanometer pores in monolayer MoS_2, resulting in ion selectivity that closely mimics biological channels. These membranes demonstrate K^+/Na^+ selectivity ratios up to 4:1, with water permeance exceeding 1000 L m^{-2} h^{-1} bar^{-1} (Xiao et al., 2016).

6.4 Comparative performance analysis: 2D nanomaterials versus commercial ones

The emergence of 2D nanomaterials has enabled transformative advances in membrane technologies across environmental, pharmaceutical, and industrial applications. Owing to their atomic-scale thickness, tunable pore structures, and unique physicochemical properties, 2D membranes offer superior selectivity, permeability, and antifouling behavior compared to conventional polymeric membranes. This section systematically examines the performance of 2D membranes in key application domains, including heavy metal and organic micropollutant removal, desalination, pharmaceutical purification, and fouling resistance. Specific attention is given to the separation mechanisms, material stability, and scalability challenges associated with graphene oxide, MXene, MoS_2, COFs, and hybrid membranes. By comparing the advantages and limitations of 2D nanomaterials against conventional membrane technologies, this section provides a comprehensive overview of their potential for next-generation water purification and process intensification.

6.4.1 Environmental applications

6.4.1.1 Heavy metal removal

The removal of heavy metals from aqueous systems presents distinct challenges that are addressed through fundamentally different mechanisms in conventional polymeric membranes versus 2D nanomaterial-based membranes. Traditional thin-film composite polyamide (TFC) membranes, such as NF270, rely primarily on size exclusion and Donnan effects for heavy metal rejection. The polyamide active layer, with a pore size distribution of 0.5–1 nm, effectively retains divalent ions like Cd^{2+} (hydrated diameter: 0.82 nm) under neutral pH conditions, achieving rejection rates of 85%–92% (Zhu et al., 2025; Peng et al., 2022). However, the hydrolytic instability of amide bonds under strongly acidic conditions ($pH < 3$) leads to a drastic decline in rejection efficiency (≤65%), while prolonged exposure to heavy metal ions induces irreversible fouling due to coordination between metal cations and carboxyl groups in the polyamide matrix (Peng et al., 2022; Tang et al., 2011). This necessitates frequent chemical cleaning (every 50–100 operational hours), significantly reducing membrane lifespan.

In contrast, 2D nanomaterial membranes, particularly MXene laminates, employ a synergistic separation mechanism combining precise size exclusion, electrostatic repulsion, and selective adsorption. The sub-nanometer interlayer spacing (0.6–0.9 nm) of MXene membranes provides steric hindrance, while their negatively charged surface functional groups (–O/–F terminations, ζ-potential ≈ –45 mV) enhance electrostatic rejection of cationic heavy metals. Furthermore, the exposed Ti sites in MXene nanosheets exhibit d-orbital vacancies that facilitate selective coordination with heavy metal ions, enabling >95% rejection of Cd^{2+} across a broad pH range (2–12) (Karthikeyan et al., 2021). Unlike polymeric membranes, MXene's inherent chemical stability allows for aggressive cleaning with 10 mM HCl while maintaining 98% flux recovery. Nevertheless, interlayer swelling in high-salinity environments remains a critical limitation, as Na^{+} intercalation expands the MXene interlayer spacing by ~0.3 nm, reducing Cd^{2+} rejection by 8%–10% (Lin et al., 2022).

To mitigate swelling, GO/MOF hybrid membranes have been developed, where the rigid framework of UiO-66-NH_2 nanoparticles restricts interlayer expansion, stabilizing the nanochannel spacing at 0.7 nm even in 1 M NaCl solutions. These membranes achieve exceptional Cr(VI) rejection (94%) but suffer from high manufacturing costs (5–8× higher than TFC membranes) and performance inconsistencies due to nanosheet stacking defects (Du et al., 2019). Recent advances in defect engineering, such as plasma-induced edge functionalization, have improved GO membrane rejection by 12% (Wang et al., 2012), offering a promising route to optimize cost-performance trade-offs.

6.4.1.2 *Organic micropollutant removal*

The removal of persistent organic micropollutants, including pharmaceuticals, pesticides, and endocrine-disrupting compounds, presents unique challenges that highlight fundamental differences between conventional polymeric membranes and emerging 2D nanomaterial-based membranes. TFC membranes, such as NF90, achieve 60%–85% rejection of neutral pharmaceuticals like carbamazepine (MW=236 Da) primarily through size exclusion, with their 200–300 Da molecular weight cutoff (MWCO) (Lin et al., 2019). However, their performance is severely compromised for smaller, polar compounds (e.g., N-nitrosodimethylamine, NDMA) where rejection drops below 30%, reflecting the intrinsic limitation of polymeric networks in discriminating sub-300Da molecules (Liu et al., 2018). Furthermore, chlorine exposure during cleaning (>0.5 ppm) degrades polyamide chains, creating larger pores that reduce rejection efficiency by 15%–20% over 100 operational hours (An et al., 2023).

In contrast, 2D MoS_2 membranes with precisely etched nanopores (0.6 ± 0.1 nm) demonstrate remarkable size selectivity, rejecting >90% of carbamazepine while allowing rapid water permeation (50–80 $L\ m^{-2}\ h^{-1}\ bar^{-1}$) due to atomic-scale pore uniformity (Sapkota et al., 2020). The hydrophobic MoS_2 basal planes minimize organic adsorption, reducing fouling propensity by 60% compared to polyamide membranes when treating wastewater containing natural organic matter (NOM) (Liu et al., 2021b). For polar micropollutants, GO membranes leverage tunable interlayer spacing (0.7–1.2 nm) and oxygen functional groups to achieve dual rejection mechanisms: steric exclusion for larger molecules (e.g., 98% rejection of tetracycline, MW=444 Da) and electrostatic repulsion/hydrogen bonding for smaller polar compounds (75% rejection of atenolol, MW=266 Da) (Hung et al., 2020). The π–π interactions between GO's sp-carbon (Xue et al., 2021) domains and aromatic pollutants further enhance the removal of endocrine disruptors like bisphenol A (rejection >95%) (Ahmed et al., 2018).

Advanced photocatalytic 2D membranes, such as g-C_3N_4/TiO_2 heterostructures, uniquely degrade pollutants during filtration. Under visible light irradiation, these membranes mineralize 80% of adsorbed sulfamethoxazole ($k=0.12\ min^{-1}$) through generated reactive oxygen species while maintaining stable water flux (Zhao et al., 2016). This contrasts sharply with conventional membranes, where accumulated organics necessitate frequent chemical cleaning, increasing operational costs by 30%–40% (Al-Amoudi and Lovitt, 2007).

6.4.2 Desalination performance

6.4.2.1 Brackish water desalination

The desalination of brackish water (TDS 1000–10,000 mg/L) presents distinct challenges that are addressed through fundamentally different transport mechanisms in conventional polymeric membranes versus 2D nanomaterial membranes. TFC polyamide membranes, exemplified by SW30XLE, achieve 99.5% salt rejection through a dense polyamide layer (thickness ~100 nm) where solution-diffusion dominates ion transport, requiring high operating pressures (55–60 bar) that incur substantial energy costs (5.3 kWh/m (Kang et al., 2019)). While effective for monovalent ion removal, these membranes exhibit poor selectivity for divalent ions (Mg^{2+}/Na^+ selectivity ~2.5) due to limited charge discrimination, and suffer from compaction-induced flux decline (20%–30% over 1000 hours) under sustained high pressure (Massons et al., 2023). Their chlorine sensitivity (>0.1 ppm causes polyamide degradation) necessitates costly dechlorination pretreatment, and organic fouling reduces water flux by 40%–60% when processing brackish water with natural organic matter (NOM) (Meng et al., 2023). In contrast, 2D graphene nanomesh membranes achieve 98.1% NaCl rejection at just 5 bar through size exclusion in their angstrom-scale pores (0.35–0.45 nm diameter), coupled with partial dehydration of ions at the hydrophobic graphene edges, enabling ultrahigh water flux (97.6 $L\ m^{-2}\ h^{-1}\ bar^{-1}$) (Yang et al., 2019). The atomic thickness (<1 nm) of graphene minimizes transport resistance while the precisely controlled pore geometry provides exceptional monovalent/divalent ion selectivity (Mg^{2+}/Na^+ >15), outperforming conventional membranes by sixfold (Rollings et al., 2016). However, graphene membranes face challenges in maintaining pore size distribution during scale-up, with current fabrication methods yielding only 60%–70% of pores within the target 0.3–1.2 nm range, leading to 10%–15% variability in salt rejection (Mi, 2019).

Alternative 2D materials like LDHs address this through tunable interlayer spacing (0.7–1.2 nm) that can be precisely adjusted with intercalating ions (e.g., CO_3^{2-} expands spacing by 0.2 nm versus Cl^-), enabling optimization for specific brackish water compositions (Altalhi et al., 2024). LDH membranes demonstrate stable operation (>6 months) in high-salinity (8000 mg/L TDS) and high-temperature (50°C) conditions where polymeric membranes degrade, though their water permeability (8–12 $L\ m^{-2}\ h^{-1}\ bar^{-1}$) remains lower than graphene counterparts (Sajid et al., 2022). MXene membranes offer a balance of performance and scalability, with 97.6% Na_2SO_4 rejection at 3 bar through a combination of size exclusion and electrostatic repulsion from their negatively charged surface terminations (-O, -F), while their metallic conductivity enables novel antifouling strategies via electrochemical oxidation of organic foulants (Xue and Zhang, 2021). Nevertheless, MXene membranes exhibit swelling in high-ionic-strength solutions (>5000 mg/L NaCl), increasing interlayer spacing by 0.3–0.4 nm and reducing rejection by 8%–12%, a limitation partially mitigated by covalent crosslinking with molecules like ethylenediamine (Sun et al., 2021; Mu et al., 2024). Hybrid 2D/polymer membranes, such as polyamide thin films incorporating MoS_2 nanosheets, combine the processability of conventional membranes with enhanced performance (40% higher flux than TFC at equivalent rejection), though long-term stability studies (>100 hours) are still needed to assess industrial viability (Wang et al., 2024).

6.4.2.2 Seawater pretreatment

Seawater pretreatment for desalination requires robust separation technologies capable of addressing the complex matrix of dissolved salts, organic matter, and biological foulants present in marine environments. Conventional seawater reverse osmosis (SWRO) systems typically employ a multistage pretreatment process combining ultrafiltration (UF) and cartridge filtration to remove particulate matter, followed by chemical dosing (e.g., antiscalants, biocides) to mitigate scaling and biofouling (Farooque, 2020). While effective, this approach suffers from several limitations: UF membranes (MWCO 10–100 kDa) demonstrate poor removal of low molecular weight organics (e.g., algal toxins, polysaccharides) that subsequently foul RO membranes, reducing water flux by 30%–50% within the first 500 operational hours (Imbrogno et al., 2025). The required chemical additives increase operational costs by 15%–20% and generate toxic byproducts (e.g., halogenated disinfection byproducts) that necessitate additional posttreatment (Amy et al., 2017). Furthermore, conventional pretreatment fails to address divalent ion scaling (Ca^{2+}, Mg^{2+}, SO_4^{2-}) at the RO stage, necessitating high-pressure operation (55–70 bar) that consumes 3.5–4.5 kWh/m (Kang et al., 2019) for seawater desalination (Subramani and Jacangelo, 2015).

In contrast, 2D nanomaterial membranes offer transformative potential for seawater pretreatment through fundamentally different separation mechanisms. GO membranes with precisely tuned interlayer spacing (0.7–1.0 nm) achieve >95% rejection of divalent ions (Mg^{2+}, SO_4^{2-}) via size exclusion and Donnan effects while allowing rapid permeation of monovalent ions (Na^+, Cl^-) (Abraham et al., 2017). The oxygen functional groups (-COOH, -OH) on GO nanosheets impart inherent antifouling properties, reducing bacterial adhesion by 70%–80% compared to conventional UF membranes (Sun and Li, 2018). MXene membranes demonstrate even greater fouling resistance through their conductive properties - applied potentials of 1.0–1.5 V in situ oxidize organic foulants and prevent biofilm formation (Qian et al., 2023). MXene membranes are susceptible to oxidative degradation in chlorine-containing seawater, with >50% performance loss occurring after 200 hours of exposure to 1 ppm free chlorine (Khalid et al., 2025). Recent advances in stabilization strategies, such as graphene oxide quantum dot (GOQD) intercalation between MXene layers, have improved stability while maintaining high water permeability (20–25 LMH/bar) (Liu et al., 2020). Another promising development involves Janus-type 2D membranes with asymmetric surface properties - hydrophilic GO on the feed side for fouling resistance and hydrophobic reduced GO on the permeate side to enhance water transport, achieving 50% higher flux than conventional membranes in seawater pretreatment (Li et al., 2024). Despite these advances, scale-up challenges persist, particularly in maintaining defect-free nanosheet alignment over large membrane areas (>1 m; Xue et al., 2021), current 2D membranes exhibit 10–20% performance variation across commercial-scale modules compared to <5% for conventional UF (Liu et al., 2023b). The higher manufacturing costs of 2D membranes ($80–150/m (Xue et al., 2021) vs. $20–50/m (Xue et al., 2021) for UF) currently limit their deployment to niche applications, though life-cycle analyses suggest break-even points may be reached within 5–7 years of operation due to energy and chemical savings (Cao et al., 2022). Emerging hybrid approaches that combine 2D materials with conventional polymers (e.g., polyamide thin-film nanocomposites with embedded MXene nanosheets) offer a pragmatic transition pathway, delivering 30%–40% performance improvements over standard SWRO pretreatment while leveraging existing manufacturing infrastructure (Wang et al., 2020).

6.4.3 Pharmaceutical/Biotechnological applications

6.4.3.1 Active pharmaceutical ingredient purification

The purification of active pharmaceutical ingredients (APIs) from complex reaction mixtures and genotoxic impurities presents distinct challenges that highlight fundamental differences between conventional polymeric membranes and 2D nanomaterial-based membranes. Traditional organic solvent nanofiltration (OSN) membranes, such as polyimide-based Starmem^{TM} series, rely primarily on solution-diffusion mechanisms with MWCO ranging from 200 to 500 Da, achieving 60%–85% API recovery in aprotic solvents like dimethylformamide (DMF) (Marchetti et al., 2014). However, these membranes suffer from significant limitations: their amorphous polymer networks exhibit swelling in polar solvents, leading to inconsistent rejection and substantial solvent losses during diafiltration. Furthermore, their lack of molecular recognition sites results in poor selectivity for structurally similar impurities, necessitating multiple purification stages that reduce overall yield to 70%–80% (Shen et al., 2022). The thermal instability of polymeric membranes (<80°C) also precludes their use in high-temperature API crystallization processes (Marchetti et al., 2014). In contrast, 2D COF membranes with precisely engineered pore apertures (0.8–1.2 nm) demonstrate exceptional size- and shape-selectivity through molecular-sieving effects, achieving >95% API recovery with impurity removal efficiencies up to 99.99% in single-pass operation (Li et al., 2020). The crystalline structure of COFs maintains pore size fidelity across solvents, showing <3% swelling variation between hexane and DMF, while their π-conjugated frameworks enable selective π-π interactions with aromatic APIs that enhance selectivity by 5–10 fold compared to polymeric membranes (Yang et al., 2024). However, 2D membranes face critical challenges in API manufacturing: the relatively low mechanical strength of COF membranes (tensile strength 20–30 MPa vs. 50–70 MPa for polyimides) limits their applicability in high-pressure (>30 bar) continuous processes (Lv et al., 2025). Recent advances in hybrid 2D-polymer membranes, such as polyamide thin films incorporating MXene nanosheets, address these limitations by combining the processability of polymers with the selectivity of 2D materials, demonstrating stable operation (>100 hours) in industrial API purification with 40% higher productivity than conventional OSN membranes (Xue et al., 2023). The development of stacked 2D heterostructures (e.g., MoS_2/h-BN/graphene) with graded pore sizes shows particular promise for multicomponent API separations, achieving simultaneous removal of genotoxic impurities (>99.9%), catalysts (>99%), and byproducts (>95%) in single-unit operations (Liu et al., 2023a).

6.4.4 Fouling behavior and stability

6.4.4.1 *Organic fouling*

Organic fouling in membrane systems arises from fundamentally different interaction mechanisms between foulants and conventional polymeric surfaces compared to 2D nanomaterial interfaces. TFC membranes experience fouling primarily through hydrophobic adsorption and pore-blocking mechanisms, where NOM components like humic acids (contact angle ~65°) adhere to the membrane's hydrophobic surface (contact angle ~50°–60°) via van der Waals forces, reducing water flux by 40%–60% within the first 100 operational hours (Zhang, 2023). The rough surface morphology (RMS roughness ~80–100 nm) of polyamide membranes further exacerbates fouling by providing nucleation sites for biofilm formation (Mi and Elimelech, 2008). In contrast, 2D GO membranes demonstrate superior antifouling properties due to their unique physicochemical characteristics: the oxygen functional groups create a hydrophilic surface (contact angle ~30°) that forms a tightly bound hydration layer (thickness ~1–2 nm), reducing organic adsorption by 70%–80% compared to polyamide membranes (Wu et al., 2023). However, 2D membranes face specific fouling challenges: GO membranes in high-ionic-strength solutions (>100 mM NaCl) experience reduced electrostatic repulsion between nanosheets, enabling foulant penetration into interlayer spaces that decreases flux recovery by 15%–20% after cleaning (Krishnamoorthi et al., 2024). Recent advances in 2D membrane design address these limitations through innovative strategies: GO membranes intercalated with zwitterionic polymers (e.g., poly(sulfobetaine methacrylate)) maintain fouling resistance across wide salinity ranges by preserving hydration layers through strong dipole interactions (Razavi et al., 2025). Hybrid 2D-polyamide thin-film nanocomposite (TFN) membranes incorporating MXene nanosheets (0.05–0.1 wt%) combine the processability of conventional membranes with enhanced antifouling properties, showing 60% lower flux decline than standard TFC membranes during long-term operation with wastewater. The higher manufacturing costs of 2D membranes ($80–150/m (Xue et al., 2021) vs. $30–50/m (Xue et al., 2021) for TFC) remain a barrier for widespread adoption, though life-cycle analyses indicate potential cost savings from reduced cleaning frequency and extended membrane lifespan (Gao et al., 2025). Emerging self-cleaning 2D membranes incorporating photocatalytic materials (e.g., g-C_3N_4/TiO_2 heterostructures) offer additional fouling control through light-activated degradation of organic foulants, maintaining stable flux (>95% recovery) under continuous operation with minimal chemical cleaning (Gao et al., 2025). These developments position 2D membranes as promising solutions for challenging feed waters containing high fouling potential, though further optimization is needed to address stability concerns and enable cost-effective large-scale deployment.

6.4.4.2 *Mechanical durability*

The mechanical durability of membranes under fouling conditions reveals fundamental differences between conventional polymeric systems and 2D nanomaterial-based membranes. TFC membranes exhibit tensile strengths of 50–70 MPa but suffer from plasticization under sustained pressure (20–30 bar), leading to irreversible compaction that reduces water flux by 20%–30% after 1000 operational hours. This compaction is exacerbated by organic fouling, where adsorbed foulants (e.g., humic acids) increase local stress concentrations by 2–3 fold, accelerating the formation of microvoids and pinhole defects (Rajendran et al., 2025). The viscoelastic nature of polyamide (elastic modulus ~2 GPa) causes creep deformation under constant load, particularly at elevated temperatures (>40°C) common in industrial operations, with studies showing 15%–20% thickness reduction after 6 months of use (Reynolds et al., 2018). In contrast, 2D GO membranes demonstrate exceptional in-plane mechanical strength (elastic modulus ~250 GPa) and maintain dimensional stability under pressures up to 50 bar, showing <5% flux decline from compaction over equivalent operation periods (Akbari et al., 2016). The covalent crosslinking of GO nanosheets with diamines (e.g., ethylenediamine) enhances interlayer bonding energy from ~0.3 eV to >1.0 eV, preventing delamination during backwashing cycles while maintaining 90%–95% of initial mechanical strength after 1000 hours in fouling conditions (Abraham et al., 2017). MXene membranes exhibit even greater robustness, with their metallic bonding providing fracture toughness (3.5–4.5 $MPa{\cdot}m^{1/2}$) 5–10 times higher than polymeric membranes, enabling stable operation in high-shear environments (shear rate >10,000 s^{-1}) where conventional membranes fail. However, 2D membranes face unique mechanical challenges: the weak out-of-plane strength of GO laminates (~50 MPa vs. >200 MPa in-plane) makes them susceptible to hydraulic damage during pressure surges, with studies showing 10%–15% performance loss after 50 rapid (0.1 s) pressure cycles from 0 to 30 bar (Gupta et al., 2024). The anisotropic mechanical properties of 2D materials also lead to complex failure modes under fouling conditions, while GO membranes resist compaction, localized organic accumulation can induce nanosheet buckling that increases hydraulic resistance by 20%–25% (Zhang et al., 2022). Recent advances in 3D-interconnected 2D architectures (e.g., graphene aerogel membranes) address these limitations by combining in-plane strength with improved z-direction stability (compressive strength ~15 MPa vs. ~5 MPa for laminates), maintaining stable performance during long-term fouling tests (Ding and Li, 2022). Emerging self-healing 2D membranes incorporating dynamic covalent bonds (e.g., boronic ester linkages) autonomously repair mechanical damage during operation, recovering >95% of initial strength after fracture—a capability unmatched by conventional membranes (Mondal and Cohen, 2022). These innovations position 2D membranes as durable solutions for harsh operating conditions, though further development is needed to optimize their mechanical reliability across all stress states and enable cost-effective manufacturing at industrial scales.

6.5 Conclusion and perspectives

2D nanomaterials have emerged as transformative building blocks for advanced membrane systems, offering exceptional opportunities to overcome the intrinsic trade-offs between permeability, selectivity, and fouling resistance in traditional polymer-based membranes. Their atomic-scale thickness, tunable interlayer spacing, unique surface chemistry, and strong interfacial interactions collectively enable high-performance separations in diverse water purification applications, including desalination, heavy metal removal, organic micropollutant rejection, and pharmaceutical separations.

This chapter comprehensively reviewed the state-of-the-art progress in 2D nanomaterial membranes, covering fundamental transport mechanisms, fabrication strategies, structural designs, and application-specific performance. Notably, the confinement of water and ions within angstrom-scale channels enables ultrafast transport coupled with precise selectivity, while surface functionalization allows tunable electrostatic and chemical interactions to enhance separation efficiency. Additionally, the integration of 2D materials into laminar, nanoporous, or mixed-matrix architectures provides versatile platforms for addressing application-specific challenges such as organic fouling, scaling, and membrane compaction.

Despite these promising developments, several critical challenges must be addressed before 2D membranes can be widely adopted in practical water treatment systems:

Scalability and manufacturing consistency. Most current fabrication methods, such as vacuum-assisted filtration or drop casting, are difficult to scale and prone to introducing structural defects or non-uniformities across large membrane areas.

High cost of raw materials and process incompatibility. The synthesis and purification of high-quality 2D nanomaterials (e.g., MXenes, COFs, and MOFs) remain costly and may not align with existing industrial membrane manufacturing processes.

Stability under harsh operational conditions. 2D membranes are susceptible to oxidation, swelling, or delamination under extreme pH, high ionic strength, or oxidizing environments, which can compromise long-term performance and durability.

Limited predictive modeling and material design tools. A lack of multiscale, data-driven models connecting nanoscale structures with macroscopic transport properties hinders rational design and optimization.

To overcome these limitations and accelerate the translation of 2D membranes from laboratory research to industrial deployment, several future research directions are proposed:

Development of scalable and robust fabrication methods. Future efforts should prioritize scalable, cost-effective, and defect-minimizing membrane fabrication techniques, such as spray coating, roll-to-roll printing, or layer-by-layer self-assembly, that are compatible with large-scale production and ensure consistent membrane quality.

Design of multifunctional hybrid membrane architectures. Incorporating 2D materials with polymers, nanoparticles, or catalytic agents into mixed-matrix or Janus-type membranes can provide synergistic functionalities including enhanced mechanical strength, fouling resistance, and active degradation of pollutants.

Construction of stimuli-responsive and self-healing membranes. Responsive 2D materials capable of reacting to light, electric fields, pH, or temperature offer opportunities to design dynamic membranes with tunable separation performance, self-cleaning behavior, and prolonged operational lifespans.

Integration of multiscale modeling and materials informatics. Coupling molecular dynamics simulations, continuum models, and machine learning approaches can facilitate high-throughput screening and rational design of 2D nanomaterials with optimized pore geometry, surface chemistry, and transport behavior.

Comprehensive lifecycle and techno-economic assessments. To ensure environmental and economic sustainability, future studies should incorporate full lifecycle assessments (LCA) and cost–benefit analyses comparing 2D membranes with commercial alternatives in various water treatment contexts.

In conclusion, 2D nanomaterials provide a powerful and versatile toolkit for constructing next-generation membranes capable of addressing global water scarcity, pollution, and treatment efficiency challenges. Continued interdisciplinary collaboration among materials scientists, membrane engineers, computational modelers, and environmental technologists will be crucial for translating these innovations into scalable, durable, and economically viable membrane technologies. With strategic advancements in fabrication, integration, and application, 2D membranes are poised to play a critical role in the sustainable management of global water resources.

References

Abraham, J., Vasu, K.S., Williams, C.D., Gopinadhan, K., Su, Y., Cherian, C.T., et al., 2017. Tunable sieving of ions using graphene oxide membranes. Nature Nanotechnology 12 (6), 546–550. Available at https://doi.org/10.1038/nnano.2017.21, http://www.nature.com/nnano/index.html.

Ahmed, M.B., Zhou, J.L., Ngo, H.H., Johir, M.A.H., Sun, L., Asadullah, M., et al., 2018. Sorption of hydrophobic organic contaminants on functionalized biochar: protagonist role of Π-Π electron-donor-acceptor interactions and hydrogen bonds. Journal of Hazardous Materials 360, 270–278. Available at https://doi.org/10.1016/j.jhazmat.2018.08.005, www.elsevier.com/locate/jhazmat.

Akbari, A., Sheath, P., Martin, S.T., Shinde, D.B., Shaibani, M., Banerjee, P.C., et al., 2016. Large-area graphene-based nanofiltration membranes by shear alignment of discotic nematic liquid crystals of graphene oxide. Nature Communications 7. Available at https://doi.org/10.1038/ncomms10891, http://www.nature.com/ncomms/index.html.

Akbari, M., Shariaty-Niassar, M., Matsuura, T., Ismail, A.F., 2018. Janus graphene oxide nanosheet: a promising additive for enhancement of polymeric membranes performance prepared via phase inversion. Journal of Colloid and Interface Science 527, 10–24. Available at https://doi.org/10.1016/j.jcis.2018.05.012, http://www.elsevier.com/inca/publications/store/6/2/2/8/6/1/index.htt.

Al-Amoudi, A., Lovitt, R.W., 2007. Fouling strategies and the cleaning system of NF membranes and factors affecting cleaning efficiency. Journal of Membrane Science 303 (1-2), 4–28. Available at https://doi.org/10.1016/j.memsci.2007.06.002.

Altalhi, A.A., Mohamed, E.A., Negm, N.A., 2024. Recent advances in layered double hydroxide (LDH)-based materials: fabrication, modification strategies, characterization, promising environmental catalytic applications, and prospective aspects. Energy Advances 3 (9), 2136–2151. Available at https://doi.org/10.1039/d4ya00272e, https://www.rsc.org/journals-books-databases/about-journals/energy-advances/.

Amy, G., Ghaffour, N., Li, Z., Francis, L., Linares, R.V., Missimer, T., et al., 2017. Membrane-based seawater desalination: present and future prospects. Desalination 401, 16–21. Available at https://doi.org/10.1016/j.desal.2016.10.002.

An, S.A., Park, C.G., Lee, J.S., Cho, S.M., Woo, Y.C., Kim, H.S., 2023. Exposure dose and temperature of chlorine on deterioration of thin-film composite membranes for reverse osmosis and nanofiltration. Chemosphere 333. Available at https://doi.org/10.1016/j.chemosphere.2023.138929, www.elsevier.com/locate/chemosphere.

Cao, Y., Taghvaie Nakhjiri, A., Ghadiri, M., 2022. Membrane desalination for water treatment: recent developments, techno-economic evaluation and innovative approaches toward water sustainability. The European Physical Journal Plus 137 (7). Available at https://doi.org/10.1140/epjp/s13360-022-02999-8.

Castelletto, S., Boretti, A., 2021. Advantages, limitations, and future suggestions in studying graphene-based desalination membranes. RSC Advances 11 (14), 7981–8002. Available at https://doi.org/10.1039/d1ra00278c.

Chen, L., Shi, G., Shen, J., Peng, B., Zhang, B., Wang, Y., et al., 2017. Ion sieving in graphene oxide membranes via cationic control of interlayer spacing. Nature 550 (7676), 380–383. Available at https://doi.org/10.1038/nature24044.

Chen, X., Qin, Y., Zhu, Y., Pan, X., Wang, Y., Ma, H., et al., 2024. Accurate prediction of solvent flux in sub-1-nm slit-pore nanosheet membranes. Science Advances 10 (17). Available at https://doi.org/10.1126/sciadv.adl1455, https://www.science.org/toc/sciadv/10/17.

Choi, W., Choi, J., Bang, J., Lee, J.H., 2013. Layer-by-layer assembly of graphene oxide nanosheets on polyamide membranes for durable reverse-osmosis applications. ACS Applied Materials and Interfaces 5 (23), 12510–12519. Available at https://doi.org/10.1021/am403790s.

Chu, C., Fu, C.-F., Zhang, P., Pan, T., Ai, X., Wu, Y., et al., 2020. Precise ångström controlling the interlayer channel of MoS_2 membranes by cation intercalation. Journal of Membrane Science 615, 118520. Available at https://doi.org/10.1016/j.memsci.2020.118520.

Ding, L., Li, L., Liu, Y., Wu, Y., Lu, Z., Deng, J., et al., 2020. Effective ion sieving with Ti3C2Tx MXene membranes for production of drinking water from seawater. Nature Sustainability 3 (4), 296–302. Available at https://doi.org/10.1038/s41893-020-0474-0.

Ding, M., Li, C., 2022. Recent advances in simple preparation of 3D graphene aerogels based on 2D graphene materials. Frontiers in Chemistry 10. Available at https://doi.org/10.3389/fchem.2022.815463.

Ding, Y., Zeng, M., Zheng, Q., Zhang, J., Xu, D., Chen, W., et al., 2021. Bidirectional and reversible tuning of the interlayer spacing of two-dimensional materials. Nature Communications 12 (1). Available at https://doi.org/10.1038/s41467-021-26139-5.

Du, X.-D., Yi, X.-H., Wang, P., Zheng, W., Deng, J., Wang, C.-C., 2019. Robust photocatalytic reduction of Cr(VI) on UiO-66-NH2(Zr/Hf) metal-organic framework membrane under sunlight irradiation. Chemical Engineering Journal 356, 393–399. Available at https://doi.org/10.1016/j.cej.2018.09.084.

Farooque, A.M., 2020. Reverse osmosis desalination: Performance and challenges. Corrosion and Fouling Control in Desalination Industry. Springer International Publishing, Saudi Arabia, pp. 49–69. Available at http://doi.org/10.1007/978-3-030-34284-5, http://doi.org/10.1007/978-3-030-34284-5_3.

Gao, T., Jin, Y., Xiao, K., 2025. Techno–economic and environmental impact assessment of membrane bioreactors for wastewater treatment: a review. Engineering. Available at https://doi.org/10.1016/j.eng.2025.05.001.

Guo, H., Zhang, J., Peng, L.E., Li, X., Chen, Y., Yao, Z., et al., 2021. High-efficiency capture and recovery of anionic perfluoroalkyl substances from water using PVA/PDDA nanofibrous membranes with near-zero energy consumption. Environmental Science and Technology Letters 8 (4), 350–355. Available at https://doi.org/10.1021/acs.estlett.1c00128, http://pubs.acs.org/page/estlcu/about.html.

Gupta, K., Heo, J.H., Im, S.H., Kim, W.S., 2024. Efficient physical delamination of Ti3C2Tx MXene under periodic and constant shear field of Taylor vortex flow. Chemical Engineering Journal 481. Available at https://doi.org/10.1016/j.cej.2024.148700, https://www.sciencedirect.com/science/journal/13858947.

Han, C., Jiang, J., Mu, L., Zhao, W., Liu, J., Lan, J., et al., 2025. Quasi-vertically asymmetric channels of graphene oxide membrane for ultrafast ion sieving. Nature Communications 16 (1). Available at https://doi.org/10.1038/s41467-025-56358-z, https://www.nature.com/ncomms/.

Hegab, H.M., Zou, L., 2015. Graphene oxide-assisted membranes: Fabrication and potential applications in desalination and water purification. Journal of Membrane Science 484, 95–106. Available at https://doi.org/10.1016/j.memsci.2015.03.011, www.elsevier.com/locate/memsci.

Huang, Z., Liu, J., Liu, Y., Xu, Y., Li, R., Hong, H., et al., 2021. Enhanced permeability and antifouling performance of polyether sulfone (PES) membrane via elevating magnetic Ni@MXene nanoparticles to upper layer in phase inversion process. Journal of Membrane Science 623. Available at https://doi.org/10.1016/j.memsci.2021.119080, www.elsevier.com/locate/memsci.

Hung, W.S., Lai, Y.L., Lee, P.H., Chiao, Y.H., Sengupta, A., Sivakumar, M., et al., 2020. Tuneable interlayer spacing self-assembling on graphene oxide-framework membrane for enhance air dehumidification. Separation and Purification Technology 239. Available at https://doi.org/10.1016/j.seppur.2019.116499, http://www.journals.elsevier.com/separation-and-purification-technology/.

Imbrogno, A., Calvo, J.I., Breida, M., Schwaiger, R., Schäfer, A.I., 2025. Molecular weight cut off (MWCO) determination in ultra- and nanofiltration: review of methods and implications on organic matter removal. Separation and Purification Technology 354, 128612. Available at https://doi.org/10.1016/j.seppur.2024.128612.

Jang, D., Bakli, C., Chakraborty, S., Karnik, R., 2022. Molecular self-assembly enables tuning of nanopores in atomically thin graphene membranes for highly selective transport. Advanced Materials 34 (11). Available at https://doi.org/10.1002/adma.202108940.

Kang, K.M., Kim, D.W., Ren, C.E., Cho, K.M., Kim, S.J., Choi, J.H., et al., 2017. Selective molecular separation on ti3c2tx-graphene oxide membranes during pressure-driven filtration: comparison with graphene oxide and MXenes. ACS Applied Materials and Interfaces 9 (51), 44687–44694. Available at https://doi.org/10.1021/acsami.7b10932, http://pubs.acs.org/journal/aamick.

Kang, Y., Xia, Y., Wang, H., Zhang, X., 2019. 2D laminar membranes for selective water and ion transport. Advanced Functional Materials 29 (29). Available at https://doi.org/10.1002/adfm.201902014.

Karahan, H.E., Goh, K., Zhang, C., Yang, E., Yıldırım, C., Chuah, C.Y., et al., 2020. MXene materials for designing advanced separation membranes. Advanced Materials 32 (29). Available at https://doi.org/10.1002/adma.201906697, http://onlinelibrary.wiley.com/journal/10.1002/(ISSN)1521-4095.

Karthikeyan, P., Elanchezhiyan, S.S., Preethi, J., Talukdar, K., Meenakshi, S., Park, C.M., 2021. Two-dimensional (2D) $Ti_3C_2T_x$ MXene nanosheets with superior adsorption behavior for phosphate and nitrate ions from the aqueous environment. Ceramics International 47 (1), 732–739. Available at https://doi.org/10.1016/j.ceramint.2020.08.183, https://www.journals.elsevier.com/ceramics-international.

Khalid, Z., Hadi, F., Xie, J., Chandrabose, V., Oh, J.M., 2025. The future of MXenes: exploring oxidative degradation pathways and coping with surface/edge passivation approach. Small (Weinheim an der Bergstrasse, Germany) 21 (6). Available at https://doi.org/10.1002/smll.202407856, http://onlinelibrary.wiley.com/journal/10.1002/(ISSN)1613-6829.

Kim, C., Koh, D.Y., Lee, Y., Choi, J., Cho, H.S., Choi, M., 2023a. Bottom-up synthesis of two-dimensional carbon with vertically aligned ordered micropores for ultrafast nanofiltration. Science Advances 9 (6). Available at https://doi.org/10.1126/sciadv.ade7871, https://www.science.org/doi/10.1126/sciadv.ade7871.

Kim, S., Choi, H., Kim, B., Lim, G., Kim, T., Lee, M., et al., 2023b. Extreme ion-transport inorganic 2D membranes for nanofluidic applications. Advanced Materials 35 (43). Available at https://doi.org/10.1002/adma.202206354.

Krishnamoorthi, R., Butt, F.S., Mazlan, N.A., Chen, S., Radacsi, N., Yang, S., et al., 2024. Tuning the interlayer spacing of graphene oxide membrane via surfactant intercalation for ultrafast nanofiltration. Journal of Membrane Science 706. Available at https://doi.org/10.1016/j.memsci.2024.122942, https://www.sciencedirect.com/science/journal/03767388.

Lee, S., Kim, E.H., Yu, S., Kim, H., Park, C., Lee, S.W., et al., 2021. Polymer-laminated $Ti_3C_2T_X$ MXene electrodes for transparent and flexible field-driven electronics. ACS Nano 15 (5), 8940–8952. Available at https://doi.org/10.1021/acsnano.1c01621, http://pubs.acs.org/journal/ancac3.

Li, S., Zhang, X., Su, J., 2024. Desalination performance in janus graphene oxide channels: geometric asymmetry vs charge polarity. ACS Applied Materials & Interfaces 16 (2), 2659–2671. Available at https://doi.org/10.1021/acsami.3c16592.

Li, Y., Wu, Q., Guo, X., Zhang, M., Chen, B., Wei, G., et al., 2020. Laminated self-standing covalent organic framework membrane with uniformly distributed subnanopores for ionic and molecular sieving. Nature Communications 11 (1). Available at https://doi.org/10.1038/s41467-019-14056-7.

Lin, Q., Liu, Y., Yang, Z., He, Z., Wang, H., Zhang, L., et al., 2022. Construction and application of two-dimensional MXene-based membranes for water treatment: a mini-review. Results in Engineering 15, 100494. Available at https://doi.org/10.1016/j.rineng.2022.100494.

Lin, Y.L., Tsai, J.Z., Hung, C.H., 2019. Using in situ modification to enhance organic fouling resistance and rejection of pharmaceutical and personal care products in a thin-film composite nanofiltration membrane. Environmental Science and Pollution Research 26 (33), 34073–34084. Available at https://doi.org/10.1007/s11356-018-3234-1, https://link.springer.com/journal/11356.

Liu, H., Zhang, X., Lv, Z., Wei, F., Liang, Q., Qian, L., et al., 2023a. Ternary heterostructure membranes with two-dimensional tunable channels for highly selective ion separation. JACS Au 3 (11), 3089–3100. Available at https://doi.org/10.1021/jacsau.3c00473.

Liu, S., Liu, G., Chen, G., Liu, G., Jin, W., 2023b. Scale-up fabrication of two-dimensional material membranes: challenges and opportunities. Current Opinion in Chemical Engineering 39, 100892. Available at https://doi.org/10.1016/j.coche.2022.100892.

Liu, L.F., Huang, X., Zhang, X., Li, K., Ji, Y.L., Yu, C.Y., et al., 2018. Modification of polyamide TFC nanofiltration membrane for improving separation and antifouling properties. RSC Advances 8 (27), 15102–15110. Available at https://doi.org/10.1039/c8ra01374h, http://pubs.rsc.org/en/journals/journal/ra.

Liu, T., Liu, X., Graham, N., Yu, W., Sun, K., 2020. Two-dimensional MXene incorporated graphene oxide composite membrane with enhanced water purification performance. Journal of Membrane Science 593, 117431. Available at https://doi.org/10.1016/j.memsci.2019.117431.

Liu, X., Zhang, L., Cui, X., Zhang, Q., Hu, W., Du, J., et al., 2021a. 2D material nanofiltration membranes: from fundamental understandings to rational design. Advanced Science 8 (23). Available at https://doi.org/10.1002/advs.202102493.

Liu, Y., Zhao, Y., Zhang, X., Huang, X., Liao, W., Zhao, Y., 2021b. MoS_2-based membranes in water treatment and purification. Chemical Engineering Journal 422, 130082. Available at https://doi.org/10.1016/j.cej.2021.130082.

Loske, L., Nakagawa, K., Yoshioka, T., Matsuyama, H., 2020. 2D nanocomposite membranes: water purification and fouling mitigation. Membranes 10 (10), 295. Available at https://doi.org/10.3390/membranes10100295.

Lu, L., Li, L., Chu, M., Chen, C., Wang, B., Wang, J., et al., 2025. Recent advancement in 2D metal–organic framework for environmental remediation: a review. Advanced Functional Materials 35 (14). Available at https://doi.org/10.1002/adfm.202419433, http://onlinelibrary.wiley.com/journal/10.1002/(ISSN)1616-3028.

Lv, H., Lai, Z., Xing, Z., Wang, S., Sun, Q., 2025. Progress in covalent organic framework membranes for water desalination: mechanistic insights, synthesis strategies, applications, and future prospects. Desalination 614, 119146. Available at https://doi.org/10.1016/j.desal.2025.119146.

Marchetti, P., Jimenez Solomon, M.F., Szekely, G., Livingston, A.G., 2014. Molecular separation with organic solvent nanofiltration: a critical review. Chemical Reviews 114 (21), 10735–10806. Available at https://doi.org/10.1021/cr500006j, http://pubs.acs.org/journal/chreay.

Massons, G., Gilabert-Oriol, G., Pandya, H., Balakrishnan, R.M., Maciá, M.Á.P., 2023. Experience with seawater reverse osmosis SW30XFR-400/34i membrane at large desalination plant in United Arab Emirates. Desalination and Water Treatment 309, 105–108. Available at https://doi.org/10.5004/dwt.2023.29921, https://www.deswater.com/readfulltextopenaccess.php?id=RFdUX2FydGljbGVzL3ZvbF8zMDlfcGFwZXJzLzMwOV8yMDIzXzEwNS5wZGY=.

Meng, Q.-W., Cheng, L., Ge, Q., 2023. Recent advances and future challenges of polyamide-based chlorine-resistant membrane. Advanced Membranes 3, 100075. Available at https://doi.org/10.1016/j.advmem.2023.100075.

Mi, B., 2019. Scaling up nanoporous graphene membranes. Science (New York, N.Y.) 364 (6445), 1033–1034. Available at https://doi.org/10.1126/science.aax3103.

Mi, B., Elimelech, M., 2008. Chemical and physical aspects of organic fouling of forward osmosis membranes. Journal of Membrane Science 320 (1-2), 292–302. Available at https://doi.org/10.1016/j.memsci.2008.04.036.

Mondal, P., Cohen, S.M., 2022. Self-healing mixed matrix membranes containing metal-organic frameworks. Chemical Science 13 (41), 12127–12135. Available at https://doi.org/10.1039/d2sc04345a, http://pubs.rsc.org/en/journals/journal/sc.

Mozafari, M., Arabi Shamsabadi, A., Rahimpour, A., Soroush, M., 2021. Ion-selective MXene-based membranes: current status and prospects. Advanced Materials Technologies 6 (10). Available at https://doi.org/10.1002/admt.202001189.

Mu, M., Graham, N., Yu, W., Sun, K., Xu, X., Liu, T., 2024. Optimal cross-linking of MXene-based membranes for high rejection and low adsorption with long-term stability for water treatment. Journal of Membrane Science 701, 122738. Available at https://doi.org/10.1016/j.memsci.2024.122738.

Pang, D., Liu, L., Zhang, X., Guo, C., Jia, J., Liu, C., et al., 2025. Ion separation with graphene oxide nanofluidic membranes: a review. Desalination 593, 118218. Available at https://doi.org/10.1016/j.desal.2024.118218.

Peng, L.E., Yang, Z., Long, L., Zhou, S., Guo, H., Tang, C.Y., 2022. A critical review on porous substrates of TFC polyamide membranes: mechanisms, membrane performances, and future perspectives. Journal of Membrane Science 641. Available at https://doi.org/10.1016/j.memsci.2021.119871, www.elsevier.com/locate/memsci.

Qian, L., Yuan, C., Wang, X., Zhang, H., Du, L., Wei, G., et al., 2023. Conductive MXene ultrafiltration membrane for improved antifouling ability and water quality under electrochemical assistance. RSC Advances 13 (23), 15872–15880. Available at https://doi.org/10.1039/d3ra01116j.

Radha, B., Esfandiar, A., Wang, F.C., Rooney, A.P., Gopinadhan, K., Keerthi, A., et al., 2016. Molecular transport through capillaries made with atomic-scale precision. Nature 538 (7624), 222–225. Available at https://doi.org/10.1038/nature19363.

Rajendran, D.S., Devi, E.G., Subikshaa, V.S., Sethi, P., Patil, A., Chakraborty, A., et al., 2025. Recent advances in various cleaning strategies to control membrane fouling: a comprehensive review. Clean Technologies and Environmental Policy 27 (2), 649–664. Available at https://doi.org/10.1007/s10098-024-03000-z, https://www.springer.com/journal/10098.

Razavi, R., Shakeri, A., Salehi, H., 2025. A novel thin film nanocomposite forward osmosis membrane modified with zwitterionic GO-g-PSBMA with high fouling resistance and heavy metal ions rejection. Journal of Environmental Chemical Engineering 13 (1), 115308. Available at https://doi.org/10.1016/j.jece.2025.115308.

Reynolds, C., Thompson, R., McLeish, T., 2018. Pressure and shear rate dependence of the viscosity and stress relaxation of polymer melts. Journal of Rheology 62 (2), 631–642. Available at https://doi.org/10.1122/1.5012969.

Rollings, R.C., Kuan, A.T., Golovchenko, J.A., 2016. Ion selectivity of graphene nanopores. Nature Communications 7. Available at https://doi.org/10.1038/ncomms11408, http://www.nature.com/ncomms/index.html.

Sajid, M., Sajid Jillani, S.M., Baig, N., Alhooshani, K., 2022. Layered double hydroxide-modified membranes for water treatment: recent advances and prospects. Chemosphere 287 (Pt 1), 132140. Available at https://doi.org/10.1016/j.chemosphere.2021.132140.

Sapkota, B., Liang, W., VahidMohammadi, A., Karnik, R., Noy, A., Wanunu, M., 2020. High permeability sub-nanometre sieve composite MoS_2 membranes. Nature Communications 11 (1). Available at https://doi.org/10.1038/s41467-020-16577-y.

Shen, J., Aljarb, A., Cai, Y., Liu, X., Min, J., Wang, Y., et al., 2025. Engineering grain boundaries in monolayer molybdenum disulfide for efficient water-ion separation. Science (New York, N.Y.) 387 (6735), 776–782. Available at https://doi.org/10.1126/science.ado7489, https://www.science.org/toc/science/387/6735.

Shen, J., Cai, Y., Zhang, C., Wei, W., Chen, C., Liu, L., et al., 2022. Fast water transport and molecular sieving through ultrathin ordered conjugated-polymer-framework membranes. Nature Materials 21 (10), 1183–1190. Available at https://doi.org/10.1038/s41563-022-01325-y, http://www.nature.com/nmat/.

Subramani, A., Jacangelo, J.G., 2015. Emerging desalination technologies for water treatment: a critical review. Water Research 75, 164–187. Available at https://doi.org/10.1016/j.watres.2015.02.032, www.elsevier.com/locate/watres.

Sun, M., Li, J., 2018. Graphene oxide membranes: functional structures, preparation and environmental applications. Nano Today 20, 121–137. Available at https://doi.org/10.1016/j.nantod.2018.04.007.

Sun, Y., Yi, F., Li, R.H., Min, X., Qin, H., Cheng, S.Q., et al., 2022. Inorganic–organic hybrid membrane based on pillararene-intercalated MXene nanosheets for efficient water purification. Angewandte Chemie - International Edition 61 (14). Available at https://doi.org/10.1002/anie.202200482, http://onlinelibrary.wiley.com/journal/10.1002/(ISSN)1521-3773.

Sun, Y., Xu, D., Li, S., Cui, L., Zhuang, Y., Xing, W., et al., 2021. Assembly of multidimensional MXene-carbon nanotube ultrathin membranes with an enhanced anti-swelling property for water purification. Journal of Membrane Science 623, 119075. Available at https://doi.org/10.1016/j.memsci.2021.119075.

Tang, C.Y., Chong, T.H., Fane, A.G., 2011. Colloidal interactions and fouling of NF and RO membranes: a review. Advances in Colloid and Interface Science 164 (1-2), 126–143. Available at https://doi.org/10.1016/j.cis.2010.10.007.

Tian, M., Liu, Y., Zhang, S., Yu, C., Ostrikov, K., Zhang, Z., 2024. Overcoming the permeability-selectivity challenge in water purification using two-dimensional cobalt-functionalized vermiculite membrane. Nature Communications 15 (1). Available at https://doi.org/10.1038/s41467-024-44699-0.

Tian, W., VahidMohammadi, A., Wang, Z., Ouyang, L., Beidaghi, M., Hamedi, M.M., 2019. Layer-by-layer self-assembly of pillared two-dimensional multilayers. Nature Communications 10 (1). Available at https://doi.org/10.1038/s41467-019-10631-0, http://www.nature.com/ncomms/index.html.

Wan, L., Shamsaei, E., Easton, C.D., Yu, D., Liang, Y., Chen, X., et al., 2017. ZIF-8 derived nitrogen-doped porous carbon/carbon nanotube composite for high-performance supercapacitor. Carbon 121, 330–336. Available at https://doi.org/10.1016/j.carbon.2017.06.017, http://www.journals.elsevier.com/carbon/.

Wang, D., Yuan, S., Zhang, N., Wang, Z., Zhu, J., Wang, Z., 2024. Thin-film composite membranes interlayered with amphiphilic MoS_2 nanosheets via controllable interfacial polymerization for enhanced desalination Performance. Environmental Science & Technology 58 (26), 11855–11863. Available at https://doi.org/10.1021/acs.est.4c04063.

Wang, J., Zhang, X., Yu, Z., Gao, Y., Lu, Q., Ma, C., et al., 2025. Fast water transport and ionic sieving in ultrathin stacked nanoporous 2D membranes. National Science Review 12 (3). Available at https://doi.org/10.1093/nsr/nwae482.

Wang, L., Boutilier, M.S.H., Kidambi, P.R., Jang, D., Hadjiconstantinou, N.G., Karnik, R., 2017. Fundamental transport mechanisms, fabrication and potential applications of nanoporous atomically thin membranes. Nature Nanotechnology 12 (6), 509–522. Available at https://doi.org/10.1038/nnano.2017.72, http://www.nature.com/nnano/index.html.

Wang, X., Li, Q., Zhang, J., Huang, H., Wu, S., Yang, Y., 2020. Novel thin-film reverse osmosis membrane with MXene Ti_3C_2T embedded in polyamide to enhance the water flux, anti-fouling and chlorine resistance for water desalination. Journal of Membrane Science 603, 118036. Available at https://doi.org/10.1016/j.memsci.2020.118036.

Wang, X., Jiao, L., Sheng, K., Li, C., Dai, L., Shi, G., 2013. Solution-processable graphene nanomeshes with controlled pore structures. Scientific Reports 3 (1). Available at https://doi.org/10.1038/srep01996.

Wang, X.L., Wei, J.F., Dai, Z., Zhao, K.Y., Zhang, H., 2012. Preparation and characterization of negatively charged hollow fiber nanofiltration membrane by plasma-induced graft polymerization. Desalination 286, 138–144. Available at https://doi.org/10.1016/j.desal.2011.11.014.

Wieszczycka, K., Staszak, K., Woźniak-Budych, M.J., Litowczenko, J., Maciejewska, B.M., Jurga, S., 2021. Surface functionalization – the way for advanced applications of smart materials. Coordination Chemistry Reviews 436. Available at https://doi.org/10.1016/j.ccr.2021.213846, http://www.journals.elsevier.com/coordination-chemistry-reviews/.

Wu, X., Zhang, J., Wang, H., Huo, Y., Xie, Z., 2023. Construction of 2D/0D Graphene Oxide/Copper(I) Oxide-incorporated titanium dioxide mixed-dimensional membranes with ultrafast water transport and enhanced antifouling properties. ACS Applied Materials & Interfaces 15 (26), 31561–31571. Available at https://doi.org/10.1021/acsami.3c06103.

Xiao, S., Xiao, P., Zhang, X., Yan, D., Gu, X., Qin, F., et al., 2016. Atomic-layer soft plasma etching of MoS 2. Scientific Reports 6. Available at https://doi.org/10.1038/srep19945, www.nature.com/srep/index.html.

Xie, L., Tang, J., Qin, R., Zhang, Q., Liu, J., Jin, Y., et al., 2023. Surface charge modification on 2D nanofluidic membrane for regulating ion transport. Advanced Functional Materials 33 (4). Available at https://doi.org/10.1002/adfm.202208959.

Xiong, X., Luo, W., Hu, X., Chen, C., Qie, L., Hou, D., et al., 2015. Flexible membranes of MoS_2/C nanofibers by electrospinning as binder-free anodes for high-performance sodium-ion batteries. Scientific Reports 5 (1). Available at https://doi.org/10.1038/srep09254.

Xue, Q., Meng, W., Fang, Q., Zhu, J., Zhang, K., 2023. Water nanochannels in mxene polyamide nanofiltration membranes: implications for permeability. ACS Applied Nano Materials 6 (13), 11282–11290. Available at https://doi.org/10.1021/acsanm.3c01337.

Xue, Q., Zhang, K., 2021. MXene nanocomposite nanofiltration membrane for low carbon and long-lasting desalination. Journal of Membrane Science 640, 119808. Available at https://doi.org/10.1016/j.memsci.2021.119808.

Xue, Y., Xia, Y., Yang, S., Alsaid, Y., Fong, K.Y., Wang, Y., et al., 2021. Atomic-scale ion transistor with ultrahigh diffusivity. Science (New York, N.Y.) 372 (6541), 501–503. Available at https://doi.org/10.1126/science.abb5144, https://science.sciencemag.org/content/372/6541/501.full.

Yang, H., Wang, N., Wang, L., Liu, H.X., An, Q.F., Ji, S., 2018. Vacuum-assisted assembly of ZIF-8@GO composite membranes on ceramic tube with enhanced organic solvent nanofiltration performance. Journal of Membrane Science 545, 158–166. Available at https://doi.org/10.1016/j.memsci.2017.09.074, www.elsevier.com/locate/memsci.

Yang, H., Zhang, H., Kang, C., Ji, C., Shi, D., Zhao, D., 2024. Solvent-responsive covalent organic framework membranes for precise and tunable molecular sieving. Science Advances 10 (51). Available at https://doi.org/10.1126/sciadv.ads0260.

Yang, L., Xiao, X., Shen, S., Lama, J., Hu, M., Jia, F., et al., 2022. Recent advances in graphene oxide membranes for nanofiltration. ACS Applied Nano Materials 5 (3), 3121–3145. Available at https://doi.org/10.1021/acsanm.1c04469, https://pubs.acs.org/journal/aanmf6.

Yang, Q., Su, Y., Chi, C., Cherian, C.T., Huang, K., Kravets, V.G., et al., 2017. Ultrathin graphene-based membrane with precise molecular sieving and ultrafast solvent permeation. Nature Materials 16 (12), 1198–1202. Available at https://doi.org/10.1038/nmat5025.

Yang, Y., Yang, X., Liang, L., Gao, Y., Cheng, H., Li, X., et al., 2019. Large-area graphene-nanomesh/ carbon-nanotube hybrid membranes for ionic and molecular nanofiltration. Science (New York, N.Y.) 364 (6445), 1057–1062. Available at https://doi.org/10.1126/science.aau5321, https://science.sciencemag.org/content/364/6445/1057/tab-pdf.

Yu, X., Ren, W., 2024. Ion and water transport in 2D nanofluidic channels. Advanced Functional Materials 34 (30). Available at https://doi.org/10.1002/adfm.202313968.

Zhang, M., Guan, K., Ji, Y., Liu, G., Jin, W., Xu, N., 2019a. Controllable ion transport by surface-charged graphene oxide membrane. Nature Communications 10 (1). Available at https://doi.org/10.1038/s41467-019-09286-8.

Zhang, X., Li, H., Wang, J., Peng, D., Liu, J., Zhang, Y., 2019b. In-situ grown covalent organic framework nanosheets on graphene for membrane-based dye/salt separation. Journal of Membrane Science 581, 321–330. Available at https://doi.org/10.1016/j.memsci.2019.03.070.

Zhang, P., Wang, Y., Li, P., Luo, X., Feng, J., Kong, H., et al., 2022. Improving stability and separation performance of graphene oxide/graphene nanofiltration membranes by adjusting the laminated regularity of stacking-sheets. Science of The Total Environment 827, 154175. Available at https://doi.org/10.1016/j.scitotenv.2022.154175.

Zhang, S., 2023. Surface roughness and membrane fouling. Nature Water 1 (2). Available at https://doi.org/10.1038/s44221-023-00045-0 132–132.

Zhao, H., Chen, S., Quan, X., Yu, H., Zhao, H., 2016. Integration of microfiltration and visible-light-driven photocatalysis on g-C 3N 4 nanosheet/reduced graphene oxide membrane for enhanced water treatment. Applied Catalysis B: Environmental 194, 134–140. Available at https://doi.org/10.1016/j.apcatb.2016.04.042.

Zhao, X., Che, Y., Mo, Y., Huang, W., Wang, C., 2021. Fabrication of PEI modified GO/MXene composite membrane and its application in removing metal cations from water. Journal of Membrane Science 640, 119847. Available at https://doi.org/10.1016/j.memsci.2021.119847.

Zhu, X., Gao, Q., Lou, M., Bai, Y., Xu, X., Li, F., et al., 2024a. GO-based membranes with enhanced stability and permeability by implanting etched-MXene nanosheets: the role of binding energy in stabilizing 2D membranes. Journal of Membrane Science 707, 122983. Available at https://doi.org/10.1016/j.memsci.2024.122983.

Zhu, Y., Jian, M., Meng, N., Ji, Y., Bai, X., Wu, L., et al., 2024b. Recent progress in developing 2D MOFs/COFs/Zeolites nanosheets membranes for water purification. Separation and Purification Technology 337, 126404. Available at https://doi.org/10.1016/j.seppur.2024.126404.

Zhu, X., Tian, E., Li, Z., Xie, D., Shao, Q., Sun, Z., et al., 2025. Graphene oxide-carbon nanotube hybrid membranes for high-pressure and high-flux nanofiltration. Advanced Functional Materials. Available at https://doi.org/10.1002/adfm.202503432.

CHAPTER

7

Exploration of 2D nanomaterials in membrane distillation for water purification

Soheil Zarghami[1], *and Toraj Mohammadi*[2]

[1]Department of Chemical Engineering, Isfahan University of Technology, Isfahan, Iran [2]Center of Excellence for Membrane Research and Technology, Department of Chemical, Petroleum and Gas Engineering, Iran University of Science and Technology (IUST), Narmak, Tehran, Iran

7.1 Introduction

Membrane distillation (MD) is an advanced membrane-based separation technology that combines both thermal and membrane processes. It employs a hydrophobic microporous membrane to separate the liquid and vapor phases. The membrane's pores allow only vapor molecules to pass through, while retaining the liquid phase and any dissolved constituents (Fig. 7.1A). This characteristic enables MD to effectively separate nonvolatile solutes from water, as only water vapor permeates the membrane. Moreover, MD is also effective for the removal of volatile organic compounds (VOCs) from aqueous solutions, as VOCs can traverse the membrane more readily than water molecules due to their higher volatility. MD operates based on vapor pressure gradient across the membrane, which serves as the driving force for vapor transport (Kujawska et al., 2015; Ye et al., 2024).

As shown in Fig. 7.1B, there are four principal configurations of MD: direct contact membrane distillation (DCMD), air gap membrane distillation (AGMD), sweep gas membrane distillation (SGMD), and vacuum membrane distillation (VMD), distinguished by their respective vapor condensation methods. MD has been investigated for applications in both desalination and wastewater treatment, owing to its unique separation mechanism and relatively mild operational conditions and moderate temperatures. Consequently, MD membranes are subject to lower mechanical stress and therefore require less structural strength compared to conventional pressure-driven membrane processes, such as reverse osmosis (RO), nanofiltration (NF), ultrafiltration (UF), and microfiltration (MF) (Samadi et al., 2023; Qasim et al., 2021).

Two-Dimensional Nanomaterials for Water Purification
DOI: https://doi.org/10.1016/B978-0-443-44530-9.00009-2

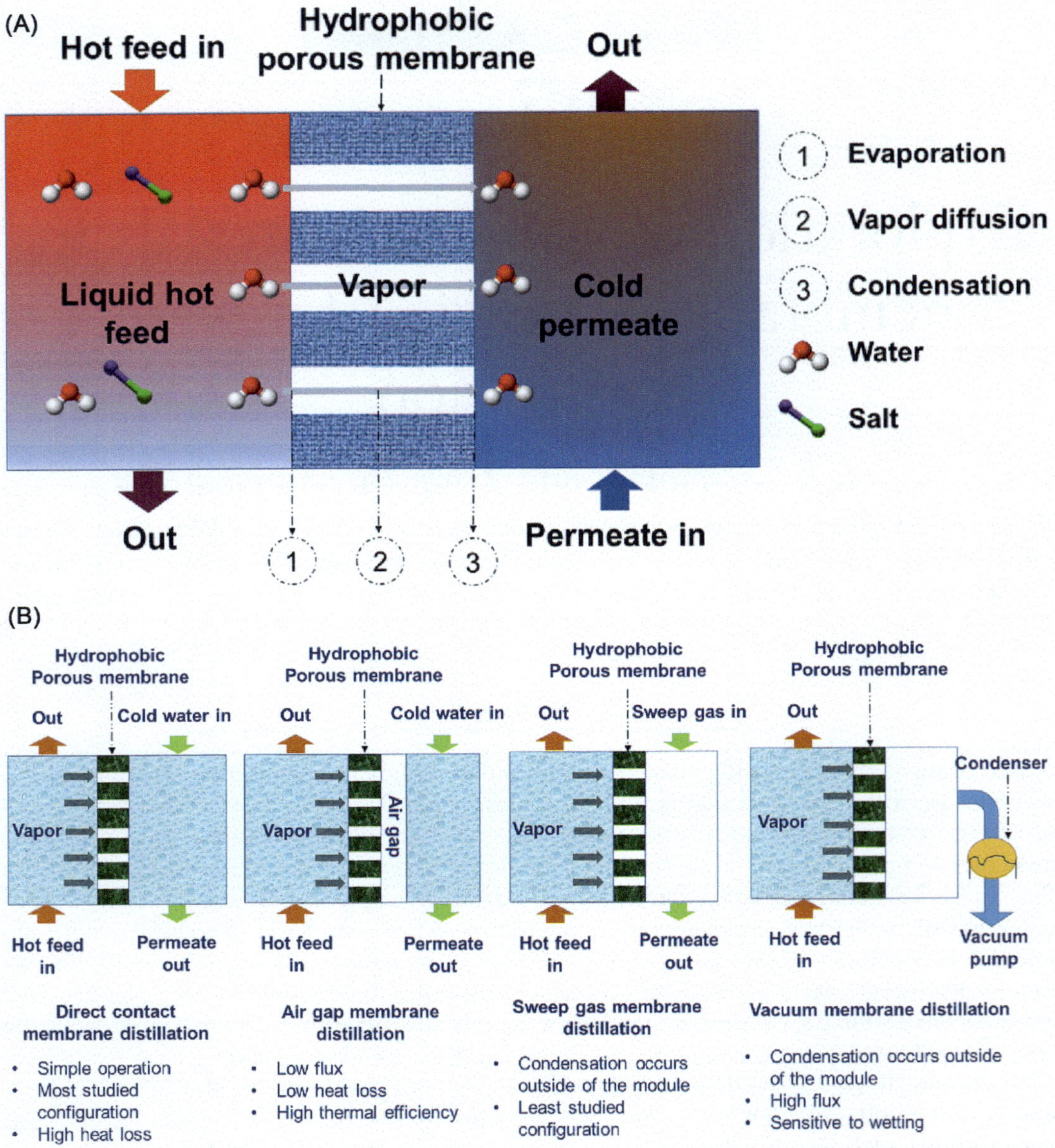

FIGURE 7.1 (A) Schematic illustration of membrane distillation (MD) process. (B) Different MD configurations and their specific features (Samadi et al., 2023). *Copyright 2023. Reproduced with permission from Elsevier Ltd.*

The integration of various nanoparticles (NPs) has been extensively investigated as a promising strategy for the development of high-performance membranes. In particular, incorporation of nanomaterials into membrane fabrication for MD processes has demonstrated potential benefits, including improved membrane properties, leading to enhanced water vapor flux, and reduced fouling tendencies.

Among the diverse range of nanomaterials, two-dimensional (2D) NPs such as graphene oxide (GO) and hexagonal boron nitride (h-BN), covalent organic frameworks (COFs), molybdenum disulfide (MoS_2), and MXene have recently garnered considerable attention due to their unique physicochemical properties, including high surface area, chemical stability, and tunable surface functionalities. These 2D nanomaterials have shown significant potential in enhancing the structural and functional characteristics of MD membranes, including hydrophobicity, porosity, and thermal conductivity. As a result, their incorporation has opened new pathways to address key challenges in MD processes, such as membrane wetting, thermal polarization, and fouling resistance.

In this context, the present review is structured into two major sections: fabrication of novel MD membranes incorporating 2D NPs, and performance evaluation of these membranes. Based on the current body of literature, studies can be broadly classified into two main categories: (1) synthesis and structural design of MD membranes using 2D nanomaterials and (2) functional modification and performance assessment of these membranes.

7.2 Mechanisms of membrane distillation using 2D nanomaterials

2D nanomaterials with sophisticated nanoconfined structures have emerged as promising candidates for enhancing membrane surface properties, thereby improving resistance to wetting in MD processes. Typically, incorporating these materials onto the membrane surface creates nanoconfined channels capable of blocking salts and foulants via an ion sieving mechanism (i.e., Donnan exclusion effect), which increases the membrane's wetting resistance. Surface engineering of membranes with 2D nanomaterials has been demonstrated as an effective strategy to prevent solute penetration (Wang et al., 2023). For example, GO has been used as a promising additive for MD membranes. Incorporating GO into polymers such as polysulfone (PSF) or polyvinylidene fluoride (PVDF) could significantly improve membranes' performance. In PSF membranes, GO interacts with sulfonyl functional groups, enhancing mechanical strength, reducing surface scaling, and decreasing the thickness of the functional layer. These effects shorten water vapor transport paths, thereby enhancing water vapor flux and membrane durability. Similarly, adding GO and its derivatives (e.g., reduced GO or functionalized GO) to PVDF membranes improves phase separation during fabrication, resulting in higher porosity and greater water vapor flux in AGMD systems. Additionally, hydrophobic functionalization of GO reduces pore wetting and increases liquid entry pressure (LEP), further enhancing long-term stability and efficiency of PVDF-based MD membranes (Camacho et al., 2020; Luque-Alled et al., 2023).

Also, h-BN nanosheets, structurally like graphene, have been incorporated into membranes, resulting in a porous and rough surface, along with finger-like macrovoids beneath the top layer due to hindered solvent and nonsolvent exchange during fabrication. Increasing h-BN content enhances nonsolvent diffusion during phase inversion, accelerating phase separation and producing a coarser structure with higher porosity. This leads to the bigger pore size and the more pore number, improving permeate vapor flux in MD processes (Zahirifar et al., 2019; Yadav et al., 2022).

In MD, LEP is a key parameter that defines the minimum pressure at which the liquid phase penetrates through the membrane pores, causing wetting. LEP is primarily influenced by pore size and shape, liquid surface tension, and contact angle between the liquid phase and the membrane surface, as described by the Laplace–Young equation. To enhance LEP, the membranes with smaller pores, higher water contact angles (WCAs), and lower surface energy are preferred (McGaughey and Childress, 2022; Chamani et al., 2020; Silva et al., 2021).

7.3 Fabrication and modification methods to develop 2D nanomaterial-based membrane distillation membranes

7.3.1 Synthesis of membrane distillation membranes using 2D nanomaterials

Two of the most important and widely studied methods for membrane fabrication are phase inversion and electrospinning, both of which offer significant potential for producing high-performance membranes. The incorporation of NPs within the polymeric solutions in these fabrication processes can play a crucial role in tailoring membrane structure and enhancing overall performance.

7.3.1.1 Phase inversion method

Phase inversion is one of the most common methods used for membrane fabrication. This technique is based on the preparation of a homogeneous, bubble-free polymer solution, in which NPs can also be dispersed to tailor the membrane characteristics (Shabani et al., 2020).

A modified wet-phase inversion method was utilized to develop GO-PSF mixed matrix membranes (MMMs) with various GO concentrations (0.25, 0.50, 1.0, and 2.0 wt%). GO was dispersed in solvent (N-methyl-2-pyrrolidone [NMP]), sonicated, and mixed with PSF polymer until dissolved. For a 2.0 wt% GO membrane, 1.5 g PSF was dissolved in 10 ml GO-NMP, with NMP added to reach 12.5 g. Lower GO concentrations were adjusted accordingly. The casting solution was spread on glass, submerged in a deionized (DI) water bath, rinsed, and dried (Camacho et al., 2020).

Another MMMs were prepared through phase inversion, utilizing a 14 wt% of PVDF polymer in solvent (Dimethylformamide (DMF)) with two types of hydrophobic GO octylamine-functionalized reduced GO (OA-rGO), and perfluorooctylamine-functionalized reduced GO (PFOA-rGO) (0.2, 0.5, 0.7, and 1 wt%). The casting solution was cast over nonwoven fabric using a doctor blade (with 200 μm clearance gap). Then, they were immersed in a DI water bath, initiating phase separation. After 5 min, they were transferred to another DI water bath for 24 h to remove residual solvents, followed by a controlled drying process (Luque-Alled et al., 2023).

PVDF/h-BN membranes were synthesized via phase inversion, using PVDF as polymer, DMF as solvent, and h-BN nanosheets (0, 0.5, 1, and 2 wt% based on the polymer weight) as the additive. The h-BN was dispersed in DMF, sonicated for 2 h, and mixed with 15 wt% PVDF under heating and stirring for 24 h. After degassing, the solution was cast (250 μm thickness) on a glass plate and immersed in a coagulation bath for 24 h. The membranes were then washed and air-dried (Zahirifar et al., 2019).

In an additional approach, ZnO nanoflowers were incorporated to introduce photocatalytic activity for dye degradation. Flat-sheet MMMs were fabricated via phase inversion using 18% w/v PVDF, 1% w/v ZnO, and 0–2% w/v h-BN (M-0, M-BZ0.5, M-BZ-1, and M-BZ-2). h-BN and ZnO were dispersed in DMF and sonicated for a uniform solution. PVDF was then added and stirred at 50°C for 8 h until fully dissolved. After overnight degassing, the solution was cast onto a nonwoven polyester fabric with a 0.12 mm knife gap (Yadav et al., 2022).

In another study, the PVDF membranes were modified by incorporating exfoliated h-BN and polyethylene glycol (PEG200) to alter their properties. The main material for all the membranes was PVDF. The first membrane was made by dissolving PVDF in DMF to form a uniform casting solution. This solution was then cast onto a glass plate and immersed in water to solidify and form the membrane structure. To enhance membrane properties, PEG200 was added to the PVDF solution, producing the second membrane. For the third membrane, exfoliated h-BN was first dispersed in DMF using ultrasonic waves to improve distribution. This dispersion was then blended with PEG200 and PVDF to create the modified casting solution. All membranes were fabricated under the same casting, coagulation, and drying conditions, with the aim of improving their performance in MD processes (Srivastava and Murthy, 2021).

Hollow fiber (HF) membranes composed of PVDF were fabricated using the dry-jet wet-phase inversion method for application in DCMD. Two types of metal-organic frameworks (MOFs) were incorporated into the PVDF matrix: pristine UiO-66 and UiO-66 modified with h-BN. To prepare the membranes, 25 wt% PVDF was dissolved in DMF at 45°C and stirred for 12–14 h to form a homogeneous dope solution. The UiO-66 and h-BN@UiO-66 powders were first dispersed in DMF using ultrasonication before being added to the dope solution. The solution was stirred for an additional 4–5 h, followed by degassing through ultrasonication. The spinning process involved the use of DI water as bore fluid and a spinneret with inner and outer diameters of 0.8 and 1.4 mm, respectively. The nascent fibers were coagulated in a water bath and then stored in water for 24 h to complete the phase inversion. Finally, the HF membranes were dried at room temperature for 48 h (Gajipara et al., 2024).

A PVDF membrane integrated with $Ti_3C_2T_x$ MXene nanosheets was developed to enhance the MD performance. Initially, an 18 wt% PVDF solution was prepared by dissolving the polymer in dimethylacetamide (DMAc) at 80°C for 8 h. In parallel, $Ti_3C_2T_x$ nanosheets were dispersed in isopropyl alcohol and ultrasonicated for 30 min to ensure uniform dispersion. A 2 mL aliquot of this dispersion was added to 100 mL of the PVDF solution, followed by stirring at 80°C for 6 h. The final casting solution was applied to a polyethylene terephthalate support with a thickness of 150 μm. Both the pristine and MXene-modified membranes were subjected to non-solvent-induced phase separation by immersion in a coagulation bath for 24 h and subsequently dried for DCMD testing (Ray et al., 2024).

The phase inversion method was also employed for the development of photothermal composite membranes using PVDF and DMF as the main polymer and solvent, respectively. Different nanomaterials, including silver NPs, carbon black, and MoS_2, were incorporated at mass fractions of 2%, 6%, 12%, and 25%. The casting solutions were applied to glass plates using a film applicator (with 200 μm gap). The samples were immersed in tap water for 2 h, followed by soaking in water to eliminate solvent traces, and then dried in an oven at 60°C overnight (Eljaddi and Cabassud, 2023).

7.3.1.2 *Electrospinning technique*

1 H, 1 H, 2 H, 2H-perfluorooctyltriethoxysilane (POTS)-modified GO was dispersed in DMF through 2 h of sonication, while PVDF was dispersed in a DMF-acetone mixture (4:1 ratio). The two solutions were combined and stirred at 60°C for 24 h to form a homogeneous spinning solution, which was then degassed under vacuum for 6 h. PVDF concentration was maintained at 14 wt%, and FTES-GO content was varied between 0.5 and 7 wt%. Electrospinning was performed using a 10 mL solution at 1.23 mL/h, applying 24.1 kV, with a 27.7 cm gap between the nozzle and the collector. The nanofibrous layers were deposited onto a polypropylene (PP) nonwoven fabric substrate for mechanical reinforcement. After 6 h of spinning, the composites were detached and dried at 50°C for 24 h (Li et al., 2020).

In another study, pristine PVDF-co-hexafluoropropylene (HFP) solution (18 wt%) was prepared in a DMF/acetone mixture (4:1 v/v) under overnight stirring. For composite solutions, graphene (1–10 wt%) was first dispersed in the solvent (DMF/acetone) via 1 h sonication, then mixed with the polymer solution under stirring at 80°C for 1 h and at room temperature for 24 h. Electrospinning was performed at 10 kV, 100 mm distance, and 1.0 mL/h flow rate. Nanofibers were collected on a rotating drum covered with aluminum foil. The process was conducted at 18°C–23°C and 40%–50% humidity. Membranes were dried at 50°C for 24 h to remove residual solvents (Woo et al., 2016).

Electrospinning was employed to fabricate micro/nano fiber membranes using polystyrene (PS) and PVDF-HFP with a 1:7 mass ratio. These membranes served as substrate for synthesizing multistage MoS_2 photothermal fiber membranes via hydrothermal reaction. In this process, $(NH_4)_6$ $Mo_7O_{24}{\cdot}4H_2O$ and CH_4N_2S were used in a 1:28 molar ratio, resulting in growth of MoS_2 nanoflowers on the fiber surface. The decomposition of CH_4N_2S generated H_2S, while $(NH_4)_6$ $Mo_7O_{24}{\cdot}4H_2O$ produced MoO_3, which was subsequently reduced under acidic hydrothermal conditions to form MoS_2. Polydimethylsiloxane (PDMS) was also introduced to modify the membrane surface due to its low surface energy and abundant methyl (-CH_3) groups, enhancing the hydrophobicity of the membranes. The preparation parameters, such as Mo/membrane ratio, reaction temperature, and reaction duration, were optimized to ensure uniform nanoflower growth and favorable structural characteristics (Ju et al., 2024).

7.3.2 Postmodification of membranes using 2D nanomaterials

In various studies, surface coating methods, vacuum filtration, and spray coating have been employed for post-modification of membranes with the aim of developing MD membranes using 2D NPs. These approaches are now being reviewed.

PVDF-HFP/GO/Octadecyl trichlorosilane (ODS) HF membranes were fabricated using the dry/wet jet spinning method. A dope solution was prepared by dissolving 12 wt% PVDF-HFP and varying concentrations of GO (0, 1, 3, and 5 wt%) in NMP. The resulting membranes were then surface-modified through silanization with ODS under acidic conditions (Dastbaz et al., 2017).

GO-immobilized membranes were fabricated by dispersing GO and the main polymer (PVDF) in solvent (Cyclohexanone). The resulting dispersions were cast onto flat composite membranes (PTFE/nonwoven PP) to obtain a uniform coating. After coating, the membranes were rinsed with acetone to remove residual PVDF (Bhadra et al., 2016).

Fuzil et al. prepared PVDF-based HF membranes with PES additive via the dry-wet spinning using NMP as solvent and ethylene glycol (EG) as pore former. After spinning at varied air gaps (10 and 20 cm), membranes were coagulated, washed, and underwent ethanol treatment to prevent shrinkage before drying. For modification, a MoS_2–TiO_2 composite (with an optimal ratio of 1:1) was synthesized hydrothermally and mixed with trichloro(octadecyl)silane (OTS) to create a dip-coating solution. The PVDF-based HF membranes were coated with this solution, enhancing their hydrophobicity and MD performance (Fuzil et al., 2023).

To enhance surface hydrophobicity, PVDF–PES HF membranes were modified using MoS_2–TiO_2 composite coating assisted by dopamine (DA) as an adhesive layer. Two dip-coating strategies were explored: the layering technique and the interlinking technique. In the layering approach, dopamine hydrochloride was polymerized to form poly (3,4-dihydroxy-l-phenylalanine) coating on the membrane surface. Subsequently, a suspension of MoS_2–TiO_2 NPs in ethanol was applied. This process was repeated in multiple dipping cycles to evaluate its impact on the coating layer thickness. In contrast, the interlinking method involved a one-step process where MoS_2–TiO_2 NPs were dispersed in a DA solution. DA was polymerized simultaneously with the composite material, creating a combined coating solution. The prewetted HF membranes were then dipped into this solution and dried at room temperature (Fuzil et al., 2024).

A Janus MXene-PVDF membrane was developed to enhance resistance to fouling and wetting in MD for the treatment of low surface tension wastewater. First, a pristine PVDF membrane was immersed in a 2 wt% PDMS/n-heptane solution (10:1 prepolymer:curing agent) for 10 s and cured at 70°C for 12 h to create a hydrophobic PDMS-modified substrate. A thin polydopamine (PDA) layer was then formed by floating the PDMS-PVDF membrane on a tris-buffer (at pH of 8.5) containing $CuSO_4$, DA, and H_2O_2 for 8 min, serving as a reactive platform for further functionalization. Subsequently, the PDA-coated membrane was floated on a 0.1 wt% polyethyleneimine (PEI) solution for 5 min, enabling cross-linking with PDA via hydrogen bonding and electrostatic interactions. A suspension of MXene (0.01 g in 30 mL ethanol) was ultrasonicated and vacuum-filtered onto the PEI-PVDF surface, forming a superhydrophilic and superoleophobic top layer. The final Janus MXene-PVDF membrane was rinsed, dried at 70°C for 3 h, and sealed for storage (Yan et al., 2022).

An electrothermal membrane for Joule heating membrane distillation (JHMD) was developed by vacuum-assisted filtration (VAF) of MXene nanosheets onto a commercial polyethylene terephthalate (PET)-PTFE membrane. Before coating, the membrane was treated with 10 wt% NaOH at 80°C for 2 h to hydrolyze ester bonds, followed by ethanol immersion to enhance surface hydrophilicity. The MXene nanosheets were then filtered onto the PET side and dried under vacuum at room temperature, yielding a composite membrane with 3.5 wt% MXene loading (Wang et al., 2023).

Porous MXene-coated PVDF membranes were fabricated to enhance MD performance by overcoming the typical trade-off between permeability and antiwetting capability. MXene nanosheets were deposited on PVDF substrates (pore size 0.22 μm) via vacuum filtration with varying MXene loadings of 16, 20, 24, 28, and 32 mg. After coating, each membrane was treated with 5 μL hydrazine monohydrate and carefully sandwiched between ceramic plates with supporting spacers to allow for volume expansion. The assembly was then cured in a vacuum drying chamber at 90°C for 12 h, resulting in a foam-like MXene structure with a hydrophobic surface and fusiform heterostructural channels (Wang et al., 2023).

A Co-MoS_2/CNT-modified PTFE membrane was prepared via vacuum filtration to enable solar-driven MD with antifouling capability. The nanocomposite was dispersed in ethanol by stirring and ultrasonication, then filtered onto the membrane under 15 kPa vacuum. Varying loadings (0–0.9 mg/cm^2) were achieved by adjusting the solution volume (Yan et al., 2022).

Vacuum filtration was also used for the development of GO composite membranes. For this purpose, GO was dispersed, centrifuged, and diluted before being vacuum-filtered onto ethanol-prewetted PVDF membranes. The GO-coated membranes were rinsed and dried at 80°C to obtain samples with different GO loadings. Reduced GO-loaded membranes were produced by placing the GO-coated membranes above hydroiodic acid solution, followed by rinsing and drying. The SiO_2-modified membranes were prepared similarly by filtering SiO_2 suspensions onto the PVDF membranes and further drying (Qiu et al., 2018).

The zeolitic imidazolate framework-67 (ZIF-67) wrapped graphene membranes (ZGM) were prepared by mixing G-ZIF dispersion with PVDF binder in methanol, ultrasonicated, and then vacuum-filtered onto PTFE membranes. The membranes were dried at 50°C to form the composites. Graphene-only modified membranes (GM) were prepared similarly without ZIF-67 (Lou et al., 2023).

A scalable strategy was used for the fabrication of membranes coated with various 2D materials, including GO, h-BN, MoS_2, and WS_2. The 2D flakes were dispersed in isopropyl alcohol, sonicated for 30 min, and then deposited via spray coating using a pyrolytic process onto commercial PVDF substrates. Prior to coating, a PDA layer was applied to enhance adhesion and prevent leaching of the nanomaterials. A final PDA layer was also coated, encapsulating the 2D nanomaterials on both sides to minimize detachment and improve stability in aqueous environments. The deposition process was conducted at 120°C, with a feed flow rate of 8000 μl/min over 10×10 cm^2 area and lasted for 2 min. The entire coating approach could be adapted into a roll-to-roll configuration, indicating potential for mass production. Membranes prepared through this process were denoted as GO-PDA/PVDF, h-BN-PDA/PVDF, MoS_2-PDA/PVDF, and WS_2-PDA/PVDF. Control samples included pristine PVDF and PDA/PVDF membranes with dual PDA layers and identical thermal exposure (Skuse et al., 2024).

Antifouling COF-LZU1/PVDF composite membranes were fabricated using an electrostatic spraying technique. In this process, TFB and PDA precursors were ultrasonically homogenized in acetic acid, loaded into syringes connected to a high-voltage supply, and sprayed onto a commercial PVDF substrate. Key parameters included 8 kV positive voltage, 3 kV negative voltage, 3 cm needle-to-collector distance, 20 rpm collector rotation, and 0.1 mL/min flow rate. These settings ensured uniform coating and controlled condensation of the COF-LZU1 layer on the membrane surface. After spraying, the membranes were vacuum dried for 8 h to remove residual solvent (Zhao et al., 2024).

A photothermal and superhydrophobic membrane was fabricated by incorporating $Ti_3C_2T_x$ MXene into a PDMS/PVDF matrix through an electrostatic spraying method. The membrane surface was engineered to self-assemble hierarchical polymeric nanospheres on MXene sheets, converting its hydrophilic nature to superhydrophobic. To prepare the membrane, an electrospray solution containing PDMS, PVDF, and tetrahydrofuran (THF)/DMF (1:1) was stirred at 65°C for 12 h, followed by dispersion of MXene. The solution was electrosprayed onto a commercial PVDF membrane at 20 kV and 1 mL/h. Membranes with varying MXene contents and thicknesses were fabricated, with the optimal configuration consisting of 10 μm MXene top layer on 100 μm PVDF base layer (Zhang et al., 2022).

7.4 Performance of 2D nanomaterial-based membrane distillation membranes

The GO-PSF MMMs showed promising performance in DCMD, with the membrane containing 1.0 wt% GO achieving the best balance of high water flux (up to 31.9 L/m^2 h (LMH)) and high salt rejection (99.85%). Increasing GO content enhanced membrane porosity and thus water flux, but excessive loading reduced salt rejection. The optimized membrane also exhibited reduced membrane fouling and improved ion retention without compromising water quality. However, incorporating GO into the membrane enhanced heat transfer from the hot feed side to the cold permeate side. The impact of GO loading on improving heat transfer aligns with the well-known high thermal conductivity of graphene-based nanomaterials. In general, as GO content increases, heat transfer through the GO-PSF membrane also increases. In MD processes, excessive heat transfer is undesirable because it reduces the temperature gradient across the membrane, thereby decreasing the vapor pressure difference that drives the separation; thus an optimum GO content should be obtained (Camacho et al., 2020).

The incorporation of hydrophobic functionalized rGO nanofillers, specifically OA-rGO and PFOA-rGO, into PVDF membranes significantly enhanced membrane hydrophobicity and porosity. Membranes with 0.7 wt% nanofillers achieved water flux values of 8.8–9.1 LMH and salt rejection above 99.9%, outperforming the pristine PVDF membrane. Both nanofiller types showed stable long-term performance, making them promising candidates for efficient and cost-effective MD applications (Luque-Alled et al., 2023).

The incorporation of 0.5 wt% h-BN into PVDF membranes significantly enhanced MD performance, achieving a water flux of 7.1 LMH and salt rejection of 99% at 80°C for over 60 h of continuous operation. This good performance was attributed to the high LEP of the membrane, which is influenced by contact angle, pore size, and pore shape, as well as its enhanced hydrophobicity, associated with low surface energy and high roughness. As noted, the addition of h-BN altered the surface roughness and pore structure of the modified membranes, leading to higher LEP and hydrophobicity. Improved porosity, surface roughness, and van der Waals interactions with water molecules also facilitated rapid and selective water vapor transport. These enhancements highlighted the potential of PVDF/h-BN membranes for efficient AGMD desalination (Zahirifar et al., 2019).

The PVDF membranes incorporated with h-BN nanosheets and ZnO nanoflowers demonstrated enhanced water flux and high removal efficiency in VMD for textile wastewater treatment. The best-performing membrane achieved a high water flux of 13.0 LMH and above 98.0% removal. Performance decline due to fouling was reversible with UV treatment, indicating good reusability and stability. SEM images were analyzed to examine the impact of incorporating h-BN nanosheets and ZnO nanoflowers on the membrane structure (Fig. 7.2). M-0 membrane exhibited a smooth and dense surface and a spongy cross-sectional structure. The incorporation of h-BN into the membrane led to the formation of a rougher and more porous surface and a cross-sectional morphology characterized by finger-like structures with macrovoids beneath the top layer. This morphological change was attributed to the restricted solvent–nonsolvent exchange, leading to the bigger pore size and higher pore density, which in turn enhanced the permeate flux (Yadav et al., 2022).

The incorporation of PEG200 and h-BN into PVDF was also found to be influential on the membrane structure and function. In particular, the PVDF/PEG200/h-BN membrane achieved higher permeate flux values, 3.4 LMH for K_2SO_4 and 2.6 LMH for KNO_3, at a feed temperature of 80°C, compared to that of the pristine PVDF membrane (2.3 LMH for K_2SO_4 and 1.9 LMH for KNO_3). Salt recovery was also greater for the modified membrane.

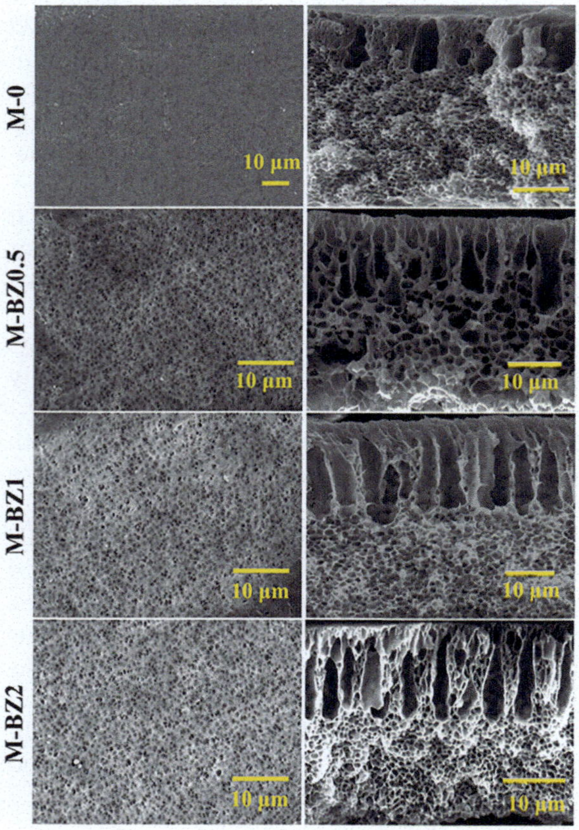

FIGURE 7.2 (A) Top surface and (B) cross-section scanning electron microscopy images of the synthesized pristine and mixed matrix membranes (Yadav et al., 2022). *Copyright 2022. Reproduced with permission from Elsevier Ltd.*

A 100 h continuous operation test demonstrated that the modified membrane exhibits stable permeate flux with only a slight increase in permeate conductivity. Enhanced water vapor transport was linked to increased porosity and favorable interactions between water vapor molecules and h-BN NPs (Srivastava and Murthy, 2021).

The performance of PVDF-based HF membranes loaded with UiO-66 and h-BN@UiO-66 MOFs was assessed in the DCMD process. Three membrane types were studied: pristine PVDF, PVDF with UiO-66, and PVDF with h-BN@UiO-66. Process optimization was carried out using response surface methodology (RSM) with Box-Behnken design to evaluate the influences of feed temperature, feed concentration, and feed flow rate on the membranes' performance. The results showed that the addition of h-BN to UiO-66 increases the MOF's surface area, which contributes to performance improvement. Among the membranes, h-BN@uio-66/PVDF demonstrated the highest permeate flux of 26. 8 LMH and salt rejection of 99.9% under the optimized conditions (65 °C, 1.5 wt.% feed concentration). It was observed that increasing feed temperature enhances vapor pressure, thus increasing permeate flux, while higher temperatures also slightly increase the risk of membrane wetting. Increased feed flow rate helped to mitigate temperature and concentration polarization effects, leading to further flux improvement. Additionally, the rejection performance of the membranes was evaluated using various salts (NaCl, KCl, $MgSO_4$, $CaSO_4$). The permeate flux followed the following trend: NaCl > KCl > $MgSO_4$ > $CaSO_4$. The h-BN@uio-66/PVDF membrane maintained high salt rejection (>99.9%) and stable permeate flux for over 80 h of operation under the optimized conditions (Gajipara et al., 2024).

The incorporation of $Ti_3C_2T_x$ MXene nanosheets into PVDF membranes led to changes in membrane surface properties and thus membrane behavior during MD processes. Compared to the pristine PVDF membrane, which suffered from significant wetting and gradual water flux decline from 6.5 to 6.0 LMH over 14 h, the MXene-modified membrane maintained consistent water flux of 8.0 LMH with negligible salt passage. This improvement was associated with the transition from hydrophilic to superhydrophobic surface characteristics, achieved by introducing MXene nanosheets that promote the formation of hierarchical micro/nanostructures. These structures contributed to WCA of 143°, along with enhanced self-cleaning and antiwetting behavior, allowing more stable operation during saline water treatment via MD processes (Ray et al., 2024).

In photothermal MD, membranes incorporating MoS_2 and carbon black NPs demonstrated superior performance compared to conventional PVDF and Ag-based membranes. These membranes showed higher vapor permeability and better wetting resistance, which are both crucial for stable desalination operation. MoS_2 and carbon black also offered cost advantages and lower wetting indicators. It was claimed that among them, carbon black stands out due to its low cost and strong potential for scalable application, though further optimization studies are still needed (Eljaddi and Cabassud, 2023).

The incorporation of FTES-functionalized GO nanosheets at 0.5, 1, 4, and 7 wt% into PVDF nanofibrous membranes, named G1, G2, G3, and G4, respectively, significantly enhanced their performance in the VMD process. Key improvements included increased surface hydrophobicity (with WCA up to 140.5°), high water flux (up to 36.4 LMH, double that of the pristine membrane), and excellent salt rejection (above 99.9%). The modified membranes also showed strong antiwetting behavior and stable operation over 60 h, making them promising candidates for efficient desalination. Fig. 7.3 presents SEM images of the

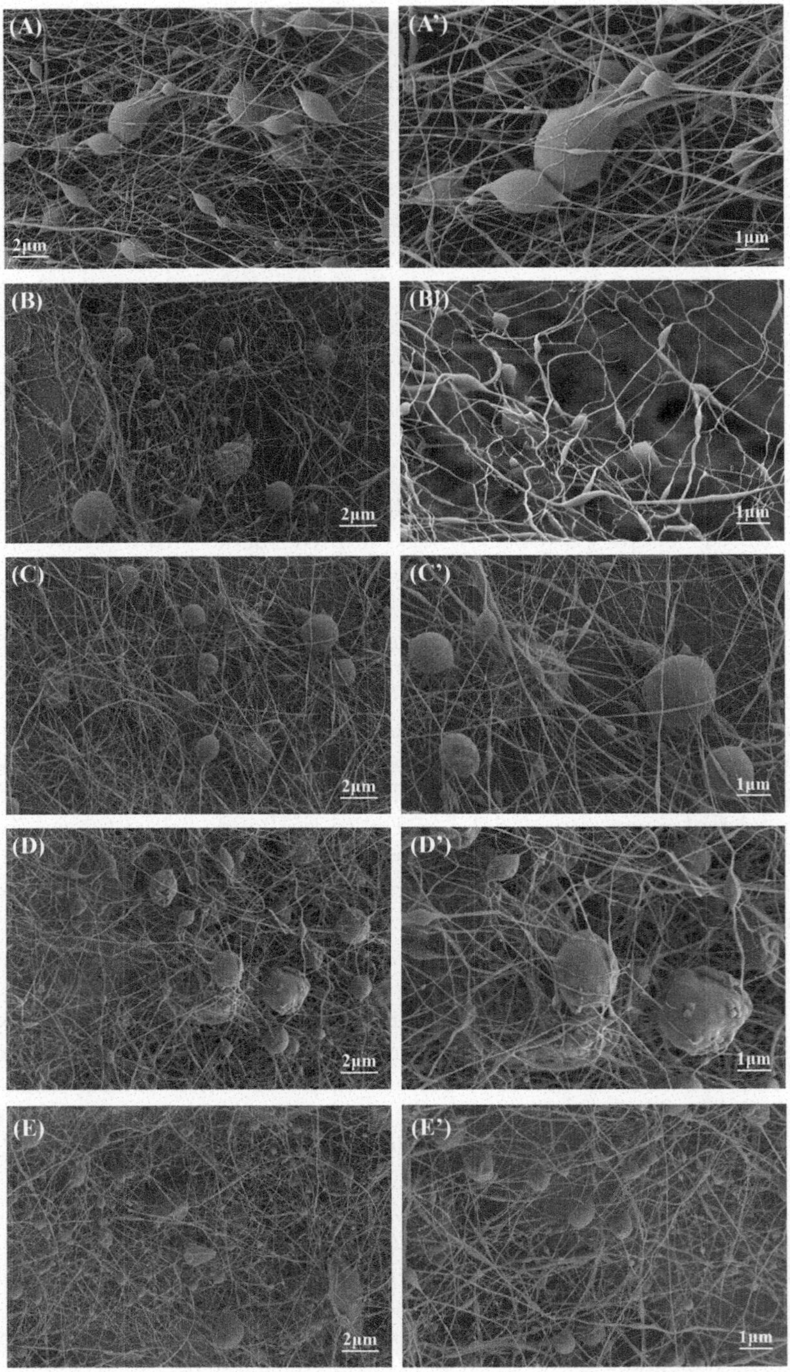

FIGURE 7.3 Scanning electron microscopy images of different PVDF electrospun nanofibrous layers (*left column* × 5.0 k) and the magnified photos (*right column* × 10.0 k). (A/A′) G0, (B/B′) G1, (C/C′) G2, (D/D′) G3, and (E/E′) G4 (Li et al., 2020). *Copyright 2020. Reproduced with permission from Elsevier Ltd.*

PVDF electrospun nanofibrous layers, revealing beaded fiber structures in the pristine membrane, which enhance surface hydrophobicity and support efficient water vapor transfer in MD. Incorporating FTES-GO nanosheets led to the formation of larger and more spherical beads, with bead density increasing alongside the nanosheets concentration (Li et al., 2020).

Electrospun graphene/PVDF-HFP membranes with 5 wt% graphene achieved stable water flux of 22.9 LMH and salt rejection of 99.99% during 60 h of AGMD operation using 3.5 wt% NaCl feed solution. Their high porosity, WCA, and LEP contributed to the superior performance of the membranes. Compared to the pristine PVDF membrane (4.7 LMH, 99.2% rejection), they showed markedly better performance. As illustrated in Fig. 7.4, the incorporation of graphene introduces multiscale surface roughness to the membrane, enhancing its hydrophobicity and consequently increasing its LEP. Due to its high aspect

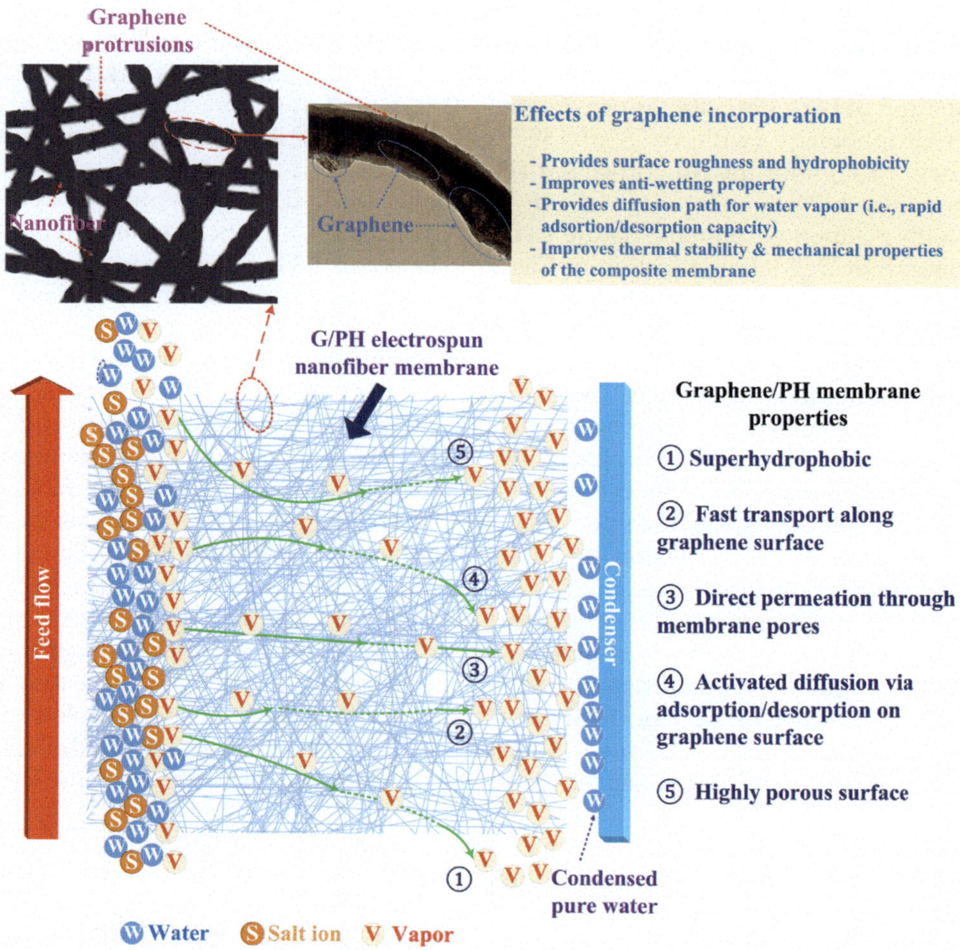

FIGURE 7.4 Schematic of the effect of graphene on the membrane performance in air gap membrane distillation process (Woo et al., 2016). *Copyright 2016. Reproduced with permission from Elsevier Ltd.*

ratio, graphene tends to protrude from the surface of nanofibers following electrospinning. Upon exposure to the feed solution (Fig. 7.4), the membrane's nano- and micro-scale surface features, combined with its superhydrophobic nature, effectively prevent water molecules from penetrating through the membrane pores. At the same time, the high volume fraction of the membrane facilitates greater water vapor transport. Furthermore, the exposed graphene nanosheets serve as diffusion pathways for water vapor molecules regarding their rapid adsorption–desorption properties, contributing to improved vapor transmission across the membrane. These properties confirm their suitability for efficient desalination (Woo et al., 2016).

The photothermal membranes, composed of PS/PVDF-HFP fibers modified with MoS_2 nanoflowers and PDMS, displayed consistent and efficient performance during the photothermal MD process. With an average light absorption of 90.9% across the 240–2600 nm range, the membrane's surface temperature increased to 74.2°C within 2 min under 1 kW/m^2 solar illumination. The system maintained a steady permeate flux of 1.2 LMH over 20 h, while achieving salt rejection of 99.96%. These results highlighted the effective solar absorption and heat localization capabilities of the membrane. The combination of high porosity (approximately 70.8%), moderate pore size (~1.82 μm), and high hydrophobicity (with WCA of 139.3°) contributed to its stability and separation efficiency throughout the MD processes (Ju et al., 2024).

The modified PVDF-HFP/GO/ODS HF membranes showed a high permeate flux of 34.1 LMH with a superhydrophobic nature (with WCA of 162°), significantly improving their performance in DCMD desalination of Caspian Sea water. LEP also increased to 181 kPa, enhancing resistance to pore wetting and fouling. Flux remained stable over 10 days, with 96% recovery after chemical cleaning. Membranes with low contact angle hysteresis, exhibiting dynamic water-repellent behavior, can effectively reduce fouling by minimizing the water-membrane contact area. The low surface energy of fluoropolymers, attributed to the high electronegativity of fluorine, results in weak attraction between fluorine and other materials, thereby enhancing hydrophobicity. PVDF-HFP copolymers typically demonstrate high WCAs, generally ranging between 110° and 130°, depending on their fabrication methods and conditions. The incorporation of GO into the PVDF-HFP matrix was found to increase surface roughness. Following surface modification with ODS, WCA significantly increased from 115° to 162°, indicating the successful development of a superhydrophobic surface. This improvement was primarily attributed to the synergistic effect of GO deposition within the membrane structure and the subsequent low-surface-energy functionalization. Overall, the modification strategy enhanced the membranes' resistance to wetting and fouling. Additionally, the near-spherical shape and mobility of water droplets on the superhydrophobic surface confirmed the presence of a self-cleaning effect, although this behavior might present certain challenges in precise contact angle measurement (Dastbaz et al., 2017).

The GO-immobilized PTFE membrane exhibited a remarkable permeate flux of up to 97.0 LMH at 80°C with complete salt rejection in the DCMD process. Its enhanced performance was attributed to selective sorption, nanocapillary effects, and reduced temperature polarization. The membrane maintained stable operation for 90 days, unaffected by salt concentrations up to 34,000 ppm. The observed increased permeate flux was primarily due to synergistic effects such as preferential sorption, enhanced transport through nanocapillaries, and reduced temperature polarization. Graphene's 2D structure of sp^2-bonded carbon atoms provides effective sorption sites for water vapor molecules while repelling bulk water molecules (Fig. 7.5). The ultrathin PVDF layer on the GO surface further enhances rejection of saltwater clusters. Additionally, polar functional groups on

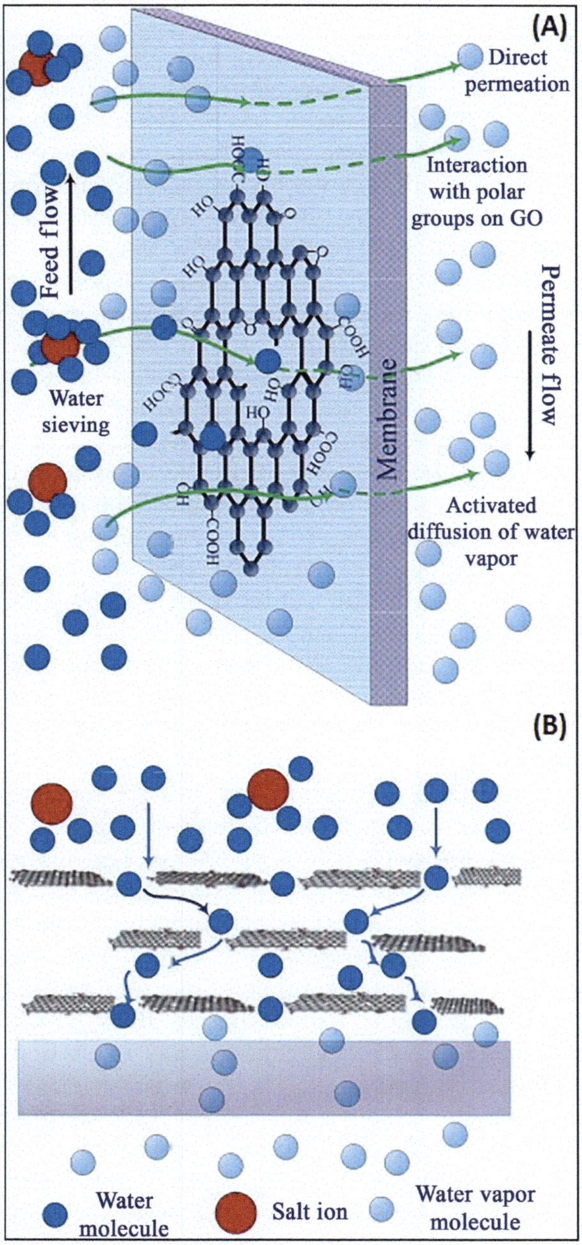

FIGURE 7.5 (A) Schematic of proposed mechanism graphene oxide immobilized membrane for (GOIM) and (B) nanocapillary action through graphene oxide layers (Bhadra et al., 2016). *Copyright 2016. Reproduced with permission from Elsevier Ltd.*

GO, such as epoxy, hydroxyl, and carboxyl, enhance membrane-vapor interactions, boosting overall vapor permeation. This aligns with previous findings using oxidized carbon nanotubes and nanodiamonds. GO also contains nanopores with adjacent hydroxylated and carboxylated carbon atoms, which aid in selectively sieving water vapor from salt clusters. The distribution of functional groups creates nanoscale wrinkles and defects that serve as transport pathways, while hydrophobic graphene regions offer nearly frictionless channels for rapid water flow (Bhadra et al., 2016).

The MoS_2–TiO_2-coated PVDF-based HF membranes demonstrated high MD performance. At 70°C, a membrane with MoS_2–TiO_2 loading of 0.2 wt% exhibited a permeate flux of 23.3 LMH, along with a salt rejection of 99.85%. WCA increased to 136.8°, due to the low surface energy of OTS, from 90.6° for the pristine membrane, and MoS_2–TiO_2 coating minimized temperature polarization. Altering the MoS_2–TiO_2 content resulted in a variation in the membrane's hydrophobicity. These results confirmed that the MoS_2–TiO_2 composite enhances surface roughness and reduces membrane pore wetting. However, further increasing the MoS_2–TiO_2 content reduced WCA to 128.7°. This decrease was likely attributed to the increased availability of exposed sites for water adsorption through Ti-OH bonds as the composite concentration rises (Fuzil et al., 2023).

The MoS_2–TiO_2/PVDF–PES membranes were evaluated for their desalination performance in MD process. The number of dipping cycles was found to affect hydrophobicity and subsequently permeate flux. Contrary to typical expectations, thicker coatings resulting from additional cycles led to higher permeate flux values. This was attributed to enhanced surface hydrophobicity, which reduces water molecule adhesion and minimizes the risk of pore wetting. Between the two coating methods, membranes produced via the layering technique demonstrated superior results. After three dipping cycles, the layered membrane showed WCA of 99.0°, delivering permeate flux of 2.1 LMH and salt rejection of 99.99%. In comparison, the interlinking method resulted in a WCA of 86°, with a permeate flux of 1.7 LMH and a salt rejection of 99.96%. The performance of the interlinking-coated membranes was still comparable to previously reported MD membranes, confirming the effectiveness of both methods in enhancing membrane characteristics through hydrophobic surface modification (Fuzil et al., 2024).

The MXene-PVDF membrane demonstrated excellent antifouling and antiwetting performance when treating low surface tension wastewaters. It maintained stable flux about 9.3 LMH with nearly complete salt rejection (~ 99.9%) for over 15 h of operation. This high-performance came from its superhydrophilic and superoleophobic MXene top layer (underwater oil contact angle: 169.9°) and high LEP (91.1 kPa), which prevents fouling and wetting even in challenging oily and surfactant-rich wastewaters. The LEP value reflects the pressure needed for a liquid to penetrate the largest pore of a membrane. Membranes with narrower pores and higher hydrophobicity generally exhibit higher LEP, which improves resistance to wetting. Consequently, the PDMS-PVDF membrane, exhibiting increased hydrophobicity, demonstrated higher LEP compared to the pristine PVDF membrane. In contrast, the MXene-PVDF membrane achieved the highest LEP despite possessing a superhydrophilic surface. This was attributed to surface agglomerations narrowing the pores and the absence of pore defects during modification of the hydrophobic substrate. Increased LEP alongside reduced in-air WCAs is likely resulted from reduced pore sizes due to the surface modification. To examine membrane surface charge, which affects fouling behavior and electrostatic interactions in MD systems, zeta potential measurements were also conducted. The PVDF and PDMS-PVDF membranes exhibited negative surface charges, with PDA-PVDF being more negative due to the presence of carboxyl groups. In contrast, the PEI-PVDF membrane exhibited a positive surface charge (+3.5 mV), allowing electrostatic assembly of a negatively charged MXene layer (−3.2 mV) on its surface (Yan et al., 2022).

Porous MXene-coated PVDF membranes exhibited enhanced performance in MD, achieving an average flux of 55.6 LMH at 60°C temperature gradient (feed: 70°C, permeate: 10°C), nearly 2.5 times higher than the pristine membrane. The fusiform heterostructural channels improved water vapor transport while selectively rejecting Na^+ and pollutants. The membranes also maintained robust antiwetting performance under salt and foulant exposure, with fluxes of 50.8 and 47.2 LMH, respectively (Wang et al., 2023).

The Co-MoS_2/CNT/PTFE membrane demonstrated effective dye wastewater treatment under solar-driven MD. Under 1 kW/m^2 light irradiation and at 35°C feed temperature, it achieved 80% degradation of RhB within 1 h on the feed side while maintaining a permeate flux of 2.4 LMH. The membrane sustained over 95% flux recovery after three fouling-cleaning cycles (Yan et al., 2022).

The GO/PVDF composite membrane demonstrated improved antifouling and antiwetting performance in the DCMD process, particularly with anionic surfactants such as sodium dodecyl benzene sulfonate (SDBS). GO NPs, deposited mainly on pore surfaces, preserved membrane porosity while reducing pore defects. Compared to the pristine PVDF and SiO_2/PVDF membranes, the GO-modified membrane maintained higher permeate flux and superior fouling resistance under anionic foulant conditions. However, its performance declined with cationic surfactants like sixteen alkyl three ammonium bromide (CTMAB), highlighting the role of surface chemistry in foulant-specific resistance. In contrast to the GO-modified membranes, those incorporating SiO_2 did not demonstrate effective anti-fouling properties, as indicated by a rapid rise in permeate conductivity within a short duration. The inclusion of SiO_2 led to an increase in the membrane's zeta potential, which diminished electrostatic repulsion against SDBS and compromised fouling resistance. The interaction between the foulant and pore surfaces enhanced membrane hydrophilicity, further contributing to the sharp increase in permeate conductivity. Unlike GO, which integrated more uniformly, SiO_2 formed a discrete surface layer, increasing mass transfer resistance and thereby reducing permeate flux (Qiu et al., 2018).

ZGM exhibited improved photothermal performance in the MD process, achieving vapor flux of 0.9 LMH and a photothermal efficiency of 62.1% under simulated sunlight. The assembly of ZIF-67 on the graphene surface formed a multilevel hierarchical structure with high surface area, enhancing light absorption and reducing reflection, which sustained a high cross-membrane temperature gradient. Additionally, ZGM effectively decreased phenol permeation; upon the addition of peroxymonosulfate to the feed, and thus the permeate phenol concentration was reduced by 88.0%, enabling VOC interception (Lou et al., 2023).

The spray membranes coated with 2D nanomaterials (GO, h-BN, MoS_2, and WS_2) were evaluated in MD experiments for over 72 h using wastewaters containing 150 ppm humic acid (HA) and 200 ppm paraffin oil, designed to simulate wastewater from oil and gas operations. Performance indicators included permeate flux stability, salt rejection, and fouling resistance. All evaluated membranes achieved high salt rejection (>99.9%). However, significant differences were observed in flux decline behavior. The pristine PVDF membrane exhibited rapid fouling, leading to complete pore wetting after just 10 h. The PDA/PVDF control membrane demonstrated a high flux decline rate (0.3%/h), while the membranes coated with GO and h-BN maintained much lower flux decline rates (0.002 and 0.028%/h, respectively). Among the evaluated membranes, only the GO-based variant maintained stable flux even in the presence of surfactants such as Triton X-100. Additionally, mechanical testing showed that the h-BN-coated membranes have tensile properties compared to the GO-coated ones. Overall, the spray coating process yielded membranes that resist both organic fouling and pore wetting under extended MD operation (Skuse et al., 2024).

The COF-LZU1/PVDF composite membranes were used in the DCMD process for treating industrial wastewater containing emerging contaminants (ECs). The COF-1.0 membrane exhibited higher permeate flux and improved salt rejection compared to the pristine PVDF substrate, particularly for sulfamethoxazole and ibuprofen, across pH values of 4–7 and temperatures of 40°C–70°C. Its hydrophilic surface and near-neutral zeta potential reduced electrostatic and hydrophobic–hydrophobic interactions between ECs and the membrane surface, leading to better fouling resistance (Zhao et al., 2024).

The $Ti_3C_2T_x$ MXene-engineered membrane exhibited enhanced MD performance by efficiently harvesting solar energy for localized photothermal heating, reducing heat loss during water vaporization. Its superhydrophobic surface, formed by hierarchical polymeric nanospheres, maintained a stable temperature gradient and strong wetting resistance. These features significantly increased the freshwater production rate and long-term operational stability under one sun illumination. This membrane design offered an energy-efficient and salt-resistant solution for sustainable desalination (Zhang et al., 2022).

7.5 Conclusion and perspectives

In this chapter, the use of 2D NPs in the MD process for water treatment (mainly desalination) was discussed. Various membrane fabrication methods, including phase inversion and electrospinning, used to produce MD membranes were examined. The nanomaterials included GO, h-BN, COFs, MoS_2, and MXene. Additionally, the properties and performance of modified membranes incorporating these 2D nanomaterials were analyzed. The enhanced MD performance achieved through the incorporation of these NPs was also presented.

The development of advanced membranes with antifouling and antiwetting properties, achieved by incorporating NPs into the membrane matrix or on the membrane surface, aims to improve MD performance, leading to higher permeate flux and salt rejection. However, the following considerations should be addressed in future research:

1. Exploring bioinspired coating strategies for the deposition of 2D NPs for developing MD membranes.
2. Comparing single-step and multistep coating methods to evaluate their effectiveness in stabilizing 2D NPs within MD systems.

3. Designing hybrid nanomaterials by combining 2D NPs with other NPs to develop MD membranes possessing high chemical and mechanical resistance as well as low thermal conductivity.
4. Developing Janus membranes with asymmetric wettability using 2D NPs to enhance separation efficiency.
5. Integrating 2D nanomaterials into MD membranes suitable for hybrid desalination systems (e.g., RO-MD, RO-MD-Crystallizer, MD-FO, MD-FO-MF, and RO-MD-PRO).

References

Bhadra, M., Roy, S., Mitra, S., 2016. Desalination across a graphene oxide membrane via direct contact membrane distillation. Desalination 378, 37–43. Available at https://doi.org/10.1016/j.desal.2015.09.026.

Camacho, L.M., Pinion, T.A., Olatunji, S.O., 2020. Behavior of mixed-matrix graphene oxide – polysulfone membranes in the process of direct contact membrane distillation. Separation and Purification Technology 240. Available at https://doi.org/10.1016/j.seppur.2020.116645, http://www.journals.elsevier.com/separation-and-purification-technology/.

Chamani, H., Yazgan-Birgi, P., Matsuura, T., Rana, D., Hassan Ali, M.I., Arafat, H.A., et al., 2020. CFD-based genetic programming model for liquid entry pressure estimation of hydrophobic membranes. Desalination 476. Available at https://doi.org/10.1016/j.desal.2019.114231, https://www.journals.elsevier.com/desalination.

Dastbaz, A., Karimi-Sabet, J., Ahadi, H., Amini, Y., 2017. Preparation and characterization of novel modified PVDF-HFP/GO/ODS composite hollow fiber membrane for caspian sea water desalination. Desalination 424, 62–73. Available at https://doi.org/10.1016/j.desal.2017.09.030, https://www.journals.elsevier.com/desalination.

Eljaddi, T., Cabassud, C., 2023. Wetting-based comparison of Ag, carbon black, and MoS2 composite membranes for photothermal membrane distillation. Membranes 13 (9). Available at https://doi.org/10.3390/membranes13090780, http://www.mdpi.com/journal/membranes.

Fuzil, N.S., Othman, N.H., Alias, N.H., Mat Shayuti, M.S., Shahruddin, M.Z., Marpani, F., et al., 2024. Effects of coating technique on the physicochemical properties and structure of MoS2-TiO2/PVDF-PES for membrane distillation application. Journal of Chemical Technology and Biotechnology 99 (8), 1754–1762. Available at https://doi.org/10.1002/jctb.7679, http://onlinelibrary.wiley.com/journal/10.1002/(ISSN)1097-4660.

Fuzil, N.S., Othman, N.H., Alias, N.H., Marpani, F., Mat Shayuti, M.S., Shahruddin, M.Z., et al., 2023. MoS2-TiO2 coated PVDF-based hollow fiber membranes for permeate flux enhancement in membrane distillation. Journal of Environmental Chemical Engineering 11 (3). Available at https://doi.org/10.1016/j.jece.2023.109866, http://www.journals.elsevier.com/journal-of-environmental-chemical-engineering/.

Gajipara, D.H., Kalla, S., Murthy, Z.V.P., 2024. Desalination performance evaluation of hBN@UiO-66 MOF incorporated hollow fiber membrane using membrane distillation. ChemistrySelect 9 (32). Available at https://doi.org/10.1002/slct.202402088, http://onlinelibrary.wiley.com/journal/10.1002/(ISSN)2365-6549.

Ju, J., Chen, Y., Huang, Y., Zhang, Y., Cheng, B., Kang, W., 2024. Designing a multi-stage MoS2 fiber membrane for photothermal membrane distillation. Journal of Membrane Science 690, 122237. Available at https://doi.org/10.1016/j.memsci.2023.122237.

Kujawska, A., Kujawski, J., Bryjak, M., Kujawski, W., 2015. ABE fermentation products recovery methods - a review. Renewable and Sustainable Energy Reviews 48, 648–661. Available at https://doi.org/10.1016/j.rser.2015.04.028, https://www.journals.elsevier.com/renewable-and-sustainable-energy-reviews.

Li, H., Shi, W., Zeng, X., Huang, S., Zhang, H., Qin, X., 2020. Improved desalination properties of hydrophobic GO-incorporated PVDF electrospun nanofibrous composites for vacuum membrane distillation. Separation and Purification Technology 230, 115889. Available at https://doi.org/10.1016/j.seppur.2019.115889.

Lou, M., Li, J., Zhu, X., Chen, J., Zhang, X., Fang, X., et al., 2023. Difunctional MOF-wrapped graphene membranes for efficient photothermal membrane distillation and VOCs interception. Journal of Membrane Science 676, 121592. Available at https://doi.org/10.1016/j.memsci.2023.121592.

Luque-Alled, J.M., Leaper, S., Abdel-Karim, A., Skuse, C., Gorgojo, P., 2023. PVDF membranes containing alkyl and perfluoroalkyl-functionalized graphene nanosheets for improved membrane distillation. Journal of Environmental Chemical Engineering 11 (3), 109898. Available at https://doi.org/10.1016/j.jece.2023.109898.

McGaughey, A.L., Childress, A.E., 2022. Wetting indicators, modes, and trade-offs in membrane distillation. Journal of Membrane Science 642. Available at https://doi.org/10.1016/j.memsci.2021.119947, www.elsevier.com/locate/memsci.

Qasim, M., Samad, I.U., Darwish, N.A., Hilal, N., 2021. Comprehensive review of membrane design and synthesis for membrane distillation. Desalination 518. Available at https://doi.org/10.1016/j.desal.2021.115168, https://www.journals.elsevier.com/desalination.

Qiu, H., Peng, Y., Ge, L., Villacorta Hernandez, B., Zhu, Z., 2018. Pore channel surface modification for enhancing anti-fouling membrane distillation. Applied Surface Science 443, 217–226. Available at https://doi.org/10.1016/j.apsusc.2018.03.004, http://www.journals.elsevier.com/applied-surface-science/.

Ray, S.S., Sharma, T.S.K., Myung, S., Choi, W.M., Kwon, Y.N., 2024. Holistic approach for fabrication of superhydrophobic MXene-based membrane for enhanced membrane distillation. Desalination 580. Available at https://doi.org/10.1016/j.desal.2024.117536, https://www.sciencedirect.com/science/journal/00119164.

Samadi, A., Ni, T., Fontananova, E., Tang, G., Shon, H., Zhao, S., 2023. Engineering antiwetting hydrophobic surfaces for membrane distillation: a review. Desalination 563, 116722. Available at https://doi.org/10.1016/j.desal.2023.116722.

Shabani, Z., Zarghami, S., Mohammadi, T., 2020. Nanomaterials for fouling-resistant RO membranes. Nanotechnology in the Beverage Industry: Fundamentals and Applications. Elsevier, Iran, pp. 151–184. Available at https://www.elsevier.com/books/nanotechnology-in-the-beverage-industry/amrane/978-0-12-819941-1, http://doi.org/10.1016/B978-0-12-819941-1.00006-7.

Silva, R.D.S., Ramlow, H., Santos, B.D.C., Madalosso, H.B., Machado, R.A.F., Marangoni, C., 2021. Membrane distillation: experimental evaluation of liquid entry pressure in commercial membranes with textile dye solutions. Journal of Water Process Engineering 44. Available at https://doi.org/10.1016/j.jwpe.2021.102339, http://www.journals.elsevier.com/journal-of-water-process-engineering/.

Skuse, C., Alberto, M., Luque-Alled, J.M., Mercadillo, V.O., Asuquo, E., Gallego-Schmid, A., et al., 2024. Spray coating of 2D materials in the production of antifouling membranes for membrane distillation. Journal of Membrane Science 711. Available at https://doi.org/10.1016/j.memsci.2024.123162, https://www.sciencedirect.com/science/journal/03767388.

Srivastava, A., Murthy, Z.V.P., 2021. Investigating the effect of PEG200 and two-dimensional h-BN on PVDF membrane performance for membrane distillation–crystallization. Materials Today Chemistry 22, 100545. Available at https://doi.org/10.1016/j.mtchem.2021.100545.

Wang, A., Chen, W., Lan, X., Ding, M., 2023. Simultaneous improvement of wetting resistance and permeability in membrane distillation using MXene-foam coating layer with heterostructural channels. Desalination 566, 116885. Available at https://doi.org/10.1016/j.desal.2023.116885.

Wang, Z., Li, Z., Yong, B., Huang, H., Yao, L., Deng, L., 2023. Depression of electro-oxidation of Ti3C2Tx MXene Joule heater by alternating current for Joule heating membrane distillation. Chemical Engineering Journal 461, 142149. Available at https://doi.org/10.1016/j.cej.2023.142149.

Woo, Y.C., Tijing, L.D., Shim, W.G., Choi, J.S., Kim, S.H., He, T., et al., 2016. W5ater desalination using graphene-enhanced electrospun nanofiber membrane via air gap membrane distillation. Journal of Membrane Science 520, 99–110. Available at https://doi.org/10.1016/j.memsci.2016.07.049, www.elsevier.com/locate/memsci.

Yadav, A., Patel, R.V., Awasthi, P., Sinha, N., Labhasetwar, P.K., 2022. Hexagonal boron nitride nanosheets incorporated photocatalytic polyvinylidene fluoride mixed matrix membranes for textile wastewater treatment via vacuum-assisted distillation. Colloids and Surfaces A: Physicochemical and Engineering Aspects 653. Available at https://doi.org/10.1016/j.colsurfa.2022.129955, www.elsevier.com/locate/colsurfa.

Yan, X., Yang, C., Ma, C., Tao, H., Cheng, S., Chen, L., et al., 2022. A novel janus membrane modified by MXene for enhanced anti-fouling and anti-wetting in direct contact membrane distillation. Chemosphere 307 (Pt 4), 136114. Available at https://doi.org/10.1016/j.chemosphere.2022.136114.

Yan, Z., Chen, X., Bao, S., Chang, H., Liu, H., Fan, G., et al., 2022. Integration of in situ Fenton-like self-cleaning and photothermal membrane distillation for wastewater treatment via Co-MoS2/CNT catalytic membrane. Separation and Purification Technology 303, 122207. Available at https://doi.org/10.1016/j.seppur.2022.122207.

Ye, Y., Zhao, L., Liu, Z., Wang, Z., Niu, Q.J., 2024. Recent advances in novel membranes for improved anti-wettability in membrane distillation: a mini review. Desalination 571. Available at https://doi.org/10.1016/j.desal.2023.117066, https://www.journals.elsevier.com/desalination.

Zahirifar, J., Hadi, A., Karimi-Sabet, J., Dastbaz, A., 2019. Influence of hexagonal boron nitride nanosheets as the additives on the characteristics and performance of PVDF for air gap membrane distillation. Desalination 460, 81–91. Available at https://doi.org/10.1016/j.desal.2019.03.004, https://www.journals.elsevier.com/desalination.

Zhang, B., Wong, P.W., Guo, J., Zhou, Y., Wang, Y., Sun, J., et al., 2022. Transforming Ti3C2Tx MXene's intrinsic hydrophilicity into superhydrophobicity for efficient photothermal membrane desalination. Nature Communications 13 (1). Available at https://doi.org/10.1038/s41467-022-31028-6, http://www.nature.com/ncomms/index.html.

Zhao, Y., Wang, C., Ren, L.-F., Zhang, B., Shao, J., 2024. Facile preparation of COF-LZU1 modified membrane via electrostatic spraying for emerging contaminants treatment in direct contact membrane distillation. Separation and Purification Technology 337, 126411. Available at https://doi.org/10.1016/j.seppur.2024.126411.

CHAPTER

8

Advancements in 2D nanomaterials for efficient solar-driven water evaporation

Syazwani Mohd Zaki[1], *Flora Serati*[1,2], *Sa'adah Ahmad Sowi*[1], *and Nadiah Mokhtar*[3]

[1]Department of Manufacturing and Materials Engineering, International Islamic University Malaysia, Kuala Lumpur, Malaysia [2]Department of Electrical and Computer Engineering, International Islamic University Malaysia, Kuala Lumpur, Malaysia [3]Faculty of Civil Engineering Technology, Universiti Malaysia Pahang Al-Sultan Abdullah, Pekan, Pahang, Malaysia

8.1 Introduction of two-dimensional nanomaterials in solar-driven water evaporation

8.1.1 Introduction

The rising global population, rapid industrialization, climate change, and the ongoing energy and natural resource crises are placing increasing pressure on the potable freshwater supplies. Freshwater shortages and water pollution currently affect nearly 300 million people living in economically disadvantaged regions worldwide (Jing et al., 2024). As a result, global water scarcity poses a severe threat to sustainable development, necessitating the advancement of innovative and scalable water purification technologies. Among emerging solutions, solar-driven evaporation technology has garnered significant attention as a promising method for producing freshwater. This is due to its ability to utilize abundant solar energy for clean water production. This approach presents a promising pathway not only for freshwater generation but also for contributing to green energy solutions. However, the inherent limitations of water, such as its poor optical absorption and substantial heat losses, lead to low natural light-to-vapor conversion efficiency, rendering it impractical for large-scale freshwater production (Fan et al., 2021). To address these challenges, extensive research has been conducted to develop advanced solar systems. These systems employ rationally designed photothermal nanomaterials and interfacial heating strategies to significantly enhance solar energy absorption and evaporation efficiency (see Fig. 8.1).

Two-Dimensional Nanomaterials for Water Purification
DOI: https://doi.org/10.1016/B978-0-443-44530-9.00008-0

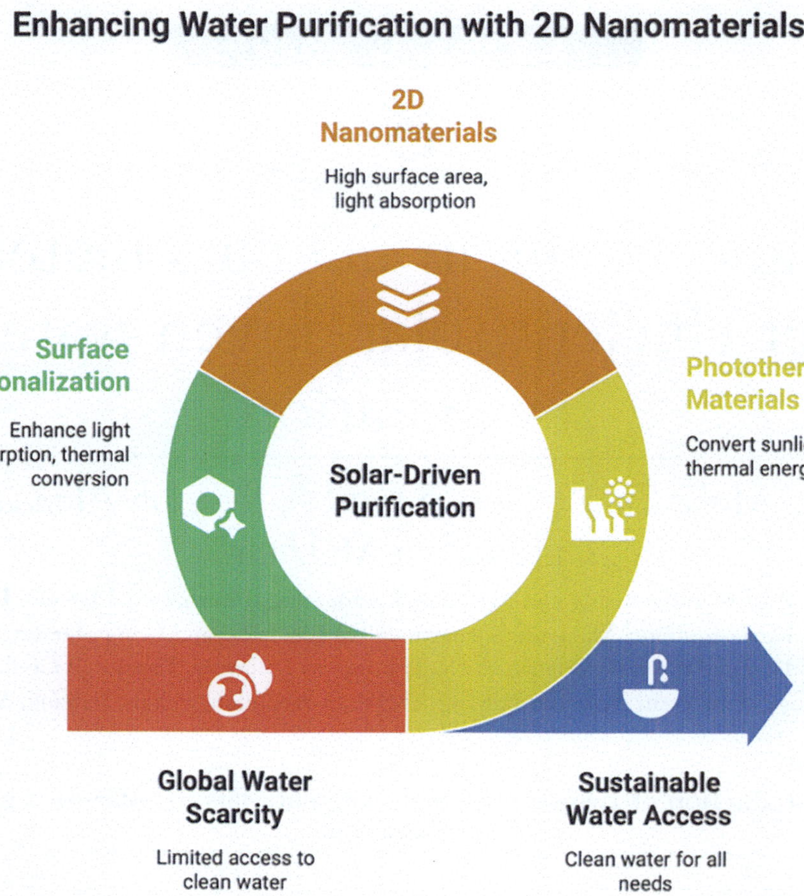

FIGURE 8.1 Enhancing solar-driven water purification by using two-dimensional nanomaterials for photothermal.

8.1.2 Photothermal materials

Photothermal materials function as light absorbers, converting light into heat. In recent years, a diverse range of photothermal materials, including polymer materials, plasmonic metals, and carbon-based compounds (Li et al., 2025), semiconductor materials, graphene (Chang et al., 2024), hybrid composites, aerogel (Ahmad et al., 2025), and hydrogel (Du et al., 2025; Zaed et al., 2025), have been actively investigated for solar-driven water purification systems (Mittal et al., 2023; Jiang et al., 2024). These materials serve as the core component in a solar vapor generation platform by adsorbing sunlight and converting it into localized heat at the water–air interface. This thermal energy facilitates water evaporation, enabling phase separation from a variety of contaminants, including dissolved salts, organic pollutants, and pathogenic microorganisms. The resulting vapor, once condensed, provides clean water suitable for drinking, sanitation, and agricultural use. To optimize the efficiency of this solar–thermal conversion process, ideal photothermal materials should possess the following attributes: exhibit broad-spectrum light absorption (Mittal et al., 2023), high photothermal conversion efficiency (Jing et al., 2024), hydrophilicity (Ahmad et al., 2025), thermal stability (Du et al., 2025), and compatibility with scalable fabrication methods (Mittal et al., 2023). The integration of 2D photothermal materials into interfacial water evaporation systems holds significant potential in addressing global freshwater scarcity through sustainable, energy-efficient, and decentralized purification solutions. Future advances in material engineering, surface functionalization, and system-level design are essential to further optimize these systems and support their large-scale implementation.

8.1.3 Two-dimensional nanomaterial solar-driven

Nanomaterials are typically defined as structures with a size in the range of 1–100 nm (Jiang et al., 2024). Among them, two-dimensional (2D) nanomaterials have gained significant attention due to their distinctive physicochemical properties, including extraordinary surface area (Noor et al., 2024), tunable electronic band gap (Jiang et al., 2024), and remarkable light absorption capabilities (Jiang et al., 2024). Characterized by their ultrathin atomic layers and simple planar structures, 2D nanomaterials are also valued for their low cost, ease of synthesis, and simple maintenance, which support their practical applicability in large-scale purification systems (Song et al., 2022). Such features enable efficient absorption of solar energy, rapid thermal response, and effective water transport, which are critical for achieving high-performance solar-driven evaporation. The planar morphology of 2D materials localizes thermal energy at the air–water interface, thereby reducing heat dissipation into the bulk water and improving overall system efficiency (Fatima et al., 2022).

Recent experimental studies have demonstrated that 2D nanomaterial-based evaporators can achieve solar-to-vapor conversion efficiencies exceeding 80% under one sun irradiation (Usman et al., 2022). Furthermore, these systems show excellent antifouling behavior and reusability (Ahmad et al., 2025), making them ideal candidates for real-world applications in brackish water desalination, wastewater treatment, and emergency water supply systems. A wide range of 2D nanomaterials has been explored for photothermal applications, including MXenes (Li et al., 2025; Zhang et al., 2025a, 2025b, 2025c; Wen et al., 2025), graphene (Khoei et al., 2023), transition metal dichalcogenides (TMDs), semiconductor materials (Du et al., 2025), polymer materials (Jing et al., 2024), and carbon-based materials (Wang et al., 2025). These materials have shown exceptional potential in improving solar steam generation and are now central to next-generation solar desalination and wastewater purification technologies (Jiang et al., 2024). As a consequence, 2D materials-based photothermal is fascinating and has garnered extensive attention in water treatment (Jiang et al., 2024).

The incorporation of 2D nanomaterials into solar evaporators has led to increased evaporation rates, higher energy conversion efficiencies, and improved long-term operational stability, making them highly attractive for sustainable and scalable water purification technologies (Fang et al., 2025). Notably, black absorptive materials, including metallic plasmonic structures, semiconducting nanoparticles, and porous carbon materials, have demonstrated superior solar energy harvesting capabilities (Liu et al., 2017). A typical MXene material, $Ti_3C_2T_x$, on the other hand, has a unique localized surface plasmon resonance (LSPR) effect (Li et al., 2025). Previous findings have reported that the hybridized graphene-based material exhibits outstanding evaporation performance, achieving a high rate of 2.08 $kg.m^{-2}\ h^{-1}$ for pure water evaporation under one-sun irradiation (Chang et al., 2024). Additionally, the employment of porous materials can significantly enhance water transport and vapor escape, contributing to improved evaporation efficiency. A super-hydrophilic and highly porous $Ti_3C_2T_x$/alginate aerogel solar evaporator has exhibited an excellent evaporation rate of 2.09 $kg\ m^{-2}\ h^{-1}$, surpassing conventional systems (Ahmad et al., 2025). However, while focusing on performance enhancement, it is crucial to consider the economic feasibility, recyclability, and scalability of such materials for real-world deployment. Further enhancement of 2D photothermal materials involves surface functionalization, defect engineering, heterojunction formation, and the development of composite materials (Fang et al., 2025). For example, integrating MXenes with carbon-based or polymeric supports has been shown to improve both photothermal stability and mechanical integrity. Likewise, the formation of heterostructures combining graphene and MoS_2 can broaden light absorption ranges and synergize thermal conductivity and hydrophilicity, resulting in higher evaporation flux and long-term durability under continuous solar illumination. In the planar surface and tight staking design, the 2D photothermal material faced significant reflected light and poor solar adsorption (Jiang et al., 2024).

8.1.4 Effective solar-driven key factor

In solar-driven water purification technologies, understanding and optimizing the factors that influence evaporation performance is critical. The efficiency of these systems is primarily governed by the capacity of the photothermal material to absorb a wide range of solar wavelengths, its photothermal conversion efficiency, and the management of thermal energy to ensure localized heating with minimal losses. These elements collectively determine the final vapor generation rate and overall system efficiency. To achieve these goals, design strategies for photothermal materials emphasize maximizing solar absorption across UV, visible, and NIR regions while minimizing reflection and transmission losses. It is equally important to suppress nonradiative energy losses by ensuring that absorbed photons are predominantly converted to thermal energy rather than reemitted as light or dissipated through conduction.

The availability and intensity of sunlight directly determine the energy input for vapor generation. Materials with broad-spectrum light absorption (UV–visible–NIR) enhance the utilization of solar energy. Materials such as reduced graphene oxide and black phosphorus exhibit excellent spectral overlap with the solar spectrum, boosting evaporation efficiency. Efficient solar-to-heat conversion is critical. 2D nanomaterials like MXenes, MoS_2, and graphene derivatives convert absorbed solar energy into localized heat with minimal radiative or conductive losses. Their inherent high photothermal conversion efficiencies (over 85% in some cases) make them ideal candidates for an interfacial heating system.

Surface functionalization refers to the modification of the surface chemistry or structure of 2D nanomaterials to enhance their performance in solar-driven water purification systems (Wieszczycka et al., 2021). These modifications are strategically applied to improve light absorption, photothermal conversion efficiency, water transport, and antifouling properties. Chemical functionalization enhances hydrophilicity, stability, and light absorption. For example, oxygen-containing groups on graphene oxide (GO) improve water affinity, while doping with nitrogen or sulfur can alter electronic structures for better photothermal response. The functional layers also support the antifouling and salt-rejection capabilities.

Minimizing heat loss to the bulk water is crucial. 2D materials are typically used in floating or layered configurations with thermal insulators beneath to ensure that generated heat remains at the air–water interface. This enhances vapor generation while reducing energy waste. Hydrophilic channels, capillary wicks, and porous structures ensure a continuous water supply to the evaporative surface. Studies show that combining hydrophilic surfaces with vertically aligned porous layers significantly increases evaporation rates by optimizing water flux. Factors such as ambient temperature, humidity, wind speed, and feedwater salinity can influence evaporation rate, condensation efficiency, and material degradation. Proper system design and protective coatings are recommended to ensure stable performance under variable conditions. Overall, advanced structural and material engineering approaches have been developed, including the incorporation of porous architectures, hydrophilic channels, and thermal insulating layers. These features support directional heat transfer to the evaporation surface, enhance capillary water transport, and maintain elevated surface temperatures necessary for sustained and efficient evaporation (Gnanasekaran and Rajaram, 2024).

8.1.5 Summary

In summary, 2D nanomaterials represent a transformative class of photothermal agents in the field of solar-driven water purification. Their tunable properties, combined with advanced design strategies, have enabled significant advancements in evaporation performance, energy efficiency, and material stability. Continued interdisciplinary research in materials science, nanotechnology, and environmental engineering is essential to overcome current limitations and scale up these systems for broader deployment in water-scarce regions. The effectiveness of solar-driven water purification systems based on 2D nanomaterials depends on the interplay between material properties, surface functionalization, system design, and environmental conditions. By strategically optimizing these key factors, we can unlock the potential for highly efficient, cost-effective, and sustainable water purification technologies.

8.2 Mechanisms of photothermal water evaporation using two-dimensional nanomaterials

8.2.1 Introduction

This topic introduces the core principles and mechanisms behind solar-driven water evaporation utilizing 2D nanomaterials. As global demand for clean water continues to rise, photothermal evaporation has emerged as a highly promising, energy-efficient, and environmentally sustainable approach to freshwater generation. By harnessing solar energy and converting it into localized heat, this method offers a viable solution for off-grid and resource-limited regions. Two-dimensional nanomaterials such as graphene, transition metal dichalcogenides (TMDs), MXenes, and black phosphorus are at the forefront of this technology due to their exceptional optical and thermal properties (Azni Farhana Mazri et al., 2023). Their ultrathin structures, large surface areas, high thermal conductivity, and broadband light absorption capabilities make them ideal for capturing solar energy and facilitating efficient evaporation. Fig. 8.2 illustrates the photothermal water evaporation mechanism using 2D nanomaterials. When sunlight strikes the porous evaporator surface, part of the light is reflected, some is transmitted, and the rest is absorbed. The absorbed light energy is then converted into heat through photothermal conversion via nonradiative relaxation processes. This conversion leads to localized heating on the surface of the porous structure (Wang et al., 2020). The generated heat undergoes several modes of transfer, such as conduction, radiation, convection, and scattering. Conducted heat travels downward into the porous layer beneath, reaching the water interface. Due to the thermal insulation properties or design of the 2D nanomaterials, the heat is confined to the water–material interface, effectively minimizing heat loss to the bulk water (Ren and Sun, 2025). As a result of this uniform localized heating, surface water rapidly evaporates, generating water vapor. The porous structure further facilitates continuous water transport from the bulk reservoir to the heated interface, ensuring a sustained evaporation process (Wang et al., 2020). This subchapter focuses on how 2D nanomaterials enhance solar-driven water evaporation through five interrelated mechanisms: solar light absorption, nonradiative relaxation, plasmonic localized heating, interfacial water evaporation, and water transport (see Fig. 8.2). A detailed understanding of these mechanisms provides a foundation for appreciating the critical role of 2D nanomaterials in the development of next-generation water purification systems.

8.2.2 Solar light absorption

2D materials have an atomic arrangement that gives them an ultrathin thickness, typically only a few nanometers. This structure also leads to a high surface-area-to-volume ratio. These materials can absorb significant portions of the solar spectrum (from ultraviolet [UV] to near-infrared [NIR]) despite their minimal thickness due to their high absorption coefficients. For instance, monolayer molybdenum disulfide (MoS_2) can absorb up to 10% of incident light at specific wavelengths. The layered crystal structures are held together by van der Waals forces, allowing easy exfoliation and integration onto various substrates (Wang et al., 2021). Electronic band structures in 2D nanomaterials are highly tunable. Their

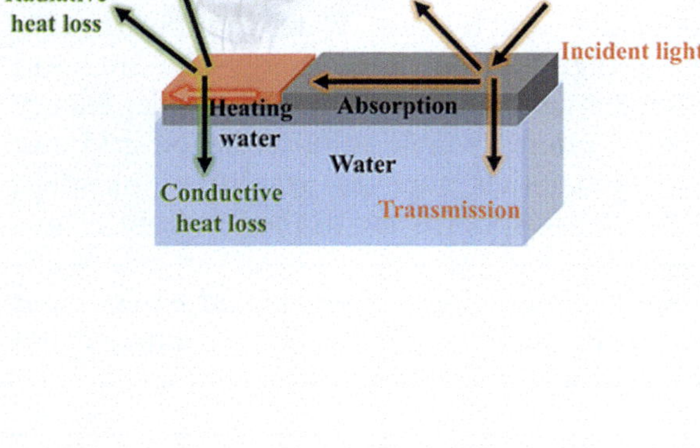

FIGURE 8.2 A schematic illustrating the variation in the light path reaching the evaporator surface and the corresponding energy balance. A schematic illustrating the variation in the light path reaching the evaporator surface and the corresponding energy balance, encompassing the stages of light-to-heat conversion, heat-to-steam generation, and steam condensation. *Data from Xiao, Y., Guo, H., Li, M., He, J., Xu, X., Liu, S., Wang, L., James, T.D., 2025. Strategies for enhancing the photothermal conversion efficiency of solar-driven interfacial evaporation. Coordination Chemistry Reviews, 527, 216378. https://doi.org/10.1016/j.ccr.2024.216378.*

bandgap can be engineered by changing the number of layers, applying strain or doping. In monolayer TMDs, for example, the material exhibits a direct bandgap, which is favorable for photon absorption and exciton generation. In contrast, the bulk forms of these materials typically have indirect band gaps, making them less efficient light absorbers (Shao et al., 2019). 2D nanomaterials possess unique optical properties that enable them to capture a significant portion of the solar spectrum. Their atomically thin structure enhances light–matter interactions, and their optical absorption can be further improved through structural modifications such as doping, defect engineering, and the creation of heterostructures (Xie et al., 2020). Interestingly, pristine graphene can absorb approximately 2.3% of incident light per atomic layer. When incorporated into composites or hybrid structures, this absorption can be significantly enhanced (Bramhaiah and Bhattacharyya, 2022).

Additionally, certain 2D materials exhibit excitonic effects and surface plasmon resonance, both of which intensify local electromagnetic fields at the material's surface, further boosting light absorption (Zhao et al., 2020). The ability of 2D materials to convert solar energy into localized thermal energy enables effective performance even under low solar intensity, thereby addressing the growing demand for sustainable water purification. Incorporating 2D materials into solar evaporator designs facilitates interfacial heating, accelerating the liquid-to-vapor phase transition and optimizing overall evaporation efficiency (Irshad et al., 2025). Solar light absorption is a fundamental process through which radiant energy from the Sun is captured and transformed into thermal energy. When photons from sunlight strike a material, they transfer their energy to atoms or molecules, exciting electrons to higher energy states or increasing the vibrational motion of these atoms or molecules. This excitation raises the internal energy of the material, often resulting in a temperature increase. The efficiency of this absorption depends on the material's color, composition, texture, and electronic structure (Nabavi Hamedani et al., 2025).

In the context of 2D materials, solar absorption is governed primarily by their electronic band structures, the presence of surface or structural defects, and plasmonic resonances. The reduced dimensionality of 2D materials enhances photon confinement and enables effective absorption of visible and NIR light, which is crucial for efficient photothermal energy conversion (Subhan and Mourad, 2025). When integrated into heterostructures, the synergistic effects between layers further optimize solar energy capture, making these materials well-suited for applications in water evaporation systems. These absorption mechanisms have significant implications for the development of sustainable water evaporation technologies. Studies have shown that 2D nanomaterials can substantially increase evaporation rates by efficiently converting sunlight into localized heat. This principle is central to the design of solar-assisted desalination systems. The ability of 2D materials to maintain performance even under low-intensity sunlight makes them particularly valuable for real-world applications in varying environmental conditions. Materials such as layered tin monoselenide (SnSe) have demonstrated remarkable photothermal conversion efficiencies, which are attributed to their direct bandgap and strong excitonic behavior. These properties enable enhanced light absorption and effective heat generation upon solar exposure, supporting both water evaporation and broader thermal management applications (Yao et al., 2018).

Carbon-based 2D materials, including graphene, carbon nanotubes, and their derivatives, are also extensively studied for their roles in photothermal systems. Their high thermal conductivity and adjustable electronic properties contribute to enhanced evaporation performance (Jayaprakash et al., 2024). When these materials are fabricated into hybrid structures such as foams, composites, or layered films, they exhibit enhanced light trapping and heat transfer at the water–material interface, further accelerating evaporation (Li et al., 2023). Emerging materials such as MXenes (metal carbides and nitrides) and engineered black photothermal sheets have also shown great promise. MXenes, with their unique layered structure and excellent electrical conductivity, demonstrate superior optical absorption and energy conversion compared to conventional materials. Their plasmonic properties enable the generation of localized heating zones, thereby improving evaporation efficiency even in suboptimal lighting conditions (Liu et al., 2017). Black photothermal sheets, designed by integrating 2D materials with specially engineered surface textures, exhibit strong broadband solar absorption and minimized thermal losses. These sheets represent a shift from traditional bulk absorbers to highly efficient 2D-based platforms, capable of delivering improved performance in solar-driven water evaporation systems (Namboorimadathil Backer et al., 2020). The integration of 2D materials into interfacial evaporation systems, sometimes referred to as "hybrid solar skins," has demonstrated transformative potential for addressing global water shortages. These systems capitalize on the unique structural and electronic characteristics of 2D materials, including localized surface plasmon resonance and excitonic behavior, to enhance photon absorption and heat localization at the evaporation interface.

Research has shown that layered designs leveraging the light-harvesting capabilities of 2D materials can dramatically increase the rate of water vapor production. Innovations in material architecture, such as porous layers or nanostructured coatings, allow for greater interaction between light and water molecules, significantly improving the performance of solar evaporation systems (Ding et al., 2021). Further exploration into the light absorption behavior of 2D materials is opening new pathways for improving water purification technologies. By tailoring their optical properties to match the solar spectrum and incorporating them into composite systems, researchers are developing multifunctional evaporators with both photothermal and photovoltaic benefits. Additionally, the scalability and versatility of 2D materials enable their integration into a wide range of environmental and industrial applications. Advances in fabrication techniques and flexible substrates support the deployment of 2D-based evaporation technologies in diverse geographic and socioeconomic settings, from urban desalination plants to remote, off-grid communities. In summary, the continued study and application of 2D materials in solar light absorption represent a critical advancement in the quest for sustainable water solutions. Their ability to efficiently harvest and convert solar energy, along with their tunability and adaptability, positions them as key enablers in developing next-generation interfacial evaporation systems that can help mitigate the impacts of global water scarcity.

8.2.3 Nonradiative relaxation

Nonradiative relaxation is the process by which photo-excited electrons in a material return to their ground state without emitting light. Instead of releasing energy as photons, the excess energy is transferred to the atomic lattice of the material and dissipated as heat in the form of phonons (quantized lattice vibrations) (Yu et al., 2025). When exposed to sunlight, these materials absorb photons, exciting electrons within their structure. The nonradiative relaxation of these excited electrons leads to the generation of localized heat. This process is highly efficient due to the materials' high surface area and strong light–matter interaction (Jeffries et al., 2022). When 2D nanomaterials absorb photons with energies close to or greater than the bandgap, the resulting electron–hole pairs (excitons) carry excess energy. A significant portion of these pairs recombines through radiative processes, releasing photons, a process known as radiative recombination. This reduces the amount of absorbed light energy converted into heat, thereby lowering photothermal conversion efficiency. On the other hand, 2D nanomaterials with narrower band gaps allow the excited carriers to relax more effectively through nonradiative mechanisms, such as phonon scattering. In this case, the excess energy is quickly released as heat via interactions with the lattice instead of being lost as light. Therefore, narrow-bandgap 2D materials tend to exhibit higher photothermal conversion efficiency, as a greater share of the absorbed solar energy is directly converted into thermal energy (Jeffries et al., 2022).

Surface engineering of 2D nanomaterials further enhances photothermal performance due to surface textures or nano-patterns. These materials can trap more light and increase the surface area available for evaporation. Such morphological modifications not only improve light absorption but also promote localized heating by enhancing light–matter interactions. This interaction between structural design and photothermal performance represents a vital research frontier, indicating that advancements in materials engineering can substantially elevate the efficiency of water evaporation technologies (Gärtner, 1961). The optimization of 2D nanomaterials for photothermal conversion also depends on material selection. Notable examples include MXenes and transition metal dichalcogenides (TMDs), which have drawn increasing interest due to their superior photothermal characteristics. MXenes, a family of 2D transition metal carbides and nitrides, exhibit excellent light absorption and heat generation due to their layered metallic structure and high electrical conductivity (Chang et al., 2024). For instance, MBene membranes, a subclass of MXenes, have demonstrated the ability to enhance water evaporation efficiency through strong solar absorption and effective heat localization at the liquid–air interface. Customized MBene membranes significantly increased water evaporation rates by optimizing light–heat conversion at the surface (Cui et al., 2023).

8.2.4 Plasmonic localized heating

In addition to conventional photothermal conversion, many 2D nanomaterials and their hybrid structures exhibit plasmonic properties that can significantly enhance solar energy utilization. Plasmonic localized heating refers to the process where incident light excites localized surface plasmons, coherent oscillations of conduction electrons in metallic or conductive nanostructures. When the frequency of incident light matches the natural oscillation frequency of surface electrons, a phenomenon known as localized surface plasmon resonance (LSPR) occurs. This resonance leads to intense confinement of electromagnetic energy near the surface of the nanostructure, resulting in strong localized heating through nonradiative processes such as electron–phonon interactions (Subhan and Mourad, 2025).

Materials like gold (Au), silver (Ag), and titanium nitride (TiN) are well-known plasmonic nanostructures, but certain 2D materials such as doped graphene, MoS_2, and MXenes can also support or be engineered to exhibit plasmonic behavior. When integrated

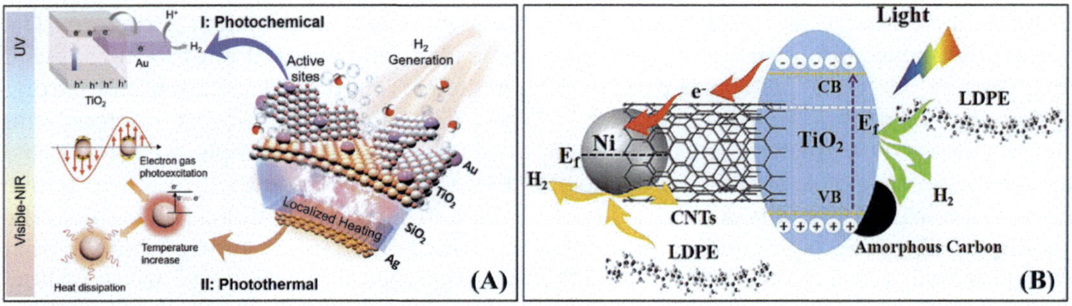

FIGURE 8.3 Schematic illustration of the general reaction mechanism involved in photothermal catalysis (Ng et al., 2021). (B) Mechanistic representation of light-driven CNT growth during the photothermal catalytic pyrolysis of LDPE (Yang et al., 2023). When light strikes the surface of a plasmonic nanostructure, surface plasmons are excited, and energy is converted into heat in the immediate vicinity of the structure, creating a localized high-temperature zone. This localized thermal energy enhances interfacial evaporation by minimizing energy losses to the bulk water. *From Yang, H., Yan, S., Yao, D., Luo, H., Zeng, K., Li, J., et al., 2023. Photo-thermal catalytic pyrolysis of waste plastics: investigation into light-induced metal-support interaction. Fuel 333, 126496. https://doi.org/10.1016/j.fuel.2022.126496.*

into solar water evaporation systems, these materials absorb solar radiation and generate heat directly at the interface between the water and the air. This contrasts with traditional solar heating systems that aim to raise the temperature of the entire water volume, often resulting in substantial heat losses. The highly localized heating effect of plasmonic materials concentrates energy at the evaporative interface, rapidly exciting water molecules and facilitating their transition from liquid to vapor, even under low-intensity sunlight (see Fig. 8.3). This effect is attributed to the layered structure of MXenes, which facilitates multiple internal reflections, enhancing the absorption and attenuation of electromagnetic waves. In addition, the abundant free charge enables tunable surface plasmon resonances (LSPRs), further improving the photothermal performance.

2D nanomaterials are particularly effective for plasmonic applications due to their high surface-to-volume ratio, tunable electronic structure, and exceptional light–matter interaction capabilities. Transition metal dichalcogenides (TMDs) such as MoS_2 and WS_2, when combined with plasmonic metals like gold, form hybrid nanostructures that leverage the benefits of both components. These systems exhibit synergistic effects in which the metallic component supports LSPR while the 2D material facilitates additional light absorption and energy conversion (Irshad et al., 2023). In these hybrid systems, plasmon-induced hot electron transfer (PHET) plays a crucial role. When plasmonic nanostructures are excited by sunlight, they generate energetic charge carriers known as hot electrons that can be injected into the conduction band of adjacent 2D materials. This not only extends the light absorption range but also increases the rate of thermal conversion. The strong coupling between the plasmonic nanostructure and the 2D material can enhance the excitation of electron–hole pairs within the semiconductor matrix, further boosting performance. For example, in a reported Au–MoS_2 hybrid system, the incorporation of plasmonic gold nanoparticles led to a nearly 40-fold increase in photocatalytic hydrogen production compared to MoS_2 alone. This performance gain underscores the importance of interface engineering and material pairing in optimizing the plasmonic photothermal effect (Huan et al., 2025).

Furthermore, the geometric configuration of plasmonic nanostructures including particle size, shape, and layer thickness can be tuned to optimize resonance with incident solar wavelengths. When the LSPR peak aligns with the optical absorption edge of the 2D material, the resulting thermal enhancement is maximized. In this configuration, the nanostructured interface acts like a nanoscale optical antenna, focusing solar energy into an extremely confined region where it can be efficiently converted into heat (Soo Joo et al., 2022). Emerging materials such as titanium nitride (TiN) and silicon nanocables have also shown great promise in plasmon-enhanced evaporation systems. TiN exhibits strong optical absorption in the near-infrared range and excellent thermal stability, making it ideal for solar thermal applications. Silicon nanocables, with their high aspect ratio and tunable morphology, support enhanced light scattering and absorption, resulting in localized heating and improved evaporation rates (Zhang et al., 2019). To enhance the efficiency of photothermal evaporation systems, it is essential to optimize plasmonic heating by tuning the SPR conditions. The resonance frequency of surface plasmons (ω) is determined by the plasma frequency of the metal (ω_p), the dielectric function of the metal (ε_m), and the dielectric constant of the surrounding medium (ε_d), as expressed by the equation (Yu et al., 2025):

$$\omega = \omega_p \left(1 + \frac{\varepsilon_m}{\varepsilon_d}\right)^{1/2} \tag{8.1}$$

This relationship demonstrates that the SPR can be tailored by selecting materials with suitable optical properties and by engineering the surrounding dielectric environment. By aligning these parameters appropriately, the material can achieve strong light-to-matter interaction at specific wavelengths, thereby maximizing localized heating and enhancing solar-driven water evaporation performance. In summary, the integration of plasmonic 2D nanomaterials into solar evaporation systems provides an advanced mechanism for maximizing energy utilization through highly localized heating. The strategic design of hybrid materials and interfaces enables precise control over light absorption, energy conversion, and thermal distribution. This not only enhances evaporation rates but also supports additional applications such as photocatalysis and hydrogen generation. With over two billion people facing water stress worldwide, these innovations offer a transformative pathway toward energy-efficient and sustainable water purification technologies.

8.2.5 Interfacial water evaporation

Traditional water evaporation methods often suffer from substantial energy losses because the heat is dispersed into the bulk water rather than being used directly at the evaporation interface (Zheng et al., 2025). To overcome this inefficiency, 2D nanomaterials are often designed to float or be positioned at the water's surface. This strategic placement ensures that the heat generated from solar absorption is concentrated right at the water–air interface, where evaporation occurs, thus significantly improving thermal efficiency. Moreover, the surface of these 2D materials can be engineered with porous structures or rough textures, which increase the effective surface area available for evaporation (Wang et al., 2025). In some cases, the materials are formed into foams, sponges, or multilayered films, further enhancing their ability to retain water and interact with sunlight. These designs enhance the interaction between the material, air, and water, thereby accelerating vapor generation. This approach, known as interfacial water evaporation, is a dominant photothermal mechanism that uses solar energy more efficiently by minimizing heat loss to the underlying water (Zhang et al., 2025a, 2025b, 2025c). By keeping the heat localized at the surface, this method maximizes the rate of evaporation, making it a powerful strategy for sustainable water purification and desalination systems.

The fundamental principle behind interfacial water evaporation involves confining solar-induced thermal energy to a thin layer of water at the evaporative surface, which reduces energy losses to the bulk and environment. When 2D nanomaterials such as MXenes, graphene, or TMDs are employed as the photothermal layer, their high-light absorption and low thermal conductivity help sustain a steep temperature gradient across the interface. For instance, $Ti_3C_2T_x$ MXenes have demonstrated excellent light-to-heat conversion properties owing to their metallic conductivity and tuneable surface terminations, making them suitable for rapid interfacial heating (Liu et al., 2019). A typical interfacial evaporation system consists of a floating photothermal structure comprising a porous matrix decorated with 2D nanomaterials. This design promotes capillary-driven water transport from the bulk reservoir to the hot surface and sustains continuous evaporation. Graphene oxide-coated foams, for example, have shown evaporation rates exceeding 1.6 kg m^{-2} h^{-1} under one sun illumination due to their optimized porosity and localized heating (Li et al., 2021). Furthermore, the surface roughness and nanoarchitecture of 2D materials enhance light scattering and trapping, thereby further improving heat localization at the interface (Zhang et al., 2025a, 2025b, 2025c). Importantly, interfacial evaporation also benefits from nonradiative relaxation processes within 2D materials, where absorbed photons generate excited charge carriers that rapidly lose energy via phonon scattering rather than radiative recombination. This results in a rapid temperature rise at the evaporative surface without significant energy dissipation (Singh et al., 2021). Such efficient photothermal conversion, when combined with a continuous water supply and optimized thermal insulation, results in enhanced evaporation performance and lower energy consumption. In summary, the synergy between interfacial heating and the unique physicochemical properties of 2D nanomaterials has opened new avenues for solar steam generation with high energy efficiency. Ongoing research continues to focus on material functionalization, structural engineering, and the integration of hybrid systems to push the boundaries of this promising green technology.

8.2.6 Water transport

An essential component of photothermal water evaporation systems is the continuous and efficient transport of water to the heated surface. Without a reliable water supply, even the most advanced photothermal materials such as 2D nanomaterials including MXenes, graphene derivatives, and TMDs cannot maintain sustained evaporation. Thus, water transport represents the final, but equally critical, mechanism in the photothermal evaporation process (see Fig. 8.4). In 2D nanomaterial-based systems, water transport is predominantly driven by capillary action within porous, hydrophilic structures. These architectures are typically designed with interconnected micro- and nanoscale pores that serve as channels for spontaneous water flow from the bulk reservoir to the photothermal interface. This capillary wicking is governed by surface tension and pressure differences,

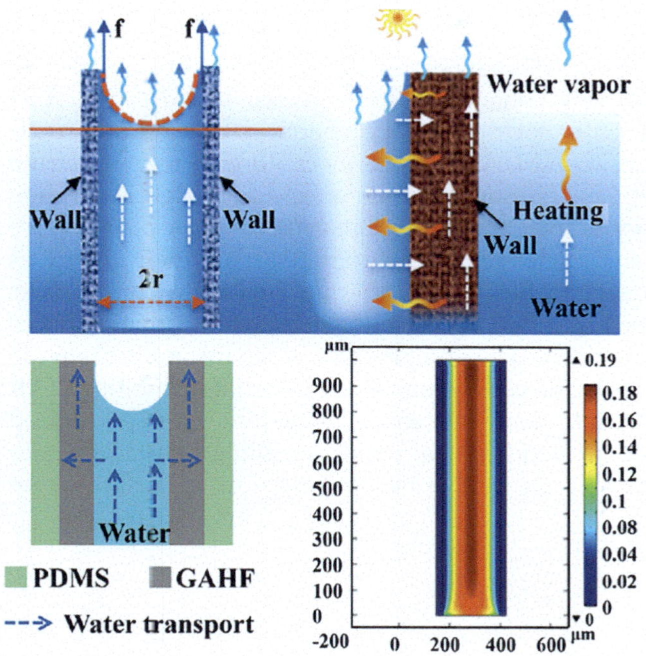

FIGURE 8.4 Water transport mechanism of graphene aerogel hollow fiber (GAHF) evaporators. *From Li, G., Fang, D., Hong, G., Eychmüller, A., Zhang, X., Song, W., 2022. Assembling graphene aerogel hollow fibres for solar steam generation. Composites Communications 35, 101302. https://doi.org/10.1016/j.coco.2022.101302.*

with smaller pore sizes generating stronger capillary forces that enhance vertical or lateral water migration to the evaporative zone (Li et al., 2022).

The nitrogen-doped graphene foam features an interconnected porous network that facilitates efficient water delivery through capillary action while confining solar-generated heat at the evaporation surface. Such vertically aligned structures promote rapid and continuous water replenishment and minimize energy loss to the bulk water that enhance evaporation efficiency (Namboorimadathil Backer et al., 2020). The intrinsic or engineered hydrophilicity of the substrate plays a crucial role. Many 2D materials such as $Ti_3C_2T_x$ MXenes and graphene oxide naturally exhibit hydrophilic surface terminations (e.g., –OH, –O, –F), or can be chemically modified to improve their water-attracting capacity. These hydrophilic interfaces enable rapid water uptake, ensuring that the surface remains wet even under intense solar irradiation (Li et al., 2021). Thermal gradients formed during photothermal heating can further contribute to water transport. The temperature difference between the heated surface and the cooler underlying bulk water can induce thermally driven flow (such as the Marangoni effect), drawing water toward the hot interface to sustain evaporation (Yan et al., 2021). In practical systems, 2D nanomaterials are often integrated into porous foams, sponges, hydrogels, or layered films that combine capillarity, hydrophilicity, and structural robustness. For instance, vertically aligned reduced graphene oxide foams have demonstrated evaporation rates exceeding 1.6 $kg\,m^{-2}\,h^{-1}$ under one sun, primarily due to their optimized water transport and heat localization capabilities (Li et al., 2023). Similarly, MXene-coated cellulose membranes and aerogels exhibit rapid wicking and high evaporation performance, proving the effectiveness of water transport integration (Singh et al., 2021).

8.2.7 Summary

Photothermal water evaporation using 2D nanomaterials is driven by a series of interconnected mechanisms that collectively enhance energy conversion and evaporation efficiency. These mechanisms begin with solar light absorption, where ultrathin 2D materials like graphene, TMDs, MXenes, and black phosphorus capture a broad spectrum of sunlight due to their tunable band gaps, strong excitonic effects, and high absorption coefficients. This absorbed solar energy is then converted into heat through nonradiative relaxation, a highly efficient process in which excited electrons dissipate their energy as phonons rather than emitting light, resulting in localized thermal buildup at the surface. Surface engineering further amplifies this effect by introducing textures that enhance light–matter interactions and heat confinement. The photothermal mechanism illustrated in Fig. 8.5 is further enhanced by plasmonic localized heating, where materials exhibiting LSPR that concentrate electromagnetic energy into nanoscale zones, generating intense localized heat (see Fig. 8.5). This heat is confined to the interfacial region, enabling interfacial water evaporation that minimizes thermal loss to the bulk water and maximizes surface vapor

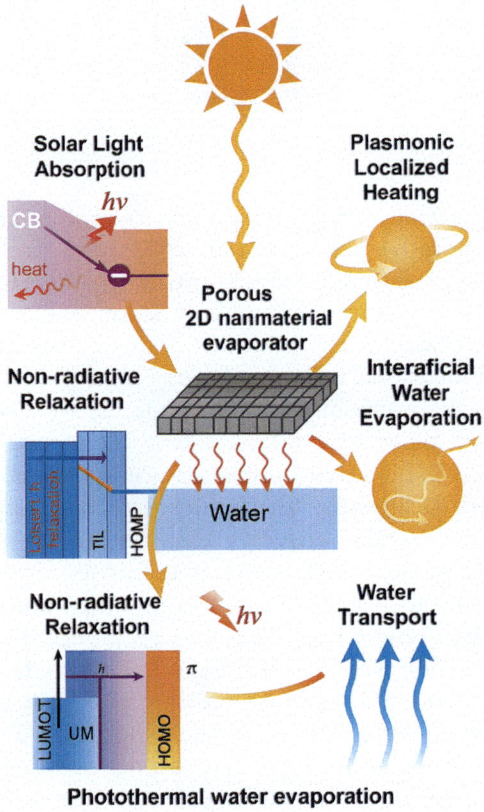

FIGURE 8.5 Mechanisms of photothermal water evaporation using two-dimensional (2D) nanomaterials. The diagram illustrates five interrelated processes: solar light absorption, nonradiative relaxation, plasmonic localized heating, interfacial water evaporation, and water transport.

generation. To sustain this process, efficient water transport is vital; 2D material-based systems are engineered with porous, hydrophilic structures that promote capillary wicking, thermal-driven flow, and continuous replenishment of water at the hot interface. Together, these five synergistic mechanisms position 2D nanomaterials as transformative platforms for high-efficiency, solar-driven water purification systems.

8.3 Photothermal performance of two-dimensional nanomaterials in solar water purification

8.3.1 Introduction to photothermal performance

Photothermal performance in solar water purification refers to a material's ability to efficiently absorb sunlight, convert it into heat, and drive water evaporation for desalination or purification. Key factors include high solar absorption, effective heat localization, and minimal energy loss, which are often measured by the evaporation rate ($kg{\cdot}m^{-2}{\cdot}h^{-1}$) and solar-to-thermal efficiency (Hou et al., 2022). Two-dimensional (2D) nanomaterials, such as graphene, MXene, and transition metal dichalcogenides (TMDs), are gaining significant attention as promising materials for solar-powered water purification (Bai et al., 2024; Patibandla et al., 2024; Peng et al., 2023). Their exceptional properties, including high surface area, tunable electronic characteristics, and efficient solar energy absorption, make them ideal for integration into purification membranes. Their atomic-scale thickness maximizes heat generation at the water interface while minimizing energy loss, and their adjustable bandgap allows precise solar spectrum utilization, enabling efficient and scalable photothermal water purification systems (Raj and Kulshrestha, 2022; Xie et al., 2020). The photothermal mechanisms in 2D nanomaterials, such as MXenes and TMDs, leverage their unique structural properties to convert light into heat efficiently. MXenes exhibit broadband light absorption due to their metallic conductivity, whereas TMDs can selectively absorb specific wavelengths (Hu et al., 2024). The absorbed photons excite charge carriers, which then thermalize through phonon interactions, effectively converting light energy into heat (Huang et al., 2020; Qu et al., 2022). This process is enhanced by interfacial heat confinement, which minimizes bulk losses and localizes thermal energy, making these materials suitable for applications in therapy and catalysis. The ultrathin structure and high surface area of these materials further optimize heat generation and retention (Qu et al., 2022). Moreover, the absorption capabilities of 2D nanomaterials span a wide range of the electromagnetic spectrum, owing to their diverse electronic properties. For instance, graphene has a zero bandgap, enabling broadband light interactions, while molybdenum disulfide (MoS_2) exhibits a larger bandgap, shifting its absorption range from the mid-infrared to the visible spectrum (Wang et al., 2015). Direct bandgap semiconductors such as monolayer MoS_2 and phosphorene are particularly distinctive, as they facilitate highly efficient energy conversion without the need for additional phonon momentum.

8.3.2 Material-specific performance in solar water purification

The photothermal performance of 2D nanomaterials in solar-driven water purification is highly dependent on their intrinsic optical, thermal, and structural properties. This section provides a detailed analysis of the key 2D materials used in photothermal applications, focusing on their efficiency, stability, and the practical trade-offs associated with them.

8.3.2.1 MXene

MXenes, a novel class of 2D materials, were first reported in 2011 by Yury Gogotsi's group (Gogotsi and Huang, 2021). These materials, primarily composed of transition metal carbides, nitrides, and carbonitrides, have since expanded into a family with over 50 compositions. Since its discovery, MXene has garnered significant research interest due to its exceptional chemical, electrical, and mechanical properties, resulting in diverse applications across various fields. In this context, 2D MXene nanomaterials, particularly $Ti_3C_2T_x$ (see Fig. 8.6), have garnered considerable attention for their exceptional photothermal conversion efficiency, making them ideal candidates for enhancing solar water purification processes. Their unique properties, including broad-spectrum solar absorption and efficient light-to-heat conversion, enable high evaporation rates in solar desalination and wastewater treatment systems.

MXenes exhibit remarkable photothermal performance due to their ability to absorb a wide range of solar radiation and convert it into thermal energy. Recent studies have demonstrated that integrating MXenes with other functional materials can further enhance their efficiency.

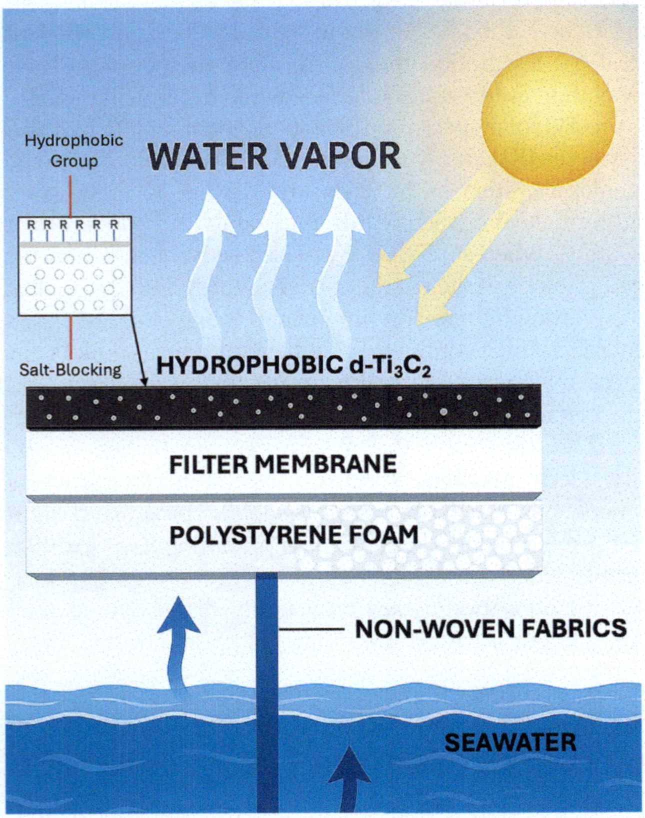

FIGURE 8.6 Illustration of the solar-driven evaporation device using a salt-resistant Ti_3C_2 membrane.

For instance, Wang's group reported that a composite membrane of MXene and copper indium selenide (CIS) nanoparticles achieved a water evaporation efficiency of 90.04%, attributed to improved light absorption and reduced reflection (Wang et al., 2022). Such composites optimize solar energy utilization, making them highly effective for practical applications (Almarzooqi et al., 2024). Structural modifications and hybrid material designs have also been explored to maximize the photothermal performance of MXene-based systems. Combining MXenes with multi-walled carbon nanotubes (MWCNTs) or perovskite oxides like $La_{0.5}Sr_{0.5}CoO_3$ (LSC) enhances water transport, vapor escape, and thermal insulation, all of which are critical for efficient solar evaporation (Fan et al., 2021; Yang et al., 2022).

Additionally, the incorporation of carbon quantum dots (CQDs) with MXenes has been demonstrated to induce localized surface plasmon resonance (LSPR) effects, thereby further enhancing photothermal conversion and evaporation rates (Jing et al., 2024). These advancements highlight the versatility of MXenes in forming high-performance composite structures. Zhao's group has explored the evaporation performance of hydrophobic Ti_3C_2 MXene membranes. The Ti_3C_2 MXene was modified with trimethoxy silane, and a commercial filter membrane was used for the water supply, as shown in Fig. 8.6 owing to its strong light-conversion capability, optimized surface engineering, and extensive water transport channels, the membrane demonstrated outstanding solar evaporation performance, achieving a high solar steam-conversion efficiency of 71% and an evaporation rate of 1.31 kg m^{-2} h^{-1} (Zhao et al., 2018).

Beyond evaporation, MXene-based photothermal materials exhibit multifunctionality, including photocatalytic degradation of pollutants. By integrating MXenes with photocatalytic components, researchers have developed systems capable of simultaneous water purification and contaminant breakdown (Fan et al., 2021; Liu et al., 2023). Advanced designs, such as the Cu_3BiS_3@V_2C/BF membrane, demonstrate the potential for integrated solar steam generation, electricity production, and water purification, showcasing the broad applicability of MXene-based materials (Xu et al., 2024). Despite these promising developments, challenges remain in the practical implementation of MXene-based photothermal systems. Researchers have addressed issues such as material stability, environmental impact, and production costs to ensure sustainable deployment (Awan et al., 2023; Balu et al., 2024). Recent efforts in developing durable and cost-effective composites, such as MXene/polyvinylidene fluoride (PVDF) membranes, offer potential solutions for long-term use in harsh environments (Ma et al., 2024). Additionally, the environmental and social implications of MXene production and disposal require careful consideration to prevent unintended consequences (Xie et al., 2020).

MXene 2D nanomaterials represent a significant advancement in solar water purification technology, offering high photothermal efficiency, multifunctionality, and adaptability in composite designs. Continued research is crucial to overcoming existing challenges and optimizing these materials for large-scale, sustainable applications. By addressing stability, cost, and environmental concerns, MXene-based systems could play a pivotal role in meeting global clean water demands through solar-driven purification.

8.3.2.2 *Graphene*

Graphene-based 2D nanomaterials have proven to be highly promising materials for solar-driven water purification due to their exceptional photothermal properties. These advanced materials harness solar energy to efficiently convert water into vapor, enabling effective desalination and purification processes. The unique structural and chemical characteristics of graphene and its derivatives contribute to their outstanding light absorption, thermal conversion, and water transport capabilities, making them particularly suitable for solar-based water treatment systems.

One of the most significant advantages of graphene-based nanomaterials is their high photothermal conversion efficiency. Their broadband light absorption and low thermal emittance allow for superior energy utilization. For instance, a composite material consisting of nickel foam, reduced graphene oxide, and carbon nanotubes achieved an impressive photothermal conversion efficiency of 96.4% under 1 kW/m^2 solar irradiation (Chen et al., 2022). Additionally, the structural versatility of graphene-based materials enables surface and interlayer modifications that enhance water transport and evaporation rates, which are critical for optimizing solar desalination performance (Aizudin et al., 2022). In practical applications, graphene-based nanomaterials have shown remarkable effectiveness in various water purification processes. For desalination, graphene oxide-based solar steam generators (GO-SSGs) exhibit high solar energy conversion efficiency and excellent salt-rejection capabilities, even in high-salinity environments (Chen et al., 2021). In wastewater treatment, graphene-based membranes have been successfully employed to remove organic pollutants and heavy metals, achieving high rejection rates for various contaminants (Yang Wang et al., 2019). Furthermore, hybrid structures incorporating materials such as V_2O_5 and polypyrrole into graphene oxide layers demonstrate multifunctional capabilities, including simultaneous water purification, disinfection, and even power generation, thereby expanding their potential applications (Marimuthu et al., 2024).

Despite their advantages, several challenges must be addressed to fully realize the potential of graphene-based nanomaterials in large-scale water purification. One major concern is the long-term stability and environmental impact of these materials. The risk of leaching and potential environmental contamination necessitates a thorough investigation to prevent unintended ecological consequences (Xie et al., 2020). Additionally, optimizing photothermal evaporation systems, particularly in terms of material localization and system design, is crucial for maximizing efficiency in real-world applications (Ghodsi and Fashandi, 2024). Another critical challenge lies in scalability and practical implementation. While laboratory-scale studies have yielded promising results, transitioning these technologies to industrial-scale production requires the development of scalable manufacturing processes and seamless integration into existing water treatment infrastructures (Yang et al., 2018).

Graphene-based 2D nanomaterials hold immense potential for revolutionizing solar water purification due to their high photothermal efficiency, structural adaptability, and cost-effectiveness. Their applications in desalination, wastewater treatment, and multifunctional systems demonstrate their versatility. However, challenges related to material stability, system optimization, and scalability must be addressed to facilitate widespread adoption. Additionally, ongoing research into alternative 2D materials may further enhance the field, paving the way for more efficient and sustainable water purification solutions in the future.

8.3.2.3 Transition metal dichalcogenides

Transition metal dichalcogenides (TMDs) have emerged as highly promising materials for solar-driven water purification due to their exceptional photothermal conversion capabilities. Recent studies have demonstrated that TMDs, such as molybdenum disulfide (MoS_2) and tungsten disulfide (WS_2), possess unique structural and optical properties that facilitate effective light absorption and heat localization, making them ideal candidates for solar interfacial evaporation applications (Xie et al., 2020; Lin et al., 2019). The large surface area and tunable electronic properties of TMDs further enhance their photothermal performance, allowing for optimized energy utilization in water purification processes (Andoshe et al., 2015; Dervin et al., 2016).

Recent advancements highlight the efficiency, design innovations, and multifunctional applications of MoS_2-based systems in addressing water purification challenges. MoS_2-based composites have achieved remarkable photothermal conversion efficiencies, making them highly effective for solar desalination and wastewater treatment. For instance, Fe_3O_4-decorated MoS_2 nanosheets exhibit an evaporation rate of 1.42 kg m^{-2} h^{-1} with 89.18% efficiency under one sun irradiation, while simultaneously removing over 98% of salt ions and nearly 100% of heavy metal ions (Bai et al., 2023). Similarly, a MoS_2/graphene oxide (GO) heterojunction film achieves an evaporation rate of 1.69 kg m^{-2} h^{-1}, with biomimetic design modifications further boosting performance to 3.20 kg m^{-2} h^{-1} through optimized heterojunction interfaces and nanoporous surfaces (Zheng et al., 2024). Another study reports that a poly(N-phenylglycine)/MoS_2 nanohybrid system attains a solar evaporation rate of 1.70 kg m^{-2} h^{-1}, underscoring the synergistic benefits of efficient photothermal conversion and water transport (Lin et al., 2022).

Design innovations have further expanded the functionality and efficiency of MoS_2-based evaporators. Functionalization of MoS_2 with surface groups such as -COOH, -OH, and -NH_2 enhances compatibility with organic matrices, yielding an evaporation rate of 1.35 kg m^{-2} h^{-1} and 83% efficiency (Yu et al., 2023). Additionally, vertically aligned MoS_2/MXene aerogels demonstrate superior light absorption and mechanical stability, achieving an evaporation rate of approximately 2.75 kg m^{-2} h^{-1} (Yang et al., 2024). Superhydrophobic MoS_2 nanoflowers coated with PFDTES exhibit an evaporation rate of 1.52 kg m^{-2} h^{-1} while preventing salt crystallization, ensuring sustained desalination performance (Yang et al., 2024).

Beyond evaporation efficiency, MoS_2-based systems have been engineered for multifunctional applications, including pollutant degradation and freshwater collection. These systems demonstrate long-term stability and durability across multiple cycles while effectively removing organic contaminants (Chen et al., 2019). Furthermore, reduced graphene oxide (RGO)-MoS_2 nanocomposites exhibit high photocatalytic activity, enabling the degradation of organic dyes under solar irradiation and highlighting their potential for integrated environmental remediation (Chen et al., 2019). Despite these advancements, challenges remain in optimizing material integration, ensuring long-term stability, and evaluating environmental safety. The ecotoxicity of MoS_2 nanomaterials requires further investigation to mitigate potential risks in water treatment applications (Amaral and Daniel-da-Silva, 2022). Addressing these challenges will be crucial for scaling up MoS_2-based solar purification systems and maximizing their real-world impact in combating global water scarcity.

8.3.2.4 *Boron nitride nanosheets*

Boron nitride (BN) nanosheets, as 2D nanomaterials, have gained recognition as one of the most promising materials for enhancing solar water purification due to their exceptional photothermal properties. Their integration into solar vapor generation (SVG) systems has demonstrated significant improvements in evaporation rates and energy conversion efficiency, making them a viable candidate for sustainable desalination and wastewater treatment technologies (Zhang et al., 2024). The unique characteristics of BN nanosheets, including their large surface area, high thermal conductivity, and chemical stability, contribute to their effectiveness in converting solar energy into heat, thereby optimizing water evaporation processes (Lei et al., 2013).

One of the key advantages of BN nanosheets is their ability to enhance solar thermal conversion efficiency. When incorporated into hydrogel-based solar vapor generators, BN nanosheets have achieved an evaporation rate of 4.03 kg m^{-2} h^{-1} under 1 sun irradiation (Zhang et al., 2024). This performance is attributed to their capacity to reduce water evaporation enthalpy (845.11 kJ kg^{-1}) and facilitate the formation of intermediate water, which lowers the energy required for vaporization. Additionally, their high thermal conductivity ensures efficient heat distribution, further improving the overall efficiency of solar-driven water purification systems.

Durability and salt resistance are critical factors in desalination applications, and BN nanosheets exhibit strong resistance to oxidation and fouling, ensuring long-term stability in saline environments. Their reusability and ease of cleaning further enhance their practicality in continuous water purification processes. Beyond thermal evaporation, BN nanosheets also demonstrate potential in photocatalytic and filtration applications. When hybridized with materials such as TiO_2, they can extend photocatalytic activity into the visible light spectrum, improving the degradation of organic pollutants (Liu et al., 2017). Moreover, BN-based membranes have shown near-complete retention of water-soluble dyes, highlighting their effectiveness in nanofiltration (Doménech et al., 2022; García Doménech et al., 2022). Despite these advantages, challenges remain in the widespread adoption of BN nanosheets. Their wide bandgap (5.5 eV) limits their standalone photocatalytic efficiency, necessitating hybridization with other semiconductors to enhance performance (Liu et al., 2017).

BN nanosheets hold significant promise for advancing solar water purification technologies due to their superior photothermal conversion, durability, and multifunctional capabilities. Future research should focus on optimizing their properties, reducing production costs, and assessing environmental impacts to facilitate their practical application in addressing global water scarcity challenges (Xie et al., 2020).

8.3.2.5 Black phosphorus

The photothermal performance of phosphorene, particularly black phosphorus (BP), has garnered significant attention in the field of solar water purification due to its exceptional ability to convert solar energy into thermal energy. This unique property, coupled with its anisotropic structure, high thermal conductivity, and superior photothermal conversion efficiency, positions BP as a promising material for enhancing solar-driven water purification processes, including desalination and disinfection. Recent advancements in BP-based systems have demonstrated notable improvements in efficiency and functionality, though challenges related to stability and photoreactivity remain critical areas of investigation. BP's high photothermal conversion efficiency is attributed to its distinctive puckered crystal structure and excellent thermal conductivity, which facilitate effective heat storage and transfer, a key requirement for solar–thermal applications (García Doménech et al., 2022). When integrated into self-floating materials, BP further enhances light-heat conversion by concentrating thermal energy at the water–air interface, thereby optimizing solar desalination efficiency (Cai et al., 2020). However, BP's susceptibility to oxidation in aqueous environments poses a significant limitation to its long-term stability. To mitigate this issue, passivation techniques, such as silica coating, have been developed to protect BP from degradation, thereby extending its operational lifespan in solar desalination devices (Li et al., 2020). Additionally, combining BP with other materials, such as bismuth vanadate, has proven effective in improving durability by preventing self-oxidation, as demonstrated in solar water splitting systems (Zhang et al., 2019).

Beyond desalination, BP has shown considerable potential in membrane-based water treatment and point-of-use disinfection systems. Its incorporation into filtration membranes enhances pollutant rejection while imparting antibacterial properties, thereby improving overall water quality (Hussain et al., 2024). Furthermore, BP nanosheets have been successfully employed in advanced disinfection systems, effectively eliminating microbial contaminants and underscoring BP's versatility in water purification applications. Despite these advantages, BP's narrow bandgap presents a challenge by limiting the efficient separation of photogenerated electron-hole pairs, which is essential for photoreactivity. To address this limitation, researchers are exploring hybrid nanostructures and heterojunctions, which can enhance charge separation and improve photocatalytic performance (Xin et al., 2021). For instance, BP-based van der Waals heterojunctions have demonstrated enhanced visible light absorption and reduced electron-hole recombination, thereby boosting photocatalytic efficiency (Wang et al., 2019). While BP holds immense promise for solar water purification, its widespread adoption hinges on overcoming existing challenges related to stability and photoreactivity. Ongoing research into hybrid material systems, passivation techniques, and novel structural designs is critical to unlocking BP's full potential. Furthermore, expanding its integration with complementary materials and exploring new applications in water treatment will be essential for advancing its role in environmental remediation. As these developments progress, BP is poised to become a cornerstone material in sustainable water purification technologies.

8.3.2.6 *Graphitic carbon nitride nanosheets*

Graphitic carbon nitride (g-C_3N_4) nanosheets have become a promising material for solar-driven water purification due to their exceptional photothermal and photocatalytic properties. With a medium band gap of approximately 2.7 eV, g-C_3N_4 exhibits strong absorption in the visible light spectrum, making it an efficient candidate for harnessing solar energy (Ding et al., 2017). The 2D nanosheet structure further enhances its performance by providing a large surface area, which improves pollutant interaction and facilitates efficient charge separation and transport (Kumar et al., 2018; Wang et al., 2024). These characteristics are critical in mitigating the common issue of electron-hole pair recombination in photocatalytic processes, thereby increasing overall efficiency (Fan et al., 2017).

To further enhance the photocatalytic performance of g-C_3N_4, researchers have explored its integration with other advanced materials. For instance, the hybridization of g-C_3N_4 with carbon quantum dots (CQDs) has been shown to increase photocatalytic degradation efficiency by fourfold compared to pure g-C_3N_4 (Wang et al., 2023). Similarly, composites such as g-C_3N_4:TiN-NFG have demonstrated a 16-fold improvement in water reduction activity (Shanker et al., 2018). Additionally, the incorporation of thermally stable materials like hexagonal boron nitride (h-BN) and gold (Au) nanoparticles has improved the recyclability and durability of g-C_3N_4-based catalysts, making them more viable for long-term applications in water purification (Wang et al., 2024).

The practical applications of g-C_3N_4 nanosheets in water purification are extensive. They have proven highly effective in degrading organic pollutants, including synthetic dyes and antibiotics, achieving removal rates exceeding 95% under visible light irradiation (Li et al., 2023). Beyond pollutant degradation, g-C_3N_4's strong photothermal conversion efficiency makes it suitable for solar-driven water evaporation, a promising approach for desalination and wastewater treatment. However, despite these advancements, challenges remain in optimizing g-C_3N_4's performance for large-scale applications. Key limitations include the high recombination rate of charge carriers and restricted light absorption beyond 460 nm, which hinders its overall efficiency (P. Wang et al., 2019; Zhang et al., 2023). Addressing these issues through advanced material engineering and composite design will be crucial for maximizing the potential of g-C_3N_4 in sustainable water purification. Moreover, ensuring that these technological solutions do not inadvertently introduce new environmental or social concerns is essential for their responsible deployment. G-C_3N_4 nanosheets represent a highly promising material for solar water purification, offering a combination of visible light absorption, efficient charge separation, and enhanced performance through composite formation. While significant progress has been made, further research is needed to overcome existing limitations and ensure the scalability and sustainability of these technologies. Advances in material modifications and hybrid systems will play a pivotal role in unlocking the full potential of g-C_3N_4 for global water purification challenges.

8.3.2.7 Covalent organic frameworks

Covalent organic frameworks (COFs), particularly in their 2D form, are increasingly recognized as suitable materials for solar-driven water purification due to their exceptional photothermal properties. These materials exhibit a combination of high surface area, tunable porosity, and broad-spectrum light absorption, making them ideal for efficient solar energy conversion and water evaporation (Liu et al., 2023). The structural and chemical adaptability of 2D COFs allows for precise optimization of their photothermal performance, enabling high efficiency in converting sunlight into heat for water purification. A key advantage of 2D COFs is their broadband light absorption, which enhances their ability to harness solar energy. For instance, thienyl-functionalized benzodithiophene-based COFs demonstrate optical absorption extending up to 600 nm, ensuring effective utilization of a significant portion of the solar spectrum (Liu et al., 2023). This property is critical for maximizing energy capture and conversion in photothermal applications. Additionally, certain COFs, such as those incorporating triazine-phenanthroline-amine-derived (TPAD) units, exhibit remarkable photothermal conversion efficiencies. Yan's group has reported that TPAD-based COFs achieve water evaporation rates of 1.42 kg m^{-2} h^{-1} with an energy conversion efficiency of 94% under 1 sun irradiation (Yan et al., 2022). This high performance is attributed to their superhydrophilic nature and efficient light-to-heat conversion mechanisms.

The structural versatility of 2D COFs further enhances their functionality in water purification systems. The formation of 2D/2D heterojunctions, such as those between COFs and SnNb2O6, has been shown to improve charge separation and light utilization efficiency (Ren et al., 2021). Such heterostructures can be tailored to optimize photothermal performance, thereby enhancing water evaporation rates. Moreover, the inherent porosity and hydrophilicity of COFs facilitate high water permeability and selective molecular transport, which are advantageous for purification processes (Kuehl et al., 2018; Yan et al., 2022). The nanoporous structure not only allows for rapid water diffusion but also ensures efficient heat localization, thereby further enhancing evaporation efficiency. From an environmental perspective, COFs offer significant advantages as they are metal-free, reducing the ecological risks associated with metal-based photocatalysts (Yan et al., 2023). Their chemical stability and recyclability make them sustainable candidates for long-term water purification applications. Furthermore, the ability to functionalize COFs with various chemical groups enables customization for specific purification needs, enhancing their adaptability to different water sources and environmental conditions (Kuehl et al., 2018).

Despite these advantages, challenges remain in scaling up COF-based photothermal systems for practical applications. Issues such as long-term stability under operational conditions, cost-effective synthesis, and large-scale manufacturing need to be addressed. Additionally, the environmental impact of COF production and disposal must be carefully evaluated to ensure that these materials do not introduce new ecological concerns. Future research should focus on developing more robust COF structures with enhanced photothermal efficiency while ensuring economic viability and minimal environmental footprint. COFs represent a highly promising class of materials for solar water purification, owing to their superior photothermal properties, structural adaptability, and environmental benefits. Continued advancements in material design and system integration will be crucial for realizing their full potential in addressing the global water scarcity challenge.

8.3.2.8 *Layered double hydroxides*

Layered double hydroxides (LDH) have demonstrated potential characteristics as 2D nanomaterials for solar water purification, due to their unique structural and electronic properties. LDH 2D nanomaterials exhibit exceptional photothermal performance, making them a candidate for efficient solar energy conversion in water treatment processes (Tan et al., 2020). Their layered structure, tunable composition, and ability to form heterostructures enhance their photocatalytic capabilities, positioning them as key materials in addressing water purification challenges (Luo et al., 2024). The structural versatility of LDH, characterized by $M(OH)_6$ octahedral layers with intercalated anions, allows for precise tuning of their photocatalytic properties (Tan et al., 2020). This layered configuration provides a large surface area, facilitating the separation of photogenerated electron-hole pairs and improving photocatalytic efficiency. Furthermore, the integration of visible-light-active LDH shells with UV-responsive cores, such as TiO_2, has been shown to enhance solar energy utilization significantly, broadening their applicability in photocatalysis (Dou et al., 2015).

In photocatalytic applications, LDH has demonstrated remarkable efficiency in degrading organic pollutants and facilitating water splitting for hydrogen and oxygen production (Wang et al., 2021). LDH adsorption-photocatalytic synergism is particularly effective in removing contaminants from aqueous environments, underscoring their potential in advanced water purification systems (Jing Bai et al., 2024). Recent advancements have focused on optimizing photocatalytic performance through strategies such as morphology control and defect engineering, further improving their selectivity and activity (Tan et al., 2020). For solar water purification, LDH stands out due to its broad-spectrum sunlight absorption and efficient conversion into thermal energy (Gama et al., 2022). Their integration with materials like graphene enhances visible light absorption and photocatalytic activity, making them even more effective in water treatment applications (Gunjakar et al., 2013). Additionally, combining LDH with filter materials, such as activated carbon, can improve system longevity by reducing the residual metal content in purified water. Developing novel synthesis methods and exploring advanced LDH composites could further enhance their photothermal properties (Gao et al., 2023). Integrating LDH with plasmonic nanoparticles or other advanced materials may also open new pathways for improving solar water purification efficiency. Continued research in these areas will be crucial for realizing the full potential of LDH in sustainable water treatment solutions.

8.3.3 Case studies of two-dimensional nanomaterials efficiency in solar water purification

Recent advancements in solar-driven water purification have been propelled by the development of high-performance 2D nanomaterials, with several landmark studies demonstrating exceptional photothermal conversion efficiencies exceeding 90%, as summarized in. Among these 2D nanomaterials, MXene-based composites and graphene-enhanced systems have established benchmarks in terms of efficiency and multifunctionality. Wang et al. reported a composite membrane integrating MXene ($Ti_3C_2T_x$) with copper indium selenide (CIS) nanoparticles, achieving a remarkable solar evaporation efficiency of 90.04% under 1 sun irradiation with water evaporation rate of 1.434 kg m^{-2} h^{-1} (Wang et al., 2022). This performance was attributed to the synergistic effects of broad-spectrum light absorption and minimized reflection losses, showcasing MXene's potential in high-efficiency desalination. Similarly, a notable breakthrough was achieved through the development of a hydrophobic MXene membrane featuring trimethoxy(1 *H*,1 *H*,2 *H*,2*H*-perfluorodecyl) silane-modified Ti_3C_2 nanosheets paired with a commercial filter membrane. This innovative design combines efficient sunlight harvesting with 71% conversion efficiency, along with remarkable salt-rejection capabilities, maintaining stable operation in high-salinity conditions for over 200 hours while achieving an evaporation rate of 1.31 kg m^{-2} h^{-1}. The system's effectiveness extends to purifying wastewater containing organic dyes and heavy metals, demonstrating the versatility of MXene-based photothermal materials for practical applications (Zhao et al., 2018).

Further, a graphene-based system comprising nickel foam, reduced graphene oxide (rGO), and carbon nanotubes (CNTs) demonstrated an unprecedented photothermal efficiency of 96.4% with a high water evaporation rate of 1.37 kg m^{-2} h^{-1}, leveraging its superior broadband absorption and thermal localization properties (Chen et al., 2022). These studies underscore the pivotal role of material design in optimizing solar energy utilization for water purification. Another notable example involves the use of covalent organic frameworks (COFs) for solar evaporation. Yan's group developed a triazine-phenanthroline-amine-derived (TPAD) COF membrane, which achieved a water evaporation rate of 1.42 kg m^{-2} h^{-1} with a photothermal conversion efficiency of 94% (Yan et al., 2022). The superhydrophilic nature and tailored porosity of the COF facilitated rapid water transport and efficient heat localization, highlighting the potential of organic frameworks in sustainable water treatment.

8.3.4 Comparative performance of two-dimensional photothermal materials

A comparative analysis of key 2D nanomaterials reveals distinct advantages and trade-offs in solar water purification applications (see Table 8.1). The following table provides a detailed summary of the performance metrics associated with these prominent materials, drawing insights from recent studies and showcasing their effectiveness in addressing water purification challenges.

This comparative overview highlights the diverse capabilities of 2D nanomaterials, with MXene and graphene leading in efficiency, while COFs and BN nanosheets offer unique structural and environmental benefits. However, challenges such as material stability, scalability, and ecological impact necessitate further research to enable large-scale deployment. Future innovations in composite design and hybrid systems may bridge these gaps, advancing solar water purification toward practical, sustainable solutions.

8.4 Summary

The photothermal performance of 2D nanomaterials in solar-driven water purification is governed by several key determinants, including their intrinsic optical absorption properties, thermal conductivity, structural stability, and interfacial interactions with water. These materials exhibit exceptional light-to-heat conversion efficiency due to their high surface area, tunable electronic properties, and capacity for localized surface plasmon resonance (LSPR) effects. Broad-spectrum solar absorption, efficient charge separation, and low thermal emittance further enhance their photothermal capabilities, making them indispensable for applications such as desalination, wastewater treatment, and pollutant degradation.

Among the most promising 2D nanomaterials, MXenes stand out for their outstanding photothermal conversion efficiency and versatility in composite designs. Their ability to integrate with materials such as copper indium selenide (CIS) and carbon quantum dots (CQDs) enhances light absorption and evaporation rates, achieving efficiencies exceeding 90%. Graphene-based nanomaterials, including reduced graphene oxide (RGO) and graphene oxide (GO), exhibit superior broadband absorption and structural adaptability, enabling high evaporation rates and multifunctional applications in desalination and the removal of organic pollutants. Transition metal dichalcogenides (TMDs), particularly MoS_2, exhibit tunable band gaps and strong light-matter interactions, facilitating efficient solar evaporation and photocatalytic degradation. Boron nitride (BN) nanosheets contribute high thermal conductivity and chemical stability, reducing water evaporation enthalpy and improving system durability. Black phosphorus (BP) leverages its anisotropic structure and high thermal conductivity for efficient photothermal conversion, though stability challenges necessitate protective modifications. Graphitic carbon nitride (g-C_3N_4) nanosheets excel in

TABLE 8.1 Comparative performance of two-dimensional (2D) nanomaterials in solar water purification systems.

System design	Materials	Evaporation rates ($kg\ m^{-2}\ h^{-1}$)	Efficiency	Key features	References
Plasmonic-enhanced composite membrane	MXene ($Ti_3C_2T_x$)/CIS nanoparticles	1.43	90.04	Broad-spectrum absorption, LSPR effects, antireflective properties	Wang et al. (2022)
Porous foam-based evaporator	rGO/CNTs/Ni foam	1.37	96.4	Hierarchical porosity, high thermal conductivity, mechanical stability	Chen et al. (2022)
Superhydrophilic COF membrane	TPAD-based COF	1.42	94	Tunable nanoporosity, metal-free, rapid water transport	Chen et al. (2022)
Magnetic photothermal evaporator	Fe_3O_4-decorated MoS_2	1.42	89.18	Multifunctional (heavy metal removal), recyclable, low-cost	Bai et al. (2023)
Hydrogel-based SVG system	BN nanosheets/ polymer hydrogel	4.03	94.6	Low evaporation enthalpy (845.11 kJ kg*'), salt-resistant	Zhang et al. (2024)
Silica-passivated BP evaporator	BP/SiO_2	1.63	-	Anisotropic heat transfer, oxidation-resistant coating	Li et al. (2020)
Hydrophobic salt-blocking membrane	Modified Ti_3C_2 MXene/ commercial filter	1.31	71	200+ hour high-salinity stability, effective dye/heavy metal removal	Zhao et al. (2018)

visible-light-driven photocatalysis, with hybrid structures enhancing charge separation and pollutant degradation. Covalent organic frameworks (COFs) offer tunable porosity and broadband absorption, achieving evaporation efficiencies up to 94%, while layered double hydroxides (LDH) combine structural versatility with photocatalytic synergism for advanced water treatment.

The unique advantages of these 2D nanomaterials lie in their multifunctionality, scalability, and potential for integration into hybrid systems. Their ability to simultaneously address water purification, energy generation, and environmental remediation underscores their transformative potential in sustainable technologies. However, challenges such as material stability, environmental impact, and large-scale production must be addressed to fully realize their practical applications. Continued research into advanced material engineering, composite optimization, and system design will be crucial in overcoming these limitations and advancing solar-driven water purification solutions to address the global water scarcity challenge.

8.5 Conclusion and perspectives

2D nanomaterials have revolutionized the field of solar-driven water purification by providing a versatile platform for high-efficiency photothermal conversion, interfacial heating, and the sustainable generation of clean water. Their intrinsic features, atomic-scale thickness, high surface-to-volume ratio, tunable band gaps, and strong excitonic effects allow them to capture a broad solar spectrum and efficiently convert absorbed light into localized heat. This photothermal effect is further enhanced through nonradiative relaxation processes, localized surface plasmon resonance (LSPR), and interfacial heat confinement. Together, these mechanisms enable rapid, energy-efficient evaporation at the water–air interface, substantially reducing thermal loss to the bulk water and improving overall system performance.

The development of 2D nanomaterial-based evaporators has advanced from fundamental proof-of-concept systems to highly engineered architectures with practical functionalities. Materials such as graphene and its derivatives (GO, RGO), MXenes ($Ti_3C_2T_x$, Nb_2C), transition metal dichalcogenides (MoS_2, WS_2), black phosphorus (BP), boron nitride (BN), graphitic carbon nitride (g-C_3N_4), covalent organic frameworks (COFs), and layered double hydroxides (LDHs) have all shown distinct advantages in light absorption, heat management, and environmental resilience. Among these, MXenes have emerged as standout candidates, achieving evaporation rates exceeding 3.6 kg m^{-2} h^{-1} and efficiencies above 90% when incorporated into hydrogels and aerogels with vertically aligned channels or dual-network structures. Graphene-based materials, with their broadband absorption and excellent thermal stability, continue to serve as the benchmark for scalable, multifunctional designs.

Recent advancements demonstrate that engineered structures such as Janus layers, vertically aligned channels, and composite hydrogels can maintain long-term operation even under high-salinity or polluted conditions. For example, MXene–CA/PAAm hydrogels have achieved stable operation for over 360 hours at 1.78 kg m^{-2} h^{-1}, while Nb_2C MXene dual-network hydrogels demonstrated 93.2% photothermal efficiency and remarkable oxidation resistance. The introduction of MXene sediment–PVA/SA aerogels with vertical pores achieved evaporation rates of 3.6 $kgm^{-2} h^{-1}$ during continuous outdoor testing without salt deposition over 14 days, highlighting real-world robustness. Moreover, MXene/CdS hybrid hydrogels have shown multifunctional capability, combining 1.80 $kg m^{-2} h^{-1}$ evaporation with simultaneous VOC and antibiotic degradation, signifying the evolution toward integrated water purification platforms.

However, several technical and practical challenges continue to limit large-scale deployment. The foremost issues include:

1. Material stability: Reactive 2D materials such as MXenes and BP are prone to oxidation and structural degradation under prolonged sunlight and water exposure.
2. Salt accumulation and scaling: Continuous operation in saline or brackish water leads to crystallization on photothermal surfaces, reducing efficiency.
3. Scalability and reproducibility: Synthesis of high-quality, defect-controlled 2D nanomaterials at a large scale remains expensive and complex.
4. Environmental and health safety: Potential leaching or persistence of nanomaterials in aquatic environments raises sustainability and regulatory concerns.

To address these challenges, innovative design strategies have emerged. Spatial decoupling between evaporation and crystallization zones, as demonstrated in radial or suspended Janus architectures, has effectively mitigated salt fouling. For instance, a horizontal disk–radial wick design redirected salt crystallization to the periphery, while suspended fibrous Janus mats allowed continuous evaporation (~1.94 $kg m^{-2} h^{-1}$) as salts crystallized below. Similarly, hydrophilic–hydrophobic interfacial engineering in porous substrates supports continuous capillary-driven water transport while maintaining localized heating. Self-healing and flexible polymeric matrices have extended device durability, while composite designs integrating carbon quantum dots (CQDs), nanoparticles, or semiconducting photocatalysts have enhanced optical absorption and improved long-term stability.

From a broader perspective, the next generation of 2D-based solar evaporators will likely move beyond single-function photothermal devices toward integrated, multifunctional systems capable of simultaneous desalination, disinfection, and pollutant degradation. The synergy between photothermal, photocatalytic, and adsorption mechanisms opens new opportunities for complete water purification under natural sunlight. In addition, interdisciplinary approaches that combine materials science, nanophotonic, computational modeling, and environmental engineering are essential for designing scalable, durable, and eco-safe water purification technologies.

Future research should focus on developing oxidation-resistant and bio-derived 2D nanomaterials for sustainable operation, establishing standardized testing protocols for performance comparison under realistic environmental conditions, integrating data-driven material design and AI-guided simulations to accelerate discovery of optimized material–structure combinations, scaling fabrication through low-cost printing, coating, or templating methods compatible with industrial production, evaluating environmental impact and life-cycle performance, ensuring safety and regulatory compliance. In summary, 2D nanomaterials have proven to be one of the most potent and versatile platforms for solar-driven water purification. Their unique combination of high-light absorption, tunable properties, and structural flexibility enables efficient, sustainable, and multifunctional water treatment systems. While significant challenges persist, particularly in stability, scalability, and safety, the rapid pace of innovation in composite engineering and device architecture is steadily bridging the gap between laboratory breakthroughs and real-world applications. With continued interdisciplinary collaboration, 2D nanomaterials will play a pivotal role in shaping the future of low-carbon, decentralized, and sustainable freshwater generation technologies for water-stressed communities worldwide.

AI disclosure

During the preparation of this work, the authors used Grammarly in order to assist grammatical errors. After using this tool/service, the authors reviewed and edited the content as needed and take full responsibility for the content of the publication.

References

Ahmad, M.M., Ma, Y., Saravanan, V., Shahid, H., Huang, L., Badshah, M., et al., 2025. Facile construction of a superhydrophilic and highly porous $Ti_3C_2T_x$/alginate aerogel solar evaporator for water evaporation and purification. Separation and Purification Technology 371, 133363. Available at https://doi.org/10.1016/j.seppur.2025.133363.

Aizudin, M., Alias, N.H., Ng, Y.X.A., Mahmod Fadzuli, M.H., Ang, S.C., Ng, Y.X., et al., 2022. Membranes prepared from graphene-based nanomaterials for water purification: a mini-review. Nanoscale 14 (48), 17871–17886. Available at https://doi.org/10.1039/d2nr05328d.

Almarzooqi, N., Alwan, R.A., AlMarzooqi, F., Ghaffour, N., Hong, S., Arafat, H.A., 2024. Solar-driven surface-heating membrane distillation using $Ti_3C_2T_x$ MXene-coated spacers. Chemosphere 351. Available at https://doi.org/10.1016/j.chemosphere.2024.141129.

Amaral, L.O., Daniel-da-Silva, A.L., 2022. MoS_2 and MoS_2 nanocomposites for adsorption and photodegradation of water pollutants: a review. Molecules 27 (20). Available at https://doi.org/10.3390/molecules27206782.

Andoshe, D.M., Jeon, J.M., Kim, S.Y., Jang, H.W., 2015. Two-dimensional transition metal dichalcogenide nanomaterials for solar water splitting. Electronic Materials Letters 11 (3), 323–335. Available at https://doi.org/10.1007/s13391-015-4402-9.

Awan, H.T.A., Kumar, L., Wong, W.P., Walvekar, R., Khalid, M., 2023. Recent progress and challenges in mxene-based phase change material for solar and thermal energy applications. Energies 16 (4). Available at https://doi.org/10.3390/en16041977.

Azni Farhana Mazri, N., Arifutzzaman, A., Kheireddine Aroua, M., Ekhlasur Rahman, M., Ali Mazari, S., 2023. Graphene and its tailoring as emerging 2D nanomaterials in efficient CO_2 absorption: a state-of-the-art interpretative review. Alexandria Engineering Journal 77, 479–502. Available at https://doi.org/10.1016/j.aej.2023.06.070.

Bai, X., Wang, Q., Tan, M., Zhang, P., Sun, S., Wu, D., et al., 2024. Spongy and anti-pollution MXene/Ag_2S/cellulose acetate membrane for sustainable solar-driven interfacial evaporation and water purification. Chemical Engineering Journal 489, 151441. Available at https://doi.org/10.1016/j.cej.2024.151441.

Bai, Z., Xu, H., Li, G., Yang, B., Yao, J., Guo, K., et al., 2023. MoS_2 nanosheets decorated with Fe_3O_4 nanoparticles for highly efficient solar steam generation and water treatment. Molecules 28 (4), 1719. Available at https://doi.org/10.3390/molecules28041719.

Balu, S.kumar, Xing, R., Liu, S., 2024. Emergent MXene evaporators in high-efficiency solar steam generation. Sustainable Materials and Technologies 40, e00941. Available at https://doi.org/10.1016/j.susmat.2024.e00941.

Bramhaiah, K., Bhattacharyya, S., 2022. Challenges and future prospects of graphene-based hybrids for solar fuel generation: moving towards next generation photocatalysts. Materials Advances 3 (1), 142–172. Available at https://doi.org/10.1039/d1ma00748c.

Cai, W., Mu, X., Pan, Y., Li, Z., Wang, J., Zhou, X., et al., 2020. Black phosphorous nanosheets: a novel solar vapor generator. Solar RRL 4 (4). Available at https://doi.org/10.1002/solr.201900537.

Chang, M., Ai, L., Yang, R., Wang, X., Xu, Y., Jiang, J., 2024. Two-dimensional layered MBene membrane towards sustainable freshwater production from solar interfacial evaporation. Chemical Engineering Journal 486, 150078. Available at https://doi.org/10.1016/j.cej.2024.150078.

Chang, M., Yang, R., Lai, X., Wang, X., Jiang, J., 2024. Porous rGO-MBene monolith: a highly efficient solar evaporator for salt-resistant desalination and multitasking water purification. Desalination 592, 118164. Available at https://doi.org/10.1016/j.desal.2024.118164.

Chen, Z., Li, X., Liu, R., Ma, K., Sang, H., Huang, Y., et al., 2022. Nickel foam@reduced graphene oxide–carbon nanotube composite as an efficient solar evaporator for water purification and electricity generation. Industrial & Engineering Chemistry Research 61 (44), 16565–16576. Available at https://doi.org/10.1021/acs.iecr.2c02646.

Chen, R., Wang, X., Gan, Q., Zhang, T., Zhu, K., Ye, M., 2019. A bifunctional MoS_2 -based solar evaporator for both efficient water evaporation and clean freshwater collection. Journal of Materials Chemistry A 7 (18), 11177–11185. Available at https://doi.org/10.1039/c9ta02002k.

Chen, L., Wei, J., Tian, Q., Han, Z., Li, L., Meng, S., et al., 2021. Dual-functional graphene oxide-based photothermal materials with aligned channels and oleophobicity for efficient solar steam generation. Langmuir 37 (33), 10191–10199. Available at https://doi.org/10.1021/acs.langmuir.1c01647.

Cui, X., Ruan, Q., Zhuo, X., Xia, X., Hu, J., Fu, R., et al., 2023. Photothermal nanomaterials: a powerful light-to-heat converter. Chemical Reviews 123 (11), 6891–6952. Available at https://doi.org/10.1021/acs.chemrev.3c00159.

Dervin, S., Dionysiou, D.D., Pillai, S.C., 2016. 2D nanostructures for water purification: graphene and beyond. Nanoscale 8 (33), 15115–15131. Available at https://doi.org/10.1039/c6nr04508a.

Ding, F., Yang, D., Tong, Z., Nan, Y., Wang, Y., Zou, X., et al., 2017. Graphitic carbon nitride-based nanocomposites as visible-light driven photocatalysts for environmental purification. Environmental Science: Nano 4 (7), 1455–1469. Available at https://doi.org/10.1039/C7EN00255F.

Ding, T., Zhou, Y., Ong, W.L., Ho, G.W., 2021. Hybrid solar-driven interfacial evaporation systems: beyond water production towards high solar energy utilization. Materials Today 42, 178–191. Available at https://doi.org/10.1016/j.mattod.2020.10.022.

Doménech, N.G., Coogan, Á., Purcell-Milton, F., Casasín García, M.L., Arjona, A.S., Cabré, M.B., et al., 2022. Partially oxidised boron nitride as a 2D nanomaterial for nanofiltration applications. Nanoscale Advances 4 (22), 4895–4904. Available at https://doi.org/10.1039/d2na00472k.

Dou, Y., Zhang, S., Pan, T., Xu, S., Zhou, A., Pu, M., et al., 2015. TiO_2@layered double hydroxide core-shell nanospheres with largely enhanced photocatalytic activity toward O_2 generation. Wiley-VCH Verlag, China. Advanced Functional Materials 25 (15), 2243–2249. Available at https://doi.org/10.1002/adfm.201404496.

Du, X., Han, L., Liu, R., Nikolai, M., Li, Y., Sun, Z., et al., 2025. Synergy design and performance optimization of hydrogel-based materials for solar driven water purification applications. Desalination 602, 118627. Available at https://doi.org/10.1016/j.desal.2025.118627.

Fang, Y., Gao, H., Cheng, K., Bai, L., Li, Z., Zhao, Y., et al., 2025. An overview of photothermal materials for solar-driven interfacial evaporation. Chinese Chemical Letters 36 (3), 109925. Available at https://doi.org/10.1016/j.cclet.2024.109925.

Fan, D., Lu, Y., Zhang, H., Xu, H., Lu, C., Tang, Y., et al., 2021. Synergy of photocatalysis and photothermal effect in integrated 0D perovskite oxide/2D MXene heterostructures for simultaneous water purification and solar steam generation. Applied Catalysis B: Environmental 295, 120285. Available at https://doi.org/10.1016/j.apcatb.2021.120285.

Fan, C., Miao, J., Xu, G., Liu, J., Lv, J., Wu, Y., 2017. Graphitic carbon nitride nanosheets obtained by liquid stripping as efficient photocatalysts under visible light. RSC Advances 7 (59), 37185–37193. Available at https://doi.org/10.1039/C7RA05732F.

Fan, Q., Wu, L., Liang, Y., Xu, Z., Li, Y., Wang, J., et al., 2021. The role of micro-nano pores in interfacial solar evaporation systems – a review. Applied Energy 292, 116871. Available at https://doi.org/10.1016/j.apenergy.2021.116871.

Fatima, J., Shah, A.N., Tahir, M.B., Mehmood, T., Shah, A.A., Tanveer, M., et al., 2022. Tunable 2D nanomaterials; their key roles and mechanisms in water purification and monitoring. Frontiers in Environmental Science 10. Available at https://doi.org/10.3389/fenvs.2022.766743.

Gama, B.M.Vd, Selvasembian, R., Giannakoudakis, D.A., Triantafyllidis, K.S., McKay, G., Meili, L., 2022. Layered double hydroxides as rising-star adsorbents for water purification: a brief discussion. Molecules 27 (15). Available at https://doi.org/10.3390/molecules27154900.

Gao, Y., Zhou, X., Fu, N., Su, S., Ma, B., Ruan, Q., et al., 2023. Superior solar water purification achieved by the excellent photothermal conversion and thermal management of porous plasmonic TiN nanoparticles. Chemical Engineering Journal 475, 146078. Available at https://doi.org/10.1016/j.cej.2023.146078.

García Doménech, N., Purcell-Milton, F., Sanz Arjona, A., Casasín García, M.-L., Ward, M., Cabré, M.B., et al., 2022. High-performance boron nitride-based membranes for water purification. Nanomaterials 12 (3), 473. Available at https://doi.org/10.3390/nano12030473.

Ghodsi, A., Fashandi, H., 2024. Influence of photothermal nanomaterials localization within the electrospun membrane structure on purification of saline oily wastewater based on photothermal vacuum membrane distillation. Journal of Environmental Management 366, 121866. Available at https://doi.org/10.1016/j.jenvman.2024.121866.

Gnanasekaran, A., Rajaram, K., 2024. Rational design of different interfacial evaporators for solar steam generation: recent development, fabrication, challenges and applications. Renewable and Sustainable Energy Reviews 192, 114202. Available at https://doi.org/10.1016/j.rser.2023.114202.

Gogotsi, Y., Huang, Q., 2021. MXenes: two-dimensional building blocks for future materials and devices. ACS Nano 15 (4), 5775–5780. Available at https://doi.org/10.1021/acsnano.1c03161.

Gunjakar, J.L., Kim, I.Y., Lee, J.M., Lee, N.S., Hwang, S.J., 2013. Self-assembly of layered double hydroxide 2D nanoplates with graphene nanosheets: an effective way to improve the photocatalytic activity of 2D nanostructured materials for visible light-induced O2 generation. Energy and Environmental Science 6 (3), 1008–1017. Available at https://doi.org/10.1039/c3ee23989f.

Gärtner, W.W., 1961. Photothermal effect in semiconductors. Physical Review 122 (2), 419–424. Available at https://doi.org/10.1103/PhysRev.122.419.

Hou, L., Wang, N., Yu, L.-J., Liu, J., Zhang, S., Cui, Z., et al., 2022. High-performance janus solar evaporator for water purification with broad spectrum absorption and ultralow heat loss. ACS Energy Letters 8 (1), 553–564. Available at https://doi.org/10.1021/acsenergylett.2c02567.

Huang, Z., Cui, X., Li, S., Wei, J., Li, P., Wang, Y., et al., 2020. Two-dimensional MXene-based materials for photothermal therapy. Nanophotonics 9 (8). Available at https://doi.org/10.1515/nanoph-2019-0571.

Huan, X., Wang, W., Wang, C., Tian, Z., Li, Y., 2025. Plasmonic nanomaterials for versatile solar energy conversion Applications. ACS Omega 10 (27), 28615–28629. Available at https://doi.org/10.1021/acsomega.5c02075.

Hussain, A.K., Hussain, R.K., Kadhom, M., Yousif, E., 2024. Two-Dimensional Black Phosphorus: Preparation, Properties, and Environmental Applications in Water Treatment. Volume 62: Hydraulic and Civil Engineering Technology IX. IOS Press BV, Iraq, pp. 1463–1485. Available at https://doi.org/10.3233/ATDE241148.

Hu, X.Y., Fan, C., Wang, S., Chen, Y., Wang, D., Chen, K., et al., 2024. Two-dimensional mxenes: innovative materials for efficient thermal management and safety solutions. Research; a Journal of Science and its Applications 7, 542. Available at https://doi.org/10.34133/research.0542.

Irshad, M.S., Arshad, N., Asghar, M.S., Hao, Y., Alomar, M., Zhang, S., et al., 2023. Advances of 2D-enabled photothermal materials in hybrid solar-driven interfacial evaporation systems toward water-fuel-energy crisis. Advanced Functional Materials 33 (51). Available at https://doi.org/10.1002/adfm.202304936.

Irshad, M.S., Arshad, N., Maqsood, G., Ahmed, I., Shakoor, B., Asghar, M.S., et al., 2025. Advancing water collection efficiency in hybrid solar evaporators: key factors, strategic innovations, and synergistic applications. Materials Science and Engineering R: Reports 165. Available at https://doi.org/10.1016/j.mser.2025.101018.

Jayaprakash, N., Elumalai, K., Manickam, S., Bakthavatchalam, G., Tamilselvan, P., 2024. Nano Materials Science. Available at https://doi.org/10.1016/j.nanoms.2024.11.004.

Jeffries, W.R., Fagan, A.M., Schaak, R.E., Knappenberger, K.L., 2022. Influence of band alignment on electronic relaxation in plasmonic metal-semiconductor hybrid nanoparticles. Journal of Physical Chemistry C 126 (19), 8384–8392. Available at https://doi.org/10.1021/acs.jpcc.2c01378.

Jiang, T., Bian, S., Wang, Y., Fan, X., Zhu, L., Song, X., et al., 2024. The nexus of solar absorption and morphology designs based on 2D photothermal materials: from rational design to advanced application of water purification. Chemical Engineering Journal 499, 156450. Available at https://doi.org/10.1016/j.cej.2024.156450.

Jing Bai, X., Zhang, C., Wang, X., Li, Z., Xu, C., Jing, T., et al., 2024. The adsorption-photocatalytic synergism of LDHs-based nanocomposites on the removal of pollutants in aqueous environment: a critical review. Journal of Cleaner Production 436, 140705. Available at https://doi.org/10.1016/j.jclepro.2024.140705.

Jing, X., Chen, L., Li, Y., Yin, H., Chen, J., Su, M., et al., 2024. Synergistic effect between 0D CQDs and 2D MXene to enhance the photothermal conversion of hydrogel evaporators for efficient solar water evaporation, photothermal sensing and electricity generation. Small (Weinheim an der Bergstrasse, Germany) 20 (50). Available at https://doi.org/10.1002/smll.202405587.

Jing, X., Liu, F., Abdiryim, T., Liu, X., 2024. Hydrogels as promising platforms for solar-driven water evaporators. Chemical Engineering Journal 479, 147519. Available at https://doi.org/10.1016/j.cej.2023.147519.

Khoei, J.K., Bafqi, M.S.S., Saeidiharzand, S., Mohammadilooey, M., Hezarkhani, M., Okan, B.S., et al., 2023. Upcycled graphene integrated fiber-based photothermal hybrid nanocomposites for solar-driven interfacial water evaporation. SSRN Electronic Journal. Available at https://doi.org/10.2139/ssrn.4355040.

Kuehl, V.A., Yin, J., Duong, P.H.H., Mastorovich, B., Newell, B., Li-Oakey, K.D., et al., 2018. A highly ordered nanoporous, two-dimensional covalent organic framework with modifiable pores, and its application in water purification and ion sieving. Journal of the American Chemical Society 140 (51), 18200–18207. Available at https://doi.org/10.1021/jacs.8b11482.

Kumar, S., Karthikeyan, S., Lee, A.F., 2018. g-C_3N_4-based nanomaterials for visible light-driven photocatalysis. Catalysts 8 (2). Available at https://doi.org/10.3390/catal8020074.

Lei, W., Portehault, D., Liu, D., Qin, S., Chin, Y., 2013. Porous boron nitride nanosheets for effective water cleaning. Nature Communications 4 (1). Available at https://doi.org/10.1038/ncomms2818.

Lin, Wu, Feng, Shi, Zhou, Zhu, et al., 2022. Poly (N-phenylglycine)/MoS_2 nanohybrid with synergistic solar-thermal conversion for efficient water purification and thermoelectric power generation. ACS Applied Materials & Interfaces 14 (1), 1034–1044. Available at https://doi.org/10.1021/acsami.1c20393.

Lin, Y., Xu, H., Shan, X., Di, Y., Zhao, A., Hu, Y., et al., 2019. Solar steam generation based on the photothermal effect: from designs to applications, and beyond. Journal of Materials Chemistry A 7 (33), 19203–19227. Available at https://doi.org/10.1039/c9ta05935k.

Liu, F., Lai, Y., Zhao, B., Bradley, R., Wu, W., 2019. Photothermal materials for efficient solar powered steam generation. Frontiers of Chemical Science and Engineering 13 (4), 636–653. Available at https://doi.org/10.1007/s11705-019-1824-1.

Liu, Y., Qu, S.Z., Zhou, Z.R., Song, X.P., Ma, L., Ding, S.J., et al., 2023. Synergistic photothermal conversion and photocatalysis in 2D/2D MXene/Bi_2S_3 hybrids for efficient solar-driven water purification. Nanoscale 15 (36), 14886–14895. Available at https://doi.org/10.1039/d3nr02848h.

Liu, Y., Wang, M., Dong, C., Yu, H., Lu, Y., Huang, X., et al., 2023. A thienyl-benzodithiophene-based two-dimensional conjugated covalent organic framework for fast photothermal conversion. Journal of Polymer Science 61 (16), 1843–1848. Available at https://doi.org/10.1002/pol.20230090.

Liu, G., Xu, J., Wang, K., 2017. Solar water evaporation by black photothermal sheets. Nano Energy 41, 269–284. Available at https://doi.org/10.1016/j.nanoen.2017.09.005.

Liu, D., Zhang, M., Xie, W., Sun, L., Chen, Y., Lei, W., 2017. Porous BN/TiO_2 hybrid nanosheets as highly efficient visible-light-driven photocatalysts. Applied Catalysis B: Environmental 207, 72–78. Available at https://doi.org/10.1016/j.apcatb.2017.02.011.

Li, G., Fang, D., Hong, G., Eychmüller, A., Zhang, X., Song, W., 2022. Assembling graphene aerogel hollow fibres for solar steam generation. Composites Communications 35, 101302. Available at https://doi.org/10.1016/j.coco.2022.101302.

Li, Z., Liu, F., Jing, X., Zhou, W., Pei, S., Abdiryim, T., et al., 2025. $Ti_3C_2T_x$/CuO composite hydrogels with low evaporation enthalpy and efficient photothermal conversion for solar-driven water purification, electricity generation and pollutant degradation. Chemical Engineering Journal 519, 165019. Available at https://doi.org/10.1016/j.cej.2025.165019.

Li, Y., Shi, Y., Wang, H., Liu, T., Zheng, X., Gao, S., et al., 2023. Recent advances in carbon-based materials for solar-driven interfacial photothermal conversion water evaporation: assemblies, structures, applications, and prospective. Carbon Energy 5 (11). Available at https://doi.org/10.1002/cey2.331.

Li, W., Tian, X., Li, X., Han, S., Li, C., Zhai, X.Z., et al., 2021. Ultrahigh solar steam generation rate of a vertically aligned reduced graphene oxide foam realized by dynamic compression. Journal of Materials Chemistry A 9 (26), 14859–14867. Available at https://doi.org/10.1039/d1ta03014k.

Li, X., Wu, J., An, S., Li, K., Zhang, J., Pei, M., et al., 2023. Ultrathin crystalline carbon nitride nanosheets for highly efficient photocatalytic pollutant removal and hydrogen production. ACS Applied Nano Materials 6 (13), 11601–11611. Available at https://doi.org/10.1021/acsanm.3c01624.

Li, M., Zhao, Q., Zhang, S., Li, D., Li, H., Zhang, X., et al., 2020. Facile passivation of black phosphorus nanosheets via silica coating for stable and efficient solar desalination. Environmental Science: Nano 7 (2), 414–423. Available at https://doi.org/10.1039/c9en01138b.

Li, J., et al., 2023. High-performance Janus evaporator based on self-healing hydrogels for salt-rejecting interfacial solar desalination. Desalination 564. Available at https://doi.org/10.1016/j.desal.2023.116812.

Luo, L., Hou, C., Wang, L., Zhang, W., Wang, C., Liu, J., et al., 2024. Layered double hydroxide-based photocatalysts for the removal of emerging contaminants: progress in past ten years. Catalysts 14 (4), 252. Available at https://doi.org/10.3390/catal14040252.

Marimuthu, A., Basu, H., Singh, S., Saha, S., Basu, R., Chandwadkar, P., et al., 2024. Intercalation of V_2O_5 and polypyrrole into a graphene oxide layer: a hybrid multifunctional photothermal structure for efficient solar evaporation, water purification, disinfection, and power production. ACS Applied Materials and Interfaces 16 (34), 45063–45077. Available at https://doi.org/10.1021/acsami.4c12476.

Ma, J., Ma, L., He, Y., Tang, X., Park, C.B., 2024. Sustainable solar-driven MXene/polyvinylidene fluoride composite nanofiber membrane for oil/saltwater separation and distillation in the complex environment. Journal of Water Process Engineering 57, 104571. Available at https://doi.org/10.1016/j.jwpe.2023.104571.

Mittal, A., Brajpuriya, R., Gupta, R., 2023. Solar steam generation using hybrid nanomaterials to address global environmental pollution and water shortage crisis. Materials Today Sustainability 21, 100319. Available at https://doi.org/10.1016/j.mtsust.2023.100319.

Nabavi Hamedani, S.A., Amjad-Iranagh, S., Dehghani, K., 2025. Photothermal high-entropy materials: a review. Sustainable Materials and Technologies 43, e01316. Available at https://doi.org/10.1016/j.susmat.2025.e01316.

Namboorimadathil Backer, S., Ramachandran, A.M., Venugopal, A.A., Mohamed, A.P., Asok, A., Pillai, S., 2020. Clean water from air utilizing black TiO_2-based photothermal nanocomposite sheets. ACS Applied Nano Materials 3 (7), 6827–6835. Available at https://doi.org/10.1021/acsanm.0c01207.

Ng, S.W.L., Gao, M., Lu, W., Hong, M., Ho, G.W., 2021. Selective wavelength enhanced photochemical and photothermal H_2 generation of classical oxide supported metal catalyst. Advanced Functional Materials 31 (38). Available at https://doi.org/10.1002/adfm.202104750.

Noor, U., Farid, M.F., Sharif, A., Saleem, A., Nabi, Z., Mughal, M.F., et al., 2024. Functionalized carbon 1D/2D nanomaterials for effective water desalination: synthesis, applications and cost issues. An overview. Elsevier B.V., Thailand Desalination 571. Available at https://doi.org/10.1016/j.desal.2023.117086.

Patibandla, S., Pramanik, A., Gao, Y., Gates, K., Shukla, M.K., Ray, P.C., 2024. Emerging 2D materials-based nanoarchitecture for water purification. In: Shukla, M., Ferguson, E., Leszczynski, J. (Eds.), Emerging Materials and Environment. Challenges and Advances in Computational Chemistry and Physics, vol 37. Springer, Cham, pp. 231–245. Available at https://doi.org/10.1007/978-3-031-39470-6_7.

Peng, H., Wang, R., Mei, L., Zhang, Q., Ying, T., Qian, Z., et al., 2023. Transition metal dichalcogenide-based functional membrane: synthesis, modification, and water purification applications. Matter 6 (1), 59–96. Available at https://doi.org/10.1016/j.matt.2022.09.019.

Qu, J., Li, S., Zhong, B., Deng, Z., Shu, Y., Yang, X., et al., 2022. Two-dimensional nanomaterials: synthesis and applications in photothermal catalysis. Nanoscale 15 (6), 2455–2469. Available at https://doi.org/10.1039/d2nr06092b.

Raj, S.K., Kulshrestha, V., 2022. Progress in 2D Nanomaterial Composites Membranes for Water Purification and Desalination. In: Khanam, Z., Gogoi, N., Srivastava, D.N. (Eds.), 2D Nanomaterials for Energy and Environmental Sustainability. Materials Horizons: From Nature to Nanomaterials. Springer, Singapore, pp. 125–148. Available at https://doi.org/10.1007/978-981-16-8538-5_6.

Ren, K., Sun, M., 2025. Thermal transport in 2D materials. Journal of Applied Physics 137 (13). Available at https://doi.org/10.1063/5.0267691.

Ren, J., Xia, Z., Luo, B., Li, Di, Shi, W., 2021. Fabrication of 2D/2D COF/$SnNb_2O_6$ nanosheets and their enhanced solar hydrogen production. Inorganic Chemistry Frontiers 8 (7), 1686–1694. Available at https://doi.org/10.1039/d0qi01443e.

Shanker, G.S., Bhosale, R., Ogale, S., Nag, A., 2018. 2D nanocomposite of g-C_3N_4 and TiN embedded N-doped graphene for photoelectrochemical reduction of water using sunlight. Advanced Materials Interfaces 5 (24), 1801488. Available at https://doi.org/10.1002/admi.201801488.

Shao, Z., Wang, C., Wu, K., Zhang, H., Chen, J., 2019. Fiber all-optical light control with low-dimensional materials (LDMs): thermo-optic effect and saturable absorption. Nanoscale Advances 1 (11), 4190–4206. Available at https://doi.org/10.1039/c9na00381a.

Singh, M.K., Shinde, P.V., Singh, P., Tyagi, P., 2021. Two-dimensional materials for advanced solar cells. In: Elseman, A.M. (Ed.), Solar Cells - Theory, Materials and Recent Advances. IntechOpen. Available at https://doi.org/10.5772/intechopen.94114.

Song, Z., Tiraferri, A., Yuan, R., Cao, J., Tang, P., Xie, W., et al., 2022. Theoretical evaluation of the evaporation rate of 2D solar-driven interfacial evaporation and of its large-scale application potential. Desalination 537, 115891. Available at https://doi.org/10.1016/j.desal.2022.115891.

Soo Joo, B., Soo Kim, I., Han, I.K., Ko, H., Gu Kang, J., Kang, G., 2022. Plasmonic silicon nanowires for enhanced heat localization and interfacial solar steam generation. Applied surface science 583, 152563. Available at https://doi.org/10.1016/j.apsusc.2022.152563.

Subhan, A., Mourad, A.H.I., 2025. Plasmonic metal nanostructures as performance enhancers in emerging solar cells: a review. Next Materials 6. Available at https://doi.org/10.1016/j.nxmate.2025.100509.

Tan, L., Wang, Z., Zhao, Y., Song, Y.F., 2020. Recent progress on nanostructured layered double hydroxides for visible-light-induced photoreduction of CO_2. Chemistry – An Asian Journal 15 (21), 3380–3389. Available at https://doi.org/10.1002/asia.202000963.

Usman, M., Zeb, Z., Ullah, H., Suliman, M.H., Humayun, M., Ullah, L., et al., 2022. A review of metal-organic frameworks/graphitic carbon nitride composites for solar-driven green H2 production, CO_2 reduction, and water purification. Journal of Environmental Chemical Engineering 10 (3). Available at https://doi.org/10.1016/j.jece.2022.107548.

Wang, Q., Feng, J., Fornasiero, P., 2021. Chapter 29 - Two-dimensional layered double hydroxide based photocatalysts for environmental clean-up and renewable energy production. In: García-López, E.I., Palmisano, L. (Eds.), Materials Science in Photocatalysis. Elsevier, China, pp. 485–503. Available at https://doi.org/10.1016/B978-0-12-821859-4.00028-3.

Wang, P., Jia, C., Huang, Y., Duan, X., 2021. Van der Waals heterostructures by design: from 1D and 2D to 3D. Matter 4 (2), 552–581. Available at https://doi.org/10.1016/j.matt.2020.12.015.

Wang, W., Klots, A., Yang, Y., Li, W., Kravchenko, I.I., Briggs, D.P., et al., 2015. Enhanced absorption in two-dimensional materials via Fano-resonant photonic crystals. Applied Physics Letters 106 (18). Available at https://doi.org/10.1063/1.4919760.

Wang, J., Lian, Y., Shao, J., Cheng, S., Zhao, H., Zhou, C., et al., 2025. A review: solar-driven water evaporation based on biomass carbon materials. Renewable and Sustainable Energy Reviews 217, 115755. Available at https://doi.org/10.1016/j.rser.2025.115755.

Wang, H., Liu, X., Niu, P., Wang, S., Shi, J., Li, L., 2020. Porous two-dimensional materials for photocatalytic and electrocatalytic applications. Matter 2 (6), 1377–1413. Available at https://doi.org/10.1016/j.matt.2020.04.002.

Wang, Y., Li, W., Yang, S., Liu, Z., Meng, G., Wu, J., et al., 2025. A hydrogel evaporator with adjustable internal water content distribution for seawater desalination. Colloids and Surfaces A: Physicochemical and Engineering Aspects 725, 137555. Available at https://doi.org/10.1016/j.colsurfa.2025.137555.

Wang, L., Lu, Y., Xu, J., Tao, F., Liang, P., 2023. Solvothermal synthesis of g-C_3N_4 nanosheets modified carbon quantum dots for enhanced photocatalytic degradation. Solid State Sciences 143, 107271. Available at https://doi.org/10.1016/j.solidstatesciences.2023.107271.

Wang, P., Meng, J., Huang, J., Wang, J.J., Li, Q.X., 2019. Phosphorene-based van der Waals heterojunction for solar water splitting. Chinese Journal of Chemical Physics 32 (4), 431–436. Available at https://doi.org/10.1063/1674-0068/cjcp1811244.

Wang, Y., Nie, J., He, Z., Zhi, Y., Ma, X., Zhong, P., 2022. $Ti_3C_2T_x$ MXene nanoflakes embedded with copper indium selenide nanoparticles for desalination and water purification through high-efficiency solar-driven membrane evaporation. ACS Applied Materials & Interfaces 14 (4), 5876–5886. Available at https://doi.org/10.1021/acsami.1c22952.

Wang, Y., Zhang, Z., Li, T., Ma, P., Zhang, H., Chen, M., et al., 2019. Photothermal-responsive graphene oxide membrane with smart gates for water purification. ACS Applied Materials & Interfaces 11 (47), 44886–44893. Available at https://doi.org/10.1021/acsami.9b15988.

Wang, D., Zhang, H., Yang, P., 2024. Microstructure driven charge carrier separation in superior thin g-C_3N_4 nanosheets towards enhanced photocatalytic activities. Colloids and Surfaces A: Physicochemical and Engineering Aspects 689, 133759. Available at https://doi.org/10.1016/j.colsurfa.2024.133759.

Wang, C., Zhou, S., Cui, Q., Zhang, M., Zheng, L., Li, S., et al., 2024. Self-floating g-C_3N_4/h-BN/Au photo-thermal catalysts with unprecedented thermostability at solar-activated heating condition for reusable degradation of organic pollutants. Applied Catalysis B: Environmental 343, 123491. Available at https://doi.org/10.1016/j.apcatb.2023.123491.

Wen, C., Li, X., Yan, S., Wen, J., Zheng, R., Wang, X., et al., 2025. Progress in MXene-based photothermal materials for solar-driven water evaporation and desalination. Chemical Engineering Journal 510, 161873. Available at https://doi.org/10.1016/j.cej.2025.161873.

Wieszczycka, K., Staszak, K., Woźniak-Budych, M.J., Litowczenko, J., Maciejewska, B.M., Jurga, S., 2021. Surface functionalization – The way for advanced applications of smart materials. Coordination chemistry reviews 436. Available at https://doi.org/10.1016/j.ccr.2021.213846.

Xie, Z., Duo, Y., Lin, Z., Fan, T., Xing, C., Yu, L., et al., 2020. The rise of 2D photothermal materials beyond graphene for clean water production. Advanced Science 7 (5). Available at https://doi.org/10.1002/advs.201902236.

Xie, Z., Peng, Y.P., Yu, L., Xing, C., Qiu, M., Hu, J., et al., 2020. Solar-inspired water purification based on emerging 2D materials: status and challenges. Solar RRL 4 (3). Available at https://doi.org/10.1002/solr.201900400.

Xin, P., Shixiang, X., Tong, Z., Hudie, Y., Chong, L., Wanying, L., 2021. Advances in two-dimensional black phosphorus-based nanostructures for photocatalytic applications. Progress in Chemistry 34 (3), 630–642.

Xu, R., Cui, H., Wei, N., Zhao, J., Song, X., Han, Y., et al., 2024. Advanced design of bimetal sulfide/V_2CT_X MXene photothermal nanocomposites for integrated solar steam generation, electricity generation, and water purification. Desalination 570, 117094. Available at https://doi.org/10.1016/j.desal.2023.117094.

Yang, Y., Han, Y., Zhao, J., Que, W., 2022. 2D/1D MXene/MWCNT hybrid membrane-based evaporator for solar desalination. Materials 15 (3), 929. Available at https://doi.org/10.3390/ma15030929.

Yang, K., Wang, J., Chen, X., Zhao, Q., Ghaffar, A., Chen, B., 2018. Application of graphene-based materials in water purification: from the nanoscale to specific devices. Environmental Science: Nano 5 (6), 1264–1297. Available at https://doi.org/10.1039/c8en00194d.

Yang, Z., Wei, N., Xue, N., Xu, R., Yang, E., Wang, F., et al., 2024. Highly efficient MoS_2/MXene aerogel for interfacial solar steam generation and wastewater treatment. Journal of Colloid and Interface Science 656, 189–199. Available at https://doi.org/10.1016/j.jcis.2023.11.110.

Yang, H., Yan, S., Yao, D., Luo, H., Zeng, K., Li, J., et al., 2023. Photo-thermal catalytic pyrolysis of waste plastics: investigation into light-induced metal-support interaction. Fuel 333, 126496. Available at https://doi.org/10.1016/j.fuel.2022.126496.

Yan, X., Lyu, S., Xu, X.Q., Chen, W., Shang, P., Yang, Z., et al., 2022. Superhydrophilic 2D covalent organic frameworks as broadband absorbers for efficient solar steam generation. Angewandte Chemie - International Edition 61 (19). Available at https://doi.org/10.1002/anie.202201900.

Yan, W., Wei, J., Chen, T., Duan, L., Wang, L., Xue, X., et al., 2021. Superstretchable, thermostable and ultrahigh-loading lithium–sulfur batteries based on nanostructural gel cathodes and gel electrolytes. Nano Energy 80, 105510. Available at https://doi.org/10.1016/j.nanoen.2020.105510.

Yan, G., et al., 2023. Metal-free 2D/2D van der waals heterojunction based on covalent organic frameworks for highly efficient solar energy catalysis. Nano-Micro Letters 15 (1), 132. Available at https://doi.org/10.1007/s40820-023-01100-x.

Yao, J., Zheng, Z., Yang, G., 2018. Layered tin monoselenide as advanced photothermal conversion materials for efficient solar energy-driven water evaporation. Nanoscale 10 (6), 2876–2886. Available at https://doi.org/10.1039/C7NR09229F.

Yu, X., Fan, S., Zhu, B., El-Hout, S.I., Zhang, J., Chen, C., 2025. Recent progress on photothermal nanomaterials: design, mechanism, and applications. Green Energy and Environment 10 (7), 1377–1436. Available at https://doi.org/10.1016/j.gee.2024.09.002.

Yu, Q., Wang, Q., Feng, T., Wang, L., Fan, Z., 2023. A novel functionalized MoS_2-based coating for efficient solar desalination. Materials 16 (8), 3105. Available at https://doi.org/10.3390/ma16083105.

Zaed, M.A., Tan, K.H., Saidur, R., Rezakazemi, M., Pandey, A.K., 2025. Crafting 2D MXene composite from recycled materials for solar-driven desalination. Separation and Purification Technology 376, 134110. Available at https://doi.org/10.1016/j.seppur.2025.134110.

Zhang, R., Guo, J., Zhou, P., Chai, Y., Xiao, X., Xu, J., 2025a. A natural choice towards water desalination: improving solar-powered wood interfacial evaporation by lignin-based carbon quantum dots. Industrial Crops and Products 224. Available at https://doi.org/10.1016/j.indcrop.2024.120289.

Zhang, K., Jin, B., Park, C., Cho, Y., Song, X., Shi, X., et al., 2019. Black phosphorene as a hole extraction layer boosting solar water splitting of oxygen evolution catalysts. Nature Communications 10 (1). Available at https://doi.org/10.1038/s41467-019-10034-1.

Zhang, C., Liang, H.Q., Xu, Z.K., Wang, Z., 2019. Harnessing solar-driven photothermal effect toward the water–energy nexus. Advanced Science 6 (18). Available at https://doi.org/10.1002/advs.201900883.

Zhang, X., Li, C., Dai, L., Si, C., Shen, Z., Qiu, Z., et al., 2023. Graphite carbon nitride photocatalytic materials: a roadmap to modification for current and future water purification. Journal of Environmental Chemical Engineering 11 (5), 110869. Available at https://doi.org/10.1016/j.jece.2023.110869.

Zhang, J., Zhang, X., Mai, B., Fu, N., Wang, S., Shang, B., et al., 2025. Mussel-inspired MXene-functionalized sponge for solar-driven water purification: Photothermal-driven crude oil absorption, solar steam generation, and photothermal sterilization. Chemical Engineering Journal 514, 163237. Available at https://doi.org/10.1016/j.cej.2025.163237.

Zhang, X.S., Mao, S., Wang, J., Onggowarsito, C., Feng, A., Han, R., et al., 2024. Boron nanosheets boosting solar thermal water evaporation. Nanoscale 16 (9), 4628–4636. Available at https://doi.org/10.1039/d3nr06146a.

Zhang, Y., Wang, H., Zhang, J., 2025c. Application of graphene-based solar driven interfacial evaporation-coupled photocatalysis in water treatment. Catalysts 15 (4), 336.

Zhao, F., Guo, Y., Zhou, X., Shi, W., Yu, G., 2020. Materials for solar-powered water evaporation. Nature Reviews Materials 5 (5), 388–401. Available at https://doi.org/10.1038/s41578-020-0182-4.

Zhao, J., Yang, Y., Yang, C., Tian, Y., Han, Y., Liu, J., et al., 2018. A hydrophobic surface enabled salt-blocking 2D Ti_3C_2 MXene membrane for efficient and stable solar desalination. Journal of Materials Chemistry A 6 (33), 16196–16204. Available at https://doi.org/10.1039/c8ta05569f.

Zheng, H., Fan, J., Chen, A., Li, X., Xie, X., Liu, Y., et al., 2024. Enhancing solar-driven water purification by multiscale biomimetic evaporators featuring lamellar MoS_2 /GO heterojunctions. ACS Nano 18 (4), 3115–3124. Available at https://doi.org/10.1021/acsnano.3c08648.

Zheng, N.C., Yuan, C.Z., Chen, Y.C., Mangindaan, D., Lee, M., Chien, H.W., 2025. Environmentally friendly chitosan/spent coffee ground composites with continuous hydrophobically gradient of porous structure for enhanced photothermal water evaporation. International Journal of Biological Macromolecules 307. Available at https://doi.org/10.1016/j.ijbiomac.2025.141821.

CHAPTER

9

Harnessing the power of 2D nanomaterials for photocatalytic water treatment

Mario J. Muñoz-Batista[1], *Manuel Peñas-Garzón*[2], *Daily Rodriguez-Padrón*[3], *and Rafael Rodríguez-Solís*[1]

[1]Department of Chemical Engineering, University of Granada, Granada, Spain [2]Department of Chemical Engineering, University of Málaga, Málagg, Spain [3]Section of Chemistry for the Technology (ChemTech), Department of Industrial Engineering, University of Padov, Padova (PD), Italy

9.1 Introduction

9.1.1 Growing concern about the presence of contaminants in surface waters

Water quality is a critical issue regarding the true availability of water resources. Depending on the intended use, water must meet specific requirements, which often necessitate treatment or conditioning. In particular, the presence of contaminants in water bodies, such as surface waters, has been shown to limit their practical accessibility (Rathi et al., 2021). Water contamination is generally defined as the harmful alteration of its physical, chemical, or biological characteristics due to the introduction of matter or energy, which affects its suitability for further use. The impacts of contaminated water (as well as air) on human health and ecosystems have been a long-standing concern. To illustrate this point, the World Economic Forum has identified the concept of 'pollution' as one of the top ten global risks to social stability and economic growth over the past decade (World Economic Forum, The Global Risks Report 2025 20th Edition, 2025). The presence of these pollutants, together with other factors such as escalating demand and distribution challenges, is increasing the pressure on water resources.

A significant challenge for current and future generations is the monitoring of potential aqueous contaminants, considering the continuous synthesis of new chemicals, as well as the modification of industrial processes. In this sense, the development of analytical

Two-Dimensional Nanomaterials for Water Purification
DOI: https://doi.org/10 1016/B978-0-443-44530-9.00005-5

techniques has fostered the identification and quantification of chemical compounds discharged from municipal and industrial wastewater streams into aquatic environments. As a consequence, a wide range of water contaminants has been detected in surface waters, even at trace levels (from $ng{\cdot}L^{-1}$ to $\mu g{\cdot}L^{-1}$) (Jolaosho et al., 2025; Boro et al., 2025; Yesilay et al., 2023; Rizzo et al., 2019). Among the vast number of likely aqueous contaminants, they can be classified into four main families:

1. contaminants of emerging concern (including pharmaceuticals such as diclofenac and carbamazepine, antibiotics like amoxicillin and sulfamethoxazole, personal care products such as parabens and microplastics [mainly polyethylene and polypropylene]);
2. endocrine disrupting compounds (i.e., heavy metals including Pb^{2+}, Cr^{6+}, and Hg^{2+}; industrial additives, such as per- and polyfluoroalkyl substances (PFAS); pesticides as atrazine and imidacloprid; and steroid hormones as estrone or 17b-estradiol);
3. pathogenic germs (e.g., *Escherichia coli, Staphylococcus aureus, Bacillus subtilis*); and
4. cyanotoxins (e.g., *Microcystins, Cylindrospermopsin*).

Despite the low concentrations detected for some of these compounds, it is important to emphasize that both the dose and the exposure period must be taken into account when evaluating the risk assessment related to aqueous contaminants. Indeed, it has been reported that the effect of long-term exposure to pollutants at trace levels can be more severe than that of short-term exposure at higher doses (Vasantha Raman et al., 2024; Fent et al., 2006). This is a matter of major interest regarding contaminants of emerging concern (whose potentially adverse impacts are not properly investigated due to their early identification), bacteria (and the related antibiotic resistance effect), or the endocrine-disrupting compounds, considered to be able to cause alterations in the metabolisms of living beings in aquatic ecosystems (Metcalfe et al., 2022), thus affecting the trophic chain. The unequivocal exposure of human beings to these contaminants, promoted by the increasing intake of pharmaceuticals, personal care products, and food packaging, among other sources, is well-documented, with some compounds detected in human tissues, plasma, and urine (Alaoui et al., 2024; Wilkinson et al., 2016).

Therefore, there is a clear need for research into strategies and technologies capable of treating surface waters contaminated with these substances, particularly those that are recalcitrant to conventional wastewater treatment processes. The detection of pollutants in the effluents of wastewater treatment plants (WWTP) (Ianes et al., 2024) confirms that most of these emerging contaminants are not completely removed being registered conversions no longer than 60%, on average, on the overall emerging contaminant load from the influent wastewater (Golovko et al., 2021). In this sense, regulation and mandatory frameworks are key aspects of environmental protection. For example, the recent Directive 2024/3019 of the European Union concerning urban wastewater treatment (The European Parliament and the Council of the European Union, 2024) requires member states to implement quaternary treatment processes (defined as *"a process which reduces a broad spectrum of micropollutants in urban wastewater"*) to treat at least the 20% of discharges from WWTPs (with a load ≥ 150,000 population equivalent) by the end of 2033 (being incremented up to 60% and 100% by 2039 and 2045, respectively). This Directive also stipulates specific requirements for samples, including those from the series of samples taken throughout the year, or the minimum removal of 80% (in relation to the load of the influent) for certain substances such as amisulpride, carbamazepine, citalopram, diclofenac, metoprolol, or candesartan, to mention a few. The monitoring of these compounds, mainly used as pharmaceuticals, is experiencing an increasing interest due to their widespread use in humans and veterinary activities, subsequently entering the aquatic ecosystem.

9.1.2 Photocatalysis as a key advanced oxidation process

Conversion methods transform contaminants into less hazardous or even harmless species, in contrast to nonconverting technologies (such as adsorption or filtration). Among the conversion approaches, advanced oxidation processes (abbreviated as AOPs) have attracted particular interest in the abatement of recalcitrant pollutants. The AOPs are characterized by the in situ generation of reactive oxygen species (ROS), including hydroxyl and hydroperoxyl radicals ($HO^{\bullet}$ and $HOO^{\bullet}$, respectively) and superoxide anion radical ($O_2^{\bullet-}$), with a strong redox capacity (Dong et al., 2022). Fenton-like processes, as well as ozonation, hydrogen peroxide-based treatments, thermal-activated persulfate, electrochemical oxidation, ultrasounds, nonthermal plasma, and radiation-driven technologies are examples of AOPs (Hübner et al., 2024). The combination of some of these approaches, together with the process intensification, appears as a promising solution for the quaternary abatement of aqueous contaminants. Among AOPs, heterogeneous photocatalysis (or just referred to as photocatalysis), within the radiation-based strategies, is highlighted by the possibility of operating under ambient pressure and temperature, so far considered an environmentally friendly technology. Essentially, photocatalysis is defined as the conversion of light into chemical energy, only requiring a solid semiconductor (photocatalyst) and a proper irradiation source. In an aqueous environment, the photocatalytic generation of ROS is mediated by redox reactions, as schematized in Fig. 9.1.

Following the schematized mechanism depicted in Fig. 9.1, the first step corresponds to the absorption of the incident light upon the irradiation of the semiconductor, driving the excitation of the electrons (e^-) located in the valence band (VB). If the energy is equal to or higher than the energy difference with respect to the conduction band (CB), the electron

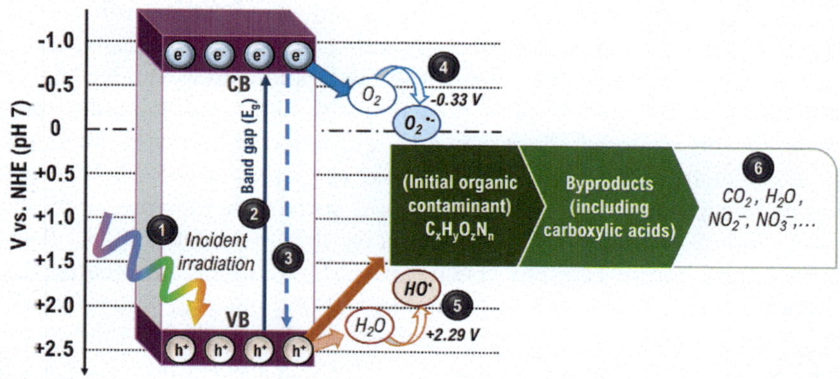

FIGURE 9.1 Scheme of the ROS generation and their action upon irradiation of a photocatalyst in aqueous solution. The diagram illustrates the formation of key ROS such as •OH, $•O_2^-$, $^1O_2$, and h^+, highlighting their interaction pathways with pollutants and their overall contribution to photocatalytic degradation processes.

successfully promotes to the CB, leaving a hole (h^+) in the VB, as represented by Step 2 of Fig. 9.1. This energy difference between CB and VB corresponds to the bandgap value (E_g) of the semiconductor, and it is considered a fundamental parameter when comparing the efficiency of different photocatalysts, as below discussed. Once the proper separation of charges (e^-/h^+) takes place, these photogenerated species might migrate to the surface of the photocatalyst, where they can interact with the adsorbed species. Conversely, the charge carriers can undergo a process of recombination (Step 3, Fig. 9.1), thereby releasing in the form of heat the energy previously absorbed as irradiation and limiting the effectiveness of the process. As indicated before, the separated electrons and holes can react with the adsorbed species at the photocatalyst surface, leading to the generation of ROS through redox processes (Li et al., 2016b). This is the case of the $O_2^{\bullet-}$, generated after the reduction of dissolved oxygen ($O_2 + e^- \rightarrow O_2^{\bullet-}$; $E_0(O_2/O_2^{\bullet-})$=-0.33 V vs. NHE), and $HO^\bullet$, formed through the oxidation of water ($H_2O + h^+ \rightarrow HO^\bullet + H^+$; $E_0(HO^\bullet/H_2O)$=+2.29 V vs. NHE), as represented in the Steps 4 and 5, respectively, of Fig. 9.1. These ROS, together with the direct interaction with h^+, can react with the pollutant, leading to its conversion to other intermediates compounds and finally to innocuous species, as schematized in step 6 (Fig. 9.1). The complete removal of the pollutant, in terms of its total conversion into CO_2 and H_2O, receives the name of mineralization. In fact, this is considered the most desirable outcome, especially with regard to the related adverse impact of the oxidation intermediate compounds (in some cases with more harmful effects than the original contaminant) (Singh et al., 2024).

Despite the increasing attention paid to photocatalysis following the contributions of Fujishima and Honda on photocatalytic water splitting in 1972 (Fujishima and Honda, 1972), the origins of photocatalysis date back to the early 20th century. Indeed, there is evidence that the terms "*photokatalytisch*"(photocatalytic) and "*photokatalyse*" (photocatalysis) already appeared in the 1910s in Prof. J. Plotnikow's book "*Photochemische Versuchstechnik*" (Photochemical experimental technique) (A, 1913). Since then, an extraordinary number of semiconductors (including metal oxides, metal sulfides, nitrides, heterojunctions between two or more compounds, metal-organic frameworks) have been examined as photocatalysts for the elimination of contaminants in aqueous solutions. Among them, titanium dioxide (TiO_2) is highlighted by its good activity and photochemical stability (Pelaez et al., 2012). Characterized by a bandgap of about 3.2 eV, TiO_2 can only harvest light from the UV region of the spectrum, restricting the suitable irradiation sources for these materials, and implying a low utilization of direct sunlight because of the negligible activity under the visible light fraction. Another important drawback is related to the high recombination rate of charge carriers shown by this material, decreasing its efficiency (Khan and Shah, 2023). As a consequence, diverse strategies have been followed to increase photocatalytic performance, including doping, metal deposition, dye-sensitization, surface engineering, or the combination with other compounds to prepare heterojunctions and heterostructures. Besides, other semiconductors different from TiO_2 have been explored to demonstrate a better development of the photocatalytic yield concerning the abatement of pollutants in water. Considering that the morphology and dimensions of the photocatalysts can have a decisive influence over their properties and activity, they can be classified as 0D (e.g., carbon dots; Yu et al., 2024b), 1D (e.g., nanotubes; Shen et al., 2025), 2D (e.g., graphitic carbon nitride sheets; Peñas-Garzón et al., 2024), and 3D (e.g., metal-organic frameworks, MOFs; Wang et al., 2024). Considering the focus of the present chapter, the following sections will be centered on the description of the properties, strengths, and drawbacks of 2D materials, and their application in the context of photocatalytic treatment of aqueous contaminants.

9.1.3 The relevance of 2D photocatalytic materials for water purification

Two-dimensional photocatalysts develop unique physicochemical properties, including high specific surface area, better availability of active sites over the surface of the materials, and strong quantum effects. These properties can be related to the improvement in the overall photocatalytic process, from enhancing the contaminant adsorption step to the light-harvesting, the migration rate of charge carriers (e^-/h^+), and the effective interfacial reactions (Wang et al., 2023a). On the other hand, these materials also present some limitations such as those related to their stability during the reaction and their preparation at a higher scale (Zhao et al., 2020b).

The isolation of graphitic carbon sheets (graphene) in 2004 (Novoselov et al., 2004) and the identification of the outstanding electronic properties attracted significant attention due to their anisotropic qualities. Concerning the use of these materials in water-related applications, graphene oxide or reduced graphene oxide (GO and rGO, respectively) containing oxygen-functional groups (-OH, -C-O-C-, etc.), provide them with a hydrophilic character, thus enhancing the interaction with water molecules. The high conductivity ability of these 2D materials makes them appropriate materials for the preparation of heterojunctions with other semiconductors (Ramalingam et al., 2022), facilitating the separation of charge carriers and preventing the recombination process. However, examples of the use of GO as a sole photocatalyst have also been reported, as in the case of the abatement of methyl 3-aminobenzoate and methyl 4-hydroxybenzoate (used as industrial additives) in aqueous solutions under simulated sunlight (Wang et al., 2023b).

Other 2D materials besides graphene have also been explored. This is the case of graphitic carbon nitride (g-C_3N_4), which exhibits a structural similarity to graphene, the nitrogen atoms being arranged in a triazine-based pattern alongside the carbon atoms. This arrangement affords the material semiconductor properties, rendering it active under visible light. The use of this material has been thoroughly examined in the last years, demonstrating its outstanding performance. For example, in the treatment of real wastewater effluents, leading to average conversions of micropollutants (including compounds of different natures as the pharmaceuticals diclofenac, carbamazepine, and sulfamethoxazole, the pesticide mecoprop, and the industrial preservative benzotriazole) higher than 80%, as well as the reduction in the activity of antibiotic-resistant bacteria and genes after few minutes under simulated solar irradiation (Zhong et al., 2022). The nitrogen-enriched carbon nitride, g-C_3N_5, displays a reduced bandgap value (around 1.8 eV instead of 2.7 eV for g-C_3N_4) as a result of the extended π conjugated electrons network due to the overlapping of azo linkage (Kumar et al., 2019). Therefore this material can benefit from a larger range of wavelengths to be activated. At this point, it is also important to remark that the materials characterized by low bandgap values are not directly linked to improved photocatalytic performance, due to the narrower energy difference between the conduction and valence bands of those semiconductors, which can also facilitate the recombination of charge carriers, thus compromising the efficacy of the photocatalyst.

Other 2D materials that are gaining interest nowadays are those composed of layered transition metal dichalcogenides (TMDs), layers of metal carbides (known as MXenes), and two-dimensional black phosphorus (BP), among the most representative materials. With respect to the former ones, TMDs are characterized by a sandwich structure of X-M-X layers (being X elements of the group VIA as S, Se, and Te; whereas M are transition metal elements from the groups IVB-VIII, as Mo or W) (Yang et al., 2023a). MoS_2 and WS_2 are examples of TMDs, being used for the preparation of heterojunctions to improve the interfacial transference of electrons, as exhibited in the degradation of the antibiotic ciprofloxacin using the MoS_2/Bi_2WO_6 system (Chen et al., 2025). A certain analogy can be drawn for MXenes (composed of noble metal-free transition metal carbides, such as Ti_3C_2), which are used as co-catalysts owing to their appropriate electron conductivity (Liu et al., 2023). In addition, Ti_3C_2 has been proven to facilitate the separation of charge carriers through the formation of energy potentials (Schottky barriers), thus blocking the backward movement of electrons (Liu et al., 2023). The monoelemental BP can exhibit even superior carrier mobility than TMDs, but its utilization as the sole photocatalyst is prevented due to some inherent limitations, such as low stability and photocatalytic activity (Hu et al., 2025). Therefore the combination of BP with other semiconductors, as described in more detail in Section 9.4.4 within this chapter, arises as a potential solution to improve the electron mobility properties through the formation of a heterojunction. To illustrate this, the combination of g-C_3N_4 and BP was successfully achieved via liquid-phase exfoliation, resulting in a 2D/2D heterostructure with an 8% BP and a superior photocatalytic performance compared to the bare components, as demonstrated by the complete elimination of the antibiotic indomethacin after 30 min under natural sunlight (He et al., 2022).

The previous examples were included to have a preliminary introduction concerning the possibilities of 2D materials for the photocatalytic elimination of aqueous contaminants. However, the research about these materials is not only circumscribed to water purification, as demonstrated by the application of some of these 2D photocatalysts in the reduction of CO_2 (Cao et al., 2024), the fixation of N_2 (Qian et al., 2021), or the evolution of hydrogen from aqueous solutions (Ganguly et al., 2019). Considering the context and scope of this book chapter, the following sections are related to the assessment of the aforementioned 2D photocatalysts and other similar materials for the abatement of pollutants in aqueous matrices. Indeed, the following section aims to provide a comprehensive description of the steps involved in the photocatalytic mechanisms (Section 9.2), followed by the description of some of the most recent synthesis methodologies (including top-down and bottom-up approaches) for their preparation (Section 9.3), and before understanding the relationship between the modification of photocatalysts and their performance (Section 9.4). The chapter is completed with the concluding remarks and the future perspectives concerning the potential use of 2D semiconductors as a promising approach for the purification of surface waters and general environmental applications.

9.2 Understanding the photocatalytic mechanisms of 2D nanomaterials

The electronic and optical behavior of 2D nanomaterials is dictated by their band structure, which controls how electrons move and results from the regular arrangement of atoms. When a material is thinned from bulk to a 2D form, the periodicity perpendicular to the plane is lost, altering the band structure. This change gives rise to unique properties, such as graphene's exceptional electrical conductivity and the visible-light photoluminescence of a MoS_2 monolayer. In 2D materials, confinement affects electron–hole interactions by reducing dielectric screening, which strengthens Coulomb forces and stabilizes excitons, shifting their absorption and emission compared with bulk. The number of layers can further tune these energy levels, with bilayers absorbing or emitting lower-energy light than monolayers, although this can also modify band structure and fluorescence. Additionally, the high surface-to-volume ratio of 2D materials provides abundant sites for adsorption and reaction, enhancing contact with pollutants and improving photocatalytic efficiency (Yang et al., 2024).

Two-dimensional nanomaterials, including g-C_3N_4, transition metal dichalcogenides, layered double hydroxides, and MXenes, combine high surface areas, adjustable band structures, and reactive edge sites, enabling efficient interaction with and transformation of diverse contaminants. In aqueous environments, conduction band electrons can reduce dissolved oxygen to form superoxide radicals ($^{\bullet}O_2^-$), while valence band holes can oxidize water or hydroxide ions to yield hydroxyl radicals ($^{\bullet}OH$) (Xie et al., 2024). Both ROS species are highly reactive, capable of mineralizing a wide range of organic contaminants into harmless end products such as CO_2 and H_2O. Additional oxidative species, such as hydrogen peroxide (H_2O_2), can form through secondary reactions of $\bullet O_2^-$ and $^{\bullet}OH$, further enhancing pollutants degradation (Fatima et al., 2022).

The intrinsic properties of 2D nanomaterials play a central role in defining their photocatalytic mechanisms. Reduced layer thickness shortens charge migration distances, minimizing electron–hole recombination and enhancing overall efficiency. The controlled introduction of defects and vacancies, such as nitrogen vacancies in g-C_3N_4 or sulfur vacancies in MoS_2, can create mid-gap states, improving visible-light absorption and facilitating charge transfer (Zeng et al., 2021). Heterojunction formation by coupling 2D semiconductors with complementary band alignments, including Type II, Z-scheme, or Schottky junctions, promotes directional charge flow and prolongs carrier lifetimes (Zhang et al., 2017). Surface functionalization with oxygen-containing groups or metal nanoparticles could further enhance pollutant adsorption and catalytic turnover. Specific material classes exhibit characteristic mechanistic features, for instance, g-C_3N_4, with its suitable bandgap (~2.7 eV) for visible-light photocatalysis, benefits from defect engineering and heterojunction formation to increase ROS generation; transition metal dichalcogenides (e.g., MoS_2, WS_2) show high edge activity, where sulfur vacancies facilitate electron transfer to adsorbed oxygen species (Liu et al., 2021); layered double hydroxides (LDHs) utilize interlayer anions as electron mediators and metal cations to enhance redox versatility (Laipan et al., 2020); and MXenes, with their exceptional electrical conductivity, enable rapid electron extraction, reducing recombination losses and boosting ROS production (Wang et al., 2022).

2D nanomaterials facilitate multiple photocatalytic pathways, with the efficiency and dominant mechanism strongly influenced by the chemical structure, redox potential, and adsorption behavior of each contaminant. Organic dyes, including cationic and anionic species such as methylene blue, rhodamine B, and methyl orange, undergo rapid degradation through oxidative cleavage of chromophoric structures (Lachheb et al., 2002; Yadav et al., 2023). Upon light irradiation, photogenerated holes (h^+) directly oxidize the dye molecules, while ROS, especially hydroxyl radicals ($\bullet OH$), further attack the conjugated π

systems, leading to decolorization and eventual mineralization to CO_2 and H_2O. The large surface area and π–π interactions between the dye molecules and the 2D catalyst surface enhance adsorption, further increasing degradation rates (Xie et al., 2024).

Zhang and coworkers compared one-dimensional CNT–TiO_2 composites with two-dimensional graphene–TiO_2, showing that the sheet–sheet contact in the graphene system provided superior interfacial coupling (Zhang et al., 2011). Similar comparative work by Yang et al. demonstrated that TiO_2 composites with graphene, CNTs, or fullerenes followed the same oxidative pathway, but differences in interfacial charge transfer influenced efficiency (Yang et al., 2013). Extending this idea, graphene combined with different TiO_2 morphologies (nanosheets, nanotubes, nanoparticles) revealed that the 2D/2D nanosheet–rGO hybrid achieved the highest activity for degrading chlorophenols and dyes due to stronger electronic coupling and faster interfacial charge transfer (Sun et al., 2013). Comparable trends were reported for CdS systems, where rGO–CdS nanosheet heterostructures outperformed nanorod and nanoparticle counterparts, again emphasizing morphology-dependent charge transport (Bera et al., 2015).

To enhance oxidation further, external oxidants such as hydrogen peroxide have been integrated with photocatalysis. In photo-Fenton systems, Fe(II) reacts with H_2O_2 to yield hydroxyl radicals, which aggressively attack pollutants. Yoon et al. exploited this by designing an Fe-doped g-C_3N_4/WO_3 hybrid, which displayed a dramatic jump in efficiency under visible light compared with pristine g-C_3N_4. Electron Paramagnetic Resonance (EPR) analysis confirmed enhanced radical production, with Fe(II) regenerated from Fe(III) under illumination, sustaining the catalytic cycle (Bera et al., 2015). Similarly, Lin et al. incorporated ferrocene into g-C_3N_4 (Fc–C_3N_4), creating strong π-conjugation that improved charge separation and stabilized Fe species in their active state. This facilitated continuous H_2O_2 activation and greatly enhanced dye degradation performance compared with unmodified g-C_3N_4 (Lin and Lin, 2017).

Pharmaceutical and personal care products (PPCPs), including antibiotics, analgesics, hormones, and UV filters, are similarly susceptible to ROS-mediated hydroxylation, dealkylation, or aromatic ring opening. For instance, fluoroquinolone antibiotics are susceptible to •OH attack at electron-rich moieties such as aromatic rings or heterocyclic groups (Li et al., 2025; Wei et al., 2025), whereas endocrine disruptors like bisphenol A are more readily oxidized through direct interaction with photogenerated holes (Mestre and Carvalho, 2019; Fu et al., 2022). Defect-rich 2D materials not only enhance visible-light absorption, promoting efficient PPCP degradation under solar irradiation, but also improve the adsorption of target molecules on the catalyst surface. This dual effect increases the likelihood of pollutant–catalyst interactions, thereby boosting photocatalytic efficiency. In the case of g-C_3N_4, the study by Wei Yan et al. showed that doping treatments did not significantly alter either the adsorption of antibiotics or the BET surface area, indicating that the increased activity in their system was primarily due to changes in electronic properties rather than surface area effects. Specifically, doping with Na or K, along with O incorporation into the triazine units, extends light absorption, improves charge separation, and prolongs carrier lifetimes, which enhances the generation of ROS ($^{\bullet}O_2^-$ and $^{\bullet}OH$) responsible for antibiotic degradation. In contrast, Ca or Mg doping introduces defects that act as recombination centers, limiting the enhancement of photocatalytic activity (Fig. 9.2A) (Yan et al., 2019).

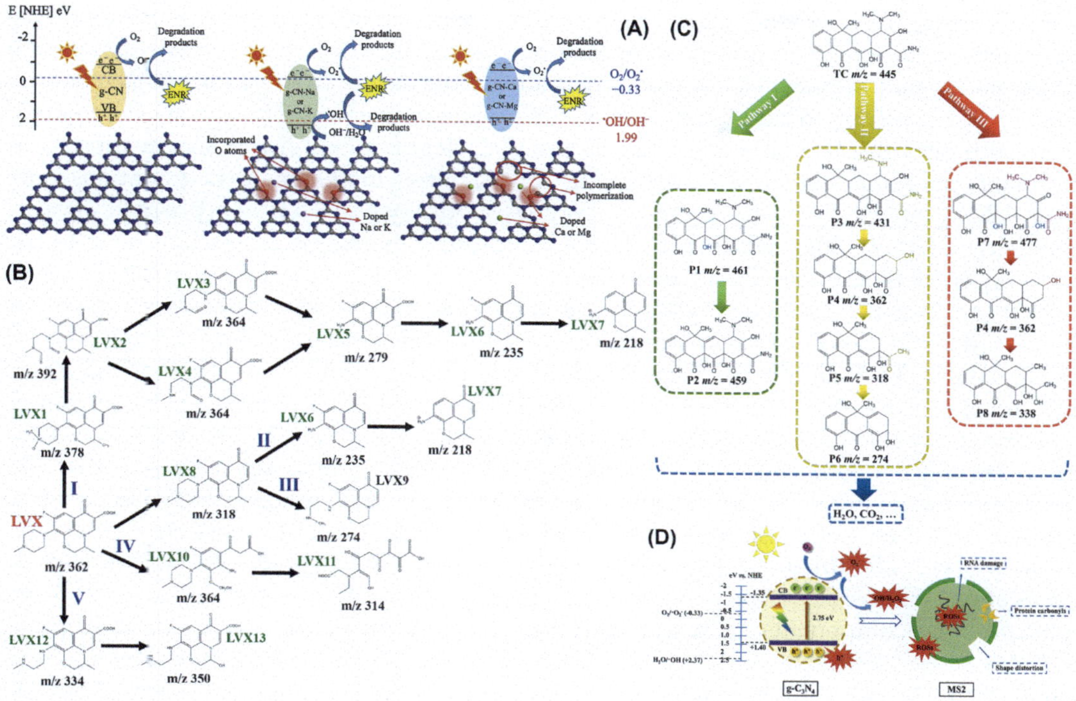

FIGURE 9.2 Proposed schematic degradation mechanism of (A) enrofloxacin, (B) levofloxacin, (C) tetracycline, and (D) MS2. *(A) Reproduced with permission from Yan, W., Yan, L., Jing, C., 2019. Impact of doped metals on urea-derived g-*C_3N_4 *for photocatalytic degradation of antibiotics: structure, photoactivity and degradation mechanisms. Applied Catalysis B: Environmental 244, 475485. https://doi.org/10.1016/j.apcatb.2018.11.0690; (B) Reproduced with permission from Lei, J., Chen, B., Zhou, L., Ding, N., Cai, Z., Wang, L., et al., 2020. Efficient degradation of antibiotics in different water matrices through the photocatalysis of inverse opal K-g-*C_3N_4*: insights into mechanism and assessment of antibacterial activity. Chemical Engineering Journal 400, 125902. https://doi.org/10.1016/j.cej.2020.125902; (C) Reproduced with permission from Duan, R., Ma, S., Xu, S., Wang, B., He, M., Li, G., et al., 2022. Soybean straw biochar activating peroxydisulfate to simultaneously eliminate tetracycline and tetracycline resistance bacteria: Insights on the mechanism. Water Research 218, 118489. https://doi.org/10.1016/j.watres.2022.118489; (D) Reproduced with permission from Li, Y., Zhang, C., Shuai, D., Naraginti, S., Wang, D., Zhang, W., 2016. Visible-light-driven photocatalytic inactivation of MS2 by metal-free g-*C_3N_4*: virucidal performance and mechanism. Water Research 106 249258. https://doi.org/10.1016/j.watres.2016.10.009.*

A related study by Juying Lei and coworkers showed that inverse opal potassium-doped carbon nitride (IO K–CN) could efficiently remove fluoroquinolone antibiotics such as levofloxacin (LVX) and norfloxacin (NOR) in diverse water matrices. The combination of a narrowed bandgap, achieved through potassium doping, and the high light-harvesting capacity of the inverse opal structure significantly enhanced photocatalytic activity compared to bulk or undoped counterparts. Degradation pathway analysis, supported by electron spin resonance and liquid chromatography–mass spectrometry, revealed multiple reaction routes involving hydroxylation, dealkylation, decarboxylation, and defluorination, leading to extensive breakdown of LVX into low-toxicity intermediates (Fig. 9.2B; Lei et al., 2020).

Furthermore, in their study, Ran Duan et al. investigated soybean straw biochars (SSBs) fabricated at different pyrolysis temperatures (600°C–1000°C) as peroxydisulfate (PS) activators for tetracycline (TC) degradation and disinfection of TC-resistant *Escherichia coli* (*E. coli*). The purification efficiency of SSBs/PS systems increased with pyrolysis temperature, with SSB1000/PS showing the highest TC degradation, *E. coli* inactivation, and simultaneous elimination of coexisting TC and bacteria. In the main degradation route (Fig. 9.2C), tetracycline (m/z 445) undergoes hydroxylation on its phenolic group, producing P1 (m/z 461), which then forms P2 through hydrogen abstraction. Due to the relatively weak N–C bond, several demethylated intermediates typically appear. In a second route (Fig. 9.2C), TC is N-demethylated to yield P3 (m/z 431), which subsequently generates P4 (via deamidation), P5 (through ring opening and dealcoholization), and P6 (by deacetylation). In a third route (Fig. 9.2C), hydroxylation produces P7 (m/z 477), which can undergo sequential demethylation, deamidation, and hydroxylation, leading again to P4. Further radical attack opens the aromatic ring, giving P8 (m/z 338). These intermediates are ultimately mineralized to CO_2 and H_2O. Mechanistic analyses revealed that TC degradation was primarily driven by surface-bound radicals, while bactericidal activity resulted from free sulfate radicals ($SO_4^{\bullet -}$) in bulk solution. The C=O functional groups on SSB1000 were identified as key active sites for both TC degradation and microbial inactivation, with reactive oxygen species damaging cell membranes and causing leakage of intracellular components. SEM and TEM imaging confirmed severe structural disruption of *E. coli* after treatment, and the correlation between C=O content and degradation/disinfection efficiency highlighted the crucial role of these groups. Moreover, SSB1000 exhibited the largest mesopore volume, highest C=O content, lowest interfacial charge transfer resistance, and strongest electron-donating capacity, explaining its superior catalytic performance (Duan et al., 2022).

Pesticides and herbicides, including organophosphates like chlorpyrifos and triazines such as atrazine, persist in agricultural runoff but can be degraded via photocatalytic cleavage of P–O, C–N, or C–Cl bonds. In these systems, heterojunction engineering within 2D materials improves charge separation, prolonging carrier lifetimes and enhancing the oxidation potential required to break these robust bonds (Pathania and Thakur, 2019). Furthermore, inorganic pollutants can undergo oxidative or reductive photocatalysis, transforming them into less harmful or even harmless forms. While heavy metals are not degraded, photocatalysis can transform toxic metal ions into less soluble or less bioavailable forms.

Hexavalent chromium (Cr^{6+}), for example, can be photoreduced to trivalent chromium (Cr^{3+}), and toxic Hg^{2+} can be reduced to elemental mercury (Hg°), facilitating removal via precipitation or adsorption (Liu et al., 2020b; Subbaiah et al., 2021; Liu et al., 2020a). Photocatalytic systems could also enable diverse redox transformations, such as the oxidation of NH_3 to N_2 and H_2S to sulfur species. For example, In_2S_3/CuS nanosheets have been shown to convert toxic H_2S into thiosulfate with concurrent H_2 evolution (Prakash et al., 2018). Exfoliated g-C_3N_4 exhibits enhanced activity for ammonia removal compared to bulk g-C_3N_4, since its band structure covers the redox window of NH_3 oxidation (Prakash et al., 2018). Similarly, flower-like MoS_2 coupled with polyaniline displays high capacity for Cr(VI) removal, explained by photocatalytic reduction followed by complexation with polyaniline (Gao et al., 2016). Phosphorus-doped porous g-C_3N_4

nanosheets have also enabled simultaneous 2,4-dichlorophenol oxidation and Cr(VI) reduction, with activity enhanced at low pH and in the presence of dissolved oxygen, maintaining stable performance over repeated cycles (Deng et al., 2017).

Moreover, pathogenic microorganisms, including bacteria, viruses, and algae, are inactivated through ROS-mediated oxidative stress, which damages cell walls, membrane proteins, and nucleic acids. High edge site density and defect engineering in 2D nanomaterials accelerate ROS generation, resulting in irreversible microbial cell death, a strategy particularly promising for decentralized disinfection without chemical additives (Fan et al., 2021; Zhao et al., 2020a). For example, Jia and co-workers developed TiO_2–Bi_2WO_6 nanosheets (TBWO) via a two-step hydrothermal synthesis. The Bi_2WO_6 component extends light absorption into the visible spectrum, allowing the composite to disrupt and kill *E. coli* through visible-light-driven photocatalysis (Jia et al., 2016). Similarly, Liu et al. engineered few-layer vertically aligned MoS_2 (FLV-MoS_2) films capable of harvesting a broad range of visible light, representing roughly half of solar energy, for water disinfection. By reducing the lateral domain size, the MoS_2 bandgap was increased from 1.3 to 1.5 eV, enhancing ROS production and achieving a >99% bacterial reduction within 20 minutes (Liu et al., 2016).

On the other hand, viruses, due to their nanoscale size, unique surface chemistry, and extremely low infectious doses (e.g., a single rotavirus particle can trigger infection), present even greater disinfection challenges than bacteria. Conventional methods, such as granular activated carbon adsorption, often fail to remove viruses effectively. In this sense, photocatalysis offers an alternative; for instance, Li et al. showed that g-C_3N_4 could deform viral morphology, oxidize capsid proteins, and damage RNA, leading to complete viral inactivation (Fig. 9.2D) (Li et al., 2016c; Zhang et al., 2018). Comparisons with other visible-light photocatalysts (TiO_2, Bi_2WO_6, Ag@AgCl) revealed that surface plasmon effects and Ag^+ release could further improve antiviral performance (Li et al., 2016c; Kumar et al., 2013; Zhang et al., 2012; Wang et al., 2015; Xia et al., 2016; Zhang et al., 2015a). Building on photocatalytic reduction strategies, Akhavan et al. incorporated graphene oxide (GO) nanosheets into WO_3 films, producing rGO–WO_3 composites that reduced virus levels from 2 x 10^6 PFU mL^{-1} to below detectable limits (<5 PFU mL^{-1}) in 3 hours of visible-light irradiation (Zhang et al., 2018; Akhavan et al., 2012).

In all cases, contaminant transformation is primarily driven by the interplay of oxidation pathways, governed by photogenerated h^+ and •OH, and reduction pathways, mediated by conduction band electrons producing $•O_2^-$ or reducing metal ions. Which pathway prevails depends on the contaminant's molecular structure, adsorption affinity, and redox potential relative to the 2D photocatalyst's band edges. Operando techniques such as EPR could pinpoint the dominant ROS involved in each pollutant class, while in situ X-ray photoelectron spectroscopy (XPS) could enable real-time monitoring of bond cleavage and oxidation state changes (Barawi et al., 2024). Complementarily, density functional theory (DFT) simulations are increasingly used to predict adsorption geometries, reaction intermediates, and activation barriers on 2D surfaces.

Although 2D nanomaterials exhibit high activity toward diverse contaminants, real wastewater introduces competing species that may block reactive sites or quench ROS. Future research should therefore prioritize selective pollutant targeting via surface functionalization, design of hybrid photocatalysts with tailored band alignments, and integration into scalable water treatment systems operable under natural sunlight.

9.3 Advanced preparation strategies for 2D nanomaterial-based photocatalysts

The synthesis of 2D nanomaterials for photocatalytic water purification has advanced significantly, driven by the need to optimize structural, electronic, and surface properties for efficient light-induced reactions. Both top-down and bottom-up strategies have been employed, each offering different advantages in terms of tunability, scalability, and ease of integration into composite systems. Below, we summarize representative methods and advances for key families of 2D materials.

Graphene and its oxidized forms are among the most extensively studied 2D materials in photocatalysis (Li et al., 2016a; Cesano et al., 2021). GO is commonly synthesized by oxidative exfoliation of graphite, typically via the modified Hummers method, which introduces oxygen-containing functional groups (e.g., hydroxyl, carboxyl, epoxy) that improve water dispersibility and facilitate integration with semiconductors (Saikam et al., 2023). Subsequent chemical or thermal reduction yields reduced graphene oxide (rGO), which partially restores the conjugated π-system and enhances electrical conductivity (Das et al., 2021; Shabil Sha et al., 2024). Liquid-phase exfoliation in organic solvents or surfactant-containing aqueous media is another scalable approach to produce few-layer graphene (Güler et al., 2021).

g-C_3N_4 is generally synthesized via thermal polymerization of nitrogen-rich precursors such as melamine, urea, or dicyandiamide. This method is straightforward, low-cost, and reproducible, making it particularly appealing for large-scale implementation (Solís et al., 2022). Recent developments have focused on enhancing the morphology and surface area of g-C_3N_4. Strategies include exfoliation via acid treatment or ultrasonication, salt-templated synthesis to generate porous structures, and doping with nonmetals or metals to adjust the electronic structure. The resulting materials exhibit improved light harvesting, higher specific surface area, and more efficient separation of photogenerated carriers (Zhu et al., 2021).

MXenes, such as $Ti_3C_2T_x$, have emerged as promising 2D materials with metallic conductivity and tunable surface terminations (–OH, –F, –O). They are produced by selectively etching the "A" layer (e.g., Al) from MAX phase precursors using fluorinated acid solutions (Murali et al., 2022). Recent advances have replaced hazardous HF with in situ etching methods using mixtures like LiF and HCl, offering safer and more controllable synthesis. After etching, mechanical agitation or intercalation steps are used to delaminate the layers into ultrathin nanosheets. MXenes are highly hydrophilic and offer abundant active sites, making them ideal as cocatalysts or conductive supports in composite photocatalysts. Their combination with visible-light semiconductors enhances charge separation and facilitates reactive species generation in water purification systems.

As aforementioned, transition metal dichalcogenides (TMDs) such as MoS_2 and WS_2 are layered materials with strong light absorption in the visible region and high surface activity, particularly at edge sites. Exfoliation techniques, either via ultrasonication in solvents or via lithium intercalation followed by aqueous exfoliation, allow for the production of few-layer nanosheets suitable for photocatalytic use (Yang et al., 2023a). Alternatively, hydrothermal and solvothermal syntheses enable the formation of nanostructured MoS_2 directly from molecular precursors, yielding morphologies such as nanoflowers or thin platelets (Yang et al., 2023a).

Other examples of emerging 2D materials include black phosphorus (BP), or phosphorene, which possesses a tunable direct bandgap and excellent carrier mobility, making it suitable for visible-light photocatalysis (Moradian et al., 2023; Yu et al., 2024a; Zheng et al., 2021). However, its high reactivity under ambient and aqueous conditions necessitates stabilization strategies during synthesis. Top-down methods such as liquid-phase exfoliation (LPE) in inert solvents (e.g., *N*-methyl-2-pyrrolidone or ionic liquids) and electrochemical exfoliation under anhydrous conditions are commonly employed to produce few-layer BP while minimizing oxidative degradation. Bottom-up techniques such as chemical vapor deposition (CVD) also allow the controlled growth of thin phosphorene layers, though typically with lower yield and limited scalability.

Two-dimensional covalent organic frameworks (COFs) and metal-organic frameworks (MOFs) offer molecularly tunable platforms with high porosity and structural regularity. COFs are generally synthesized via solvothermal reactions involving dynamic covalent chemistry (e.g., boronate ester or imine linkages) between organic monomers (Qing et al., 2025; Xiao et al., 2023; Wu et al., 2025). These reactions require elevated temperatures (typically 100–150 °C) and long reaction times to ensure crystallinity. However, novel synthetic alternatives such as microwave-assisted polymerization, interfacial synthesis, and mechanochemical methods (e.g., ball milling under ambient conditions) have recently gained traction as greener and more scalable routes (Wu et al., 2025; Trentin et al., 2025). After synthesis, liquid-phase exfoliation in polar solvents is often used to isolate ultrathin COF layers with enhanced photocatalytic activity.

Layered MOFs, such as those based on Zr, Zn, or Cu paddlewheel nodes, are likewise obtained via solvothermal synthesis, but recent developments in ultrasound-assisted, room-temperature, or solid-state synthesis have facilitated their production under milder and more sustainable conditions (Cao and Wang, 2020). Exfoliation of MOFs into 2D nanosheets is achieved via solvent-assisted ultrasonication, sometimes aided by surfactants or ionic liquids to preserve the structural integrity. Post-synthetic modifications, including linker functionalization and metal exchange, further allow tuning light absorption and redox properties for targeted photocatalytic applications.

Hexagonal boron nitride (h-BN) is another 2D material increasingly integrated into photocatalytic systems for its exceptional chemical inertness, thermal stability, and insulating properties (Shu et al., 2024; Yang et al., 2023b). These features make h-BN an excellent support material in hybrid systems, where it enhances charge separation, morphological stability, and dispersion of active phases. h-BN nanosheets are commonly synthesized via chemical vapor deposition, ball milling, or liquid-phase exfoliation in polar organic solvents (Gautam and Chelliah, 2021; Ali et al., 2025). A viable alternative for this type of system could be the in situ growth of photocatalytic nanostructures (e.g., metal oxides or sulfides) on exfoliated h-BN to form heterostructures with improved interfacial charge dynamics and long-term durability in water purification applications.

The comparative Table 9.1 outlines the principal synthesis routes for 2D nanomaterials used in photocatalytic water purification. Top-down approaches, such as chemical oxidation and ultrasound-assisted exfoliation, are commonly applied to materials like GO, rGO, and TMDs. These methods are generally scalable and low-cost, though they often yield small lateral dimensions and introduce structural defects. Bottom-up strategies such as thermal condensation of nitrogen-rich precursors (for g-C_3N_4) and hydrothermal/solvothermal

TABLE 9.1 Summary of key synthesis methods for 2D materials used in photocatalytic water purification.

Synthesis method	2D Materials Obtained	Approach	Key features	Advantages	Limitations
Chemical oxidation and exfoliation (Hummers + reduction)	GO, rGO	Top-down	Functionalization with oxygenated groups, good dispersion in water	Scalable, low-cost, compatible with semiconductors	Acidic waste generation, structural defects
Thermal condensation of nitrogen-rich precursors	g-C_3N_4	Bottom-up	Conjugated heptazine-based network, visible-light activity	Simple, metal-free, tunable porosity and doping	Low surface area in bulk form
Selective etching of MAX phases (HF or LiF/HCl)	MXenes (e.g., $Ti_3C_2T_x$)	Top-down	Metallic layers with -OH, -F, -O surface terminations, high conductivity	Promotes charge separation, hydrophilic, active supports	Reactive in water, safety precautions required
Liquid-phase exfoliation (ultrasound-assisted)	Graphene, MoS_2, WS_2, BN	Top-down	Mechanical separation of layers using ultrasonic energy	Versatile, applicable to many layered materials	Low yield, small lateral size, poor control
Hydrothermal /solvothermal synthesis	MoS_2, WS_2, SnS_2, etc.	Bottom-up	Controlled growth of hierarchical structures (e.g., nanosheets, nanoflowers)	Enables doping, morphology control, mild conditions	Time-consuming, moderate scalability
Chemical Vapor Deposition (CVD)	Graphene, MoS_2 (continuous films)	Bottom-up	Atomic-level deposition on substrates, high-crystallinity monolayers	Precise thickness control, high structural quality	Expensive, limited to thin films, not ideal for powder-based systems

methods (for MoS_2, WS_2, SnS_2) enable more precise control over composition and morphology. While these techniques allow for doping and porosity tuning, they may suffer from limited scalability. MXenes, obtained via selective etching of MAX phases, offer high conductivity and surface functionalization beneficial for charge transport, but raise safety and stability concerns due to the use of fluorinated etchants and susceptibility to oxidation. Chemical vapor deposition (CVD) stands out for producing high-quality, atomically thin films of graphene and MoS_2 with precise structural control, though its use in water purification photocatalysis is generally limited to hybrid systems due to high cost and substrate constraints. Ultimately, the choice of synthesis method depends on the desired material architecture, compatibility with photocatalytic mechanisms, and the feasibility of scaling for environmental applications.

9.4 Optimizing energy bands in 2D nanomaterials for enhanced photocatalytic performance

The 2D materials have raised attention in the photocatalysis field due to their outstanding properties, not only regarding their textural properties and high surface area, but also because of the potential tuning of the bandgap energy and the bands alignment, good electrical conductivity and the plausible modification of the charge transport, minimizing the undesirable recombination effect that make them more effective in the uptake and transformation of radiation into chemical energy. The thin nature of 2D material enables a large surface area, which implies the formation of multiple active sites in which the oxidation and reduction reactions can be triggered, driving the photocatalytic degradation mechanism. The 2D dimension usually provides opportunities to modify the bandgap energy, allowing them to absorb at different wavelength ranges. Furthermore, the charges formed inside the 2D photocatalysts can be transferred to the surface more quickly than the bulk catalysts since the ultrathin thickness shortens the transfer distance (Li et al., 2023).

The most popular developed 2D photocatalysts include graphene, graphitic carbon nitride (g-C_3N_4), phosphorene, transition metal dichalcogenides like MoS_2 and WS_2, and some metals such as TiO_2 that can be potentially prepared as nanosheets (Garg et al., 2021). The 2D materials can harvest a wide range of the electromagnetic spectrum depending on their electronic properties. The indirect semiconductors, such as graphitic carbon nitride (Rono et al., 2021), metal dichalcogenides such as MoS_2 (Splendiani et al., 2010) and WS_2 (Roy and Bermel, 2018), or phosphorene, do not require extra energy conversion and can harvest the solar and UV radiation. Graphene, instead, is a material with zero bandgap that can achieve a broadband light interaction. Although single layered graphene is recognized as a semi-metal with zero bandgap that is not suitable for light absorption, many researchers have demonstrated that functionalized graphene based analogs like graphene oxide (GO) might be potential candidates for nonmetal photocatalysts as the band structure of GO is related to its oxidation degree, which can be tuned by selecting suitable synthesis methods (Yeh et al., 2010). Diverse strategies have been deeply explored to enhance the harvesting of radiation, tailoring the bandgap and the position of the bands, or overcoming the limitations of a certain material by combining with other materials to build a heterostructure. Fig. 9.3 schematizes the most popular strategies developed to date.

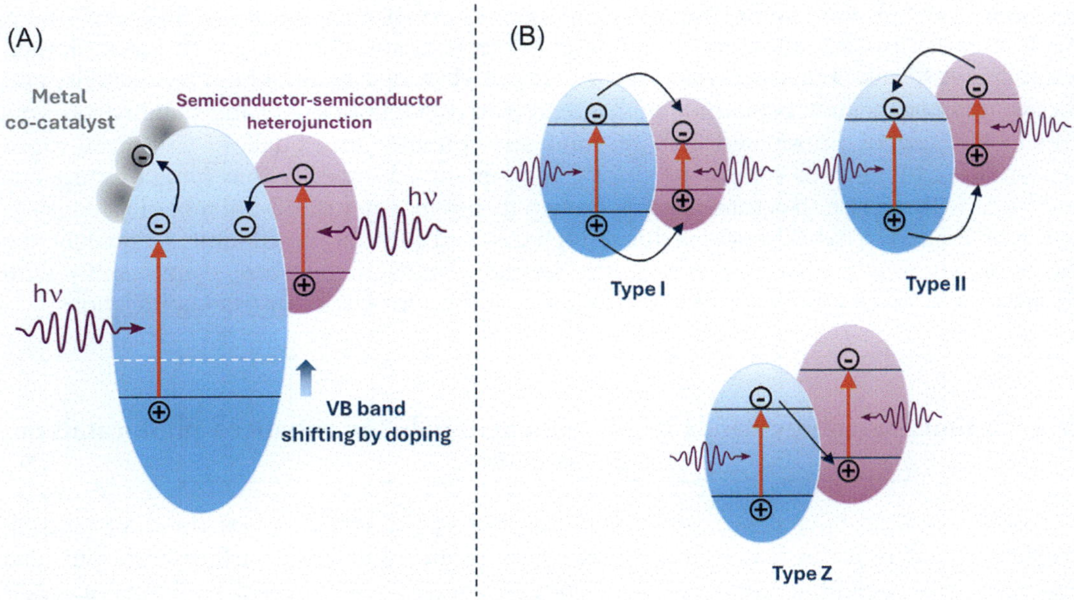

FIGURE 9.3 Strategies pursued for bandgap engineering (A) and heterojunction types (B). The schemes depict common approaches designed to optimize charge separation and enhance photocatalytic efficiency.

9.4.1 Bands' energy by thickness-dependency

The band gaps of many 2D materials are highly sensitive to their thickness or number of layers. Consequently, a wider absorption range can be defined by modifying the bandgap. Usually, the single-layered version behaves as a direct semiconductor rather than an indirect one, as happens with multilayered materials.

The exfoliation of g-C_3N_4 has been demonstrated to alter the electronic properties and, therefore the photocatalytic activation process. The single-layer g-C_3N_4 has a larger bandgap than the bulk multilayer version because of a quantum confinement effect. Moreover, the multilayer disposition performs as an indirect semiconductor while the single-layered one behaves as a direct one (Hayat et al., 2022).

Black phosphorus (BP) is a 2D layered material that can be exfoliated into thin sheets to form phosphorene. The bandgap of BP can be tailored depending on the number of layers, obtaining values of 0.3–1.88 eV (Mishra et al., 2024). The single-layer version, that is, phosphorene, displays a bandgap of 2.0 eV. These bandgap values allow the photoactivation under the solar spectrum (Cai et al., 2014). In the application of tetracycline degradation, few-layer phosphorene showed good response and stability with visible light under deoxidation conditions (Yao et al., 2022). The holes of the VB displayed a redox potential of +1.73 V, lower than the required value for the production of hydroxyl radicals. The degradation mechanism is attributed to the reduction of tetracycline by the photogenerated charges due to the deoxidation conditions.

Transition metal dichalcogenides (TMDs) hold narrow band gaps, usually <2.4 eV, endowing them with wider spectral absorption cutoff wavelengths ranging from near-infrared to visible spectrum (Gan et al., 2021). TMDs possess a thickness-dependent electronic bandgap that can be tuned, resulting from the quantum confinement effect (Yang et al., 2023a). For instance, with MoS_2, when the thickness of MoS_2 decreases to a monolayer, it changes from an indirect bandgap semiconductor to a direct bandgap semiconductor, along with a dramatic jump in luminescence (Splendiani et al., 2010) The valence band (VB) energy splits for one to five layers of MoS_2 (Zhang et al., 2015b), being the redox potential of the holes in the VB reduced with the lower number of layers (Yang et al., 2023a). For these reasons, the disinfection rate considerably improved by a few-layered MoS_2 if compared to the bulk version (Liu et al., 2016).

9.4.2 Elemental doping

The insertion of metallic and nonmetallic elements in the surface of 2D material may impact the photocatalytic performance due to a modification of the band energy scheme by shifting the valence band energy and reducing the bandgap.

The nonmetallic polymeric structure of g-C_3N_4 is susceptible to doping with nonmetals of similar size and electronegativity to C and N atoms. This polymeric structure displays a bandgap around 2.7 eV and a valence band with ~1.6 eV of redox potential (Molaei, 2023), which is insufficient for the oxidation of water molecules to hydroxyl radicals. Therefore by the insertion of nonmetal dopants such as O, S, P, or B, the bandgap can be tuned, and the conduction band can be shifted (Sahoo et al., 2024); however, the main improvement in the mechanism relies on the improvement of the charge migration and separation (Quintana et al., 2025). Nonmetals with strong electronegativity promote covalent bonds in the polymeric structure by acquiring electrons in the course of chemical reactions (Quintana et al., 2025). The g-C_3N_4 network can be doped with O at the interstitial by substituting a C or N atom. The presence of O improves the hydrophilicity surface, which helps the adsorption of hydrophilic contaminants onto the catalyst's surface (Quintana et al., 2025). The doping with O enhances the photocatalytic activity by optimizing the electronic and band structure of the g-C_3N_4. O-doped g-C_3N_4 narrows the bandgap from 2.70 to 1.98 eV, harvesting more light in the visible range. This reduction is generated by approaching the conduction and valence bands. Although this effect implies the formation of charges with lower redox potential, the exchange of O with N positions promotes the delocalization and better migration of charges (Quintana et al., 2025). The self-doping of C in the g-C_3N_4 structure has revealed the preferential replacement of N with C atoms, leaving larger delocalized large p bonds between the substituted carbons, which enhances the conductivity of g-C_3N_4 due to a better electron transfer via π delocalization (Sahoo et al., 2024). The reduction of the bandgap is also achieved. The lower reduction of the valence band does not favor the formation of hydroxyl radical; therefore the oxidation mechanism of oxidation takes place via photogenerated holes and superoxide radicals (Bao et al., 2017). Sulfur is normally inserted in the g-C_3N_4 structure by replacing N atoms. The interaction of 3p states of S and the 2p states of N leads to a substantial enhancement in the bandgap and band energies. The better performance is attributed to a better separation charge and boosted light harvesting as observed in the degradation of diverse contaminants such as ciprofloxacin (Gupta et al., 2020), tetracycline (Wang et al., 2019), or spiramycin (Dou et al., 2021) by doping with S from different precursors. Similar effects have been attributed to doping with B, for example, decreased bandgap and enhanced charge migration (Alejandra Quintana et al., 2022), leading to better degradation rates of dyes (Narkbuakaew and Intasanta, 2022; Yan et al., 2010). The doping with P has been less explored, but the insertion of P atoms in the structure aids the bandgap narrowing and the charge separation. These modifications, through holes and superoxide radical oxidation, explain better degradation in water solution (Li et al., 2019).

Both cationic and anionic elements are substitutionally replaced in TMDs by foreign atoms that have comparable radii; however, anion doping has provided better efficiencies (Balasubramaniam et al., 2019). The doping of metal-disulfides with halogens (F, Cl, Br) as well as group V elements (N, P, As) creates vacancies that favor positions that are always preferential at anion sites. The doping with metals has been reported as a popular strategy to improve the visible light absorption and boost charge migration. The 2D structure allows a uniform deposition of metallic doping, which produces better performance than that observed in bulk materials. The doping of K atoms in the interlayers of g-C_3N_4 enhances charge separation and ROS generation. It has been proved to be a better photocatalytic degradation of 1,3-diphenylguanidine (Aoudi et al., 2024). The bandgap and valence band energy decrease with K dosage, along with better migration of charges, resulting in a higher formation of superoxide radical. The insertion of 3d transition metals, for example, Ti, Mn, Fe, Co, Ni, Cu, Zn, etc., in the structure also alters the optical properties, in terms of bandgap and band edge. Generally, the narrowing of the bandgap by metal doping makes the VB edges shift upward while the CB edges shift downward (Kalidasan et al., 2024).

The substitution of metal position is an interesting strategy for defect building and tunable bandgap in TMDs, leading to better charge separation (Balasubramaniam et al., 2019). Mn-doped MoS2 nanostructures are promising for improving their charge separation and photo-stimulated dye degradation activity (Sahoo et al., 2023). The bandgap decreases with an increase in the Mn content, which reduces the distance between the VB and CB.

9.4.3 Metal co-catalyst

Metal decoration has been identified as an effective strategy to promote charge delocalization, as the presence of noble metal nanoparticles acts as a sink for electrons. In the case of water treatment, Ag has been the most studied noble metal due to the antimicrobial properties of Ag° nanoparticles.

The deposition of Ag° onto g-C_3N_4 has been demonstrated to significantly enhance photocatalytic activity in the degradation of pollutants (Viet et al., 2019). Under radiation, the VB electrons of g-C_3N_4 are excited and moved to CB, generating photo-induced holes in the VB with strong oxidation potential. The presence of Ag° acts as an electron sink, facilitating the migration of charges, enhancing the charge separation, and reducing recombination. Hence, these electrons are captured by adsorbed O_2 molecules, triggering the formation of larger superoxide radicals than in the absence of Ag° (Garg et al., 2024).

The deposition of Ag° on WS_2 has been reported to improve the photodegradation performance by collecting photoinduced electrons from the CB of WS_2 to prevent the recombination of the electrons and holes due to the Schottky barrier formed at WS_2–Ag interface (Wu et al., 2022). A similar effect has been reported in Ag nanoparticles decorating MoS_2, which facilitates efficient charge separation, reducing electron–hole recombination rates (Gogoi and Deb, 2024).

9.4.4 Heterojunctions

The combination of $AgIO_3$ and g-C_3N_4 as 2D/2D nanosheets has provided a good photocatalytic response under visible radiation, producing a synergistic effect on the degradation rate of dyes (Li et al., 2015). Under visible radiation, the electrons of g-C_3N_4 are excited to its CB, where they migrate to the CB of $AgIO_3$, promoting a better charge separation. However, these electrons are not able to reduce O_2 to superoxide radical, so this ROS does not play an important role in the degradation mechanism. BiOCl is another 2D semiconductor that has been combined with C_3N_4 nanosheets through a simple calcination route (Liu et al., 2017). Although BiOCl displays a large bandgap that prevents activation with visible light, its presence aids the charge separation. Bi_2WO_6 has also been combined with g-C_3N_4 for boosting the degradation of ibuprofen in water (Wang et al., 2017). The formation of ultrathin 2D/2D heterojunctions provokes an effective charge transfer, which explains the enhanced photocatalytic activity under visible light irradiation. Graphene has successfully attached to g-C_3N_4 at low proportions (1–5 wt.%) to insert a surface plasmon resonance-induced effect (Al Marzouqi and Selvaraj, 2023). The visible absorption was 50 times higher when compared to the pure sample, which benefited the degradation kinetics of nizatidine and ranitidine in water. In another example, reduced graphene oxide (rGO) and rod-like g-C_3N_4 also support the migration of charges of photo-generated electrons from the CB of g-C_3N_4 to the rGO support (Vinesh et al., 2020). The holes generated in the VB of g-C_3N_4 launch the creation of ROS for the degradation of tetracycline. The combination of graphene oxide, g-C_3N_4, and ZnO in a ternary composite (Ma et al., 2023). As the bandgap of g-C_3N_4 is lower than ZnO, under visible radiation, electrons are photo-excited in the CB of g-C_3N_4 and migrate to the CB of ZnO. Conversely, holes tend to be transferred from the VB of ZnO to the VB of g-C_3N_4. Summarizing both semiconductor forms a Type II heterojunction in which the presence of GO aids the charge transfer and delocalization. Recently, MXene has been postulated as an alternative electron charge transfer like graphene. A 2D/2D heterojunction of Ti_3C_2/g-C_3N_4 performed a higher diclofenac oxidation rate than the pure g-C_3N_4 with visible light irradiation (He et al., 2020). The reason behind this enhancement is the excellent electrical conductivity of Ti_3C_2, the efficient migration and separation of photo-generated charge carriers in the heterojunction. A Z-scheme heterojunction of g-C_3N_4, Ti_3C_2, and black phosphorus has been reflected thanks to the P-bridging effect between the P of black phosphorus and the N of carbon nitride (Zhou et al., 2021). The resulting heterostructure improved the visible light harvesting, delocalizing the photo-generated charges that displayed a stronger redox potential, accelerating the degradation of ciprofloxacin.

MoS_2 has been extensively used for the preparation of diverse heterojunctions (Rahman et al., 2022). The 2D/2D g-C_3N_4/MoS_2 heterojunction has provided improved photocatalytic kinetics than the pure phases. The combination led to effective charge transfer, which inhibits the recombination of electron–hole pairs (Wu et al., 2021). The heterojunction TiO_2/MoS_2 can be described as a Type II heterounion (Gopal et al., 2020). MoS_2, with a bandgap of 1.89 eV, provided a source of photo-excitation in the visible range, which cannot be harvested by TiO_2, whose bandgap is 3.28 eV. TMDs have also been combined with bismuth oxyhalides (Yousef et al., 2022). A novel hetero-structured photocatalyst WS_2/BiOBr was characterized by the presence of oxygen vacancies, which contribute to harvesting the visible light as well as the separation of photo-induced electron–hole pairs, leading to over twofold pseudo-first order rate constant than the bare semiconductors (Fu et al., 2020). A Type II $WS_2/BiOCl$ heterojunction has been reported to enhance the photocatalytic response for the degradation of dyes (Ashraf et al., 2020). The electrons reaching the conduction band of WS_2 can easily be transferred to the conduction band of BiOCl, and the holes can easily move from BiOCl to the WS_2 nanosheets. As a result, the better radiation harvesting and lower recombination rate explain the improved kinetics of the process. Other bismuth species, such as $BiVO_4$, with high photocatalytic response, have also been combined with 2D TMDs, conforming to Type II heterostructures. The combination of $MoS_2/BiVO_4$ has a suitable band edge for pollutant degradation compared to the $WS_2/BiVO_4$ heterostructure. This improvement was ascribed to the role of the MoS_2 monolayer as an electron donor, the many reactive sites on the MoS_2 surface, and the better separation of electric charges (Opoku et al., 2017). The presence of rGO helps to delocalize electrons and the adsorption of pollutants due to the higher surface area (Ramalingam et al., 2022). For instance, the effectiveness of rGO/CdS over Rhodamine B was explained based on improved adsorption uptake, a broader visible light response range, and enhanced electron–hole pair separation (Jiang et al., 2014). $NiCo_2S_4$ combined with MXene has demonstrated a larger photocatalytic activity than the bare sulfide (Vigneshwaran et al., 2021). The sulfide incorporated into 2D MXene layers enhances the adsorption and degradation of the dye Rhodamine B. The large surface area of 2D MXene and the efficient electron transfer promote accelerated electron migration and remarkable electroconductivity of the nanocomposite. Black phosphorus has been combined with CdS_2 to build a composite with boosted activity for the degradation of tetrabromobisphenol A (Zhang et al., 2024). The better photocatalytic response of the resulting Type II composite was attributed to larger radiation harvesting in the visible region, better contaminant adsorption uptake, improved charge separation, and higher ROS generation rate.

9.5 Conclusion and perspectives

Two-dimensional nanomaterials have emerged as highly promising candidates for photocatalytic water purification systems due to their large specific surface area, tunable band structures, and highly efficient charge transport properties. This chapter has examined the underlying photocatalytic mechanisms in 2D materials, with particular attention to charge separation and migration processes, surface redox reactions, and interfacial dynamics influenced by heterostructure formation and defect engineering. A thorough understanding of these aspects is essential for the rational design of photocatalysts capable of degrading persistent pollutants and emerging contaminants in aqueous media.

Advances in the synthesis and engineering of 2D nanomaterials, through top-down exfoliation, bottom-up chemical strategies, and hybrid assemblies, have expanded the possibilities for tuning structural and electronic features that govern photocatalytic behavior. Methods that allow precise control over layer thickness, lateral dimensions, crystallinity, and surface functionality have shown to significantly influence photocatalytic activity in water purification applications. Nevertheless, many of these synthetic routes remain confined to laboratory-scale procedures, often limited by poor scalability, high cost, and environmental concerns. Overcoming these constraints is vital for real-world implementation.

The optimization of energy band alignments in 2D materials represents a key strategy for enhancing visible-light-driven photocatalysis. By employing approaches such as doping, interface engineering, or strain modulation, it is possible to shift band positions and promote more efficient charge separation, suppress recombination, and activate redox reactions relevant to the degradation of contaminants in water matrices. Still, to realize their full potential, these materials must demonstrate long-term stability, resistance to fouling, and selectivity under variable water quality conditions.

Despite encouraging progress, the transition from proof-of-concept demonstrations to operational technologies faces several barriers. The design of photocatalysts that remain active and robust under real water conditions, containing natural organic matter, inorganic ions, and microorganisms, requires further investigation. Moreover, integrating 2D materials into scalable photoreactors and ensuring uniform light penetration, photocatalyst recovery, and minimal environmental impact are ongoing technical challenges. In parallel, 2D materials must prove their superiority over conventional technologies already in use for water treatment, such as activated carbon adsorption, chlorination, ozonation, or membrane processes. Demonstrating clear advantages in terms of efficiency, selectivity, and energy demand will be essential for their industrial adoption.

A promising route for accelerating development involves the integration of artificial intelligence and machine learning in material design. These tools allow the prediction of photocatalytic activity based on structural descriptors, optimization of synthesis protocols, and identification of favorable compositions and architectures with minimal experimental effort. Coupled with high-throughput experimentation and in situ characterization, data-driven approaches are poised to significantly shorten discovery cycles and guide the development of next-generation photocatalysts tailored for water purification.

In the coming years, the field is expected to evolve toward the design of multifunctional and regenerable 2D photocatalysts that can simultaneously address multiple classes of contaminants, including pharmaceuticals, pesticides, heavy metals, and microplastics. Progress will rely on a tighter coupling between fundamental insights, scalable manufacturing, and digital design tools. The development of integrated photocatalytic systems powered by solar energy and capable of decentralized operation will also be a critical step toward sustainable and accessible water treatment technologies. While 2D nanomaterials have already reshaped the conceptual landscape of photocatalysis, their ultimate impact on water purification will depend on our ability to translate these advanced materials into reliable, cost-effective, and environmentally sound solutions.

References

A, H.S., 1913. Photochemische Versuchstechnik. Nature 91 (2269), 186–187. Available at https://doi.org/10.1038/091186b0.

Akhavan, O., Choobtashani, M., Ghaderi, E., 2012. Protein degradation and RNA efflux of viruses photocatalyzed by graphene–tungsten oxide composite under visible light irradiation. The Journal of Physical Chemistry C 116 (17), 9653–9659. Available at https://doi.org/10.1021/jp301707m.

Al Marzouqi, F., Selvaraj, R., 2023. Surface plasmon resonance induced photocatalysis in 2D/2D graphene/g-C_3N_4 heterostructure for enhanced degradation of amine-based pharmaceuticals under solar light illumination. Catalysts 13 (3), 560. Available at https://doi.org/10.3390/catal13030560.

Alaoui, A., Christ, F., Silva, V., Vested, A., Schlünssen, V., González, N., et al., 2024. Identifying pesticides of high concern for ecosystem, plant, animal, and human health: a comprehensive field study across Europe and Argentina. Science of the Total Environment 948. Available at https://doi.org/10.1016/j.scitotenv.2024.174671.

Alejandra Quintana, M., Solís, R.R., Ángeles Martín-Lara, M., Blázquez, G., Mónica Calero, F., Muñoz-Batista, M.J., 2022. Enhanced boron modified graphitic carbon nitride for the selective photocatalytic production of benzaldehyde. Separation and Purification Technology 298, 121613. Available at https://doi.org/10.1016/j.seppur.2022.121613.

Ali, S., Ismail, P.M., Humayun, M., Bououdina, M., 2025. Hexagonal boron nitride: from fundamentals to applications. Desalination 599, 118442. Available at https://doi.org/10.1016/j.desal.2024.118442.

Aoudi, B., Kahkeci, J., Boluk, Y., Gamal El-Din, M., 2024. Enhanced g-C_3N_4 for sustainable solar degradation of 1,3-diphenylguanidine (DPG) in wastewater: investigating the effects of precursor, temperature, and potassium doping. Chemical Engineering Journal 487. Available at https://doi.org/10.1016/j.cej.2024.150665. 150665.

Ashraf, W., Bansal, S., Singh, V., Barman, S., Khanuja, M., 2020. BiOCl/WS2hybrid nanosheet (2D/2D) heterojunctions for visible-light-driven photocatalytic degradation of organic/inorganic water pollutants. RSC Advances 10 (42), 25073–25088. Available at https://doi.org/10.1039/d0ra02916e.

Balasubramaniam, B., Singh, N., Kar, P., Tyagi, A., Prakash, J., Gupta, R.K., 2019. Engineering of transition metal dichalcogenide-based 2D nanomaterials through doping for environmental applications. Molecular Systems Design and Engineering 4 (4), 804–827. Available at https://doi.org/10.1039/c8me00116b.

Bao, N., Hu, X., Zhang, Q., Miao, X., Jie, X., Zhou, S., 2017. Synthesis of porous carbon-doped g-C_3N_4 nanosheets with enhanced visible-light photocatalytic activity. Applied Surface Science 403, 682–690. Available at https://doi.org/10.1016/j.apsusc.2017.01.256.

Barawi, M., Mesa, C.A., Collado, L., Villar-García, I.J., Oropeza, F., de la Peña O'Shea, V.A., et al., 2024. Latest advances in in situ and operando X-ray-based techniques for the characterisation of photoelectrocatalytic systems. Journal of Materials Chemistry A 12 (35), 23125–23146. Available at https://doi.org/10.1039/d4ta03068k.

Bera, R., Kundu, S., Patra, A., 2015. 2D hybrid nanostructure of reduced graphene oxide-CdS nanosheet for enhanced photocatalysis. ACS Applied Materials and Interfaces 7 (24), 13251–13259. Available at https://doi.org/10.1021/acsami.5b03800.

Boro, D., Chirania, M., Verma, A.K., Chettri, D., Verma, A.K., 2025. Comprehensive approaches to managing emerging contaminants in wastewater: identification, sources, monitoring and remediation. Environmental Monitoring and Assessment 197 (4). Available at https://doi.org/10.1007/s10661-025-13809-w.

Cai, Y., Zhang, G., Zhang, Y.W., 2014. Layer-dependent band alignment and work function of few-layer phosphorene. Scientific Reports 4. Available at https://doi.org/10.1038/srep06677.

Cao, L., Wang, C., 2020. Metal-organic layers for electrocatalysis and photocatalysis. ACS Central Science 6 (12), 2149–2158. Available at https://doi.org/10.1021/acscentsci.0c01150.

Cao, J., Wang, J., Wang, Z., Zubairu, S.M., Ding, Y., Zhu, G., 2024. In-situ construction of Z-scheme 2D/2D g-C_3N_4/BiOBr heterojunction with enhanced photocatalytic CO_2 reduction. Surfaces and Interfaces 45, 103875. Available at https://doi.org/10.1016/j.surfin.2024.103875.

Cesano, F., Boffa, V., Bortot Coelho, F.E., Magnacca, G., 2021. Graphene and graphene-oxide for enhancing the photocatalytic properties of materials. Materials Science in Photocatalysis. Elsevier, Italy, pp. 385–396. Available at https://doi.org/10.1016/B978-0-12-821859-4.00015-5.

Chen, Q., Liu, C., Liu, R., Hou, Y., Bi, J., Yu, J.C., et al., 2025. Ultrathin 2D/2D MoS2/Bi_2WO_6 S-scheme heterojunction for boosting photocatalytic degradation of ciprofloxacin. Separation and Purification Technology 355, 129768. Available at https://doi.org/10.1016/j.seppur.2024.129768.

Das, P., Deoghare, A.B., Maity, S.R., 2021. A novel approach to synthesize reduced graphene oxide (RGO) at low thermal conditions. Arabian Journal for Science and Engineering 46 (6), 5467–5475. Available at https://doi.org/10.1007/s13369-020-04956-y.

Deng, Y., Tang, L., Zeng, G., Zhu, Z., Yan, M., Zhou, Y., et al., 2017. Insight into highly efficient simultaneous photocatalytic removal of Cr(VI) and 2,4-diclorophenol under visible light irradiation by phosphorus doped porous ultrathin g-C_3N_4 nanosheets from aqueous media: Performance and reaction mechanism. Applied Catalysis B: Environmental 203, 343–354. Available at https://doi.org/10.1016/j.apcatb.2016.10.046.

Dong, C., Fang, W., Yi, Q., Zhang, J., 2022. A comprehensive review on reactive oxygen species (ROS) in advanced oxidation processes (AOPs). Chemosphere 308 (Pt 1), 136205. Available at https://doi.org/10.1016/j.chemosphere.2022.136205.

Dou, Y., Yan, T., Zhang, Z., Sun, Q., Wang, L., Li, Y., 2021. Heterogeneous activation of peroxydisulfate by sulfur-doped g-C_3N_4 under visible-light irradiation: implications for the degradation of spiramycin and an assessment of N-nitrosodimethylamine formation potential. Journal of Hazardous Materials 406. Available at https://doi.org/10.1016/j.jhazmat.2020.124328. 124328.

Duan, R., Ma, S., Xu, S., Wang, B., He, M., Li, G., et al., 2022. Soybean straw biochar activating peroxydisulfate to simultaneously eliminate tetracycline and tetracycline resistance bacteria: Insights on the mechanism. Water Research 218, 118489. Available at https://doi.org/10.1016/j.watres.2022.118489.

Fan, X., Yang, F., Nie, C., Ma, L., Cheng, C., Haag, R., 2021. Biocatalytic nanomaterials: a new pathway for bacterial disinfection. Advanced Materials 33 (33). Available at https://doi.org/10.1002/adma.202100637.

Fatima, J., Shah, A.N., Tahir, M.B., Mehmood, T., Shah, A.A., Tanveer, M., et al., 2022. Tunable 2D nanomaterials; their key roles and mechanisms in water purification and monitoring. Frontiers in Environmental Science 10. Available at https://doi.org/10.3389/fenvs.2022.766743.

Fent, K., Weston, A.A., Caminada, D., 2006. Ecotoxicology of human pharmaceuticals. Aquatic Toxicology 76 (2), 122–159. Available at https://doi.org/10.1016/j.aquatox.2005.09.009.

Fu, S., Yuan, W., Liu, X., Yan, Y., Liu, H., Li, L., et al., 2020. A novel 0D/2D WS2/BiOBr heterostructure with rich oxygen vacancies for enhanced broad-spectrum photocatalytic performance. Journal of Colloid and Interface Science 569, 150–163. Available at https://doi.org/10.1016/j.jcis.2020.02.077.

Fu, X., Yang, R., Zhou, G., Chen, X., Liu, Y., Chi, J., et al., 2022. New progress in photocatalytic degradation of bisphenol A as representative endocrine disrupting chemicals. Current Opinion in Green and Sustainable Chemistry 35, 100629. Available at https://doi.org/10.1016/j.cogsc.2022.100629.

Fujishima, A., Honda, K., 1972. Electrochemical photolysis of water at a semiconductor electrode. Nature 238 (5358), 37–38. Available at https://doi.org/10.1038/238037a0.

Gan, X., Lei, D., Ye, R., Zhao, H., Wong, K.Y., 2021. Transition metal dichalcogenide-based mixed-dimensional heterostructures for visible-light-driven photocatalysis: dimensionality and interface engineering. Nano Research 14 (6), 2003–2022. Available at https://doi.org/10.1007/s12274-020-2955-x.

Ganguly, P., Harb, M., Cao, Z., Cavallo, L., Breen, A., Dervin, S., et al., 2019. 2D nanomaterials for photocatalytic hydrogen production. ACS Energy Letters 4 (7), 1687–1709. Available at https://doi.org/10.1021/acsenergylett.9b00940.

Gao, Y., Chen, C., Tan, X., Xu, H., Zhu, K., 2016. Polyaniline-modified 3D-flower-like molybdenum disulfide composite for efficient adsorption/photocatalytic reduction of Cr(VI). Journal of Colloid and Interface Science 476, 62–70. Available at https://doi.org/10.1016/j.jcis.2016.05.022.

Garg, A., Basu, S., Shetti, N.P., Reddy, K.R., 2021. 2D materials and its heterostructured photocatalysts: synthesis, properties, functionalization and applications in environmental remediation. Journal of Environmental Chemical Engineering 9 (6). Available at https://doi.org/10.1016/j.jece.2021.106408.

Garg, S., Choudhary, M.K., Kataria, J., 2024. Unlocking the potential of biogenic Ag@g-C_3N_4 in sustainable water purification: a kinetic and photocatalytic study. Environmental Pollution 363 (Pt 1), 125126. Available at https://doi.org/10.1016/j.envpol.2024.125126.

Gautam, C., Chelliah, S., 2021. Methods of hexagonal boron nitride exfoliation and its functionalization: covalent and non-covalent approaches. RSC Advances 11 (50), 31284–31327. Available at https://doi.org/10.1039/d1ra05727h.

Gogoi, P., Deb, S., 2024. 1T/2H mixed phase MoS2 nanosheets decorated with Ag nanoparticles for effective photocatalysis of organic dyes for wastewater treatment. Colloids and Surfaces A: Physicochemical and Engineering Aspects 703. Available at https://doi.org/10.1016/j.colsurfa.2024.135357. 135357.

Golovko, O., Örn, S., Sörengård, M., Frieberg, K., Nassazzi, W., Lai, F.Y., et al., 2021. Occurrence and removal of chemicals of emerging concern in wastewater treatment plants and their impact on receiving water systems. Science of The Total Environment 754, 142122. Available at https://doi.org/10.1016/j.scitotenv.2020.142122.

Gopal, R., Chinnapan, M.M., Bojarajan, A.K., Rotte, N.K., Ponraj, J.S., Ganesan, R., et al., 2020. Facile synthesis and defect optimization of 2D-layered MoS_2 on TiO_2 heterostructure for industrial effluent, wastewater treatments. Scientific Reports 10 (1). Available at https://doi.org/10.1038/s41598-020-78268-4.

Güler, Ö., Tekeli, M., Taşkın, M., Güler, S.H., Yahia, I.S., 2021. The production of graphene by direct liquid phase exfoliation of graphite at moderate sonication power by using low boiling liquid media: the effect of liquid media on yield and optimization. Ceramics International 47 (1), 521–533. Available at https://doi.org/10.1016/j.ceramint.2020.08.159.

Gupta, B., Gupta, A.K., Ghosal, P.S., Tiwary, C.S., 2020. Photo-induced degradation of bio-toxic Ciprofloxacin using the porous 3D hybrid architecture of an atomically thin sulfur-doped g-C_3N_4/ZnO nanosheet. Environmental Research 183. Available at https://doi.org/10.1016/j.envres.2020.109154.

Hayat, A., Al-Sehemi, A.G., El-Nasser, K.S., Taha, T.A., Al-Ghamdi, A.A., Ali Shah Syed, J., et al., 2022. Graphitic carbon nitride (g–C_3N_4)–based semiconductor as a beneficial candidate in photocatalysis diversity. International Journal of Hydrogen Energy 47 (8), 5142–5191. Available at https://doi.org/10.1016/j.ijhydene.2021.11.133.

He, J., Yang, J., Jiang, F., Liu, P., Zhu, M., 2020. Photo-assisted peroxymonosulfate activation via 2D/2D heterostructure of Ti_3C_2/g-C_3N_4 for degradation of diclofenac. Chemosphere 258, 127339. Available at https://doi.org/10.1016/j.chemosphere.2020.127339.

He, D., Jin, D., Cheng, F., Zhang, T., Qu, J., Zhou, Y., et al., 2022. Development of a metal-free black phosphorus/graphitic carbon nitride heterostructure for visible-light-driven degradation of indomethacin. Science of the Total Environment 804. Available at https://doi.org/10.1016/j.scitotenv.2021.150062.

Hu, R., Chen, W., Lai, J., Li, F., Qiao, H., Liu, Y., et al., 2025. Heterogeneous interface engineering of 2D black phosphorus-based materials for enhanced photocatalytic performance. Small (Weinheim an der Bergstrasse, Germany) 21 (6). Available at https://doi.org/10.1002/smll.202409735.

Hübner, U., Spahr, S., Lutze, H., Wieland, A., Rüting, S., Gernjak, W., et al., 2024. Advanced oxidation processes for water and wastewater treatment – guidance for systematic future research. Heliyon 10 (9), e30402. Available at https://doi.org/10.1016/j.heliyon.2024.e30402.

Ianes, J., Cantoni, B., Polesel, F., Remigi, E.U., Vezzaro, L., Antonelli, M., 2024. Monitoring (micro-)pollutants in wastewater treatment plants: comparing discharges in wet- and dry-weather. Environmental Research 263 (Pt 2), 120132. Available at https://doi.org/10.1016/j.envres.2024.120132.

Jia, Y., Zhan, S., Ma, S., Zhou, Q., 2016. Fabrication of TiO_2-Bi_2WO_6 binanosheet for enhanced solar photocatalytic disinfection of *E. coli*: insights on the mechanism. ACS Applied Materials and Interfaces 8 (11), 6841–6851. Available at https://doi.org/10.1021/acsami.6b00004.

Jiang, N., Xiu, Z., Xie, Z., Li, H., Zhao, G., Wang, W., et al., 2014. Reduced graphene oxide–CdS nanocomposites with enhanced visible-light photoactivity synthesized using ionic-liquid precursors. New Journal of Chemistry 38 (9), 4312–4320. Available at https://doi.org/10.1039/C4NJ00152D.

Jolaosho, T.L., Rasaq, M.F., Omotoye, E.V., Araomo, O.V., Adekoya, O.S., Abolaji, O.Y., et al., 2025. Microplastics in freshwater and marine ecosystems: occurrence, characterization, sources, distribution dynamics, fate, transport processes, potential mitigation strategies, and policy interventions. Ecotoxicology and Environmental Safety 294. Available at https://doi.org/10.1016/j.ecoenv.2025.118036.

Kalidasan, K., Mallapur, S., Munirathnam, K., Nagarajaiah, H., Reddy, M.B.M., Kakarla, R.R., et al., 2024. Transition metals-doped g-C_3N_4 nanostructures as advanced photocatalysts for energy and environmental applications. Chemosphere 352. Available at https://doi.org/10.1016/j.chemosphere.2024.141354.

Khan, H., Shah, M.U.H., 2023. Modification strategies of TiO_2 based photocatalysts for enhanced visible light activity and energy storage ability: a review. Journal of Environmental Chemical Engineering 11 (6), 111532. Available at https://doi.org/10.1016/j.jece.2023.111532.

Kumar, N., Maitra, U., Hegde, V.I., Waghmare, U.V., Sundaresan, A., Rao, C.N.R., 2013. Synthesis, characterization, photocatalysis, and varied properties of TiO_2 cosubstituted with nitrogen and fluorine. Inorganic Chemistry 52 (18), 10512–10519. Available at https://doi.org/10.1021/ic401426q.

Kumar, P., Vahidzadeh, E., Thakur, U.K., Kar, P., Alam, K.M., Goswami, A., et al., 2019. C_3N_5: a low bandgap semiconductor containing an azo-linked carbon nitride framework for photocatalytic, photovoltaic and adsorbent applications. Journal of the American Chemical Society 141 (13), 5415–5436. Available at https://doi.org/10.1021/jacs.9b00144.

Lachheb, H., Puzenat, E., Houas, A., Ksibi, M., Elaloui, E., Guillard, C., et al., 2002. Photocatalytic degradation of various types of dyes (Alizarin S, Crocein Orange G, Methyl Red, Congo Red, Methylene Blue) in water by UV-irradiated titania. Applied Catalysis B: Environmental 39 (1), 75–90. Available at https://doi.org/10.1016/S0926-3373(02)00078-4.

Laipan, M., Yu, J., Zhu, R., Zhu, J., Smith, A.T., He, H., et al., 2020. Functionalized layered double hydroxides for innovative applications. Materials Horizons 7 (3), 715–745. Available at https://doi.org/10.1039/c9mh01494b.

Lei, J., Chen, B., Zhou, L., Ding, N., Cai, Z., Wang, L., et al., 2020. Efficient degradation of antibiotics in different water matrices through the photocatalysis of inverse opal K-g-C_3N_4: insights into mechanism and assessment of antibacterial activity. Chemical Engineering Journal 400, 125902. Available at https://doi.org/10.1016/j.cej.2020.125902.

Li, Y., Li, K., Yang, Y., Li, L., Xing, Y., Song, S., et al., 2015. Ultrathin g-C_3N_4 nanosheets coupled with $AgIO_3$ as highly efficient heterostructured photocatalysts for enhanced visible-light photocatalytic activity. Chemistry - A European Journal 21 (49), 17739–17747. Available at https://doi.org/10.1002/chem.201502945.

Li, X., Yu, J., Wageh, S., Al-Ghamdi, A.A., Xie, J., 2016a. Graphene in photocatalysis: a review. Small (Weinheim an der Bergstrasse, Germany) 12 (48), 6640–6696. Available at https://doi.org/10.1002/smll.201600382.

Li, X., Yu, J., Jaroniec, M., 2016b. Hierarchical photocatalysts. Chemical Society Reviews 45 (9), 2603–2636. Available at https://doi.org/10.1039/c5cs00838g.

Li, Y., Zhang, C., Shuai, D., Naraginti, S., Wang, D., Zhang, W., 2016c. Visible-light-driven photocatalytic inactivation of MS2 by metal-free g-C_3N_4: virucidal performance and mechanism. Water Research 106, 249–258. Available at https://doi.org/10.1016/j.watres.2016.10.009.

Li, J., Tian, C., Zhao, H., Mei, J., Zhang, J., Yang, S., 2019. Controllable fabrication of a red phosphorus modified g-C_3N_4 photocatalyst with strong interfacial binding for the efficient removal of organic pollutants. Journal of Alloys and Compounds 810, 151885. Available at https://doi.org/10.1016/j.jallcom.2019.151885.

Li, M., Zhang, H., Zhao, Z., Wang, P., Li, Y., Zhan, S., 2023. Inorganic ultrathin 2D photocatalysts: modulation strategies and environmental/energy applications. Accounts of Materials Research 4 (1), 4–15. Available at https://doi.org/10.1021/accountsmr.2c00172.

Li, Y., Yang, J., Wang, J., Feng, Z., Jing, K., Wu, W., et al., 2025. Oxidative degradation of fluoroquinolone antibiotics by ferrate(VI): kinetics, reaction mechanism, and theoretical calculations. Chemical Engineering Journal 513, 162970. Available at https://doi.org/10.1016/j.cej.2025.162970.

Lin, K.Y.A., Lin, J.T., 2017. Ferrocene-functionalized graphitic carbon nitride as an enhanced heterogeneous catalyst of Fenton reaction for degradation of Rhodamine B under visible light irradiation. Chemosphere 182, 54–64. Available at https://doi.org/10.1016/j.chemosphere.2017.04.152.

Liu, C., Kong, D., Hsu, P.C., Yuan, H., Lee, H.W., Liu, Y., et al., 2016. Rapid water disinfection using vertically aligned MoS_2 nanofilms and visible light. Nature Nanotechnology 11 (12), 1098–1104. Available at https://doi.org/10.1038/nnano.2016.138.

Liu, W., Qiao, L., Zhu, A., Liu, Y., Pan, J., 2017. Constructing 2D BiOCl/C_3N_4 layered composite with large contact surface for visible-light-driven photocatalytic degradation. Applied Surface Science 426, 897–905. Available at https://doi.org/10.1016/j.apsusc.2017.07.225.

Liu, D., Li, B., Wu, J., Liu, Y., 2020a. Photocatalytic oxidation removal of elemental mercury from flue gas. A review. Environmental Chemistry Letters 18 (2), 417–431. Available at https://doi.org/10.1007/s10311-019-00957-y.

Liu, N., Zhang, Y., Xu, C., Liu, P., Lv, J., Liu, Y.Y., et al., 2020b. Removal mechanisms of aqueous Cr(VI) using apple wood biochar: a spectroscopic study. Journal of Hazardous Materials 384, 121371. Available at https://doi.org/10.1016/j.jhazmat.2019.121371.

Liu, Y., Wang, H., Yuan, X., Wu, Y., Wang, H., Tan, Y.Z., et al., 2021. Roles of sulfur-edge sites, metal-edge sites, terrace sites, and defects in metal sulfides for photocatalysis. Chem Catalysis 1 (1), 44–68. Available at https://doi.org/10.1016/j.checat.2021.01.002.

Liu, B., Zhang, X., Chu, J., Li, F., Jin, C., Fan, J., 2023. 2D/2D $BiOIO_3/Ti_3C_2$ MXene nanocomposite with efficient charge separation for degradation of multiple pollutants. Applied Surface Science 618, 156565. Available at https://doi.org/10.1016/j.apsusc.2023.156565.

Ma, C.Y., Liu, P.H., Wang, R.T., Zhao, G.H., Zhou, N., Zhang, Q.Y., 2023. Synthesis and characterization of enhanced visible light-driven C-doped ZnO-2D GO/g-C_3N_4 heterojunction photocatalyst. Diamond and Related Materials 139. Available at https://doi.org/10.1016/j.diamond.2023.110364. 110364.

Mestre, A.S., Carvalho, A.P., 2019. Photocatalytic degradation of pharmaceuticals carbamazepine, diclofenac, and sulfamethoxazole by semiconductor and carbon materials: a review. Molecules (Basel, Switzerland) 24 (20). Available at https://doi.org/10.3390/molecules24203702.

Metcalfe, C.D., Bayen, S., Desrosiers, M., Muñoz, G., Sauvé, S., Yargeau, V., 2022. An introduction to the sources, fate, occurrence and effects of endocrine disrupting chemicals released into the environment. Environmental Research 207, 112658. Available at https://doi.org/10.1016/j.envres.2021.112658.

Mishra, R.K., Sarkar, J., Chianella, I., Goel, S., Nezhad, H.Y., 2024. Black phosphorus: the rise of phosphorene in 2D materials applications. Next Materials 4. Available at https://doi.org/10.1016/j.nxmate.2024.100217.

Molaei, M.J., 2023. Graphitic carbon nitride (g-C_3N_4) synthesis and heterostructures, principles, mechanisms, and recent advances: a critical review. International Journal of Hydrogen Energy 48 (84), 32708–32728. Available at https://doi.org/10.1016/j.ijhydene.2023.05.066.

Moradian, S., Badiei, A., Mohammadi Ziarani, G., Mohajer, F., Varma, R.S., Iravani, S., 2023. Black phosphorus-based photocatalysts: synthesis, properties, and applications. Environmental Research 237 (Pt 1), 116910. Available at https://doi.org/10.1016/j.envres.2023.116910.

Murali, G., Reddy Modigunta, J.K., Park, Y.H., Lee, J.H., Rawal, J., Lee, S.Y., et al., 2022. A review on MXene synthesis, stability, and photocatalytic applications. ACS Nano 16 (9), 13370–13429. Available at https://doi.org/10.1021/acsnano.2c04750.

Narkbuakaew, T., Intasanta, V., 2022. Modified g-C_3N_4 with boron doping for efficient simultaneous catalytic reduction of Ag+ and organic pollutants. Materials Today Sustainability 20. Available at https://doi.org/10.1016/j.mtsust.2022.100258. 100258.

Novoselov, K.S., Geim, A.K., Morozov, S.V., Jiang, D., Zhang, Y., Dubonos, S.V., et al., 2004. Electric field effect in atomically thin carbon films. Science (New York, N.Y.) 306 (5696), 666–669. Available at https://doi.org/10.1126/science.1102896.

Opoku, F., Govender, K.K., Van Sittert, C.G.C.E., Govender, P.P., 2017. Role of MoS_2 and WS_2 monolayers on photocatalytic hydrogen production and the pollutant degradation of monoclinic $BiVO_4$: a first-principles study. New Journal of Chemistry 41 (20), 11701–11713. Available at https://doi.org/10.1039/c7nj02340e.

Pathania, D., Thakur, M., Sharma, A., 2019. Photocatalytical degradation of pesticides. Nano-Materials as Photocatalysts for Degradation of Environmental Pollutants: Challenges and Possibilities. Elsevier, India, pp. 153–172. Available at https://doi.org/10.1016/B978-0-12-818598-8.00009-2.

Pelaez, M., Nolan, N.T., Pillai, S.C., Seery, M.K., Falaras, P., Kontos, A.G., et al., 2012. A review on the visible light active titanium dioxide photocatalysts for environmental applications. Applied Catalysis B: Environmental 125, 331–349. Available at https://doi.org/10.1016/j.apcatb.2012.05.036.

Peñas-Garzón, M., Sampaio, M.J., Chávez, A.M., Silva, A.M.T., Silva, C.G., Faria, J.L., 2024. Film-immobilized carbon nitride for the degradation of pharmaceuticals: a continuous flow photoreactor performance upheld by excitation-emission matrix fluorescence spectroscopy. Chemical Engineering Journal 500. Available at https://doi.org/10.1016/j.cej.2024.157384.

Prakash, A., Dan, M., Yu, S., Wei, S., Li, Y., Wang, F., et al., 2018. In_2S_3/CuS nanosheet composite: an excellent visible light photocatalyst for H_2 production from H_2S. Solar Energy Materials and Solar Cells 180, 205–212. Available at https://doi.org/10.1016/j.solmat.2018.03.011.

Qian, J., Zhao, S., Dang, W., Liao, Y., Zhang, W., Wang, H., et al., 2021. Photocatalytic nitrogen reduction by Ti_3C_2 MXene derived oxygen vacancy-rich C/TiO_2. Advanced Sustainable Systems 5 (4). Available at https://doi.org/10.1002/adsu.202000282.

Qing, Q., Luo, J., Liu, S., Wang, J., Wang, Z., Xiong, X.G., et al., 2025. General synthesis of covalent organic frameworks under ambient condition within minutes via microplasma electrochemistry approach. Nature Communications 16 (1). Available at https://doi.org/10.1038/s41467-025-57892-6.

Quintana, M.A., Rodriguez-Padrón, D., Jiménez-Calvo, P., Calero, M., Solís, R.R., Muñoz-Batista, M.J., 2025. Selective production of aldehydes: from traditional alternatives to alcohol photo-oxidation using g-C_3N_4 -based materials. Materials Advances 6 (12), 3760–3784. Available at https://doi.org/10.1039/d5ma00229j.

Rahman, A., Jennings, J.R., Tan, A.L., Khan, M.M., 2022. Molybdenum disulfide-based nanomaterials for visible-light-induced photocatalysis. Darussalam ACS Omega 7 (26), 22089–22110. Available at https://doi.org/10.1021/acsomega.2c01314.

Ramalingam, G., Perumal, N., Priya, A.K., Rajendran, S., 2022. A review of graphene-based semiconductors for photocatalytic degradation of pollutants in wastewater. Chemosphere 300, 134391. Available at https://doi.org/10.1016/j.chemosphere.2022.134391.

Rathi, B.S., Kumar, P.S., Vo, D.V.N., 2021. Critical review on hazardous pollutants in water environment: occurrence, monitoring, fate, removal technologies and risk assessment. Science of the Total Environment 797. Available at https://doi.org/10.1016/j.scitotenv.2021.149134.

Rizzo, L., Malato, S., Antakyali, D., Beretsou, V.G., Đolić, M.B., Gernjak, W., et al., 2019. Consolidated vs new advanced treatment methods for the removal of contaminants of emerging concern from urban wastewater. Science of the Total Environment 655, 986–1008. Available at https://doi.org/10.1016/j.scitotenv.2018.11.265.

Rono, N., Kibet, J.K., Martincigh, B.S., Nyamori, V.O., 2021. A review of the current status of graphitic carbon nitride. Critical Reviews in Solid State and Materials Sciences 46 (3), 189–217. Available at https://doi.org/10.1080/10408436.2019.1709414.

Roy, S., Bermel, P., 2018. Electronic and optical properties of ultra-thin 2D tungsten disulfide for photovoltaic applications. Solar Energy Materials and Solar Cells 174, 370–379. Available at https://doi.org/10.1016/j.solmat.2017.09.011.

Sahoo, D., Tyagi, S., Agarwal, S., Shakya, J., Ali, N., Yoo, W.J., et al., 2023. Cost-effective and highly efficient manganese-doped MoS_2 nanosheets as visible-light-driven photocatalysts for wastewater treatment. Langmuir: The ACS Journal of Surfaces and Colloids 39 (20), 7109–7121. Available at https://doi.org/10.1021/acs.langmuir.3c00390.

Sahoo, S., Mahamallik, P., Das, R., Panigrahi, S., 2024. A critical review on non-metal doped g-C_3N_4 based photocatalyst for organic pollutant remediation with sustainability assessment by life cycle analysis. Environmental Research 258, 119390. Available at https://doi.org/10.1016/j.envres.2024.119390.

Saikam, L., Arthi, P., Senthil, B., Shanmugam, M., 2023. A review on exfoliated graphite: synthesis and applications. Inorganic Chemistry Communications 152, 110685. Available at https://doi.org/10.1016/j.inoche.2023.110685.

Shabil Sha, M., Anwar, H., Musthafa, F.N., Al-Lohedan, H., Alfarwati, S., Rajabathar, J.R., et al., 2024. Photocatalytic degradation of organic dyes using reduced graphene oxide (rGO). Scientific Reports 14 (1). Available at https://doi.org/10.1038/s41598-024-53626-8.

Shen, Y., Shen, Q., Zhu, C., Lin, Y., Fang, Q., Liu, R., et al., 2025. Pyridinic nitrogen-modulated carbon nanotubes facilitate dual-electron transfer for polymerization-enhanced pollutant removal. Applied Catalysis B: Environment and Energy 378, 125579. Available at https://doi.org/10.1016/j.apcatb.2025.125579.

Shu, J., Zhang, Y., Wang, K., Wang, J., Ying, J., Wang, H., 2024. Membrane hybrid system assembled with MXene@h-BN for efficient typical pharmaceuticals removal and self-cleaning: synergetic effect on adsorption-photocatalytic process. Chemical Engineering Journal 494, 152889. Available at https://doi.org/10.1016/j.cej.2024.152889.

Singh, A., Majumder, A., Saidulu, D., Bhattacharya, A., Bhatnagar, A., Gupta, A.K., 2024. Oxidative treatment of micropollutants present in wastewater: a special emphasis on transformation products, their toxicity, detection, and field-scale investigations. Journal of Environmental Management 354, 120339. Available at https://doi.org/10.1016/j.jenvman.2024.120339.

Solís, R.R., Quintana, M.A., Martín-Lara, M.Á., Pérez, A., Calero, M., Muñoz-Batista, M.J., 2022. Boosted activity of g-C_3N_4/UiO-66-NH_2 heterostructures for the photocatalytic degradation of contaminants in water. International Journal of Molecular Sciences 23 (21). Available at https://doi.org/10.3390/ijms232112871.

Splendiani, A., Sun, L., Zhang, Y., Li, T., Kim, J., Chim, C.Y., et al., 2010. Emerging photoluminescence in monolayer MoS_2. Nano Letters 10 (4), 1271–1275. Available at https://doi.org/10.1021/nl903868w.

Subbaiah, M.P., Kalimuthu, P., Jung, J., Jeon, B.H., 2021. Recent advances in effective capture of inorganic mercury from aqueous solutions through sulfurized 2D-material-based adsorbents. Journal of Materials Chemistry A 9 (34), 18086–18101. Available at https://doi.org/10.1039/d1ta01018b.

Sun, J., Zhang, H., Guo, L.H., Zhao, L., 2013. Two-dimensional interface engineering of a titania-graphene nanosheet composite for improved photocatalytic activity. ACS Applied Materials and Interfaces 5 (24), 13035–13041. Available at https://doi.org/10.1021/am403937y.

The European Parliament and the Council of the European Union, 2024. Directive (EU) 2024/3019 of the European Parliament and of the Council of 27 November 2024 concerning urban wastewater treatment (recast). Available from: https://eur-lex.europa.eu/eli/dir/2024/3019/oj/eng.

Trentin, O., Muñoz-Batista, M.J., Perosa, A., Selva, M., Rodriguez-Padron, D., 2025. Mechanochemically engineered functional materials: advancing photocatalysis for sustainable fuels. Advanced Functional Materials. Available at https://doi.org/10.1002/adfm.202506860.

Vasantha Raman, N., Gebreyohanes Belay, B.M., South, J., Botha, T.L., Pegg, J., Khosa, D., et al., 2024. Effect of an antidepressant on aquatic ecosystems in the presence of microplastics: a mesocosm study. Environmental Pollution 357. Available at https://doi.org/10.1016/j.envpol.2024.124439.

Viet, N.M., Trung, D.Q., Giang, B.L., Tri, N.L.M., Thao, P., Pham, T.H., et al., 2019. Noble metal -doped graphitic carbon nitride photocatalyst for enhancement photocatalytic decomposition of antibiotic pollutant in wastewater under visible light. Journal of Water Process Engineering 32. Available at https://doi.org/10.1016/j.jwpe.2019.100954.

Vigneshwaran, S., Park, C.M., Meenakshi, S., 2021. Designed fabrication of sulfide-rich bi-metallic-assembled MXene layered sheets with dramatically enhanced photocatalytic performance for Rhodamine B removal. Separation and Purification Technology 258, 118003. Available at https://doi.org/10.1016/j.seppur.2020.118003.

Vinesh, V., Ashokkumar, M., Neppolian, B., 2020. rGO supported self-assembly of 2D nano sheet of (g-C_3N_4) into rod-like nano structure and its application in sonophotocatalytic degradation of an antibiotic. Ultrasonics Sonochemistry 68. Available at https://doi.org/10.1016/j.ultsonch.2020.105218. 105218.

Wang, D., Li, Y., Li Puma, G., Wang, C., Wang, P., Zhang, W., et al., 2015. Mechanism and experimental study on the photocatalytic performance of Ag/AgCl @ chiral TiO_2 nanofibers photocatalyst: the impact of wastewater components. Journal of Hazardous Materials 285, 277–284. Available at https://doi.org/10.1016/j.jhazmat.2014.10.060.

Wang, J., Tang, L., Zeng, G., Deng, Y., Liu, Y., Wang, L., et al., 2017. Atomic scale g-C_3N_4/Bi_2WO_6 2D/2D heterojunction with enhanced photocatalytic degradation of ibuprofen under visible light irradiation. Applied Catalysis B: Environmental 209, 285–294. Available at https://doi.org/10.1016/j.apcatb.2017.03.019.

Wang, W., Zeng, Z., Zeng, G., Zhang, C., Xiao, R., Zhou, C., et al., 2019. Sulfur doped carbon quantum dots loaded hollow tubular g-C_3N_4 as novel photocatalyst for destruction of *Escherichia coli* and tetracycline degradation under visible light. Chemical Engineering Journal 378. Available at https://doi.org/10.1016/j.cej.2019.122132.

Wang, Y., Guo, T., Tian, Z., Bibi, K., Zhang, Y.Z., Alshareef, H.N., 2022. MXenes for energy harvesting. Advanced Materials 34 (21). Available at https://doi.org/10.1002/adma.202108560.

Wang, H., Duan, Y., Fei, G., Yan, T., Kang, Y.M., Dionysiou, D.D., 2023a. Design, synthesis and modification of 2D nanomaterials-based photocatalysts for pollutant degradation and photodegradation experiments from lab-scale to grand-scale. Chemical Engineering Journal 477. Available at https://doi.org/10.1016/j.cej.2023.147219.

Wang, H., Zou, Y., Wang, W., Zhang, Y., Mailhot, G., Li, J., et al., 2023b. Quantitative structure-activity relationship for the photooxidation of aromatic micro-pollutants induced by graphene oxide in water. Chemosphere 315, 137781. Available at https://doi.org/10.1016/j.chemosphere.2023.137781.

Wang, J.H., Kong, F., Liu, B.F., Zhuo, S.N., Ren, N.Q., Ren, H.Y., 2024. Enhanced photocatalytic degradation of diclofenac by UiO-66/MgAl-LDH: excellent performances and mechanisms. Environmental Science: Nano 11 (8), 3286–3293. Available at https://doi.org/10.1039/d4en00266k.

Wei, Y., Fu, H., Zhao, C., Zhang, L., Wang, J.-F., Liu, W., et al., 2025. Accurate targeted detoxification of fluoroquinolone antibiotics via fenton-like reaction catalyzed by asymmetric oxygen vacancy on cobalt-doped In_2O_3. Applied Catalysis B: Environment and Energy 372, 125290. Available at https://doi.org/10.1016/j.apcatb.2025.125290.

Wilkinson, J.L., Hooda, P.S., Barker, J., Barton, S., Swinden, J., 2016. Ecotoxic pharmaceuticals, personal care products, and other emerging contaminants: a review of environmental, receptor-mediated, developmental, and epigenetic toxicity with discussion of proposed toxicity to humans. Critical Reviews in Environmental Science and Technology 46 (4), 336–381. Available at https://doi.org/10.1080/10643389.2015.1096876.

World Economic Forum, 2025. The Global Risks Report 2025: 20th Edition. Available from: https://www.weforum.org/publications/global-risks-report-2025/

Wu, L., Zhang, L., Liu, R., Ge, H., Tao, Z., Meng, Q., et al., 2021. 2D/2D g-C_3N_4/1T-MoS_2 nanohybrids as schottky heterojunction photocatalysts for nuclear wastewater pretreatment. ACS ES and T Water 1 (10), 2197–2205. Available at https://doi.org/10.1021/acsestwater.1c00097.

Wu, S.C., Cheng, P., Han, J.J., Chen, Y., Yan, X., Guo, X.J., et al., 2022. Construction of two-dimensional Ag/WS_2 hybrid membranes with self-cleaning ability by photocatalysis for efficient water filtration. Journal of Membrane Science 641. Available at https://doi.org/10.1016/j.memsci.2021.119865.

Wu, D., Gu, N., Yao, J., Cao, Y., Wang, L., Shakir, I., et al., 2025. Recent advances in room-temperature synthesis of covalent organic frameworks. Chemical Science 16 (13), 5447–5463. Available at https://doi.org/10.1039/d5sc00109a.

Xia, D., An, T., Li, G., Wang, W., Zhao, H., Wong, P.K., 2016. Synergistic photocatalytic inactivation mechanisms of bacteria by graphene sheets grafted plasmonic AgAgX (X=Cl, Br, I) composite photocatalyst under visible light irradiation. Water Research 99, 149–161. Available at https://doi.org/10.1016/j.watres.2016.04.055.

Xiao, J., Chen, J., Liu, J., Ihara, H., Qiu, H., 2023. Synthesis strategies of covalent organic frameworks: an overview from nonconventional heating methods and reaction media. Green Energy & Environment 8 (6), 1596–1618. Available at https://doi.org/10.1016/j.gee.2022.05.003.

Xie, W., Liu, G., Liu, Y., Bai, Y., Liao, Y., Li, T., et al., 2024. Multidimensional TiO_2 photocatalysts for the degradation of organic dyes in wastewater treatment. Journal of Porous Materials 31 (5), 1655–1681. Available at https://doi.org/10.1007/s10934-024-01619-3.

Yadav, S., Shakya, K., Gupta, A., Singh, D., Chandran, A.R., Varayil Aanappalli, A., et al., 2023. A review on degradation of organic dyes by using metal oxide semiconductors. Environmental Science and Pollution Research 30 (28), 71912–71932. Available at https://doi.org/10.1007/s11356-022-20818-6.

Yan, S.C., Li, Z.S., Zou, Z.G., 2010. Photodegradation of Rhodamine B and Methyl Orange over Boron-Doped g-C_3N_4 under visible light irradiation. Langmuir: The ACS Journal of Surfaces and Colloids 26 (6), 3894–3901. Available at https://doi.org/10.1021/la904023j.

Yan, W., Yan, L., Jing, C., 2019. Impact of doped metals on urea-derived g-C_3N_4 for photocatalytic degradation of antibiotics: structure, photoactivity and degradation mechanisms. Applied Catalysis B: Environmental 244, 475–485. Available at https://doi.org/10.1016/j.apcatb.2018.11.069.

Yang, M.Q., Zhang, N., Xu, Y.J., 2013. Synthesis of fullerene-, carbon nanotube-, and graphene-TiO_2 nanocomposite photocatalysts for selective oxidation: a comparative study. ACS Applied Materials and Interfaces 5 (3), 1156–1164. Available at https://doi.org/10.1021/am3029798.

Yang, R., Fan, Y., Zhang, Y., Mei, L., Zhu, R., Qin, J., et al., 2023a. 2D transition metal dichalcogenides for photocatalysis. Angewandte Chemie - International Edition 62 (13). Available at https://doi.org/10.1002/anie.202218016.

Yang, S., Zhang, F., Shang, Y., Luo, L., Liu, Z., 2023b. Highly efficient photocatalytic degradation of refractory organic pollutants onto designed boron nitride: morphology control and oxygen doping. Journal of Cleaner Production 429, 139532. Available at https://doi.org/10.1016/j.jclepro.2023.139532.

Yang, F., Hu, P., Yang, F., Hua, X.J., Chen, B., Gao, L., et al., 2024. Photocatalytic applications and modification methods of two-dimensional nanomaterials: a review. Tungsten 6 (1), 77–113. Available at https://doi.org/10.1007/s42864-023-00229-x.

Yao, L., Chen, X., Yu, C., Chen, J., Lai, Y., Li, N., 2022. Metal-free photocatalyst of few-layer phosphorene with excellent activity under different light conditions. Materials Letters 306, 130884. Available at https://doi.org/10.1016/j.matlet.2021.130884.

Yeh, T.-F., Syu, J.-M., Cheng, C., Chang, T.-H., Teng, H., 2010. Graphite oxide as a photocatalyst for hydrogen production from water. Advanced Functional Materials 20 (14), 2255–2262. Available at https://doi.org/10.1002/adfm.201000274.

Yesilay, G., dos Santos, O.A.L., A, B.R., Hazeem, L.J., Backx, B.P., J, J.V., et al., 2023. Impact of pathogenic bacterial communities present in wastewater on aquatic organisms: application of nanomaterials for the removal of these pathogens. Aquatic Toxicology 261. Available at https://doi.org/10.1016/j.aquatox.2023.106620.

Yousef, A., Thiehmed, Z., Abdul Shakoor, R., Altahtamouni, T., 2022. Recent progress in WS_2-based nanomaterials employed for photocatalytic water treatment. Catalysts 12 (10), 1138. Available at https://doi.org/10.3390/catal12101138.

Yu, W., Wang, K., Li, H., Ma, T., Wu, Y., Shang, Y., et al., 2024a. An updated review of few-layer black phosphorus serving as a promising photocatalyst: synthesis, modification and applications. Nanoscale 16 (41), 19131–19173. Available at https://doi.org/10.1039/d4nr02567a.

Yu, Z., Li, F., Xiang, Q., 2024b. Carbon dots-based nanocomposites for heterogeneous photocatalysis. Journal of Materials Science and Technology 175, 244–257. Available at https://doi.org/10.1016/j.jmst.2023.08.023.

Zeng, M., Chen, M., Huang, D., Lei, S., Zhang, X., Wang, L., et al., 2021. Engineered two-dimensional nanomaterials: an emerging paradigm for water purification and monitoring. Materials Horizons 8 (3), 758–802. Available at https://doi.org/10.1039/d0mh01358g.

Zhang, Y., Tang, Z.R., Fu, X., Xu, Y.J., 2011. Engineering the unique 2D mat of graphene to achieve graphene-TiO_2 nanocomposite for photocatalytic selective transformation: what advantage does graphene have over its forebear carbon nanotube? ACS Nano 5 (9), 7426–7435. Available at https://doi.org/10.1021/nn202519j.

Zhang, Z., Wang, W., Wang, L., Sun, S., 2012. Enhancement of visible-light photocatalysis by coupling with narrow-band-gap semiconductor: a case study on Bi_2S_3/Bi_2WO_6. ACS Applied Materials and Interfaces 4 (2), 593–597. Available at https://doi.org/10.1021/am2017199.

Zhang, C., Li, Y., Wang, D., Zhang, W., Wang, Q., Wang, Y., et al., 2015a. Ag@helical chiral TiO_2 nanofibers for visible light photocatalytic degradation of 17α-ethinylestradiol. Environmental Science and Pollution Research 22 (14), 10444–10451. Available at https://doi.org/10.1007/s11356-015-4251-y.

Zhang, Y., Li, H., Wang, H., Liu, R., Zhang, S.L., Qiu, Z.J., 2015b. On valence-band splitting in layered MoS2. ACS Nano 9 (8), 8514–8519. Available at https://doi.org/10.1021/acsnano.5b03505.

Zhang, J., Xie, W., Zhao, J., Zhang, S., 2017. Band alignment of two-dimensional lateral heterostructures. 2D Materials 4 (1), 015038. Available at https://doi.org/10.1088/2053-1583/aa50cc.

Zhang, C., Li, Y., Zhang, W., Wang, P., Wang, C., 2018. Metal-free virucidal effects induced by g-C_3N_4 under visible light irradiation: statistical analysis and parameter optimization. Chemosphere 195, 551–558. Available at https://doi.org/10.1016/j.chemosphere.2017.12.122.

Zhang, Z., He, D., Zhou, Y., Bai, E., Qu, J., Zhang, Y.-nan, 2024. Fabrication of black phosphorus/CdS heterostructure with enhancement photocatalytic degradation activity for tetrabromobisphenol A and toxicity prediction of intermediates. Environmental Research 256, 119060. Available at https://doi.org/10.1016/j.envres.2024.119060.

Zhao, J., Huang, S., Ravisankar, P., Zhu, H., 2020a. Two-dimensional nanomaterials for photoinduced antibacterial applications. ACS Applied Bio Materials 3 (12), 8188–8210. Available at https://doi.org/10.1021/acsabm.0c00950.

Zhao, Y., Zhang, S., Shi, R., Waterhouse, G.I.N., Tang, J., Zhang, T., 2020b. Two-dimensional photocatalyst design: a critical review of recent experimental and computational advances. Materials Today 34, 78–91. Available at https://doi.org/10.1016/j.mattod.2019.10.022.

Zheng, Y., Chen, Y., Gao, B., Lin, B., Wang, X., 2021. Phosphorene-based heterostructured photocatalysts. Engineering 7 (7), 991–1001. Available at https://doi.org/10.1016/j.eng.2021.06.004.

Zhong, J., Ahmed, Y., Carvalho, G., Wang, Z., Wang, L., Mueller, J.F., et al., 2022. Simultaneous removal of micropollutants, antibiotic resistant bacteria, and antibiotic resistance genes using graphitic carbon nitride under simulated solar irradiation. Chemical Engineering Journal 433, 133839. Available at https://doi.org/10.1016/j.cej.2021.133839.

Zhou, Y., Yu, M., Liang, H., Chen, J., Xu, L., Niu, J., 2021. Novel dual-effective Z-scheme heterojunction with g-C_3N_4, Ti_3C_2 MXene and black phosphorus for improving visible light-induced degradation of ciprofloxacin. Applied Catalysis B: Environmental 291, 120105. Available at https://doi.org/10.1016/j.apcatb.2021.120105.

Zhu, B., Cheng, B., Fan, J., Ho, W., Yu, J., 2021. g-C_3N_4-based 2D/2D composite heterojunction photocatalyst. Small Structures 2 (12). Available at https://doi.org/10.1002/sstr.202100086.

CHAPTER

10

Antibacterial mechanisms of 2D nanomaterials: pathways to safer water purification

Hamyal Awais[1], *Usman Liaqat*[1], *Afira Khan*[1], *Mudassar Azam*[2], *Waheed Miran*[1,3], *and Sajjad Miran*[4]

[1]School of Chemical and Materials Engineering, National University of Sciences and Technology (NUST), Islamabad, Pakistan [2]Institute of Chemical Engineering and Technology, University of the Punjab, Lahore, Pakistan [3]SDU Biotechnology, Department of Green Technology, University of Southern Denmark, Odense, Denmark [4]Department of Mechanical Engineering Technology, University of Gujrat, Gujrat, Pakistan

10.1 Introduction to antibacterial properties of 2D nanomaterials

Water is a key resource of life and a pillar of civilization and growth of humanity, and even space travel. Nonetheless, only approximately 3% of water resources on the planet have the form of freshwater, and most of them are concentrated in groundwater, icecaps, and glaciers (Fu et al., 2016; Van Engelenburg, Van Slobbe and Hellegers, 2019). Freshwater shortage and pollution are becoming matters of emergency since the world is growing in population and largely industrializing at an alarming rate. Traditional methods of water treatment, including coagulation, flocculation, electrolysis, and sedimentation, have had a long history of success, but can be large, labor-intensive, and expensive. They can no longer overcome the complexity of contemporary water pollutants, that is, heavy metals, pathogens, pesticides, pharmaceuticals, and volatile organic compounds (VOCs) (Bi et al., 2020; Chen et al., 2020). Nanotechnology has been introduced into the world of water purification and, in particular, the incorporation of two-dimensional nanomaterials (2DNMs) into the developed water treatment systems in the 21st century.

2DNMs are atomically thin with enormous lateral size providing unique features of high surface-to-volume ratio and tunable surface functionalities, fast mass transport, and high mechanical and chemical stability (Anichini et al., 2018; Chimene, Alge and Gaharwar, 2015). They can also be widely used in water purification in the form of multifunctional

Two-Dimensional Nanomaterials for Water Purification
DOI: https://doi.org/10.1016/B978-0-443-44530-9.00010-9

membrane filters, adsorption, photocatalysis, and their sensors due to their structural anisotropy and high activity (Wang et al., 2016a). Some of the typical 2DNMs applied in treating water are graphene, graphene oxide (GO), graphitic carbon nitride (g-C_3N_4), hexagonal boron nitride (h-BN), transition metal dichalcogenides (TMDs), transition metal oxides (TMOs), MXenes, Metal-organic frameworks (MOFs), covalent organic frameworks (COFs), and natural clays such as montmorillonite and layered double hydroxides (LDHs).

One of them is graphene and its derivatives, which is characterized by ultrathin structure, electrical conductivity, and amazing mechanical strength. They can be used as desalination and molecular sieve membranes because they can selectively transport ions due to their distinct atomic structure and a distinct honeycomb lattice (Surwade et al., 2015; Yu et al., 2017). Likewise, GO membranes are highly hydrophilic and controllably increase their interlayer spaces that allow selective filtration. GO membranes have problems like swellings in watery conditions, which change the interlayer spacing and decrease the efficiency in filtration. To address these drawbacks, it has been suggested to use developments, such as cross-linking agents and vertical stacking on lamellar liquid-crystal phases (Zeng et al., 2021).

In addition to being membranes, 2DNMs have also proved to be high-performance adsorbents. They are suitable to adsorb a great diversity of pollutants due to their extensive surface area and adjustable surface attainments, such as oil, dyes, heavy metals such as Pb^{2+}, Hg^{2+}, and Cr^{6+}. (Ge et al., 2017; Kumar et al., 2017). The 2D surface functional groups coordinate recruits selectively with the contaminants via electrostatic attractions, pi–pi interactions, and coordination bonding. Nevertheless, the reuse of adsorbent, the aggregation in water, and recovery are also main issues that prevent the utility in practice (Premarathna et al., 2019; Zeng et al., 2021).

The newest photocatalysts in photocatalysis are ultrathin 2DNMs (e.g., g-C_3N_4) and TMDs (e.g., MoS_2), as well as MXenes, which utilize visible light in the degradation of pollutants. Their band gaps are narrow, and surface states are active, which allows performance of charge separation and production of reactive oxygen species (ROS) through solar irradiation, which effectively breaks down persistent organic pollutants (Liu et al., 2016). The incorporation of these photocatalysts into the membrane structure has also brought the concept of creating self-cleaning and antifouling membranes where the biofilms grow minimally and increase the service lifespan of the membranes (Oh et al., 2017).

2DNMs are also central to developing smart sensing platforms for real-time pollutant detection. Pathogens, heavy metals, and organic toxins can also be detected at a trace level due to the change in electric, electrochemical, or optical signal, which is owed to their surface sensitivity (Dehghani et al., 2016). Field-effect transistors (FETs) based on graphene, for example, have been used to realize ultralow detection thresholds in heavy metal detection applications. Even though these sensors are very sensitive, there is a problem of signal drift, long-term stability, and regeneration, which has slowly been solved through the advent of second-generation hybrid material systems and wireless integration.

Even with this development, 2DNM-based systems to purify water have not achieved much in terms of market translation. The primary challenges entail scalability in production, environmental friendliness, stability under extreme conditions, and affordability. Top-down synthesis in the form of exfoliation is more scalable than bottom-up synthesis, such as chemical vapor deposition (CVD), producing many 2DNMs. However, purity, consistency, and defect management are of concern. Moreover, their activities at the aqueous colloidal

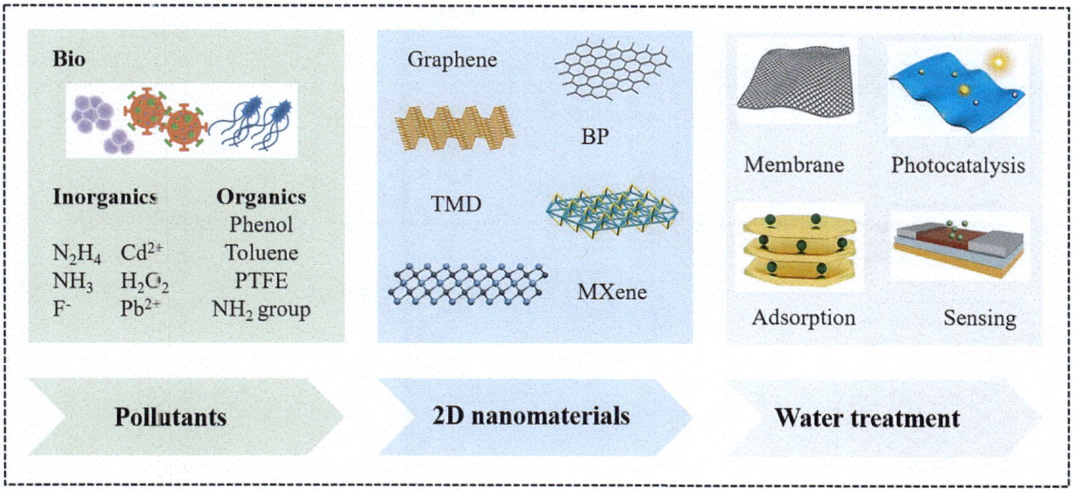

FIGURE 10.1 Schematic diagrams of water treatment through the use of the versatile nature of 2D nanomaterials in membrane separation, adsorption, photocatalysis, and sensor reactions to pollutants.

stages, such as their aggregation and leaching, have to be well recognized in order to mitigate ecological and health risks.

By determining which areas of future research are to be conducted, it is possible to list rational structure-property design, defect engineering, multifunctional integration, and environmentally benign synthesis methods. An example is to have a programmed superlattice and modes where a single nanoplatform can deploy adsorption, sensing, and catalysis using a programmed response. Manufacture by robot assembly and machine learning directed materials discovery is also emerging to enable tailoring of specific physical properties and predictive optimization of nanomaterials to a given contaminant (Zeng et al., 2021). 2DNMs have disrupted water purification due to their diverse physicochemical feature, yet great challenges will have to be overcome to progress the stage between laboratory and actual practice. The partnership between different disciplines and policy will play a major role in narrowing down this divide, which, in time, will guarantee access to clean, safe, and cheap water to all. shows water treatment through the use of the versatile nature of 2D nanomaterials in membrane separation, adsorption, photocatalysis, and sensor reactions to pollutants (Fig. 10.1).

10.2 Forms of 2D nanoparticles in water management

The high level of physicochemical properties and ultrathin structure of 2DNMs with large surface areas, distinctive electronic properties, and multiple, multifunctional applications have resulted in the great interest in these materials in environmental science and material research over the past few years (Huang et al., 2020). Since the path breaking exfoliation of monolayer graphene by the group of Geim at the University of Manchester, that heralded

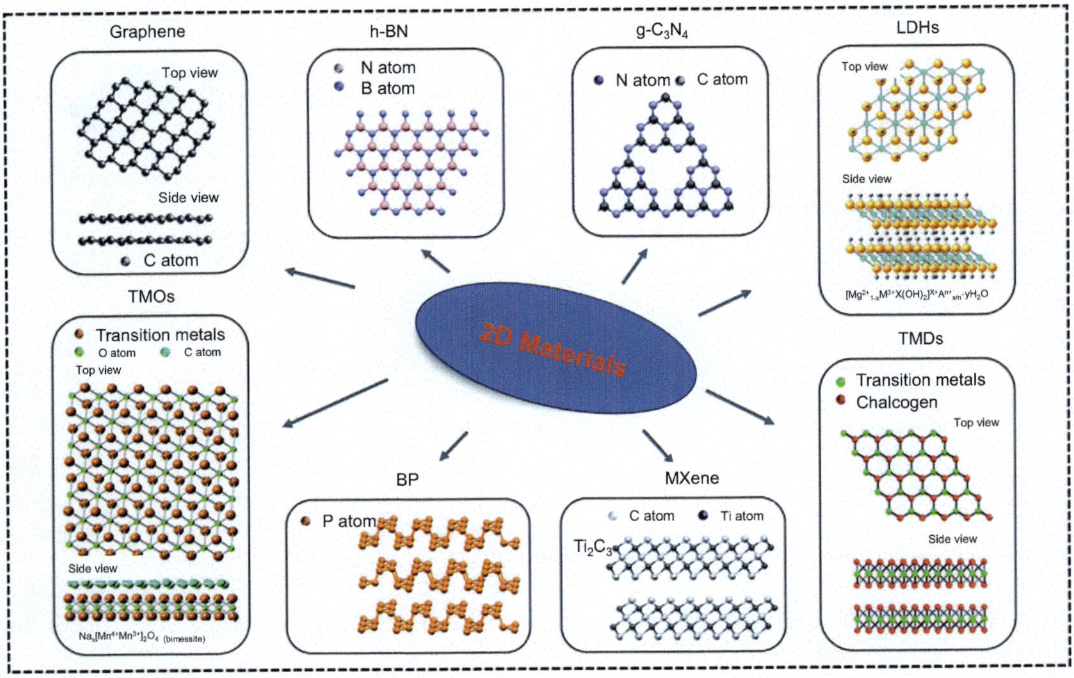

FIGURE 10.2 Classification and atomic structures of various 2D materials. This schematic illustrates the atomic configurations and structural diversity of key classes of two-dimensional (2D) materials. Represented categories include graphene, hexagonal boron nitride (h-BN), graphitic carbon nitride (g-C_3N_4), layered double hydroxides (LDHs), transition metal oxides (TMOs), black phosphorus (BP), MXenes, and transition metal dichalcogenides (TMDs). *From Raza, S., Ghasali, E., Orooji, Y., Lin, H., Karaman, C., Dragoi, E.N., Erk, N., 2023. Two dimensional (2D) materials and biomaterials for water desalination: structure, properties, and recent advances. Environmental Research, 219. https://doi.org/10.1016/j.envres.2022.114998.*

the emergence of 2D materials, other 2DNMs have been discovered, including g-C_3N_4 (Yu et al., 2017), MoS_2 (Seyler et al., 2019), and tungsten disulfide (WS_2), titanium carbide-based MXenes (Shahzad et al., 2017; Zhao et al., 2018; Huang et al., 2018) bismuth-based materials (Tian et al., 2013), and boron nitride (BN) (Gonzalez-Ortiz et al., 2020; Khan et al., 2017). The different materials show unique optical, electronic, and chemical characteristics such that they can be well applied in a variety of functions like biosensing, photothermal therapy, and photocatalysis. 2DNMs such as graphene, TMOs, TMDs, LDHs, and BNMs have shown enormous potential in adsorptive, catalytic, and separation-based types of environmental remediation applications (Huang et al., 2020). Various 2D nanomaterials and schematic nanostructures are illustrated in Fig. 10.2.

10.3 Mechanisms of antibacterial action in 2D nanomaterials

The 2DNMs have a thin atomic structure with a high surface area that enables direct contact with microbial cells and pollutants and further increases the purification reaction efficacy (Chen et al., 2016). The high surface area and tunable chemistry enhance adsorption, separation, and disinfection (Liu et al., 2019; Manna et al., 2015). These materials treat water by virtue of various antibacterial pathways involving physical disturbance, photothermal interactions, the production of reactive oxygen species, and Fenton-type oxidation (Koul et al., 2011), as shown in Fig. 10.3. These various functionalities render 2DNMs equally suitable as the next-generation water treatment systems (Li et al., 2015).

10.3.1 Stress on bacterial cells through physical contact

One of the primary antibacterial actions of 2DNMs in water purification is their ability to physically interact with microbial membranes, leading to mechanical damage. This is primarily attributed to their atomically sharp edges and high aspect ratio, allowing materials like graphene and MXene to act as nano-knives that disrupt the integrity of bacterial membranes upon contact (Chen et al., 2016). For instance, GO's wrinkled structure enables direct wrapping and entrapment of bacteria, disrupting cellular functions through mechanical cutting and oxidative stress (Koul et al., 2011). Similarly, MXene (e.g., $Ti_3C_2T_x$) has been shown to adhere to bacterial surfaces through electrostatic attraction and hydrogen bonding, followed by insertion of its nanosheet edges into membranes, causing leakage of cytoplasmic content and eventual cell lysis. The lateral dimensions and surface chemistry of

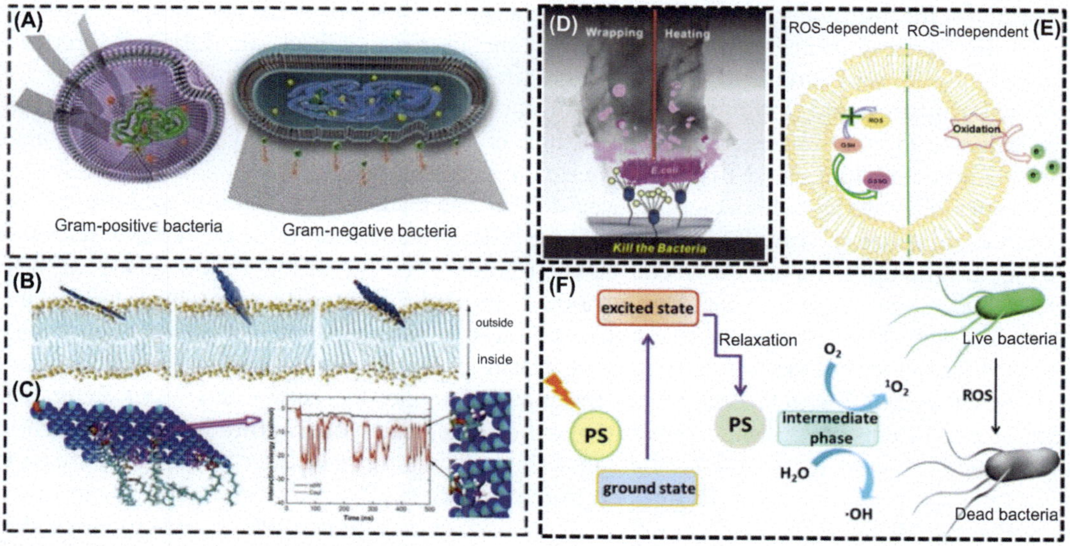

FIGURE 10.3 Antibacterial mechanisms of nanomaterials against bacteria. (A) Scheme of the various antibacterial actions of graphene-coated materials. (B) Terminal images of jectories in g-C_3N_4. (C) g-C_3N_4 nanosheet snapshot is interacting with the lipids of the bacterial membrane in three individual lipid heads at its pore site, and the interaction energies between one typical lipid and g -$C_{63}Ni_4$.(D) Antibacterial action of antibacterial graphene system (E) Diagram of oxidative stress of 2DM. (F) Photocatalyst antibacterial diagram. *(A), (B), (C), (E), and (F) Reproduced by permission from Li, B., Luo, Y., Zheng, Y., Liu, X., Tan, L., Wu, S., 2022. Two-dimensional antibacterial materials. Progress in Materials Science 130, 100976. https://doi.org/10.1016/j.pmatsci.2022.100976, Copyright 2022, Elsevier; (D) Adapted by permission Seeberger, P.H., 2017. Glycan arrays and other tools produced by automated glycan assembly. Perspectives in Science, 11, 11–17. https://doi.org/10.1016/j.pisc.2016.06.085, Copyright 2017, Elsevier.*

these materials significantly influence the efficiency of this mechanism. Small-sized nanosheets with oxygen-containing functional groups exhibit higher bactericidal performance due to enhanced interaction with bacterial walls (Shannon et al., 2008). In Fig. 10.3A, the interaction varies with Gram-positive and Gram-negative bacteria due to differences in cell wall composition, where the outer membrane in Gram-negative strains offers additional resistance, as shown by Fig. 10.3B,C. Nonetheless, the nano-blade action of 2DNMs proves effective across diverse microbial strains, supporting their use in portable water filters and antimicrobial coatings on filtration membranes (Chen et al., 2016).

10.3.2 Photothermal effects

The antibacterial activity of the 2DNMs in water purification systems is very significant because of photothermal effects. In Fig. 10.3D, shows photothermal disinfection involves bacteria killing caused by heat produced by nanomaterials in light irradiation (usually as near-infrared light (NIR) radiation). The applied power density is low. More specifically, NIR proves to be beneficial because it penetrates deeper into biological tissue (Wang et al., 2013). Some examples of nanomaterials that have shown to be effective during photothermal antibacterial activities are gold nanoparticles (NPs), magnetic NPs, and nanomaterials made of carbon (Chae et al., 2019). Graphene-based structures, in particular, reduced graphene oxide (rGO) is also of prominence among the 2DNMs (Qi et al., 2015). rGO nanosheets (NSs) decorated with multivalent sugar ligands have a high light-to-heat conversion performance under IR light and are capable of destroying bacteria (Qi et al., 2015). Moreover, glutaraldehyde functionalized rGO and glutaraldehyde Fe_3O_4 increase the number of bacterial captures and killing by virtue of the combined effect of glutaraldehyde cell attachment and rGO heat generation. At that, chitosan-functionalized magnetic GO also presents a bactericidal effect through synergistic performance.

Another 2D material of high significance, MoS_2 is also being engineered into multifunctional 2D nanocomposites such as CFM that can absorb NIR light and lead to the breakage of bacterial cell membranes upon hyperthermia effects. These composites interestingly show greater affinity to Gram-positive bacteria probably attributed to their stronger nonspecific electrostatic interaction with their negatively charged membranes (Inphonlek et al., 2010). By broadening the material range, photocatalysts harvesting plasmons in plasmonic semiconductor nanoparticles, such as MoO_3@cdots Broadening the material range, the compound MoO_3@xAg has demonstrated dual-mode disinfection. These nanosheets effectively capture NIR light which is transformed to local heat, and at the same time delivering antibacterial Ag ions, thereby improving the performance of disinfection (Xu, Yao and Xu, 2019). Those studies suggest that photothermal disinfection holds hope as a method of next-generation antimicrobial water treatment, and among uninvestigated 2DNMs, one might find many similar or better properties. Further research on these materials is needed in order to create effective, low-energy, and chemical-free approaches to water disinfection.

10.3.3 Photogeneration of reactive oxygen species

One of the strongest antibacterial effects of 2DNMs is their photocatalytic production of ROS, shown by Fig. 10.3E containing hydroxyl radicals ($^{\bullet}OH$), superoxide anions ($O_2^{\bullet -}$), and singlet oxygen ($^1O_2$). Such ROS have powerful oxidative characteristics and therefore destroy microbial proteins, lipids, and nucleic acids irreversibly and lead to the death of the bacteria (Ibrahim et al., 2016). Composite g-C_3N_4, BP, and MXene were found to be highly effective as photocatalysts even under visible and solar light, as their bandgap structure and electron mobility are quite auspicious (Li et al., 2015; Wang et al., 2009). Photoexcited electrons in these materials are capable of reducing molecular oxygen when irradiated with light, and the holes are symbolized by oxidizing water to produce ROS, which penetrate and oxidize a membrane (Krasner et al., 2006; Muellner et al., 2007). The graphene g-C_3N_4 composites can be used as a photocatalyst, which makes it possible to implement meaningful photocatalytic disinfection at low light power (Li et al., 2015). Equally, Ti_3C_2 MXene hybrids with TMDs or MOFs have longer light absorption and multichannel ROS production. Also, variation of the substitution of heterojunctions, oxygen vacancy, and metal loading provides high ROS yields and longer ROS lifetimes (Chen et al., 2016). Contrary to the conventional disinfecting agents, such as chlorine, which generate hazardous by-products, ROS-based disinfection is natural and selective, since it does not introduce chemical residuals but cleans pathogens. Such a photocatalytic antibacterial property has been significantly incorporated in multipurposed membranes and coatings to encompass point-of-use water to treat and clean-self-system based filters (Ibrahim et al., 2016).

10.3.4 Fenton reactions, Fenton-like reactions

Besides physical and photocatalytic processes, some 2DNMs are also involved in Fenton and Fenton-like reactions to produce hydroxyl radicals to exert their antibacterial effect to treat the water. The classic Fenton process entails reaction between hydrogen peroxide (H_2O_2) and Fe^{2+} ions to form a strong oxidizing agent called $^{\bullet}OH$. The transition metals (e.g., Fe, Co, Cu) doped 2DNMs or combined with other catalysts are useful in mimicking or enhancing this reaction (Chen et al., 2016). An example is that of Fe-based MXenes and graphene composites as catalysts in heterogeneous Fenton-like systems having better control of the kinetics of reaction as well as radical generation. The catalysts are particularly useful when they are employed in the treatment of complex water matrices where there is coexistence of microbial contaminants and organic pollutants. Production of ROS at localized sites of these systems is not only able to kill the microbes but also detoxify remaining antibiotics and pesticides to avoid resistance (Ibrahim et al., 2016), shown by Fig. 10.3F. In the reports in the recent years, Co-/Cu-TMDs have demonstrated the ability to produce ROS in the absence of H_2O_2 as a result of peroxidase-like activity, thereby rendering themselves as independent sources of antibacterial activity in aqueous settings. The major benefits of the 2DNM-assisted Fenton processes are their ability to be more stable, recyclable, and have a high surface activity that is appropriate to adopt the advanced oxidation processes (AOPs) in wastewater treatment facilities (Chen et al., 2016).

10.4 Fabrication techniques for antibacterial 2D nanomaterials

The synthesis of 2DNM is achieved by two broadly defined approaches: top-down and bottom-up, as shown in Fig. 10.4 (Wang et al., 2016a). The size of precursor material used, in

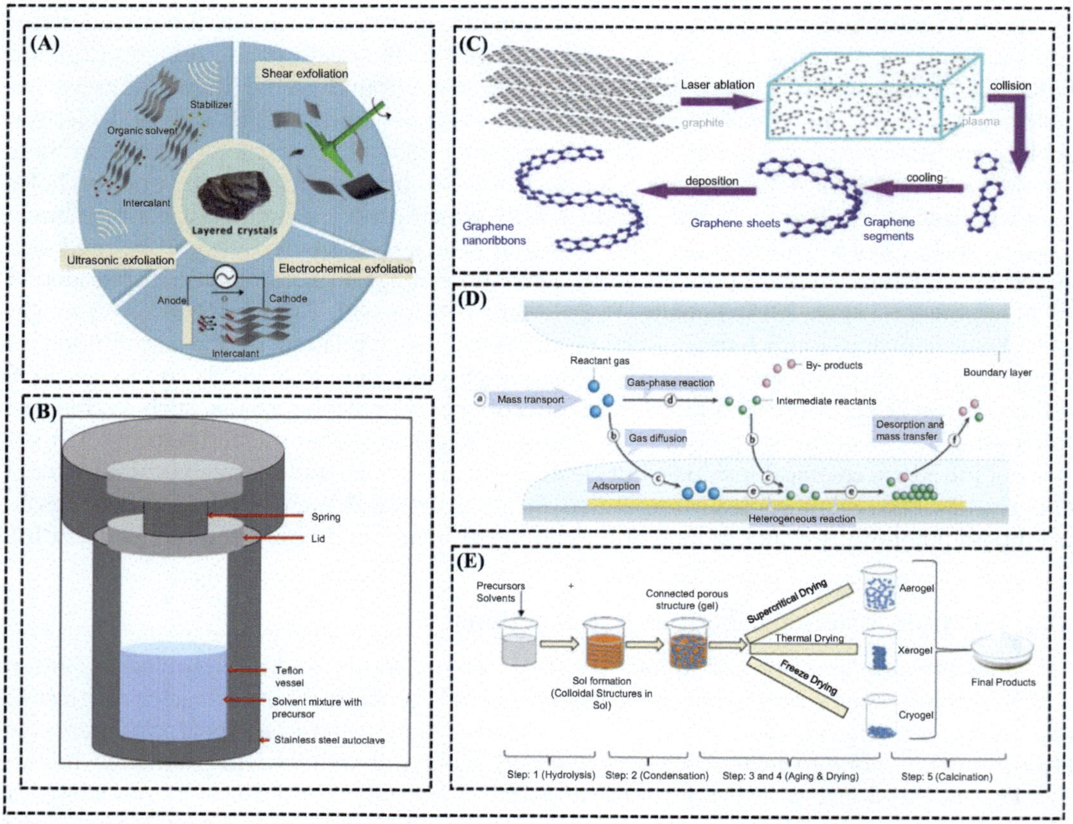

FIGURE 10.4 Overview of common synthesis methods for 2D materials. This figure presents a variety of synthesis techniques employed for the fabrication of two-dimensional (2D) materials. (A) 2D nanomaterials through the intercalation of bulk layered crystals. (B) A picture of an autoclave of solvo or hydrothermal synthesis (C) The synthesis of Graphene nanoribbons with focused laser radiation (D) Growth of 2D nanofilms through the CVD process. (E) Schematic of different steps involved in sol–gel method. *(A) Credit from Ren, Y., Xu, Q., 2018. Building close ties between CO_2 and functional two-dimensional nanomaterials with green chemistry strategy. Energy and Environmental Materials 1 (2), 46–60. https://doi.org/10.1002/eem2.12005, Copyright 2018, Wiley; (B) Credit from Mutalib, A.A.A., Jaafar, N.F., 2022. Potential of deep eutectic solvent in photocatalyst fabrication methods for water pollutant degradation: a review. Journal of Environmental Chemical Engineering 10 (3). https://doi.org/10.1016/j.jece.2022.107422, Copyright 2022, Elsevier; (C) Credit from Ye, F., Musselman, K.P., 2024. Synthesis of low dimensional nanomaterials by pulsed laser ablation in liquid. APL Materials 12 (5). https://doi.org/10.1063/5.0199104, Copyright 2024, American Institute of Physics; (D) Credit from Edward, K., Mamun, K., Narayan, S., Assaf, M., Rohindra, D., Rathnayake, U., 2023. State-of-the-art graphene synthesis methods and environmental concerns. Applied and Environmental Soil Science 2023. https://doi.org/10.1155/2023/8475504, Copyright 2023, Wiley; (E) Credit from Bokov, D., Turki Jalil, A., Chupradit, S., Suksatan, W., Javed Ansari, M., Shewael, I.H., Valiev, G.H., Kianfar, E., 2021. Nanomaterial by sol-gel method: synthesis and application. Advances in Materials Science and Engineering 2021. https://doi.org/10.1155/2021/5102014, Copyright 2021, Wiley.*

either of the two methods, distinguishes between the both Han et al. (2013). Based on the type of NM under fabrication and its required properties, the method is selected. This topic broadly covers extensively employed techniques.

10.4.1 Top-down technique

It involves controlled exfoliation or breakdown of a bulk material into one or several atoms thick sheets. Both chemical and physical methods can be employed in the construction of 2DNMs by top-down synthesis.

10.4.1.1 *Exfoliation*

As indicated from Fig. 10.3A, it is a straightforward method where mechanical force or high frequency sound waves (sonication) are targeted on a bulk material to agitate the particles and break Van der Waal forces between the layers, micro reduction of graphite to graphene (Novoselov et al., 2004; Yi and Shen, 2015), for instance upon repeated application, a single or multilayer thick material is obtained. For strong ionic solids, Ion-change exfoliation is applied, wherein ions are intercalated between the material layers, widening the space, facilitating exfoliation of the material.

10.4.1.2 *Laser ablation*

It refers to process whereby a targeted material is irradiated with high energy laser to remove the desired NM from the bulk. The melted or vaporized part is then solidified or condensed, respectively, to produce nanoparticles forming a sheet upon aggregation (Ye and Musselman, 2024) (Fig. 10.3C).

10.4.1.3 *Plasma treatment*

Ions and radicals from ionized gases break the interlayer forces to facilitate exfoliation. Strong oxidizing agents can also be used to weaken the Van der Waal force and break down the bulk material. The exfoliated particles then arrange themselves on a substrate to form thin nanosheets. Microwaves can also be used to excite the gas for plasma formation and then carry out the entire process.

10.4.2 Bottom-up technique

It is a comparatively challenging method requiring precision and accuracy, wherein atoms and molecule are aggregated or grown to achieve complex structures (Wang et al., 2016a). It results in the manufacturing of premium-grade NMs with reduced thickness and larger lateral dimension.

10.4.2.1 *Chemical vapor deposition*

It is a widely adopted strategy owing to its flexibility of adjusting process parameters, resulting in the production of high-quality pure NMs with controlled properties and defects (Li et al., 2015). CVD involves the deposition of gaseous precursors by adsorption or diffusion on specific substrates at elevated temperatures via a chemical reaction. Nucleation occurs, and atoms then reorganize to form a continuous 2D film (Fig. 10.3D).

10.4.2.2 *Hydro and solvo thermal*

They are wet chemical strategies known for optimized production rate and low cost. Hydro and solvothermal process starts by preparing a precursor solution using water or an organic solvent, respectively. The solute depends on the type of NM being synthesized. The solution is heated at relatively low temperatures of 100–200°C after being poured into a sealed stainless steel chamber, as illustrated in Fig. 10.3B. The solvent, under high pressure, vaporizes, and atoms arrange into thin 2D nanosheets.

10.4.2.3 *Sol–gel method*

A similar precursor solution, as in the hydro and solvothermal method, is prepared. Upon condensation and hydrolysis, the solution thickens to form a gel-like network which, upon drying, forms a dry porous material (Bokov et al., 2021). It is then heated to increase stability and remove residuals to achieve ultrathin 2D nanosheets. Refer to Fig. 10.3E for a schematic demonstration. This method proves to be cost-effective and provides precise control over structural properties.

10.4.2.4 *Surface functionalization*

It implies surface treatment to improve physiochemical traits of an NM. Similarly, to stabilize and enhance antimicrobial activity, NM surface is modified using various substances and strategies. Many research proves that Ag, Cu, and ZnO nanoparticles (NPs) tend to show high toxicity against bacteria and are effective in preventing bacterial attacks (Perveen et al., 2020; Shirzadi-Ahodashti et al., 2021; Suba et al., 2021). Ag and Cu ions are first reduced to NPs and then exposed to the NM surface for metal interaction. Ag-ZnO composite also proves synergistic in photocatalytic antibacterial action. The contact between a particular NM and NP produces ROS, which increase soxidative stress, causing bacterial cellular damage and ultimately cell death. Once NPs are synthesized, surface modification can be achieved by two methods having their own merits and demerits:

10.4.2.5 *Covalent functionalization*

It involves chemical bonding between the NM surface and NP by strong covalent attractions. It provides permanent attachment and a strengthened effect; however, the complexity involved in surface modification and structural destabilization via covalent functionalization serves as an obstacle.

10.4.2.6 *Noncovalent functionalization*

Nanoparticles attach to the nanomaterial surface via weak electrostatic or Van der Waals interactions. The physical bonding preserves the structure of the NM, but demonstrates less stability, and NPs may detach over time.

10.4.2.7 *Hybridization of 2DNMs*

2DNMs provide powerful applications in a vast array of fields owing to their diverse physicochemical properties, such as anisotropy, quantum confinement, large surface-to-volume ratio, and antibacterial action. Multiple research studies have been conducted to improve the existing and discover new properties of NMs, which have uncovered a promising branch of nanotechnology hybridized 2DNMs (Guan and Han, 2019). In contrast to surface modification, where only surface chemistry is altered, hybridization involves creating a completely different NM with new and improved properties than its parent components. The multifunctional applications and synergistic effects, hybridized NMs provide, make it a demanding branch of science. TiO_2, ZnO, and Ag NPs are considered functional for hybridization with Graphene (Pan et al., 2012). The hybrid structures and composites show potential photocatalytic ROS generation against bacterial attacks, which is integral in water purification (Bai et al., 2017; Xu et al., 2011). Graphene, a zero bandgap conductor, inhibits charge recombination by capturing the electrons transferred from the combining material (Guan and Han, 2019). This charge separation efficiency produces more ROS, strengthening the antibacterial effect. Moreover, interspecies 2DNM hybridization also proves instrumental in refining bacterial inhibition. Transition metal dichalcogenides (TMDs), capable of membrane damage due to their sharp edges, and MXenes with enhanced conductivity can be hybridized for transformative results. Depending on the required properties for antimicrobial effect, respective materials can be selected for hybridization. With scalability and environmental concerns presented as challenges, there is still a wide gap to bridge in achieving targeted hybrid structures.

Despite being a cutting-edge technology with vast applications, nanotechnology faces serious challenges during fabrication and large-scale production (Hachhach et al., 2025). The requirement of precision, prevention in reproducibility at the surface of NMs, accurately controlled process parameters, and intensive energy demand still challenge the cost effectiveness and optimized production of 2DNMs at an industrial scale. Most importantly, the preservation of original structure at the time of laboratory synthesis is a major obstacle; for example, MoS_2 and WS_2 or carbon nanotubes only prove beneficial in their application when their properties are conserved throughout (Grobert, 2006; Gulzar et al., 2016; Sengupta, 2018). The generation of toxins and high energy consumption in bottom-up approach and large amounts of waste produced in top-down strategy are also considered hazards to the environment (Akram et al., 2014; Dhingra et al., 2010). Advanced research on the synthesis of environment-friendly yet cost-effective 2DNMs continues to expand.

10.5 Integration of 2D nanomaterials in water purification systems

10.5.1 Membranes filtration of desalination mediated with 2D nanomaterials

Membrane-based filtration remains a cornerstone of modern water purification and desalination, fundamentally relying on a barrier that separates unwanted particles from water based on size exclusion or permeability. However, conventional membranes often face a critical trade-off between selectivity and permeability, and their limited performance has driven the search for advanced materials (Fatima et al., 2022). In this context, 2DNMs, owing to their atomic thickness typically ranging from 0.3 to a few nanometers, these materials enable highly selective and permeable pathways for water transport. Their integration into membrane structures has opened new avenues for both liquid-state filtration, as in reverse osmosis (RO), and gas-state purification, as in solar membrane distillation (Liu et al., 2017; Zhu et al., 2018). Traditional desalination methods such as multistage flash distillation and RO (Xu et al., 2016) dominate current infrastructure, but RO suffers from limitations including membrane fouling, sensitivity to chlorine, low ammonium rejection (Tałałaj et al., 2019), and high operational costs, especially in developing regions. The prevalent polyamide thin-film composite membranes, in use for over three decades (Cohen-Tanugi and Grossman, 2015), show limited improvements in permeability and are easily degraded by chlorine, hindering effective disinfection. In contrast, graphene-based membranes, with an atomic thickness of approximately 0.34 nm, not only surpass polyamide layers (typically 100 nm) (Goh and Ismail, 2015) in strength and permeability but also exhibit higher tolerance to chlorine, significantly reducing fouling (Cohen-Tanugi and Grossman, 2015). Moreover, GO's inherent antibacterial properties inhibit biofouling, thereby prolonging membrane life and lowering energy demands (Mahmoud et al., 2015). Nanoporous monolayer membranes, such as perforated graphene or MoS_2 achieve excellent water flux and salt rejection through electrostatic and size-selective interactions. Techniques like electron beam irradiation and plasma etching are employed to create cation-selective nanopores in these membranes, enabling them to outperform traditional RO membranes. However, challenges such as mechanical fragility and scale-up limitations persist. To overcome these issues, stacked multilayer laminates using 2D nanosheets are being developed via vacuum-assisted filtration or layer-by-layer (LbL) assembly. GO membranes shown by Fig. 10.5A, in particular, offer tunable interlayer spacing that can be engineered to selectively transport water molecules while blocking hydrated ions like Na^+ and Cl^-. Cross-linking agents are incorporated to enhance their structural integrity under high-pressure conditions. Adjusting interlayer distances further allows for the selective removal of organic contaminants or biomolecules in specialized applications like dialysis. The inclusion of metal ions such as Ca^{2+} and Mg^{2+} improves ion selectivity via cation–π interactions (Presumido et al., 2021). Additionally, 2D materials like g-C_3N_4 with artificial nanopores or self-supporting spacers enhance membrane durability and performance across varying chemical environments. MXenes, with their high electrical conductivity and surface area, demonstrate rapid salt removal capabilities, as seen in capacitive deionization using $Ti_3C_2T_x$, which achieved removal rates of 9.4 mg g^{-1} min^{-1} (Guo et al., 2018). Despite these advancements, hurdles remain in terms of scalable manufacturing, long-term stability, and pore size control, particularly for graphene membranes, where grain boundaries and imperfections may affect selectivity. Future development focuses on creating large-area, defect-free membranes, integrating antibacterial agents like silver nanoparticles, and incorporating self-cleaning or healing functionalities. Thus the fusion of membrane engineering and 2D nanotechnology holds exceptional promise for efficient, scalable, and sustainable desalination and water purification systems.

10.5.2 2D Nanomaterials for water remediation systems as photocatalysts

The incorporation of 2DNMs as photocatalysts in the purification of water is a very promising and sustainable method of solving the global water pollution through the use of the sun (Wang et al., 2014). The formation of the reactive oxygen species (ROS), able to oxidize the pollutants or to split the water or to kill the microorganisms, is triggered by

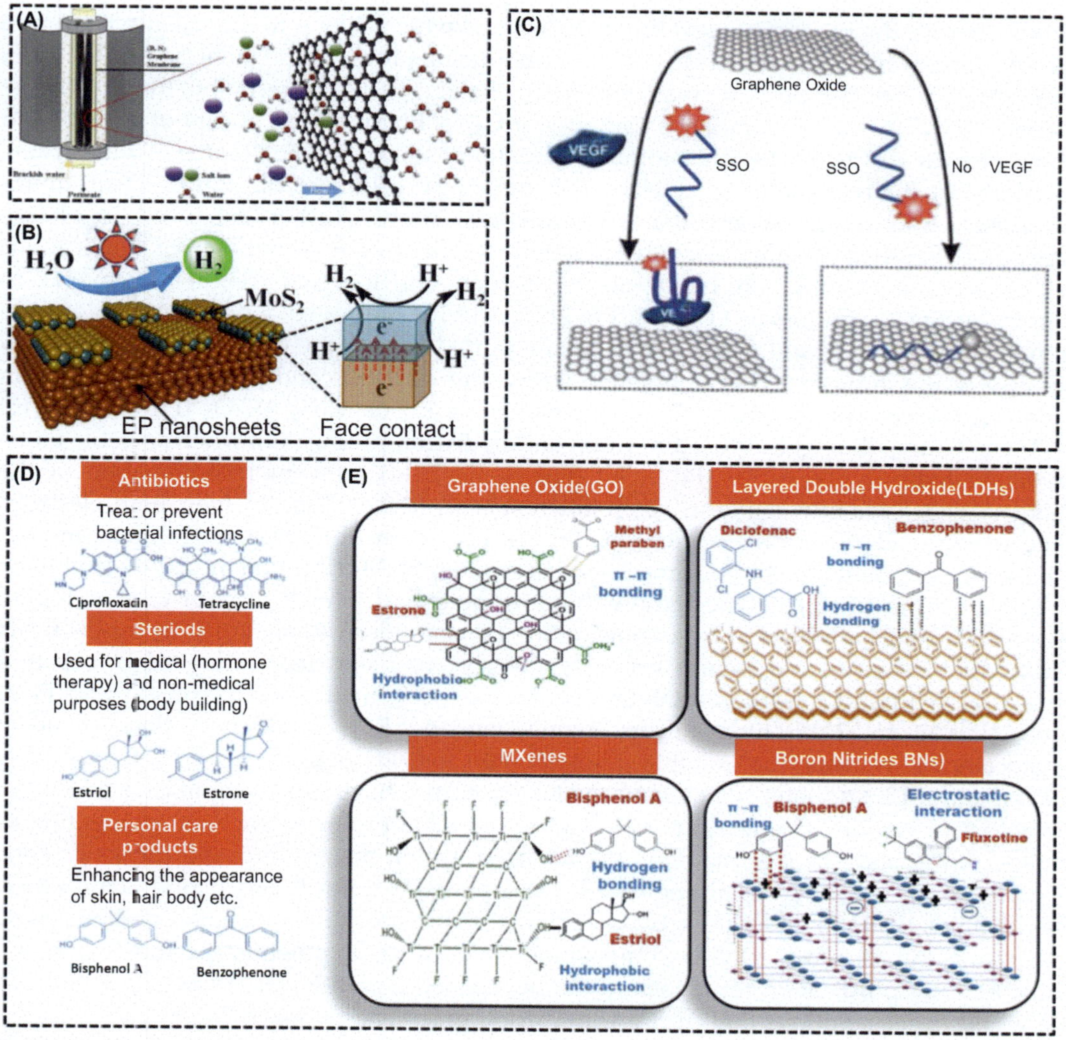

FIGURE 10.5 (A) Schematics of Ag^+ detection are presented by fluorescent silver specific oligonucleotide (SSO) and high-quenching GO nanosheets. (B) Schematics of 2D-2DBP photocatalysts 10,000/MoS_2. (C) Fluorescent sensors based on high-quenching 2D nanomaterials. (D) Graphical presentation of various categories of microcontaminants and their representatives as well as their structures. (E) 2D nanomaterials used in the adsorption of as well as relationships supporting adsorption. *(A) Reproduced by permission Presumido, P.H., Primo, A., Vilar, V.J.P., Garcia, H., 2021. Large area continuous multilayer graphene membrane for water desalination. Chemical Engineering Journal 413, 127510. https://doi.org/10.1016/j.cej.2020.127510, Copyright 2021, Elsevier; (B) Reproduced by permission Yuan, Y.J., Wang, P., Li, Z., Wu, Y., Bai, W., Su, Y., Guan, J., Wu, S., Zhong, J., Yu, Z.T., Zou, Z., 2019. The role of bandgap and interface in enhancing photocatalytic H2 generation activity of 2D-2D black phosphorus/MoS2 photocatalyst. Applied Catalysis B: Environmental 242, 1–8. https://doi.org/10.1016/j.apcatb.2018.09.100, Copyright 2019, Elsevier; (C) Reproduced by permission Cao, C., Zhang, F., Goldys, E.M., Gao, F., Liu, G., 2018. Advances in structure-switching aptasensing towards real time detection of cytokines. TrAC - Trends in Analytical Chemistry 102, 379–396. https://doi.org/10.1016/j.trac.2018.03.002, Copyright 2018, Elsevier; (D) and (E) Reproduced by permission Sikri, N., Behera, B., Kumar, A., Kumar, V., Pandey O., Mehta, J., et al., 2025. Recent advancements on 2D nanomaterials as emerging paradigm for the adsorptive removal of microcontaminants. Advances in Colloid and Interface Science 340, 103441. https://doi.org/10.1016/j.cis.2025.103441, Copyright 2025, Elsevier.*

photocatalysis, the process of elevating the electrons which belong to the valence band into the conduction band of a catalyst with semiconductor properties (Turchi and Ollis, 1990; Zhao et al., 2019). Although they are subject to challenges due to recombination of charge carriers, 2DNMs, such as graphene, MXenes, g-C_3N_4, BNMs, black phosphorus (BP), and MoS_2, provide distinctive benefits, for example, atomic thickness, high surface-volume ratio, reactivity to many systems, and electronic tunability (Zhou et al., 2013). Their multi-functional architecture allows their performances to be further enhanced in charge separation, light harvesting, and photocatalytic efficiencies in a number of applications such as organic degradation, biological disinfection, and inorganic pollutant removal, as shown by Fig. 10.5B (Yuan et al., 2019) Photoactivity improved with graphene and its composites with TiO_2 and g-C_3N_4 have been reported to be more efficient in photoactivity because of the improved charge transfer ability and electron conduction pathway. As an example, the catalyst graphene/TiO_2 (Bera et al., 2015) or MoS_2 @TiO_2 (Fu et al., 2015) visibly enhances the light-induced reaction and hydrogen production, whereas noble metal mediated hybrids, like Cu_2O/Pd/RGO (Bai et al., 2014), prevent recombination of the charges. TiO_2 is highly stable in heterojunction with Mxenes due to their high electrical conductivity and anisotropic carrier mobility, which enhances hydrogen evolution and the breakdown of pollutants. In the same line of reasoning, BNMs (Ihsanullah, 2020) result in an improvement in photocatalytic activity, binding potentiation with TiO_2 or carrier doping with conducting polymers such as polyaniline (Shahabuddin et al., 2018) to result in the degradation of dyes and pharmaceuticals. Nearly complete disinfection of E. coli and viruses also occurred in photocatalytic disinfection performed with 2D g-C_3N_4 and FLV-MoS_2 (Liu et al., 2016) by bandgap engineering and ROS production. Besides, novel photocatalytic systems, for example, photo-Fenton hybrids (e.g., Fe-doped g-C_3N_4/WO_3) (Guo et al., 2012) and intercalated nanocluster structures, have shown significant selectivity and efficiency under visible light. Nonetheless, possible problems such as aggregation of photocatalyst and its low stability in unstirred water masses are still present. Thus approaches such as interfacial engineering, morphology control, and ligand stabilization play a pivotal role in ensuring activity within realistic conditions. In total, photocatalytic water purification with 2DNMs composition holds a great potential of the high-yield, scalable water purification route toward the next generation of water cleaning solutions, as the world needs solutions that purify water using green energy technologies.

10.5.3 2D Nanomaterials as adsorbents in water cleaning systems

The use of 2DNMs as water purification adsorbents has become a formidable technique owing to their vast specific surface area, surface chemistry that can be altered, and excellent physicochemical properties. These materials, such as graphene, MXenes, MoS_2, LDHs, MOFs, and COFs, are perfect materials that allow effective removal of a broad range of pollutants, including oils, dyes, heavy metals, and anions. As shown in Fig. 10.5E and reported by Sikri et al. (2025) graphene and its derivatives, especially graphene oxide (GO), have a large adsorption capacity due to the strong adsorptive ability attributed to hydrogen bonding, electrostatic force, and the stacking of pi (pi) phases, which is increased through functionalization of the material to contain oxygen-containing groups (Han et al., 2013). The ability of MXenes to adsorb both cationic and anionic species, such as Cr^{6+} and Cu^{2+} was made possible because of their negative surface charge property, in-situ reduction, and ion-intercalation nonspecificity (Shahzad et al., 2018). The interaction forces that BNMs get involved with to trap organic substances are the van der Waals forces and hydrophobic association, whereas LDHs depend on the ion-exchange procedure and interlayer sorbing to remove anionic pollutants (Song et al., 2018; Zou et al., 2016). Adsorption of oil in oil spill response. In oil spill cleanups, 2DMs such as fluorinated GO fluorinated h-BN composites are synthesized to selectively adsorb oil within seconds, even at the worst marine conditions. However, another form of 2D COFs, the hierarchical porosity and surface roughness, is outstanding in adsorbing hydrophilic dye and stability in different pH and temperature. Besides this, the incorporation of 2D sheets on 3D scaffolds, making hierarchical flower-like nanostructures, or using magnetic Fe_3O_4 nanoparticles further enhances reusability, separation, and adsorption efficiency. The mechanisms of heavy metal removal with 2DNMs include Lewis acid-base interactions, intercalation, and surface complexation, which are shown by MoS_2, fluorinated BN, and intercalated MXenes (Wang et al., 2016b). With oxyanion pollutants such as arsenates or chromates, there can be optimized interlayer distances and surface functional groups to allow high-capacity adsorption. All in all, the adaptability, expandability, and customizable interplay mechanisms of 2DNMs make them among the leaders of the next-generation technologies of water purification, to confront various environmental challenges efficiently and sustainably.

10.5.4 Electrochemical sensors of 2DNMs in water purification systems

The design of 2DNMs into water purification system devices as SD-based sensors has provided a paradigm shift in the real-time detection of pollutants with ultrasensitive procedures. As well as graphene, MXenes, MoS_2, BP, and TMDs have extraordinary surface-to-volume ratios, conductivities, and manipulable surface chemistries, which enable them to be highly responsive platforms to sensing applications (Hassan et al., 2023; Zeng et al., 2021). The operating principle of such sensors usually consists of adsorption or interaction of molecules of the analyte to the surface of 2D nanosheets, as shown in Fig. 10.5C, and this subsequently produces detectable electrical, electrochemical, optical, or mechanical signals (Mao et al., 2017). As an example, sensors based on graphene have high sensitivity to heavy metal ions, organic contaminants, and pathogens since the chemical can be functionalized easily because of its high electrical conductivity (Novoselov et al., 2012; Sharma and Malviya, 2023). A common application of MoS_2 and BP in a FET-based sensor is sensitive molecule detection, for example, lead, mercury, or antibiotics, through a semiconductor current response and defect-mediated adsorption mode (Abbas et al., 2015). Having an abundant surface chemistry with its hydrophilicity, MXenes can be used to detect toxic ions and organic by-products in real-time via capacitive changes and resistive changes (Anasori, Lukatskaya and Gogotsi, 2023; Naguib et al., 2012). Remarkably, the insertion of functional groups, including -OH, -COOH, or $-NH_2$ on the surfaces of nanosheets, promotes the specificity to target contaminants through selective binding (Verma et al., 2024). Sensing methods based on optical techniques and 2D materials, such as surface-enhanced Raman scattering and fluorescence quenching, provide a nondestructive, label-free detection route to biological and chemical pollutants at trace levels. Moreover, integration of such sensors with microfluidic or handheld systems provides an opportunity of on-site point-of-use monitoring, particularly crucial in rural and remote locations where infrastructure is scarce. In spite of the mentioned benefits, there is still a problem of signal drift, the stability of the environment, and sensor regeneration. Nevertheless, these challenges are more recently being resolved by continuous developments in the fabrication of hybrid materials, on-surface passivation, and coupling with wireless platforms (Shao et al., 2010). Therefore 2DNM-based sensors cannot simply be considered as tools of detection, but as a part of intelligent water purification processes that allow the early warning, feedback control, and improved effectiveness of the processes.

10.6 Conclusion and perspectives

Incorporating 2DNMs in water remediation can be a revolutionary step in resolving the water crisis in the world by using the advantages of material science. Although the common traditional purification efforts seem to be fairly used, such as boiling, sedimentation, oxidation, and solar disinfection, in many instances, they are deficient in scale, cost, practicality, and reliability. Thus the urgent task to work out the next-generation technologies of water treatment characterized by economic efficiency, sustainability, and technological robustness arises. The 2DNMs, due to their ultrathin atomic-scale dimensions, large surface area-to-volume ratio, configurable surface properties, and outstanding mechanical, optical, and electrical qualities, have become a prospective choice in numerous treatment processes: membrane filtration, adsorption, photocatalysis, and sensing.

Nevertheless, the full commercialization and large-scale adoption of such technologies remain several issues that are to be dealt with in a systematic manner. The cost of production is one of the greatest obstacles. Most of the 2DNMs continue to be produced using either expensive or inefficient schemes. Top-down exfoliation based on the solutions ranks as a promising, large-scale production method because of low-cost and high-abundance layered starting materials such as graphite. Unlike most generally manufactured 1D nanomaterials, which have traditionally been produced by bottom-up methods such as CVD, exfoliation processes enable the large-scale, low-cost synthesis of nanomaterials by sonication or shear force, providing significant economic and processing time (Stafford et al., 2020; Varrla et al., 2015).

Another important issue is sustainability in its production as well as the use phase. The synthesis paths have to be environmentally friendly, employing minimal amounts of toxic reagents and energy, and should guarantee ecological sustainability and viability (Zeng et al., 2021). Layered double hydroxides (LDHs) made of earth-abundant and nontoxic elements also deserve special attention because of their low environmental impact and high water decontamination adaptability (Hu et al., 2020; Xiong et al., 2018). Operationally, the constant and extreme water matrices, extremely high or low pH, salinity, and chemical conditions, necessitate the construction of long-lasting, resistant to degradation, and low secondary pollution potential of materials.

In addition, a high level of application sustainability should ensure that the 2DNMs can sustain longer periods of structural integrity as well as functionality. For example, a well-used GO laminate in membrane separations can easily swell in water due to its hydrophilic character, and this results in alteration of interlayer spacing, impairing performance. New design principles have to be developed to ensure average nanochannels for selective filtration. The self-assembly and directed assembly techniques have proved some potential in the organization of ordered structures, producing vertical alignment and high throughput. To give an example, rapid ion transport within 2D MXene nanosheets vertically stacked with a lamellar liquid-crystal phase has been demonstrated (Zhang et al., 2019).

Fine-tuning of water treatment will also be done through defect engineering to increase selectivity, adsorption, as well as catalytic performance. According to the hard-soft acid-base (HSAB) theory, chemicals like MoS_2 and $MoSe_2$, which have soft sulfur or selenium-based dangling groups, form strong bonds with soft acids like Hg^{2+} and Pb^{2+}. However, on the flip side, the surface-terminated MXenes (e.g., U[VI], which has a termination of ^{-}OH and ^{-}F) are relatively more favorable in binding hard acids (Hu et al., 2020). Modification of functional groups is critical in a broad range of destructions of contaminants that include heavy metals as well as complicated organic pollutants.

There are also exciting prospects using chemical programmability to allow multifunctional materials by combining diverse components on a single nanostructured substrate. As an example, the integration of photocatalysts with membranes allows organic foulants and filtration processes to occur concurrently, thus solving the problem of biofouling, which is one of the problems of conventional membranes. Furthermore, the multicomponent hybrid structures and superlattice assemblies are currently designed to provide the elimination of several types of contaminants at the same time to increase the functionality and efficiency of the purification systems (Fatima et al., 2022).

New technologies of fabrication, like 3D printing and robotic assembly, are driving the proliferation of rapid prototyping of such nanostructures. The stacking of 2D materials such as graphene and h-BN into superlattices now has the potential to be accurately stacked via robot-assisted synthesis, which minimizes the human input. These functions, coupled with the additive manufacturing process, offer an unsurpassed control of the membrane architecture and layer thickness, which is important in designing customized separation properties (Wu et al., 2023; Zheng et al., 2024).

Machine learning and data-driven methods are even now being applied to materials science, forming the frontier of smart materials development with a predictive ability to determine elements such as bandgap, adsorption capacity, and permeability of materials utilizing preexisting sets of data (Zhou et al., 2023). These simulation instruments can speed up screening and optimization of 2DNMs with appropriate water treatment activities by cost-saving experimentation, trial-and-error, and logically based design (Yang et al., 2022).

The possibilities of 2DNMs in purifying the water are enormous and come with complicated issues. The solution to these needs to be multidisciplinary in nature and will involve the expertise of materials science, chemistry, environmental engineering, and computational modeling. In the future, more studies need to be conducted to come up with multifunctional/programmable, scalable, and more sustainable 2DNMs that can be used in a real-world water treatment setting. The eventual manifestation of intelligent, integrated systems, which can deliver clean, safe, and affordable water to everyone, will be brought about by bridging the existing gap between laboratory innovation and deployment on a field scale.

References

Abbas, A.N., Liu, B., Chen, L., Ma, Y., Cong, S., Aroonyadet, N., et al., 2015. Black phosphorus gas sensors. ACS Nano 9 (5), 5618–5624. Available at https://doi.org/10.1021/acsnano.5b01961.

Akram, H., Achak, O., Haffane, S., Elmoujahid, C., Elmesbahi, A., Elmessoudi, D., et al., 2014. Comparison of synthesis routes of inorganic fullerene-like nano-additives for wind turbine lubrication: Application of Life Cycle Assessment approach. 2014 International Renewable and Sustainable Energy Conference (IRSEC), Ouarzazate, Morocco. IEEE, pp. 788–790. Available at https://doi.org/10.1109/IRSEC.2014.7059811.

Anasori, B., Lukatskaya, M.R., Gogotsi, Y., 2023. 2D Metal Carbides and Nitrides for Energy Storage. MXenes: From Discovery to Applications of Two-Dimensional Metal Carbides and Nitrides. Jenny Stanford Publishing, United States, pp. 677–722.

Anichini, C., Czepa, W., Pakulski, D., Aliprandi, A., Ciesielski, A., Samorì, P., 2018. Chemical sensing with 2D materials. Chemical Society Reviews 47 (13), 4860–4908. Available at https://doi.org/10.1039/c8cs00417j.

Bai, S., Ge, J., Wang, L., Gong, M., Deng, M., Kong, Q., et al., 2014. A unique semiconductor-metal-graphene stack design to harness charge flow for photocatalysis. Advanced Materials 26 (32), 5689–5695. Available at https://doi.org/10.1002/adma.201401817.

Bai, X., Sun, C., Liu, D., Luo, X., Li, D., Wang, J., et al., 2017. Photocatalytic degradation of deoxynivalenol using graphene/ZnO hybrids in aqueous suspension. Applied Catalysis B: Environmental 204, 11–20. Available at https://doi.org/10.1016/j.apcatb.2016.11.010.

Bera, R., Kundu, S., Patra, A., 2015. 2D hybrid nanostructure of reduced graphene oxide-CdS nanosheet for enhanced photocatalysis. ACS Applied Materials and Interfaces 7 (24), 13251–13259. Available at https://doi.org/10.1021/acsami.5b03800.

Bi, H., Wang, X., Liu, H., He, Y., Wang, W., Deng, W., et al., 2020. A universal approach to aqueous energy storage via ultralow-cost electrolyte with super-concentrated sugar as hydrogen-bond-regulated solute. Advanced Materials 32 (16). Available at https://doi.org/10.1002/adma.202000074.

Bokov, D., Turki Jalil, A., Chupradit, S., Suksatan, W., Javed Ansari, M., Shewael, I.H., et al., 2021. Nanomaterial by Sol-Gel method: synthesis and application. Federation Advances in Materials Science and Engineering 2021. Available at https://doi.org/10.1155/2021/5102014.

Chae, A., Jo, S., Choi, Y., Ryu, B., Choi, C.A., Park, S.Y., et al., 2019. Enhanced photothermal bactericidal activity of chemically reduced graphene oxide stabilized by tripodal amphiphile. Applied Surface Science 474, 111–117. Available at https://doi.org/10.1016/j.apsusc.2018.04.119.

Chen, L., Cao, L., Ji, X., Hou, S., Li, Q., Chen, J., et al., 2020. Enabling safe aqueous lithium ion open batteries by suppressing oxygen reduction reaction. Nature Communications 11 (1). Available at https://doi.org/10.1038/s41467-020-16460-w.

Chen, L., He, F., Huang, Y., Meng, Y., Guo, R., 2016. Hydrogenated nanoporous TiO_2 film on Ti-25Nb-3Mo-2Sn-3Zr alloy with enhanced photocatalytic and sterilization activities driven by visible light. Journal of Alloys and Compounds 678, 5–11. Available at https://doi.org/10.1016/j.jallcom.2016.03.216.

Chimene, D., Alge, D.L., Gaharwar, A.K., 2015. Two-dimensional nanomaterials for biomedical applications: emerging trends and future prospects. Advanced Materials 27 (45), 7261–7284. Available at https://doi.org/10.1002/adma.201502422.

Cohen-Tanugi, D., Grossman, J.C., 2015. Nanoporous graphene as a reverse osmosis membrane: recent insights from theory and simulation. Desalination 366, 59–70. Available at https://doi.org/10.1016/j.desal.2014.12.046.

Dehghani, M.H., Mostofi, M., Alimohammadi, M., McKay, G., Yetilmezsoy, K., Albadarin, A.B., et al., 2016. High-performance removal of toxic phenol by single-walled and multi-walled carbon nanotubes: kinetics, adsorption, mechanism and optimization studies. Journal of Industrial and Engineering Chemistry 35, 63–74. Available at https://doi.org/10.1016/j.jiec.2015.12.010.

Dhingra, R., Naidu, S., Upreti, G., Sawhney, R., 2010. Sustainable nanotechnology: through green methods and life-cycle thinking. Sustainability 2 (10), 3323–3338. Available at https://doi.org/10.3390/su2103323.

Fatima, J., Shah, A.N., Tahir, M.B., Mehmood, T., Shah, A.A., Tanveer, M., et al., 2022. Tunable 2D nanomaterials; their key roles and mechanisms in water purification and monitoring. Frontiers in Environmental Science 10. Available at https://doi.org/10.3389/fenvs.2022.766743.

Fu, H., Yu, K., Li, H., Li, J., Guo, B., Tan, Y., et al., 2015. Enhanced field emission and photocatalytic performance of MoS_2 titania nanoheterojunctions via two synthetic approaches. Dalton Transactions 44 (4), 1664–1672. Available at https://doi.org/10.1039/c4dt03035d.

Fu, Y., Li, L., Xie, B., Dong, C., Wang, M., Jia, B., et al., 2016. How to establish a bioregenerative life support system for long-term crewed missions to the moon or mars. Astrobiology 16 (12), 925–936. Available at https://doi.org/10.1089/ast.2016.1477.

Ge, J., Shi, L.A., Wang, Y.C., Zhao, H.Y., Yao, H.B., Zhu, Y.B., et al., 2017. Joule-heated graphene-wrapped sponge enables fast clean-up of viscous crude-oil spill. Nature Nanotechnology 12 (5), 434–440. Available at https://doi.org/10.1038/nnano.2017.33.

Goh, P.S., Ismail, A.F., 2015. Graphene-based nanomaterial: the state-of-the-art material for cutting edge desalination technology. Desalination 356, 115–128. Available at https://doi.org/10.1016/j.desal.2014.10.001.

Gonzalez-Ortiz, D., Salameh, C., Bechelany, M., Miele, P., 2020. Nanostructured boron nitride–based materials: synthesis and applications. Materials Today Advances 8 (12), 100107. Available at https://doi.org/10.1016/j.mtadv.2020.100107.

Grobert, N., 2006. Nanotubes – grow or go? Materials Today 9 (10), 64. Available at https://doi.org/10.1016/S1369-7021(06)71680-7.

Guan, G., Han, M.Y., 2019. Functionalized hybridization of 2D nanomaterials. Advanced Science 6 (23). Available at https://doi.org/10.1002/advs.201901837.

Gulzar, M., Masjuki, H.H., Kalam, M.A., Varman, M., Zulkifli, N.W.M., Mufti, R.A., et al., 2016. Tribological performance of nanoparticles as lubricating oil additives. Journal of Nanoparticle Research 18 (8). Available at https://doi.org/10.1007/s11051-016-3537-4.

Guo, J., Li, Y., Zhu, S., Chen, Z., Liu, Q., Zhang, D., et al., 2012. Synthesis of WO_3@Graphene composite for enhanced photocatalytic oxygen evolution from water. RSC Advances 2 (4), 1356–1363. Available at https://doi.org/10.1039/c1ra00621e.

Guo, L., Wang, X., Leong, Z.Y., Mo, R., Sun, L., Yang, H.Y., 2018. Ar plasma modification of 2D MXene $Ti_3C_2T_x$ nanosheets for efficient capacitive desalination. FlatChem 8, 17–24. Available at https://doi.org/10.1016/j.flatc.2018.01.001.

Hachhach, M., Bayou, S., El Kasmi, A., Saidi, M.Z., Akram, H., Hanafi, M., et al., 2025. Towards sustainable scaling-up of nanomaterials fabrication: current situation, challenges, and future perspectives. Eng 6 (7). Available at https://doi.org/10.3390/eng6070149.

Han, Y., Xu, Z., Gao, C., 2013. Ultrathin graphene nanofiltration membrane for water purification. Advanced Functional Materials 23 (29), 3693–3700. Available at https://doi.org/10.1002/adfm.201202601.

Hassan, J.Z., Raza, A., Din Babar, Z.U., Qumar, U., Kaner, N.T., Cassinese, A., 2023. 2D material-based sensing devices: an update. Journal of Materials Chemistry A 11 (12), 6016–6063. Available at https://doi.org/10.1039/d2ta07653e.

Hu, L., Li, M., Wei, X., Wang, H., Wu, Y., Wen, J., et al., 2020. Modulating interfacial electronic structure of CoNi LDH nanosheets with $Ti_3C_2T_x$ MXene for enhancing water oxidation catalysis. Chemical Engineering Journal 398. Available at https://doi.org/10.1016/j.cej.2020.125605.

Huang, K., Li, Z., Lin, J., Han, G., Huang, P., 2018. Two-dimensional transition metal carbides and nitrides (MXenes) for biomedical applications. Chemical Society Reviews 47 (14), 5109–5124. Available at https://doi.org/10.1039/c7cs00838d.

Huang, T., Jiang, Y., Shen, G., Chen, D., 2020. Recent advances of two-dimensional nanomaterials for electrochemical capacitors. ChemSusChem 13 (6), 1093–1113. Available at https://doi.org/10.1002/cssc.201903260.

Ibrahim, R.K., Hayyan, M., AlSaadi, M.A., Hayyan, A., Ibrahim, S., 2016. Environmental application of nanotechnology: air, soil, and water. Environmental Science and Pollution Research 23 (14), 13754–13788. Available at https://doi.org/10.1007/s11356-016-6457-z.

Ihsanullah, I., 2020. MXenes (two-dimensional metal carbides) as emerging nanomaterials for water purification: progress, challenges and prospects. Chemical Engineering Journal 388. Available at https://doi.org/10.1016/j.cej.2020.124340.

Inphonlek, S., Pimpha, N., Sunintaboon, P., 2010. Synthesis of poly(methyl methacrylate) core/chitosan-mixed-polyethyleneimine shell nanoparticles and their antibacterial property. Colloids and Surfaces B: Biointerfaces 77 (2), 219–226. Available at https://doi.org/10.1016/j.colsurfb.2010.01.029.

Khan, M.H., Liu, H.K., Sun, X., Yamauchi, Y., Bando, Y., Golberg, D., et al., 2017. Few-atomic-layered hexagonal boron nitride: CVD growth, characterization, and applications. Materials Today 20 (10), 611–628. Available at https://doi.org/10.1016/j.mattod.2017.04.027.

Koul, A., Arnoult, E., Lounis, N., Guillemont, J., Andries, K., 2011. The challenge of new drug discovery for tuberculosis. Nature 469 (7331), 483–490. Available at https://doi.org/10.1038/nature09657.

Krasner, S.W., Weinberg, H.S., Richardson, S.D., Pastor, S.J., Chinn, R., Sclimenti, M.J., et al., 2006. Occurrence of a new generation of disinfection byproducts. Environmental Science and Technology 40 (23), 7175–7185. Available at https://doi.org/10.1021/es060353j.

Kumar, P., Pournara, A., Kim, K.H., Bansal, V., Rapti, S., Manos, M.J., 2017. Metal-organic frameworks: challenges and opportunities for ion-exchange/sorption applications. Progress in Materials Science 86, 25–74. Available at https://doi.org/10.1016/j.pmatsci.2017.01.002.

Li, G., Nie, X., Chen, J., Jiang, Q., An, T., Wong, P.K., et al., 2015. Enhanced visible-light-driven photocatalytic inactivation of *Escherichia coli* using g-C_3N_4/TiO_2 hybrid photocatalyst synthesized using a hydrothermal-calcination approach. Water Research 86, 17–24. Available at https://doi.org/10.1016/j.watres.2015.05.053.

Liu, C., Kong, D., Hsu, P.C., Yuan, H., Lee, H.W., Liu, Y., et al., 2016. Rapid water disinfection using vertically aligned MoS_2 nanofilms and visible light. Nature Nanotechnology 11 (12), 1098–1104. Available at https://doi.org/10.1038/nnano.2016.138.

Liu, C., Jia, F., Wang, Q., Yang, B., Song, S., 2017. Two-dimensional molybdenum disulfide as adsorbent for high-efficient Pb(II) removal from water. Applied Materials Today 9, 220–228. Available at https://doi.org/10.1016/j.apmt.2017.07.009.

Liu, X., Yan, Z., Zhang, Y., Liu, Z., Sun, Y., Ren, J., et al., 2019. Two-dimensional metal-organic framework/enzyme hybrid nanocatalyst as a benign and self-activated cascade reagent for in vivo wound healing. ACS Nano 13 (5), 5222–5230. Available at https://doi.org/10.1021/acsnano.8b09501.

Mahmoud, K.A., Mansoor, B., Mansour, A., Khraisheh, M., 2015. Functional graphene nanosheets: the next generation membranes for water desalination. Desalination 356, 208–225. Available at https://doi.org/10.1016/j.desal.2014.10.022.

Manna, J., Goswami, S., Shilpa, N., Sahu, N., Rana, R.K., 2015. Biomimetic method to assemble nanostructured Ag@ZnO on cotton fabrics: application as self-cleaning flexible materials with visible-light photocatalysis and antibacterial activities. ACS Applied Materials and Interfaces 7 (15), 8076–8082. Available at https://doi.org/10.1021/acsami.5b00633.

Mao, S., Chang, J., Pu, H., Lu, G., He, Q., Zhang, H., et al., 2017. Two-dimensional nanomaterial-based field-effect transistors for chemical and biological sensing. Chemical Society Reviews 46 (22), 6872–6904. Available at https://doi.org/10.1039/c6cs00827e.

Muellner, M.G., Wagner, E.D., Mccalla, K., Richardson, S.D., Woo, Y.T., Plewa, M.J., 2007. Haloacetonitriles vs. regulated haloacetic acids: are nitrogen-containing DBFs more toxic? Environmental Science and Technology 41 (2), 645–651. Available at https://doi.org/10.1021/es0617441.

Naguib, M., Mashtalir, O., Carle, J., Presser, V., Lu, J., Hultman, L., et al., 2012. Two-dimensional transition metal carbides. ACS Nano 6 (2), 1322–1331. Available at https://doi.org/10.1021/nn204153h.

Novoselov, K.S., Fal'ko, V.I., Colombo, L., Gellert, P.R., Schwab, M.G., Kim, K., 2012. A roadmap for graphene. Nature 490 (7419), 192–200. Available at https://doi.org/10.1038/nature11458.

Novoselov, K.S., Geim, A.K., Morozov, S.V., Jiang, D., Zhang, Y., Dubonos, S.V., et al., 2004. Electric field effect in atomically thin carbon films. Science (New York, N.Y.) 306 (5696), 666–669. Available at https://doi.org/10.1126/science.1102896.

Oh, Y., Armstrong, D.L., Finnerty, C., Zheng, S., Hu, M., Torrents, A., et al., 2017. Understanding the pH-responsive behavior of graphene oxide membrane in removing ions and organic micropollulants. Journal of Membrane Science 541, 235–243. Available at https://doi.org/10.1016/j.memsci.2017.07.005.

Pan, X., Zhao, Y., Liu, S., Korzeniewski, C.L., Wang, S., Fan, Z., 2012. Comparing graphene-TiO_2 nanowire and graphene-TiO_2 nanoparticle composite photocatalysts. ACS Applied Materials and Interfaces 4 (8), 3944–3950. Available at https://doi.org/10.1021/am300772t.

Perveen, R., Shujaat, S., Qureshi, Z., Nawaz, S., Khan, M.I., Iqbal, M., 2020. Green versus sol-gel synthesis of ZnO nanoparticles and antimicrobial activity evaluation against panel of pathogens. Journal of Materials Research and Technology 9 (4), 7817–7827. Available at https://doi.org/10.1016/j.jmrt.2020.05.004.

Premarathna, K.S.D., Rajapaksha, A.U., Sarkar, B., Kwon, E.E., Bhatnagar, A., Ok, Y.S., et al., 2019. Biochar-based engineered composites for sorptive decontamination of water: a review. Chemical Engineering Journal 372, 536–550. Available at https://doi.org/10.1016/j.cej.2019.04.097.

Presumido, P.H., Primo, A., Vilar, V.J.P., Garcia, H., 2021. Large area continuous multilayer graphene membrane for water desalination. Chemical Engineering Journal 413. Available at https://doi.org/10.1016/j.cej.2020.127510.

Qi, Z., Bharate, P., Lai, C.H., Ziem, B., Böttcher, C., Schulz, A., et al., 2015. Multivalency at interfaces: supramolecular carbohydrate-functionalized graphene derivatives for bacterial capture, release, and disinfection. Nano Letters 15 (9), 6051–6057. Available at https://doi.org/10.1021/acs.nanolett.5b02256.

Sengupta, J., 2018. Carbon nanotube fabrication at industrial scale: opportunities and challenges. Handbook of Nanomaterials for Industrial Applications. Elsevier, India, pp. 172–194. Available at http://doi.org/10.1016/B978-0-12-813351-4.00010-9.

Seyler, K.L., Rivera, P., Yu, H., Wilson, N.P., Ray, E.L., Mandrus, D.G., et al., 2019. Signatures of moiré-trapped valley excitons in $MoSe_2$/WSe2 heterobilayers. Nature 567 (7746), 66–70. Available at https://doi.org/10.1038/s41586-019-0957-1.

Shahabuddin, S., Khanam, R., Khalid, M., Sarih, N.M., Ching, J.J., Mohamad, S., et al., 2018. Synthesis of 2D boron nitride doped polyaniline hybrid nanocomposites for photocatalytic degradation of carcinogenic dyes from aqueous solution. Arabian Journal of Chemistry 11 (6), 1000–1016. Available at https://doi.org/10.1016/j.arabjc.2018.05.004.

Shahzad, A., Rasool, K., Miran, W., Nawaz, M., Jang, J., Mahmoud, K.A., et al., 2017. Two-dimensional $Ti_3C_2T_x$ MXene nanosheets for efficient copper removal from water. ACS Sustainable Chemistry and Engineering 5 (12), 11481–11488. Available at https://doi.org/10.1021/acssuschemeng.7b02695.

Shahzad, B., Tanveer, M., Rehman, A., Cheema, S.A., Fahad, S., Rehman, S., et al., 2018. Nickel; whether toxic or essential for plants and environment - a review. Plant Physiology and Biochemistry 132, 641–651. Available at https://doi.org/10.1016/j.plaphy.2018.10.014.

Shannon, M.A., Bohn, P.W., Elimelech, M., Georgiadis, J.G., Mariñas, B.J., Mayes, A.M., 2008. Science and technology for water purification in the coming decades. Nature 452 (7185), 301–310. Available at https://doi.org/10.1038/nature06599.

Shao, Y., Wang, J., Wu, H., Liu, J., Aksay, I.A., Lin, Y., 2010. Graphene based electrochemical sensors and biosensors: a review. Electroanalysis 22 (10), 1027–1036. Available at https://doi.org/10.1002/elan.200900571.

Sharma, R., Malviya, R., 2023. Utilization of bioactive silk protein in the development of optical devices: recent advancements and applications. Current Protein and Peptide Science 24 (5), 404–422. Available at https://doi.org/10.2174/1389203724666230412092734.

Shirzadi-Ahodashti, M., Mizwari, Z.M., Hashemi, Z., Rajabalipour, S., Ghoreishi, S.M., Mortazavi-Derazkola, S., et al., 2021. Discovery of high antibacterial and catalytic activities of biosynthesized silver nanoparticles using C. *fruticosus* (CF-AgNPs) against multi-drug resistant clinical strains and hazardous pollutants. Environmental Technology and Innovation 23. Available at https://doi.org/10.1016/j.eti.2021.101607.

Sikri, N., Behera, B., Kumar, A., Kumar, V., Pandey, O., Mehta, J., et al., 2025. Recent advancements on 2D nanomaterials as emerging paradigm for the adsorptive removal of microcontaminants. Advances in Colloid and Interface Science 340, 103441. Available at https://doi.org/10.1016/j.cis.2025.103441.

Song, S., Yin, L., Wang, X., Liu, L., Huang, S., Zhang, R., et al., 2018. Interaction of U(VI) with ternary layered double hydroxides by combined batch experiments and spectroscopy study. Chemical Engineering Journal 338, 579–590. Available at https://doi.org/10.1016/j.cej.2018.01.055.

Stafford, J., Uzo, N., Farooq, U., Favero, S., Wang, S., Chen, H.H., et al., 2020. Universal scaling and real-time monitoring of the production of liquid exfoliated graphene. ChemRxiv. Available at https://doi.org/10.26434/chemrxiv.11628267.v1.

Suba, S., Vijayakumar, S., Vidhya, E., Punitha, V.N., Nilavukkarasi, M., 2021. Microbial mediated synthesis of ZnO nanoparticles derived from *Lactobacillus* spp: characterizations, antimicrobial and biocompatibility efficiencies. Sensors International 2, 100104. Available at https://doi.org/10.1016/j.sintl.2021.100104.

Surwade, S.P., Smirnov, S.N., Vlassiouk, I.V., Unocic, R.R., Veith, G.M., Dai, S., et al., 2015. Water desalination using nanoporous single-layer graphene. Nature Nanotechnology 10 (5), 459–464. Available at https://doi.org/10.1038/nnano.2015.37.

Tałałaj, I.A., Biedka, P., Bartkowska, I., 2019. Treatment of landfill leachates with biological pretreatments and reverse osmosis. Environmental Chemistry Letters 17 (3), 1177–1193. Available at https://doi.org/10.1007/s10311-019-00860-6.

Tian, J., Sang, Y., Yu, G., Jiang, H., Mu, X., Liu, H., 2013. A Bi_2WO_6 -based hybrid photocatalyst with broad spectrum photocatalytic properties under UV, visible, and near-infrared irradiation (Adv. Mater. 36/2013). Advanced Materials 25 (36), 5074. Available at https://doi.org/10.1002/adma.201370230.

Turchi, C.S., Ollis, D.F., 1990. Photocatalytic degradation of organic water contaminants: mechanisms involving hydroxyl radical attack. Journal of Catalysis 122 (1), 178–192. Available at https://doi.org/10.1016/0021-9517(90)90269-P.

Van Engelenburg, J., Van Slobbe, E., Hellegers, P., 2019. Towards sustainable drinking water abstraction: an integrated sustainability assessment framework to support local adaptation planning. Journal of Integrative Environmental Sciences 16 (1), 89–122. Available at https://doi.org/10.1080/1943815x.2019.1636284.

Varrla, E., Backes, C., Paton, K.R., Harvey, A., Gholamvand, Z., McCauley, J., et al., 2015. Large-scale production of size-controlled MoS_2 nanosheets by shear exfoliation. Chemistry of Materials 27 (3), 1129–1139. Available at https://doi.org/10.1021/cm5044864.

Verma, S., Pathak, A.K., Rahman, B.M.A., 2024. Review of biosensors based on plasmonic-enhanced processes in the metallic and meta-material-supported nanostructures. Micromachines 15 (4). Available at https://doi.org/10.3390/mi15040502.

Wang, H., Zhang, L., Chen, Z., Hu, J., Li, S., Wang, Z., et al., 2014. Semiconductor heterojunction photocatalysts: design, construction, and photocatalytic performances. Chemical Society Reviews 43 (15), 5234–5244. Available at https://doi.org/10.1039/c4cs00126e.

Wang, J., Li, G., Li, L, 2016. Synthesis strategies about 2D materials. In: Nayak, P.K. (Ed.), Two-Dimensional Materials – Synthesis, Characterization Potential Applications. InTech, London, pp. 1–20. Available at https://doi.org/10.5772/63918.

Wang, Y., Hu, J., Zhuang, Q., Ni, Y., 2016b. Label-free fluorescence sensing of lead(II) ions and sulfide ions based on luminescent molybdenum disulfide nanosheets. ACS Sustainable Chemistry and Engineering 4 (5), 2535–2541. Available at https://doi.org/10.1021/acssuschemeng.5b01639.

Wang, X., Cao, L., Lu, F., Meziani, M. J., Li, H., Qi, G., Sun, Y. P. (2009). Photoinduced electron transfers with carbon dots. Chemical Communications, (25), 3774–3776.

Wang, Y.W., Fu, Y.Y., Wu, L.J., Li, J., Yang, H.H., Chen, G.N., 2013. Targeted photothermal ablation of pathogenic bacterium, *Staphylococcus aureus*, with nanoscale reduced graphene oxide. Journal of Materials Chemistry B 1 (19), 2496–2501. Available at https://doi.org/10.1039/c3tb20144a.

Wu, Y., An, C., Guo, Y., 2023. 3D printed graphene and graphene/polymer composites for multifunctional applications. Materials 16 (16), 5681. Available at https://doi.org/10.3390/ma16165681.

Xiong, J., Di, J., Xia, J., Zhu, W., Li, H., 2018. Surface defect engineering in 2D nanomaterials for photocatalysis. Advanced Functional Materials 28 (39). Available at https://doi.org/10.1002/adfm.201801983.

Xu, J.W., Yao, K., Xu, Z.K., 2019. Nanomaterials with a photothermal effect for antibacterial activities: an overview. Nanoscale 11 (18), 8680–8691. Available at https://doi.org/10.1039/c9nr01833f.

Xu, K., Feng, B., Zhou, C., Huang, A., 2016. Synthesis of highly stable graphene oxide membranes on polydopamine functionalized supports for seawater desalination. Chemical Engineering Science 146, 159–165. Available at https://doi.org/10.1016/j.ces.2016.03.003.

Xu, T., Zhang, L., Cheng, H., Zhu, Y., 2011. Significantly enhanced photocatalytic performance of ZnO via graphene hybridization and the mechanism study. Applied Catalysis B: Environmental 101 (3-4), 382–387. Available at https://doi.org/10.1016/j.apcatb.2010.10.007.

Yang, R.X., McCandler, C.A., Andriuc, O., Siron, M., Woods-Robinson, R., Horton, M.K., et al., 2022. Big Data in a nano world: a review on computational, data-driven design of nanomaterials structures, properties, and synthesis. ACS Nano 16 (12), 19873–19891. Available at https://doi.org/10.1021/acsnano.2c08411.

Ye, F., Musselman, K.P., 2024. Synthesis of low dimensional nanomaterials by pulsed laser ablation in liquid. APL Materials 12 (5). Available at https://doi.org/10.1063/5.0199104.

Yi, M., Shen, Z., 2015. A review on mechanical exfoliation for the scalable production of graphene. Journal of Materials Chemistry A 3 (22), 11700–11715. Available at https://doi.org/10.1039/c5ta00252d.

Yu, H., Shi, R., Zhao, Y., Bian, T., Zhao, Y., Zhou, C., et al., 2017. Alkali-assisted synthesis of nitrogen deficient graphitic carbon nitride with tunable band structures for efficient visible-light-driven hydrogen evolution. Advanced Materials 29 (16). Available at https://doi.org/10.1002/adma.201605148.

Yuan, Y.J., Wang, P., Li, Z., Wu, Y., Bai, W., Su, Y., et al., 2019. The role of bandgap and interface in enhancing photocatalytic H2 generation activity of 2D-2D black phosphorus/MoS_2 photocatalyst. Applied Catalysis B: Environmental 242, 1–8. Available at https://doi.org/10.1016/j.apcatb.2018.09.100.

Zeng, M., Chen, M., Huang, D., Lei, S., Zhang, X., Wang, L., et al., 2021. Engineered two-dimensional nanomaterials: an emerging paradigm for water purification and monitoring. Materials Horizons 8 (3), 758–802. Available at https://doi.org/10.1039/d0mh01358g.

Zhang, Z., Yang, S., Zhang, P., Zhang, J., Chen, G., Feng, X., 2019. Mechanically strong MXene/Kevlar nanofiber composite membranes as high-performance nanofluidic osmotic power generators. Nature Communications 10 (1). Available at https://doi.org/10.1038/s41467-019-10885-8.

Zhao, L., Deng, J., Sun, P., Liu, J., Ji, Y., Nakada, N., et al., 2018. Nanomaterials for treating emerging contaminants in water by adsorption and photocatalysis: systematic review and bibliometric analysis. Science of the Total Environment 627, 1253–1263. Available at https://doi.org/10.1016/j.scitotenv.2018.02.006.

Zhao, X., Pan, D., Chen, X., Li, R., Jiang, T., Wang, W., et al., 2019. g-C_3N_4 photoanode for photoelectrocatalytic synergistic pollutant degradation and hydrogen evolution. Applied Surface Science 467-468, 658–665. Available at https://doi.org/10.1016/j.apsusc.2018.10.090.

Zheng, B., Xie, Y., Xu, S., Meng, A.C., Wang, S., Wu, Y., et al., 2024. Programmed multimaterial assembly by synergized 3D printing and freeform laser induction. Nature Communications 15 (1). Available at https://doi.org/10.1038/s41467-024-48919-5.

Zhou, M., Lou, X.W., Xie, Y., 2013. Two-dimensional nanosheets for photoelectrochemical water splitting: possibilities and opportunities. Nano Today 8 (6), 598–618. Available at https://doi.org/10.1016/j.nantod.2013.12.002.

Zhou, Y., Wang, Y., Peijnenburg, W., Vijver, M.G., Balraadjsing, S., Fan, W., 2023. Using machine learning to predict adverse effects of metallic nanomaterials to various aquatic organisms. Environmental Science and Technology 57 (46), 17786–17795. Available at https://doi.org/10.1021/acs.est.2c07039.

Zhu, J., Hou, J., Uliana, A., Zhang, Y., Tian, M., Van Der Bruggen, B., 2018. The rapid emergence of two-dimensional nanomaterials for high-performance separation membranes. Journal of Materials Chemistry A 6 (9), 3773–3792. Available at https://doi.org/10.1039/c7ta10814a.

Zou, Y., Wang, X., Ai, Y., Liu, Y., Li, J., Ji, Y., et al., 2016. Coagulation behavior of graphene oxide on nanocrystallined Mg/Al layered double hydroxides: batch experimental and theoretical calculation study. Environmental Science and Technology 50 (7), 3658–3667. Available at https://doi.org/10.1021/acs.est.6b00255.

CHAPTER

11

Impact of 2D nanomaterials on adsorption performances

Sameeha Adil[1], Sufiyan Syed[1], Wajahat Khan Mohammed[1], Ashly Thomas[1], Murchana Changmai[1], and Swarnalatha Rajaguru[2]

[1]Department of Chemical Engineering, BITS Pilani, Dubai, United Arab Emirates [2]Department of Electrical & Electronics Engineering, BITS Pilani, Dubai, United Arab Emirates

11.1 Introduction

Environmental pollution is a major issue in both developing and developed countries. Anthropogenic activity causes the environmental pollution to increase (Ajith and James, 2016). Living organisms are at risk from both organic and inorganic pollutants, which have become more common and persistent in recent years. Oceans, rivers, and other inland waterways are being strained by human activity in some ways, causing significant harm to their quality. Toxic pollutants such as heavy metals, pathogens, pesticides, dyes, polycyclic aromatic hydrocarbons (PAHs), and other harmful chemicals enter water bodies by dissolving, drifting, or settling, leading to contamination (Ajith et al., 2021). Heavy metal ions, nonmetallic salts, acids, bases, dyes, phenols, pharmaceuticals, pesticides, PAHs, and volatile organic compounds (VOCs) are among the various pollutants commonly found in wastewater. These pollutants may affect parameters such as pH, TOC, COD, and BOD, as well as introduce dangerous toxins (Thines et al., 2017; Gaur et al., 2018).

Many wastewater treatment techniques have demonstrated promise in eliminating contaminants from wastewater, including the activated sludge process (ASP), adsorption using materials like zeolites, alumina, biochar, activated carbon (AC), clay minerals (Nasar and Mashkoor, 2019), and chemical treatments like chlorination, ozonation, permanganate oxidation, UV irradiation, and membrane filtration (Zhou et al., 2019). However, because of their low-concentration persistence, resistance to degradation, and high management costs, these techniques are inefficient in getting rid of microcontaminants and their byproducts. Nanotechnology-based strategies have become effective substitutes for the following problems, allowing for the quicker and more efficient removal of microcontaminants through improved adsorption capabilities (Gogoi et al., 2018).

Two-Dimensional Nanomaterials for Water Purification
DOI: https://doi.org/10.1016/B978-0-443-44530-9.00011-0

One novel and promising strategy for tackling the present problems in water treatment is nanotechnology (Lu and Astruc, 2018; Shukla et al., 2018a). It facilitates the efficient use of nontraditional water sources and opens up new possibilities for current treatment procedures. In this sector, particles with sizes ranging from a few nanometers to several hundred nanometers are used. Because of their distinct surface and physicochemical characteristics, these nanoscale particles are ideal for cutting-edge applications. Proponents of nanotechnology contend that it may offer crucial answers to urgent global problems, like ensuring a steady supply of clean drinking water for a growing population. The surface area-to-volume ratio of nanomaterials is significantly higher than that of conventional micrometer-sized adsorbents, which results in enhanced photocatalytic performance, increased adsorption capacity, greater efficiency in pollutant removal, and quicker removal rates (Ajith et al., 2020). In the context of water treatment, nanotechnology plays a role in three main areas: pollution monitoring using nanosensors and detectors built for particular contaminants, remediation through whole or partial pollutant removal, and purification. The main procedures utilized in treatment and remediation are antibacterial activity, photocatalysis, adsorption, and nanomembranes. Because of its high removal effectiveness, cheap operating costs, ease of design, and sustainability, adsorption employing nano-based materials has drawn interest (Ajith et al., 2021).

Water remediation uses a diverse range of nanoparticles, which can be generically categorized as organic, inorganic, and polymeric (Anjum et al., 2019). In order to solve the drawbacks of conventional wastewater treatment techniques, nanotechnology is emerging as a highly effective new-age solution. It can be identified and modified to remove microcontaminants from water more quickly and efficiently through absorption (Gehrke et al., 2015). Nanomaterials of different dimensions including 0D (quantum dots, metal oxides, fullerenes, polymeric particles, etc.), 1D (nanowires, carbon nanotubes, nanorods, nanobelts, nanofibres, etc.) (Garnett et al., 2019), 2D (graphene oxide [GO], layered double hydroxides [LDHs], MXene, hexagonal boron nitride [h-BN], transition metal dichalcogenides [TMDs], etc.) (Novoselov et al., 2016; Pantano and Kuljanishvili, 2020), 3D (metal organic framework [MOF], covalent organic framework [COF], nanocomposites, etc.) as summarized in Table 11.1 have been extensively investigated as efficient adsorbents of microcontaminants (Verma et al., 2023).

2D nanoparticles possess unique properties that make them highly effective adsorbents for water treatment applications. Their large specific surface area provides numerous active sites for catalytic reactions and adsorption, and their ease of surface functionalization enables precise tailoring of surface characteristics for particular applications. These materials are environmentally safe due to their biocompatibility and hydrophilicity, which enhance their interaction with water molecules and contaminants. They also feature an adjustable bandgap, allowing for the modulation of electronic properties for efficient photocatalysis, along with activated metallic hydroxide sites that facilitate catalytic degradation processes. Their semiconductor nature enhances photocatalytic activity under light irradiation, while their strong metallic conductivity facilitates rapid charge transfer and effective separation of photogenerated carriers. Moreover, the formation of Schottky junctions through surface functional groups further promotes charge separation and transfer in composite materials, and their high abundance enhances pollutant adsorption and catalytic performance (Xuanying et al., 2020). The presence of various oxygen-rich functional groups, such as

TABLE 11.1 Classification of 0D, 1D, 2D, and 3D nanomaterials for water treatment application.

Classification of nanomaterials based on dimensionality for water treatment applications			
0D	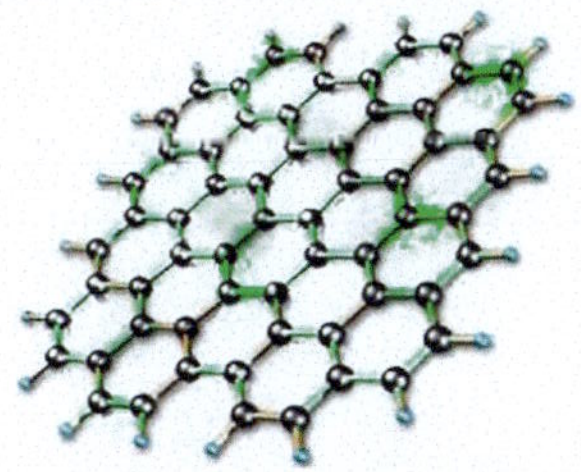 Graphene quantum dot (Graphene Quantum Dots: Introduction and Market News and Graphene-Info, 2025)	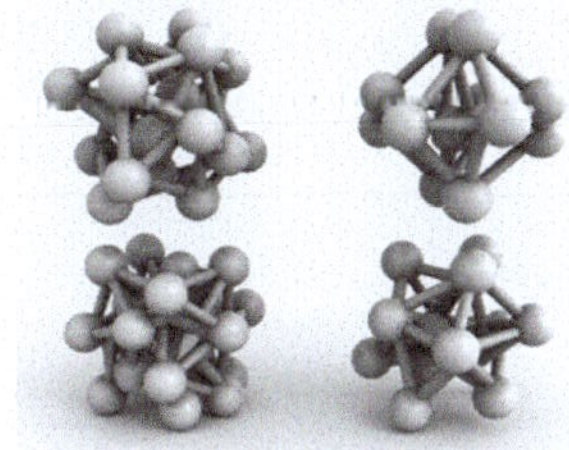 Anatase and rutile TiO_2 nanoparticles.	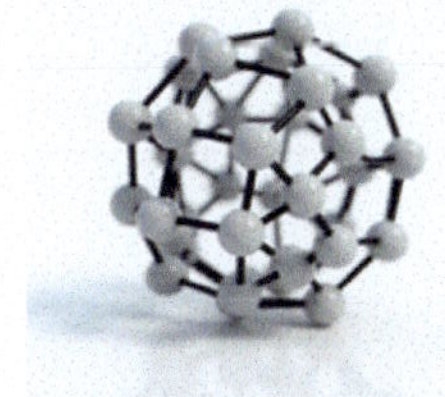 C_{60} fullerene
1D	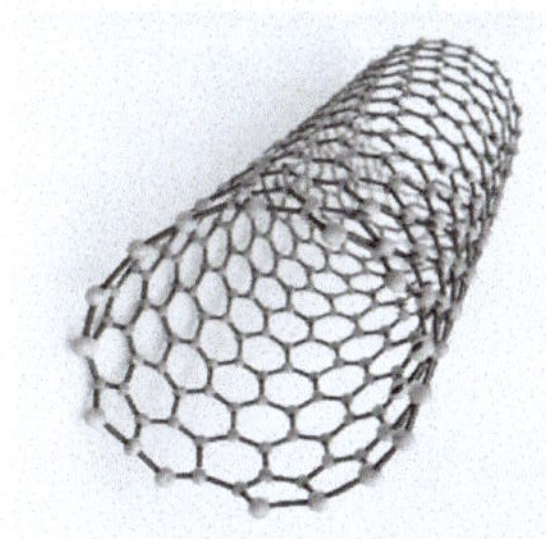 Carbon nanotube	 Electrospun carbon nanofibers	 Carbon nanobelt

(*Continued*)

TABLE 11.1 (Continued)

Classification of nanomaterials based on dimensionality for water treatment applications			
2D	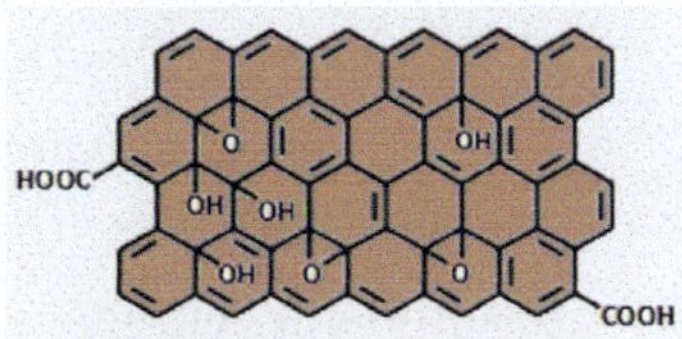 Graphene oxide	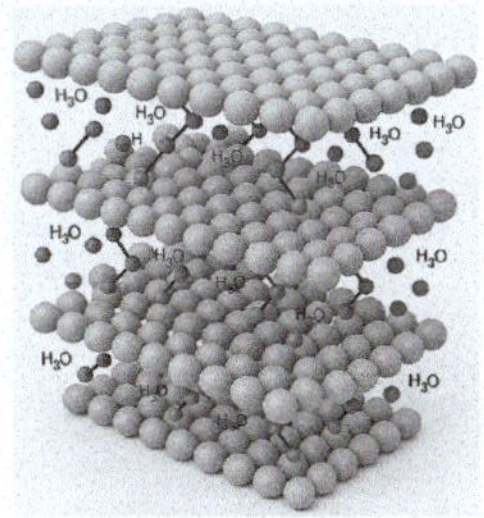 Layered double hydroxides (LDHs)	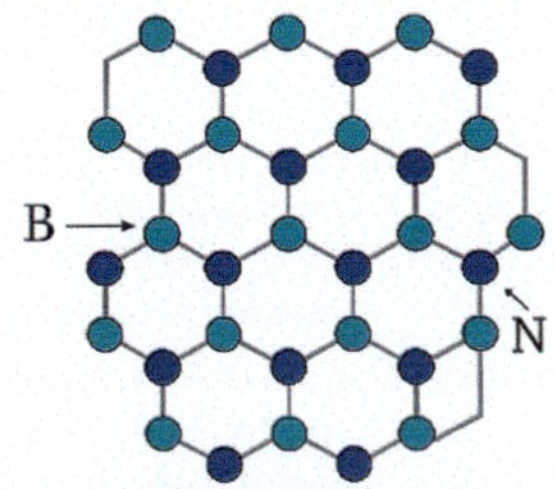 Hexagonal—boron nitride
3D	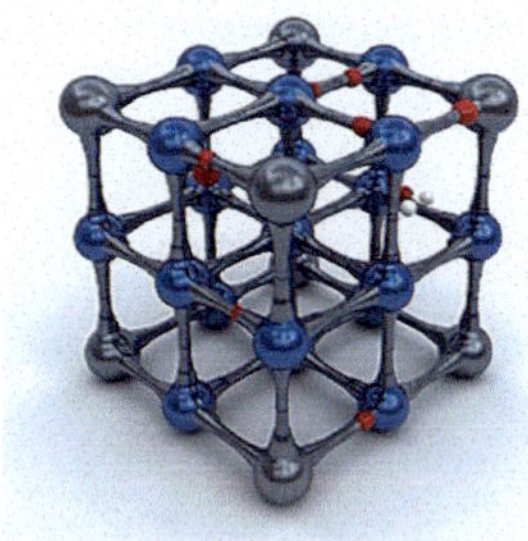 Metal-organic framework	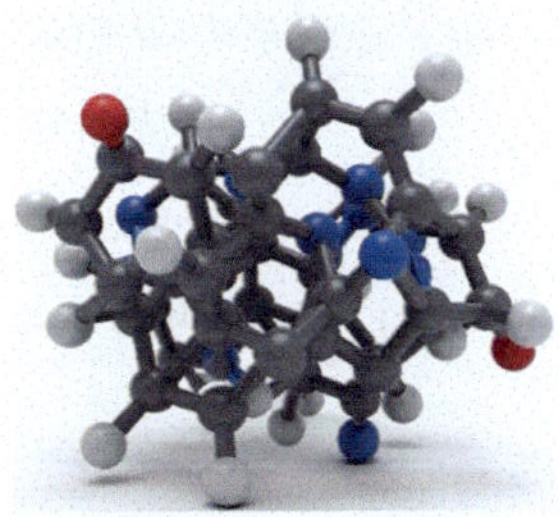 Covalent organic framework	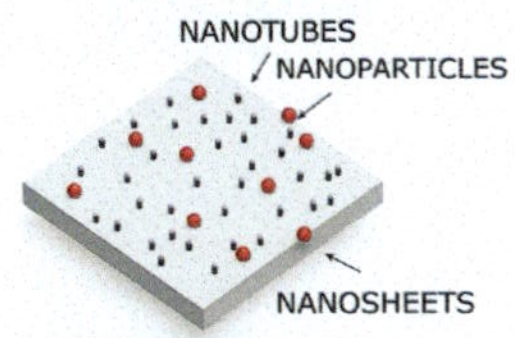 Nanocomposites

carboxyl and hydroxyl groups, makes it easier for pollutants to be adsorbed by ion exchange, hydrogen bonds, Van der Waals forces, electrostatic forces, and π–π interactions (Saini et al., 2024; Samuel and Shivaraju, 2024). These characteristics allow for fast regeneration, quick processing, minimal resource consumption, and highly efficient treatment outcomes (Zeng et al., 2021). Thus 2D nanomaterials are incorporated into water treatment systems for the efficient adsorptive removal of microcontaminants such as heavy metals, dyes, pharmaceuticals, and other emerging pollutants due to their high surface area, tunable surface chemistry, and layered structures (Khalid et al., 2020).

This chapter provides a comprehensive overview of the impact of 2D nanomaterials on adsorption-based water purification systems. It highlights how 2D nanomaterials provide higher surface area, reactivity, and selectivity compared to traditional adsorption materials. The chapter explains the fundamental principles and adsorption mechanisms unique to these materials, while addressing current challenges, possible solutions, and future paths for real-world applications. To demonstrate their benefits, a comparison with conventional adsorbents is also presented.

11.2 Role of 2D nanomaterials in adsorption systems

Numerous natural materials have been used as adsorbents traditionally, namely red mud, coconut husks, and bagasse ash, due to their wide availability and low cost (Zhao et al., 2019). However, these adsorbents have low adsorption efficiency and poor selectivity. To combat this, various nanomaterials were developed for higher performance, like zero-dimensional, one-dimensional, and two-dimensional nanomaterials. Zero-dimensional (0D) nanoparticles like quantum dots, carbon quantum dots, and fullerenes are nanomaterials with all three dimensions confined to below 100 nm, characterized by their quasi-spherical nature and high surface-to-volume ratio (Paras et al., 2022). These nanomaterials demonstrate remarkable capabilities for removing heavy metals and organic pollutants through mechanisms like electrostatic interactions, hydrogen bonding, π-π stacking, and van der Waals forces (Kong et al., 2024). They show high promise for dye and heavy metal adsorption due to their porous structure and active surfaces, and fullerenes have applications in molecular adsorption (Aziz et al., 2024). Despite its advantages, it has limited adsorption at high contaminant concentrations and is prone to self-agglomeration, which reduces adsorption capacity. It also faces degradation, which affects stability and reduces effective surface area. Its challenges with recovery and reuse play an environmental impact (Satyam and Patra, 2024).

One-dimensional (1D) nanomaterials are structures with two dimensions less than 100 nm and one dimension extending beyond 100 nm, creating extensive surfaces with high aspect ratios, coupled with anisotropic properties (Cao, 2022). Nanomaterials like Carbon Nanotubes (CNTs), nanowires, nanorods, and nanofibers exhibit exceptional adsorption capabilities due to their large specific area, high length-to-diameter ratios, and surface functional groups (Fatima et al., 2022). Carbon nanotubes, specifically, have applications in heavy metal removal, and remove Cu(II), Pb(II), and Cr(VI) from wastewater with capacities reaching 150–547 mg/g (Ji et al., 2022). Tungsten oxide nanowires show selectivity for organic dyes, with adsorption capacities of 547.32 mg/g for methylene blue. However, these nanomaterials also face issues like aggregation, which reduces surface area, stability issues that lead to degradation, and difficulty in recovery after adsorption. They also have high manufacturing costs and show difficulty in scalability. Some of these nanomaterials also show poor selectivity and release toxic products on degradation (Chrystie, 2023).

Three-dimensional (3D) nanomaterials are structures where none of the dimensions is confined under 100 nm, and are composed of nanostructures of 0D, 1D, or 2D arranged to form these 3D nanomaterials. Materials like CNT foams, graphene aerogels, and metal-organic frameworks (MOF) exhibit exceptional adsorption properties due to their large surface area and extensive porous networks (Zhou et al., 2020). Graphene aerogels remove oils with an absorption capacity of over 40 times their own mass, and 3D carbon-boron nitride structures have an adsorption capacity of 408 mg/g for methylene blue and remove heavy metals like Cr^{3+} within 2 minutes (Liu et al., 2018). MOFs have high selectivity, and have detection limits lower than 2 μg/L for palladium ions and have excellent recyclability (Liu et al., 2022). Despite this, 3D nanomaterials also face significant drawbacks. They have complex synthesis methods, high manufacturing costs, and are difficult to scale up. Additionally, their agglomeration reduces surface area and degradation due to lack of stability. Due to their macroscopic nature, there is difficulty in recovery and separation. Their heterogeneous nature leads to inconsistent performance, and there is difficulty in establishing a baseline protocol for large-scale implementation (Kalaitzidou and Merachtsaki, 2023).

The emergence of 2D nanomaterials has transformed adsorption-based water purification (Zakir et al., 2025). Properties like high surface-to-volume ratios, thin structure, and variable surface chemistry offer an advantage over conventional adsorbents for removing contaminants like heavy metals, dyes, to name a few, by offering high adsorption capacity, fast adsorption rate, and high regeneration. These features enable 2D nanomaterials to have superior adsorption capacities, selectivity, and kinetics compared to traditional nanomaterials. The large lateral size and nanometer thickness impart 2D nanomaterials with a high specific surface area, and have applications in multiple environmental remediation applications, like catalysis, sensing, and adsorption. 2D nanomaterials like graphene membranes with thin atomic layers have applications in the development of water-treatment systems due to their high adsorptive, catalytic, and separation capacity (Surwade et al., 2015). This section explores the role of 2D nanomaterials in adsorption, highlighting their application. Several 2D nanomaterials have been discussed in the literature: graphene and graphene oxide (GO), transition metal dichalcogenides (TMDs), MXenes, and boron nitride nanosheets (BNNSs), to name a few, which will be discussed in further detail in the upcoming sections. Fig. 11.1 shows the schematic representation of the said 2D nanoparticles. Some of the 2D nanomaterials used in absorption systems are dissussed as follows.

11.2.1 Graphene and graphene oxide

Graphene is the most extensively studied nanomaterial in adsorption applications, and shows great promise in environmental remediation and water treatment. Graphene consists of a single atomic layer of sp^2 hybridized carbon atoms arranged in a honeycomb lattice.

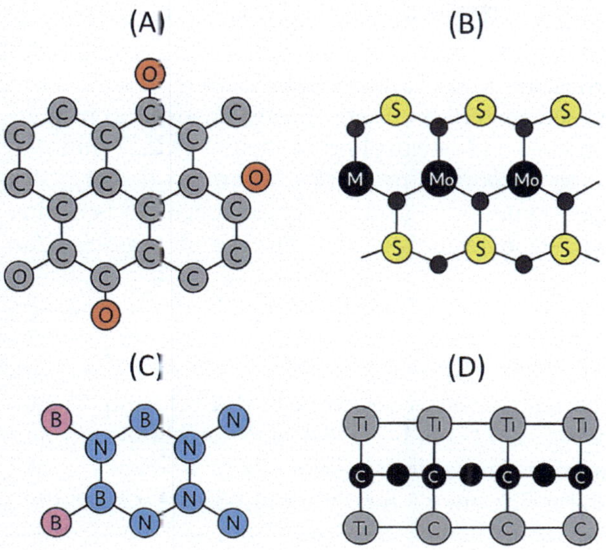

FIGURE 11.1 Schematic representations of 2D materials: (A) graphene oxide, (B) MoS_2, (C) hexagonal boron nitride (h-BN), and (D) MXene ($Ti_3C_2T_x$), illustrating atomic arrangements and layer structures.

A single atom is covalently bonded to further three carbon atoms, and has a distance of about 0.142 nm from adjacent carbon atoms. Individual layers are packed to form graphite with Van der Waal forces, with a 0.335 nm spacing between each layer (Huang et al., 2020).

Graphene's adsorption properties are due to its unique two-dimensional structure with its π-electron systems and high surface-to-volume ratios. Pure graphene is primarily hydrophobic, and its selectivity is primarily for polar contaminants. Graphene oxide has hydrophilic characteristics due to the oxygen functional groups on its surface (Lian et al., 2018). In ideal conditions, it exhibits exceptional mechanical strength (~1 TPa Young's modulus), electrical conductivity (~10^6 S/m), and specific surface area (2630 m^2/g) (Chenab et al., 2020). Its derivative, graphene oxide (GO), contains oxygen-derived functional groups including epoxy (-O-), hydroxyl (-OH), and carboxyl (-COOH) on the planes and edges. This addition allows the structure to be hydrophilic, increasing its reactivity (Han and Hong, 2025). This is done by disrupting the conjugated pi-system, enhancing the dispersibility in aqueous media (Jiao et al., 2025).

Graphene-based materials demonstrate exceptional performance in the removal of organic contaminants from aquatic environments. Graphene shows an adsorption capacity of 358.62 mg/g for rhodamine B and 547.81 mg/g for methylene blue. It removes most dyes in under 10 minutes (Jia et al., 2022). GO shows superior performance over reduced graphene oxide (rGO), with GO removing 98.67% methylene blue within 4 hours compared to rGO, which removes 93.47% in 7 hours (Jahan et al., 2022).

N-doped graphene nanosheets have exceptional capabilities for heavy metal removal through the process of capacitive deionization, achieving 90%–100% removal efficiency for cations, including Pb^{2+}, Cd^{2+}, Cu^{2+}, and Fe^{2+} within 30 minutes, with concentrations of 0.05–200 ppm (Liu et al., 2017).

Graphene oxide demonstrates high performance in removing perfluoroalkyl substances (PFASs) from tap water, with rGO microfiltration removing 138 μg/g of total PFASs in

30 mins (Khaliha et al., 2022). Alginate-graphene composite hydrogels have removal efficiency of ofloxacin that is 3–4 times higher than granular activated carbon, with adsorption reaching up to 178 mg/g for rhodamine B (Tunioli et al., 2023).

Surface functionalization of graphene-based materials enhances their adsorption properties. Amine-modified graphene aerogel achieves CO_2 adsorption capacities of 2.27 mmol/g at standard temperature and 0.1 bar, due to its porous network and amino density (Wu et al., 2022). Functionalization with oxygen-containing groups creates specific adsorption sites—carboxyl-modified graphene shows stronger adsorption for boron species (Yao et al., 2023).

Graphene oxide (GO) membranes represent a breakthrough in membrane technology, offering exceptional performance across diverse applications through their unique two-dimensional structure and tunable properties. GO membranes excel in water treatment with remarkable salt rejection capabilities, achieving 95–99% rejection of Na_2SO_4 and $MgSO_4$ while maintaining high water permeance up to 71–144 L m^{-2} h^{-1} bar^{-1} (Alkhouzaam and Qiblawey, 2021). These membranes demonstrate superior performance in heavy metal removal, with adsorption capacities reaching 535 mg g^{-1} for Fe^{3+} and 680 mg g^{-1} for Pb^{2+}, while achieving over 98% rejection of toxic metals like arsenic, cadmium, and lead (Shukla et al., 2018b). In desalination applications, GO membranes show promise with Na_2SO_4 rejection rates exceeding 98.5% and water permeance of 6.6 L m^{-2} h^{-1} bar^{-1} for reduced graphene oxide membranes (Chen et al., 2020). The membranes demonstrate excellent stability under harsh conditions, maintaining performance for over 72 hours in continuous operation and showing resistance to fouling due to their hydrophilic nature. For gas separation, GO membranes achieve exceptional selectivity with H_2/CO_2 ratios reaching 240 and CO_2/N_2 selectivity of 70–104, while maintaining high permeance values (Alen et al., 2019). In organic solvent nanofiltration, GO membranes provide over 99% rejection of dyes and organic molecules with molecular weight cutoffs of 320–500 g mol^{-1}, combined with a high methanol permeance of 155 L m^{-2} h^{-1} bar^{-1} (Jatoi et al., 2023). The tunable interlayer spacing (0.4–1.25 nm) enables precise molecular sieving, while surface functionalization with polymers, cross-linkers, and metal ions enhances stability and separation performance (Tiwary et al., 2024). These combined properties position GO membranes as next-generation materials for water purification, desalination, gas separation, and organic solvent processing applications.

Despite their exceptional properties in adsorption, graphene and its derivatives face multiple limitations that constrain their practical implementation. This includes aggregation and stacking issues due to π-π interactions and van der Waals forces that reduce the surface area, reducing its performance (Ali et al., 2023). Scalability is also an issue due to environmentally unfriendly synthesis methods like the modified Hummers process, utilizing expensive chemicals and energy-intensive procedures. Thermal degradation is also a major issue for graphene oxide, as it undergoes significant degradation at low temperatures (210°C) due to its functional groups, which limits the operating temperature range (Olowojoba et al., 2016). Environmental risks include cytotoxic effects that vary with surface structure and size, with highly mobile nanoparticles posing bioaccumulation threats and pulmonary toxicity by inhalation. Recovery difficulties arise due to its nanoscale size, making separation expensive. Repeated cycling of the nanoparticle leads to material degradation and loss of functional groups (Ghulam et al., 2022).

11.2.2 Boron nitride materials

Boron nitride materials represent a promising class of adsorbents, gaining significant attention for water treatment and environmental remediation. They are isostructural to carbon, used due to their large surface area, exceptional thermal stability, conductivity, corrosion and oxidation resistance, low density, and high bandgap (Feng et al., 2016). They exist in four forms: hexagonal BN (h-BN), which has a structure analogous to graphite, rhombohedral BN (r-BN), wurtzite BN (w-BN), and cubic BN (c-BN). The c-BN and w-BN are low-density phases with sp3 hybridized bonds, while the h-BN and r-BN are dense phases with sp^2 hybridized BeN bonds (Yu et al., 2018).

Hexagonal boron nitride nanosheets possess a layered structure like graphite, but with Boron and Nitrogen atoms that alternate in a hexagonal lattice. This imparts unique electronic and chemical properties compared to graphite (Chkhartishvili et al., 2016). The electronegativity difference creates a partial ionic character in h-BN, allowing it to have a polarizing surface that interacts with ionic and polar molecules. The structural properties depend highly on the synthesis methods: activated boron nitride achieves specific surface areas of up to 2078 m^2/g and pore volumes of 1.66 cm^3/g (Li et al., 2013).

Boron nitride demonstrates exceptional performance in removing heavy metals from aqueous media. Activated BN achieves maximum adsorption capacities of 215, 225, 235, and 282 mg/g for Co^{2+}, Pb^{2+}, Ni^{2+}, and Ce^{3+}, respectively (Gai et al., 2023). BN fibers that are synthesized using template methods have higher capacities for the mentioned cations. Phosphorus-doped boron nitride (P-BN) shows selective adsorption behavior, with maximum capacities of 69.45 mg/g for zinc and 53.80 mg/g for copper in ideal conditions (Li et al., 2020).

BN nanosheets exhibit remarkable adsorption performance for organic pollutants, like pharmaceutical compounds and dyes. For pharmaceutical drugs, BN nanosheets show thermodynamically favorable, strong adsorption for levofloxacin, tetracycline, and curcumin (Goyal et al., 2020). For dye removal, hexagonal BN has selectivity for cationic dyes over anionic ones, due to its electrostatic interactions (Tarhan, 2020). h-BN removes 90% of cationic Victoria blue B compared to 42.6% removal of anionic Metanil Yellow using 10 mg of adsorbent. It achieves adsorption capacities of 895.2 mg/g for cationic dyes and 211 mg/g for anionic dyes.

Boron Nitride Materials are utilized particularly for hydrocarbon separation. Nanoporous BN with surface areas of 1360 m^2/g demonstrates stability up to 800°C in air and 1050°C in nitrogen, significantly higher than most conventional adsorbents. It has a high selectivity for paraffins over olefins and has a high performance in CH_4/N_2, C_2H_6/CH_4, and C_3H_8/C_3H_6 separation (Saha et al., 2017). Porous BN materials show nitrogen adsorption 150–200% higher than carbon alternatives. This is due to its heterogeneous surface and unique electronic properties of BN materials (Kimura et al., 2021).

Surface functionalization enhances the adsorption properties of BN through the addition of specific functional groups and defect sites. Acid treatment using phosphoric acid and sulfuric acid induces exfoliation and functionalization of sulfur, creating stable suspensions in water, ethanol, and acetone (Bhimanapati et al., 2014).

Composite materials incorporating BN show advanced adsorption applications. Functionalized h-BN integrated into polyvinylidene fluoride and polydopamine interpenetrating network membranes achieves over 97% rejection for both methylene blue and methyl orange dyes over multiple cycles. These membranes retain a 90% recovery ratio even after continuous operation with protein solutions, showing their enhanced antifouling properties (Avasarala et al., 2024).

BN-based membranes show exceptional performance in water purification, with water-exfoliated BN membranes achieving retention values close to 100% for various water-soluble dyes, including methyl orange, methylene blue, and Evans blue (García Doménech et al., 2022). Activated BN membranes demonstrate methylene blue removal that exceeds that of activated carbon (Li et al., 2015).

Boron nitride (BN) membranes represent a highly promising class of materials for advanced separation applications, offering exceptional performance through their unique two-dimensional structure and outstanding physicochemical properties exceling in water treatment applications with remarkable performance metrics, achieving water permeance up to 1834 $L\ m^{-2}\ h^{-1}$ for boron nitride nanotube (BNNT) membranes and demonstrating near-perfect salt rejection capabilities of 98%–100% for various ionic species including NaCl, Na_2SO_4, and heavy metals (García Doménech et al., 2022). The superior performance stems from BN's unique structural features, including high surface area, excellent chemical and thermal stability (up to 1000°C), and tunable nanopore architectures that enable precise molecular sieving (Nam et al., 2023). In desalination applications, hexagonal boron nitride (h-BN) membranes show exceptional promise with water permeabilities exceeding conventional reverse osmosis membranes by several orders of magnitude while maintaining 100% salt rejection for concentrations up to 1 M, achieved through size exclusion and electrostatic interactions (Davoy et al., 2018). For organic contaminant removal, BN membranes demonstrate outstanding efficiency with near-complete (99%–100%) rejection of organic dyes, including methyl orange, methylene blue, and Evans blue, while maintaining high flux rates of 600–1310 $L\ m^{-2}\ h^{-1}\ bar^{-1}$ (Coogan et al., 2023). The membranes also exhibit remarkable stability across wide pH ranges and various organic solvents, making them suitable for diverse industrial applications (Chen et al., 2018). In gas separation applications, BN membranes show excellent selectivity for CO_2/N_2 separation with permeance values reaching 10^7 GPU and high selectivity ratios, while functionalized BN nanosheets enable tunable separation performance through controlled pore size and edge functionalization (Maurya et al., 2020). Additionally, BN membranes demonstrate unique proton conductivity properties, achieving values up to 107 $mS\ cm^{-1}$ for fuel cell applications, with exceptional stability under harsh operating conditions (Das et al., 2023). The combination of high permeability, excellent selectivity, chemical inertness, and mechanical robustness positions BN membranes as next-generation materials for water purification, desalination, gas separation, and energy applications, with particular advantages in harsh chemical environments where conventional membranes fail (Nam et al., 2023).

Despite its adsorption properties, boron nitride materials face significant limitations that inhibit practical implementation. Colloidal stability is a major challenge, as BN nanosheets are incredibly unstable in aqueous solutions, aggregating and precipitating over time in water, which reduces effective surface area (Da Silva et al., 2025). Due to its complex synthesis procedures, including high-temperature processing (>1300°C) and controlled atmospheres, it faces production and scalability challenges (Xu et al., 2019). Selectivity issues occur in complex environmental matrices, as they show poor distinction between target contaminants and other organic compounds, reducing efficiency due to competitive adsorption (Marchesini et al., 2019).

BN nanoparticles exhibit a "Trojan horse" effect, enhancing the toxicity of pollutants like cadmium by increasing their uptake, and demonstrate varying cytotoxic effects depending on particle length and surface properties (Kodali et al., 2022). Recovery and regeneration difficulties arise due to its nanoscale size, making separation technically challenging and expensive. Repeated regeneration leads to material degradation and loss of functional groups that are important for adsorption performance (Kotyakova et al., 2022). Temperature-dependent performance limitations show that adsorption capacity decreases with increasing temperature, and while BN exhibits high thermal stability, practical operating temperatures may be limited by the thermal decomposition of surface functional groups that are essential for enhanced adsorption (Gai et al., 2023).

11.2.3 Transition metal dichalcogenides

Transition metal dichalcogenides (TMDs) represent a family of layered materials with the general formula where M is a transition metal atom (such as Mo, W, Ti, Ta, or Re) and X is a chalcogen atom (S, Se, or Te), that have recently emerged as promising adsorbents (Yang et al., 2023). TMD monolayers stack collectively with the help of Van Der Waal forces. Each TMD monolayer is made up of three layers of atoms: a layer of transition metal, two layers of chalcogen, and a layer of transition metal between them. The ultrathin structure of TMDs (0.6–0.7 nm thick) combined with their high surface-area-to-mass ratio creates adsorption sites that make them attractive for contaminant removal applications (Mei et al., 2024).

TMDs have a diversity in structures with different polymorphs, including the thermodynamically stable 2 H phase and the metastable 1 T/1 T' phases, each offering electronic and chemical properties that affect adsorption behavior. The 2 H phase typically exhibits semiconducting properties with band gaps ranging from 1.2 to 2.4 eV for materials like MoS_2 (2.16 ± 0.04 eV) and WS_2 (2.38 ± 0.06 eV), while the metallic 1 T/1 T' phases provide enhanced conductivity and activated chemisorption sites (Hill et al., 2016). The layers create multiple adsorption sites, like basal planes, edge sites, and defect sites such as chalcogen vacancies, with edge sites typically exhibiting higher reactivity toward adsorbates (Wang et al., 2017). Surface areas vary depending on synthesis methods and structural modifications, and reach over 1000 m^2/g for optimized materials (Mei et al., 2024). The functional groups and defects on the surface enhance adsorption capabilities by providing binding sites and modifying the electronic structure of materials (Regis et al., 2023). The surface properties can be tuned and enhanced by doping, defect engineering, and phase transitions.

Transition metal dichalcogenides (TMDs) have outstanding efficiency in the elimination of heavy metals from aqueous solutions, with the metallic phase 1 T/1 T' TMD nanosheets attaining lead ion (Pb^{2+}) removal capabilities as high as 758 mg/g (Mei et al., 2024). These materials can decrease residual Pb^{2+} concentrations from 2 mg/L to 2 μg/L in 0.5 minutes, complying with World Health Organization drinking water guidelines. Point-of-use devices constructed with metallic phase MoS_2 nanosheets demonstrate treatment capacity of 55 L-water/g-adsorbent at feed Pb^{2+} concentrations of 1 mg/L, indicating performance that is 1–3 orders of magnitude superior to other 2D materials and commercial activated carbon. The sulfur chemisorption sites enable strong metal-chalcogen bond formation and account for

TMD's superior performance in the metal phase. Different TMD materials show varying affinities for specific heavy metals, with MoS_2, WS_2, TaS_2, and TiS_2 all demonstrating excellent removal capabilities for various heavy metal ions, including Cu^{2+}, Cd^{2+}, and other toxic species.

TMDs exhibit excellent adsorption performance for organic pollutants, including pharmaceutical compounds, dyes, and industrial chemicals (Regis et al., 2023). TMD nanosheets show strong thermodynamically favorable adsorption for levofloxacin, tetracycline, and curcumin. The adsorption mechanism involves multiple interactions like hydrogen bonding, π–π stacking, and electrostatic attractions. For dye removal, TMDs demonstrate effective removal of cationic and anionic dyes, and the efficiencies vary depending on electrostatic interactions (He et al., 2022).

TMDs show promising applications in gas-phase adsorption, and their performance depends highly on the nature of gas molecules and TMD material properties (Khan Musa et al., 2020). Studies show effective adsorption of hydrazine gas with layer-dependent behavior. Monolayer and bilayer WS_2 show complete reversibility upon desorption, while multilayer structures exhibit only partial recovery. Adsorption tendency increases with an increase in the number of WS_2 layers, while interlayer separation decreases, making desorption more difficult. MoS_2 and WS_2 heterostructures exhibit enhanced gas adsorption performance compared to individual materials, with MoS_2-WS_2 heterostructures achieving adsorption energies of −0.15 and −0.25 eV for CO and C_2H_2, respectively. This shows stronger adsorption than isolated surfaces and excellent sensing properties, with -99.56% and -99.95% for CO and C_2H_2 detection. This shows a sharp reduction in electrical resistance on adsorption (Tian et al., 2024).

Direct covalent functionalization of TMD surfaces is achieved through electrochemical methods using diaryliodonium salts, enabling attachment of various functional groups such as $-CF_3$ or $-NO_2$ that tune surface properties for specific applications (Michael, 2021). TMD-based heterostructures offer synergizing effects that limit individual materials while retaining their properties. Mixed-dimensional heterostructures combining TMDs with other nanomaterials (0D, 1D, or 3D) create enhanced photocatalytic systems with enhanced properties like charge separation, increased surface area, and light absorption ranges (Gan et al., 2021). These materials have application in flexible ultrathin devices that offer high energy density with long cycle lives (Ji and Choi, 2022).

Beyond pure adsorption, TMDs offer additional applications through adsorption-photocatalysis mechanisms that play a role in simultaneous contaminant removal and degradation (Yousef et al., 2022). They possess direct band gaps that enable efficient absorption of visible light, making them efficient photocatalysts for the degradation of pollutants (Hunge et al., 2024). Its efficient charge transfer and separation are due to its layered structure with strong in-plane covalent bonds and weak Van Der Waal interactions. WS_2-based nanomaterials have shown exceptional promise for photocatalytic water treatment, with their unique nanostructure and properties including high surface area and high carrier mobility enabling effective removal of organic contaminants, pharmaceuticals, heavy metals, and infectious microorganisms (Lu et al., 2023). Unlike traditional photocatalysts like TiO_2 and ZnO that are active only under UV light, TMDs can respond to visible light, significantly expanding their practical applicability.

TMDs also offer exceptional membrane performance for water treatment, combining ultrafast permeation with high selectivity. TMD membranes, particularly MoS_2 and WS_2

variants, achieve remarkable water permeance of 245–918 L m^{-2} h^{-1} bar^{-1} while maintaining over 92% rejection of salts and near-complete (99%) rejection of organic contaminants (Sun et al., 2013). These two-dimensional materials excel in desalination applications, demonstrating 92% Na_2SO_4 rejection with flux rates of 1.5 L m^{-2} hr^{-1} bar^{-1}, and show promise for heavy metal removal with ultrahigh adsorption capacities up to 758 mg g^{-1}. The tunable interlayer spacing (6.2–12 Å) enables precise molecular sieving, while their superior chemical stability and hydrophilic surface properties provide excellent fouling resistance compared to conventional membranes (Ang and Chew, 2019). TMD membranes also demonstrate versatility in gas separation, achieving H_2 permeance of 47,000 GPUs with high selectivity ratios, and show potential for organic solvent nanofiltration with almost 100% rejection of organic dyes in acetone while maintaining ultrafast permeance. The combination of high flux, excellent rejection performance, and operational stability positions TMD membranes as promising candidates for next-generation water treatment, desalination, and speciality separation applications.

Despite their adsorption capabilities, TMDs face several challenges, which hinder their widespread application in water treatment. Major synthesis and scalability issues arise due to high temperature requirements and costly precursors, making large-scale production impractical (Hui et al., 2020). Quality control remains challenging, with synthesis methods like chemical vapor deposition introducing defects such as chalcogen vacancies and polycrystalline films with imperfect grain boundaries, affecting performance. Toxicity varies across TMDs: WS_2 and MoS_2 show low cytotoxicity, while WSe_2 displays dose-dependent toxicity (Shi et al., 2022). Under light and aqueous conditions, TMDs showcase stability issues: WSe_2 degrades via oxidation, whereas MoS_2 is comparatively stable (Zhang et al., 2019). Recovery and regeneration are complicated by nanoscale dimensions, requiring costly separation techniques and facing degradation over cycles (Jiříček et al., 2015). TMDs also face environmental sensitivity, such as reduced effectiveness under high humidity, and performance inefficiency due to structural defects. TMDs also lack selectivity in matrices, leading to competitive adsorption, which undermines efficiency in targeting specific contaminants (Khan Musa et al., 2020).

11.2.4 MXenes

Two-dimensional MXenes, a family of transition metal carbides, nitrides, and carbonitrides with the general formula $M_{n+1}X_nT_x$ (where M is a transition metal, X is carbon/nitrogen, and T_x represents surface functional groups), have emerged as promising adsorbents since 2011 (Zhang et al., 2018). These materials possess a unique layered structure, where transition metals are interlaced with carbon or nitrogen atoms, and can have various functional groups like hydroxyl (–OH), oxygen (–O), fluorine (–F), and chlorine (–Cl) that are formed during the synthesis process. Its combination of large specific surface areas (exceeding 1000 m^2/g for certain structures), high electrical conductivity (>1000 S/m), hydrophilic nature, and abundant surface functional groups makes MXenes favorable for adsorption (Amani et al., 2025).

MXenes showcase a diversity in structures, with over 30 different compositions synthesized to date, and there is potential for hundreds more (Anasori et al., 2015). The most extensively studied MXene, $Ti_3C_2T_x$, demonstrates specific surface areas ranging from

9 to 93.08 m^2/g depending on the synthesis and processing methods employed. The layered accordion-like structure of MXenes, formed through selective etching of the A-layer atoms from MAX phase precursors, creates extensive interlayer spaces that are modified as per their application (Kumar et al., 2023). The properties of MXenes are imparted by their functional groups, introduced by wet etching (Kagalkar et al., 2025). These functional groups (–O, –OH, –F, and –Cl) create polarized surfaces that facilitate strong interactions with various adsorbates (Ibragimova et al., 2021). The distribution of functional groups can be controlled by synthesis conditions, with studies showing that oxygen-containing functional groups dominate under most preparation conditions, regardless of MXene composition. Surface functionalization exhibits similar patterns across different types of nanomaterials, which indicates that the interactions are dominated by the geometry of the triangular lattice rather than the chemical nature of the nanomaterial.

MXenes have exceptional performance in heavy metal removal, with efficiencies exceeding 95% in multiple studies (Linda et al., 2024). $Ti_3C_2T_x$ MXenes achieve remarkable removal rates for lead ions, reducing concentrations from 2 mg/L to 2 μg/L within 0.5 minutes, meeting World Health Organization drinking water standards. Vanadium carbide (V_2CT_x) nanosheets exhibit a high uptake capacity of 174 mg/g for uranium U(VI) with fast sorption kinetics and desirable selectivity. This is due to the formation of strong coordination bonds between metal ions and functional groups. Composite MXene materials demonstrate even higher capacities, with MXene/polydopamine (PDA) composite particles achieving ultrahigh adsorption capacity of 862.3 mg/g for chromium hexavalent Cr(VI) removal, while maintaining 96.7% of their original capacity after 2 months of storage in aqueous solution (Qing et al., 2023). The composite also exhibits reduction properties, converting more than half of toxic Cr(VI) to harmless chromium trivalent Cr(III). For specific metals, $Ti_3C_2T_x$ achieves maximum adsorption capacities of 169.8 mg/g for Cr(VI) at pH=2.0, 475.6 mg/g for Cd^{2+}, and demonstrates 96% lead removal efficiency in the first adsorption cycle (Lv et al., 2021). For dye removal, $Ti_3C_2T_x$ exhibits high capacity for methylene blue (~140 mg/g) due to electrostatic attractions when used at appropriate concentrations (Alkhudaydi et al., 2024). The material demonstrates high selectivity, making it suitable for positively charged organic pollutants- studies show that MXenes achieve over 80% removal of methylene blue under ideal conditions. In pharmaceutical removal applications, MXenes demonstrate superior performance compared to conventional adsorbents (Iravani and Varma, 2022). MXene-based photocatalysts show excellent degradation of dyes compared to graphene-based materials, with some compositions achieving degradation efficiencies of 96%, 94%, and 75% for Nb_2C, Ti_2C, and Ti_3C_2, respectively, within 120–160 minutes. For uremic toxins, $Ti_3C_2T_x$ achieves adsorption capacities of 45.7 mg/g for creatinine and 17.0 mg/g for uric acid (Zhao et al., 2020a).

MXenes have exceptional selectivity for radioactive waste elimination, namely, for actinide capture in aqueous solutions (Wang et al., 2020a). V_2CT_x MXene demonstrates competitive performance as an inorganic adsorbent for uranium capture, as uranyl species can bond strongly with hydroxylated MXene. Uranium adsorption capacity goes as high as 1890 mg/g for Ti_2CO_2, and its binding energy and adsorption capacity depend on its functional group.

MXenes have high capacity for gas separation and carbon capture applications, and their performance depends on their interlayer spacing and surface functionalization

(Massoumılari and Velioğlu, 2023). CO_2 can be effectively trapped within MXene nanochannels through electrostatic and van der Waals interactions, with shorter interlayer distances improving gas adsorption capacity. Nitride MXenes show promise for CO_2 capture, with compositions like Cr-, Mo-, and W-derived MXenes showing appropriate interaction strengths for carbon capture and storage applications.

The functional groups on MXene surfaces enable polymers to be ex situ blended or in situ polymerized, creating composite materials with advanced properties (Kagalkar et al., 2025). Functionalized MXene composites, like ZnO–MXene nanocomposites, demonstrate enhanced metal ion adsorption performance.

Biomimetic modification, such as the polydopamine (PDA) functionalization of MXenes, has promise to improve adsorption capacity and stability. PDA facilitates MXene exfoliation while providing additional binding sites, resulting in composite particles with ultrahigh adsorption capacity and enhanced stability against oxidation, which is usually the cause of restriction of MXene applications (Qing et al., 2023). Magnetite/MXene (Fe_3O_4/Ti_3C_2) nanocomposites achieve removal capacities of 4.68 mg/g for malachite green dye with 96% removal efficiency within 60 minutes (Alkhudaydi et al., 2024).

Fe_2O_3–SiO_2/MXene ternary nano-photo composites achieve 98% tetracycline removal efficiency under optimized conditions, through electron transfer from iron-based components to the MXene surface. The high electron mobility of MXenes facilitates the production of superoxide radicals ($\cdot O_2^-$), which contribute to contaminant degradation (Linda et al., 2024). MXene-derived oxide nanoheterostructures demonstrate exceptional photocatalytic activity, with TiNbO-based materials achieving 83% sulfamethoxazole degradation after 2 hours under UVA light-ternary oxide structures show superior performance compared to binary systems. MXene-derived oxide nanoheterostructures demonstrate exceptional photocatalytic activity. Specifically, TiNbO-based materials demonstrate 83% degradation of sulfamethoxazole after 2 hours under UVA light thus indicating that ternary oxide structures show superior performance compared to the binary systems. They show good reusability, with photocatalytic efficiency decreasing from 83% to 72% in 5 consecutive cycles (Qamar and Ali, 2025).

Due to its ultrathin structure and tunable interlayer spacing, MXenes demonstrate exceptional potential in membrane-based separation applications. Laminated MXene membranes constructed from exfoliated nanosheets achieve ultrahigh permeance for H_2 and excellent H_2/CO_2 selectivity through size exclusion mechanism. These membranes enable molecular separation based on size difference due to their well-ordered laminar structures (Huang et al., 2021). The versatility of MXene membranes extends to various applications, including gas separation, water treatment, organic solvent purification, nanofluidic ion transport, and osmotic energy conversion, due to its hydrophilicity, abundant functional groups, superb flexibility, and excellent conductivity. In organic solvent applications, MXene-embedded polymer membranes show 30% improvement for PEI membranes and 162% improvement for PDMS membranes in isopropanol flux with only 2 wt% MXene embedment.

Despite their adsorption properties, MXenes face several critical challenges that hinder their widespread use as adsorbents. Oxidation stability is a major limitation, as MXenes, especially HF-etched ones, degrade readily in the presence of water, oxygen, and elevated temperatures. This degradation converts them into metal oxides such as TiO_2 and carbon, altering their structure and nullifying their functionality (Zheng et al., 2021). Scalability is also a major barrier, as current synthesis involves multi-step processes with hazardous reagents like high-concentration HF, requiring stringent conditions and specialized equipment. Precursor cost, reactor design, energy input, and purification steps all contribute to high production expenses, complicating the transition to industrial-scale manufacturing (Bao et al., 2025).

Environmental and toxicity concerns add further constraints; studies on zebrafish embryos indicate composition-dependent toxicity, with delaminated $Nb_4C_3T_x$ showing higher mortality rates than Nb_2CT_x, and fluorinated MXenes potentially releasing harmful HF (Szuplewska et al., 2022). The nanoscale size and aggregation behavior of MXenes complicate their separation and recovery, reducing their reusability. Additionally, their broad-spectrum adsorption, while useful, can lead to poor selectivity in complex matrices where competitive adsorption lowers effectiveness for specific pollutants. Surface terminations greatly influence both adsorption performance and biocompatibility, yet controlling them during synthesis remains difficult. Certain functional groups are more prone to oxidation, and improper termination can compromise long-term stability and safety, further limiting real-world applications (Nahirniak et al., 2023). Table 11.2 shows a summarized application of 2D nanomaterials.

11.3 Principles of adsorption using 2D nanomaterials: fundamentals and mechanisms

2D nanomaterials are sheets with the presence of a few atoms. It has different materials like transition-metal dichalcogenides, hexagonal boron nitride, and MXenes. Their ultrathin geometry delivers enormous surface-to-volume ratios and lattice structures that can be turned layer by layer. 2D nanomaterials exhibit unique characteristics compared to the bulk 3D materials as they are composed of a few thick atoms. Their large lateral dimensions relative to thickness create physical, chemical, and mechanical behaviors (Xie and Zhao, 2022). For example, graphene with its electrical conductivity but lack of a semiconducting bandgap makes it unsuitable for semiconductor applications. Some of the properties are mentioned below.

11.3.1 Optical behavior

2D nanomaterials exhibit extraordinary light-related behaviors, which have driven a surge of research into their optical characteristics. When the materials are pared down to just one or a few atomic layers, they begin to show effects that will not be seen in thicker or bulk forms. Properties like distinctive emission bands, surface plasmons, heightened sensitivity to light, and strong absorption features (Zheng et al., 2020a). In fact, some atomically thin sheets are almost invisible to the naked eye in the visible spectrum. For example, single-layer graphene transmits about 97% of visible light, and freestanding $Ti_3C_2T_Z$ (form of MXene) films reach roughly 97% transparency like graphene (Dillon et al., 2016; Devi, 2020). Their pronounced response to environmental changes (temperature, chemical surroundings,

TABLE 11.2 Summary of 2D nanomaterials for adsorption applications.

Nanomaterial class	Major 2D materials	Key applications	Target pollutants	References
Graphene-based materials	Graphene, graphene oxide (GO), reduced GO, N-doped graphene	Adsorption of organic dyes and heavy metals; pharmaceutical removal; water desalination and gas-separation membranes	Dyes (Rhodamine B, methylene blue); heavy metals (Pb^{2+}, Cd^{2+}, Cu^{2+}); pharmaceuticals (e.g., ofloxacin, PFAS); gases (CO_2, H_2 via GO membranes)	Jia et al. (2022), Jahan et al. (2022), Liu et al. (2017), Khaliha et al. (2022), Tunioli et al. (2023), Alkhouzaam and Qiblawey (2021), Shukla et al. (2018b), Chen et al. (2020), Alen et al. (2019)
Boron nitride materials (BNMs)	Hexagonal BN (h-BN), rhombohedral BN (r-BN), activated BN nanosheets, BN fibers, P-doped BN	Adsorption of heavy metals and organic pollutants (dyes, pharmaceuticals); gas-phase hydrocarbon separation	Heavy metals (Co^{2+}, Pb^{2+}, Ni^{2+}, Ce^{3+}; Zn^{2+}, Cu^{2+}); pharmaceuticals (levofloxacin, tetracycline, curcumin); dyes (Victoria Blue B, Metanil Yellow); gases (CH_4, N_2, C_2H_6, C_3H_6)	Gai et al. (2023), Li et al. (2020), Goyal et al. (2020), Tarhan (2020), Saha et al. (2017), Kimura et al. (2021), García Doménech et al. (2022), Li et al. (2015), Nam et al. (2023), Davoy et al. (2018), Coogan et al. (2023)
Transition metal dichalcogenides (TMDs)	MoS_2, WS_2, TiS_2, TaS_2, $MoSe_2$, WSe_2	Adsorption of heavy metals; removal of organic dyes and pharmaceuticals; gas adsorption/sensing	Heavy metals (Pb^{2+} up to 758 mg/g; Cu^{2+}, Cd^{2+}); pharmaceuticals (levofloxacin, tetracycline, curcumin); dyes (cationic and anionic dyes); Gases (hydrazine, CO, C_2H_2)	Mei et al. (2024), Regis et al. (2023), He et al. (2022), Khan Musa et al. (2020), Tian et al. (2024)
MXenes	$Ti_3C_2T_x$, V_2CT_x, Nb_2C, Mo_2CO_2, Ti_2C	Adsorption of heavy metals; removal of dyes and pharmaceuticals; radioactive ion capture; gas (CO_2) separation; membrane separation	Heavy metals (Pb^{2+}, Cr(VI), Cd^{2+}); dyes (methylene blue, malachite green); pharmaceuticals (tetracycline); radioactive (UO_2^{2+}/U(VI)); gas (CO_2)	Linda et al. (2024), Qing et al. (2023), Lv et al. (2021), Alkhudaydi et al. (2024), Iravani and Varma (2022), Zhao et al. (2020a), Wang et al. (2020a), Massoumılari and Velioğlu. (2023)

strain, etc.) makes 2D materials effective for optical sensors. The interaction of nanomaterials with X-rays and other wavelengths is being explored if it can improve cancer phototherapy or radiotherapy, and advanced bioimaging techniques (Cheng et al., 2020; Khan et al., 2020).

11.3.2 Surface modification

Certain 2D nanomaterials carry reactive sites on their surfaces, which makes it possible to modify their behavior for pollutant adsorption, targeted drug delivery, catalysis, and molecular separation. New chemical species can be attached via two main routes of functionalization:

1. Noncovalent functionalization, where guest molecules, ions, or polymers adhere through weaker Van der Waals forces of attraction, electrostatic (ionic) interactions, hydrogen bonding, π–π stacking, etc.

   The functionalization method is illustrated in Fig. 11.2.

2. Covalent functionalization, in which the additive forms true chemical bonds with existing functional groups on the 2D sheet (Yu et al., 2020). For instance, graphene oxide has oxygen-based groups like carboxyls, hydroxyls, and epoxides. This can be further linked to polymers, nanoparticles, or biomolecules to adjust solubility, reactivity, or biocompatibility (Park et al., 2021). Likewise, MXenes naturally terminate in –F, –OH, and -O groups, for covalent grafting of dyes, catalysts, or ionic species (Cui et al., 2021). By choosing the groups to attach, multifunctional composites can be built like enabling next-generation sensors, membranes, catalysts, or energy-storage electrodes. Because of their high surface area, electronic and mechanical properties, and ultrathin geometry, 2D materials have become a

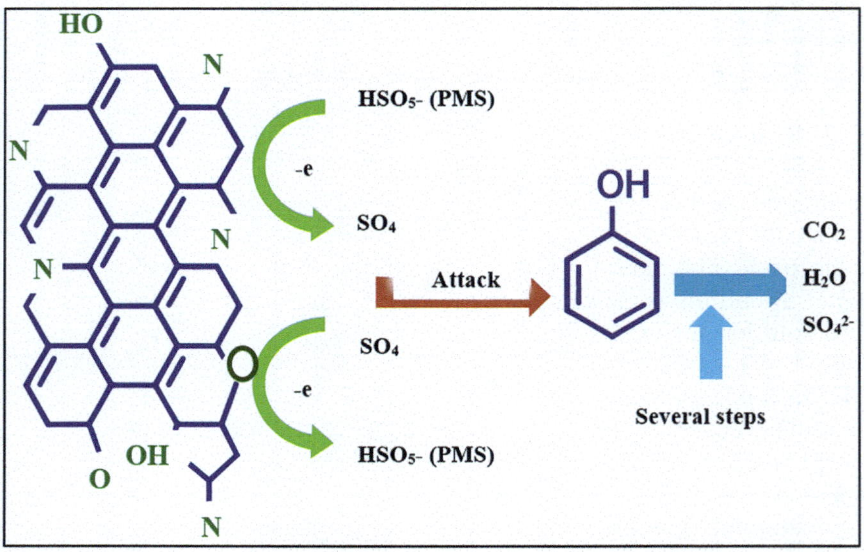

FIGURE 11.2 Carboxylate functionalized graphene oxide sheet dispersed by electrostatic repulsion, enabling uniform dispersion in water (Li et al., 2008).

central focus for research into separation technologies, advanced membrane fabrication, chemical and biological sensing, catalysis, and battery or supercapacitor designs.

11.3.3 Host-guest interaction mechanism

Layered 2D nanomaterials provide ideal "host" structures into which various "guest" ions, atoms, or even small molecules can reversibly slip between their sheets. This process, known as intercalation, involves the insertion and removal of foreign species into the gaps of the host's crystal lattice (Jun et al., 2016). For example, layered manganese dioxide (MnO_2) nanosheets can take up and release metal cations such as K^+, Li^+, Na^+, and Mg^{2+} (Xiong et al., 2017) while MXene materials have demonstrated efficient intercalation of Na^+ and K^+ ions, making them efficient for rechargeable batteries (Aslam and Xu, 2020). Insertion of species can be tuned as they enter and diffuse, intercalation in 2D hosts finds uses across energy storage, catalysis, drug-delivery systems, and environmental remediation.

11.3.4 High carrier conductivity

2D nanomaterials have several advantages, like transporting electric charges (along with heat) rapidly, which is needed for photovoltaics and high-speed electronics. It has been noted that graphene, black phosphorous, transition-metal dichalcogenides (TMDs), and 2D perovskites exhibit extremely high carrier intrinsic mobilities. For example, graphene can transport carriers at around 200,000 $cm^2\ V^{-1}\ s^{-1}$. Few-layered black phosphorus shows electron mobilities near 1000 $cm^2\ V^{-1}\ s^{-1}$ and hole mobilities around 286 $cm^2\ V^{-1}\ s^{-1}$ (Xiao et al., 2015). Beyond electrical transport, certain 2D materials boost thermal conduction; for example, a nanocomposite paper made of boron nitride (BN) nanosheets and cellulose nanofibrils reaches about 145.7 W/mK, which is comparable to aluminum alloys (Xiao et al., 2020). In general, charge carriers to a planar sheet speed up both electron and ion movement (Guo et al., 2020). Moreover, by introducing the right dopants into these ultrathin sheets, the electrical and thermal conductivities for specific devices can be fine-tuned (Joy et al., 2020).

11.3.5 High surface to volume ratio

All 2D nanomaterials have a surface area relative to their mass. For example, 1 g of graphene offers around 2630 m^2/g of exposed surface area (Zeng et al., 2018). This vast interface is a key factor in judging a material's suitability for various uses. In drug-delivery systems, for instance, such a high surface area means more space to load therapeutic molecules, making ultrathin 2D sheets ideal scaffolds for next-generation nanocarriers (Qiu et al., 2018). Compared with their bulk materials, 2D materials give devices an atomically thin contact layer, gaining fine control over how substances interact at the surface and, consequently, how the device behaves (Hu et al., 2018).

11.3.6 Structural adaptability

2D Nanomaterials are preferred because they are highly flexible and fit well in the development of flexible electronic devices and sensors. On the other hand, semiconductor devices and conventional metal films, which are used for strain gauges, break or crack during bending or flexing, which makes them unsuitable for wearable gadgets. Due to their thin, sheet-like structure, 2D materials can endure significant bending and twisting without losing performance, making them perfect for flexible sensors (Zheng et al., 2020b) and other bendable electronic components (Peng et al., 2014). 2D nanomaterials such as Xenes (e.g., silicene), transition-metal dichalcogenides (TMDs), borophene, and MXenes have all demonstrated potential in flexible-device applications (Li et al., 2019).

11.3.7 Tunable bandgap for semiconductors

2D nanomaterials exhibit a variety of electronic behaviors, with some traits common across many 2D sheets. Graphene has high conductivity but lacks an inherent bandgap and the properties of a semiconductor. So, it's not ideal for digital switching. In contrast, numerous 2D materials offer adjustable band gaps and semiconducting performance (Zhao et al., 2020b; Wang et al., 2021). Those with both high carrier mobility and moderate-to-wide band gaps (typically between 1 and 2.4 eV) are especially attractive for next-generation transistors and logic devices (Kang et al., 2020). As the gap increases, mobility decreases (Schwierz et al., 2015). An example is the family of transition-metal dichalcogenides (TMDs) in bulk form, which have indirect band gaps, but once thinned to a single or few layers, they transition to direct bandgap semiconductors, enhancing light emission and charge separation (Wang et al., 2012). Overall, 2D semiconductors are used over 3D counterparts by combining an atomically smooth interface, greater operational stability, and mobility, whereas many performances of bulk materials degrade over time (Su et al., 2021; Sharma et al., 2024) (Fig. 11.3).

11.3.8 Enhanced structural stability

2D nanomaterials are strong and can withstand great mechanical stress. Their stiffness, often quantified by Young's modulus, typically falls in the 150–400 GPa range for most 2D

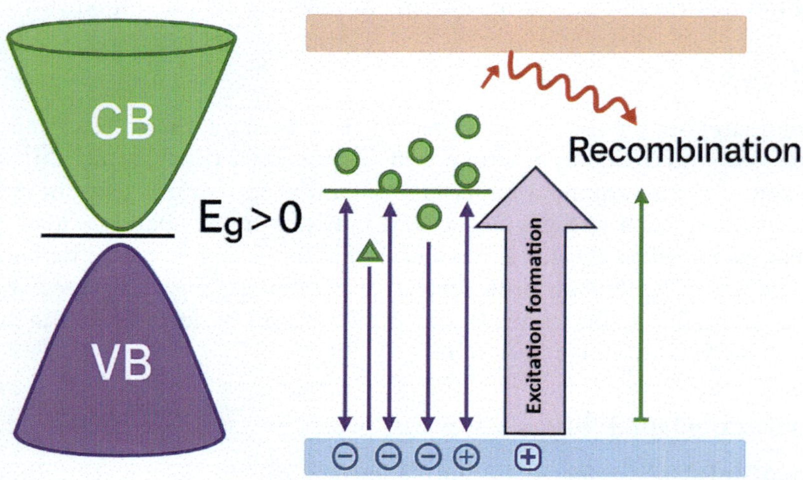

FIGURE 11.3 Schematic structure of a 2D semiconductor showing a finite bandgap (Eg>0), which separates the valence band and the conduction band, showing high carrier mobility (Sharma et al., 2024).

sheets (Jiang et al., 2020). For instance, Titanium Carbide (Ti_3C_2) exhibits an experimental Young's modulus around 330 ± 30 GPa, which is amongst the stiffest solution-processed 2D materials, while the material studio-molecular dynamics simulations predict it could reach about 502 GPa under ideal conditions (Carey and Barsoum, 2021). The atoms that terminate the surface of a 2D sheet also influence its rigidity. MXenes capped with –O groups tend to be stiffer than those with –F or –OH groups (Zhan et al., 2020). Adding 2D nanomaterials as fillers to composites or structural components can boost the mechanical stability of devices and machinery because of theirts high intrinsic strength. Table 11.3 shows a summary of the discussed nanomaterials with their properties and applications

TABLE 11.3 Features and applications of 2D nanomaterials.

2D material	Description	Main features	Example applications
Graphene	- Carbon-based 2D materials with sp^2 hybridized atoms in a hexagonal structure (Geim and Novoselov, 2007).	- High mechanical strength (Lee et al., 2008) - High thermal and electrical conductivity - Large surface area (Stoller et al., 2008) - Functionalized forms (e.g., GO, rGO) for water treatment (Baig et al., 2019)	- Batteries (Kamisan et al., 2022) - Supercapacitors (Stoller et al., 2008) Sensors (Baig et al., 2022) - Desalination (Gao et al., 2022)
Xenes	- Single-element 2D nanomaterials (Wang et al., 2020b)	- Tunable Bandgap via 2D layer adjustment (Venkateshalu et al., 2022) - Direct-gap semiconductors possible (Kamal and Ezawa, 2015) - Biocompatibility (Dongre S et al., 2022)	- Energy storage (Ren et al., 2017) - Field-effect transistors (Sultana et al., 2022) - Sensors (Kong et al., 2021) - Deionization (Gao et al., 2021)
Mxenes	- 2D transition metal carbides/nitrides with surface functional groups.	- Hydrophilic nature (Naguib et al., 2012) - High electrical conductivity - Flexible film-forming ability (Li et al., 2021)	- Supercapacitors (Chen et al., 2021) - Antifouling membranes (Zhang and Zhang, 2022) - Gas separation (Petukhov et al., 2021) - Water treatment (Hojjati-Najafabadi et al., 2022)
Transition-metal dichalcogenides	- TMDs like (MX_2) where M is a metal and X is a chalcogen atom.	- Direct and tunable bandgap - Strong light-matter interaction (Manzeli et al., 2017) - Atomic thickness - High catalytic activity	- Semiconductors - Flexible electronics (Jiang et al., 2022) - Gas sensors—energy storage (Kamisan et al., 2022)

11.4 Comparative performance analysis: 2D nanomaterials versus other adsorbents

Dimensional classification separates nanomaterials into 0D, 1D, 2D, and 3D structures, where each dimension of the same nanomaterial exhibits specific properties (Kallumottakkal et al., 2021). Although nanomaterials of diverse dimensions are found to be capable adsorbents in the removal of contaminants, 2D nanomaterials, being versatile atomically thinnest amongst the other groups, are interestingly better in the field of water treatment as they possess a broad range of properties and promising features. In this, the design of single or several atomic-thick layers can help gain exceptional materials with tremendous performance, thus enabling new perspectives in the field of nanotechnology and nanoscience. Their exclusive physicochemical, structural characteristics, and presence of oxygen functional groups promote their performance to adsorb and eliminate contaminants efficiently by different mechanisms. While traditional adsorbents are cost-effective and feasible for specific pollutants, 2D nanomaterials possess multifunctional ability and surpass in terms of adaptability, capacity, and sustainability. Certain 2D nanomaterials like Mxene, graphene, graphene oxide, LDHs, and BN are used as adsorbents for the adsorptive removal of contaminants (Sikri et al., 2025). Table 11.4 provides a comparative performance analysis of different types of commonly used 2D nanomaterials for adsorption.

2D nanomaterials possess rapid adsorption kinetics, tunable surface for the adsorption of target pollutants are effective solutions in the field of water treatment. In terms of performance, GO exhibits good efficiency over five regeneration cycles, demonstrating potent recyclability (Gomes et al., 2022) in the removal of toxic contaminants from industrial wastewater, although scalability remains a challenge. Mxene, a great adsorbent of heavy metals from water, is exothermic where removal of ions is optimum at reduced temperatures of approximately 35°C. At high-concentration ranges, MXenes perform with lower efficiency. In the case of BNNs, efficiency is dependent on the hydrophobicity of the pollutant. Since most pollutants are hydrophobic in nature, BNNs could be designed such that they target pollutants of similar hydrophobic nature, aiding increased efficiency, particularly in oil-water separation processes (Matsuzawa et al., 2010).

Adsorption capacity, removal efficiency, surface area of the nanomaterial, reusability, environmental impacts, etc, could be analyzed for comparative analysis of 2D nanomaterials with other adsorbents. 2D nanomaterials have a high surface area that enables high adsorption capacity with more than 90% removal efficiency, displaying lesser environmental impact and excellent efficiency over 5+ recyclability cycles, and are used in advanced adsorption techniques. However, with regard to traditional adsorbents like activated carbon are low cost, possess moderate surface area enabling moderate adsorption capacity, and variable removal efficiency with respect to specific pollutants displaying higher environmental impacts and lesser recyclability cycles used for separation and purification processes. Unlike 2D materials, traditional adsorbents are deficient in multifunctional efficiency.

2D nanomaterials outperform other 0D,1D,3D nanomaterials in adsorption processes. 2D nanomaterials have an unmatched, ultrahigh surface area where almost all atoms have surface active sites, creating more adsorption sites than other dimensional nanomaterials, thus providing contaminants with direct entry of pollutants to active sites, thereby intensifying adsorption kinetics. 0D nanomaterials, although they have a high surface

TABLE 11.4 Comparative performance analysis of different types of commonly used 2D nanomaterials.

Adsorbent type	Synthesis of adsorbent	Adsorption mechanism	Target pollutant removed	Properties responsible for removal	Industries involved
Graphene/ functionalized graphene/ hybridize-graphene with polymers (Li et al., 2024)	–	The mechanism of stacking, electrostatic attraction associated with the type of pollutant.	Organic and Inorganic pollutants	Intrinsic features like unique 2D structure, high surface area, tune-able surface chemistry.	Wastewater treatment
GO (Gomes et al., 2022)	Chemical oxidation (Modified Hummer) was used for the synthesis of GO	The mechanism of stacking, Electrostatic attraction, surface complexation is involved in the removal of contaminants from water.	More than 90% of Cd (II) and Pb (II) present in water was adsorbed onto GO.	High surface area, high density of negative charges from oxygen functional groups that helps in the interaction of ions.	Mining, petrochemicals, electroplating, fertilizer, etc.
Graphene and GO-based composites with hydrogels (Eraković and Stefanović, 2022)	–	Adsorption is essentially used for purification by graphene-based hydrogel composites.	Adsorption of diverse class of colored pollutants like paints, pesticides that are both organic and inorganic.	Porous structure of hydrogel composites	Wastewater
GO, granulated activated carbon (GAC), and zeolites (Marnani et al., 2024)	A modified Hummer method was used for GO synthesis.	Adsorption process. Physicochemical interaction is analyzed using isotherm models.	Removal of cationic dyes like Rhodamine B, methylene blue (MB)	Layered structure of GO, presence of large oxygen functional groups, resulting in enhanced adsorption capacity	Dye wastewater
Functionalized graphene-based composites (Iqbal et al., 2023)	Chemical vapor deposition and chemical exfoliation	Adsorption process	Removal of heavy metals and organic pollutants	Big and highly active surface, huge surface area, distinctive physicochemical properties	Wastewater treatment systems

(*Continued*)

TABLE 11.4 (Continued)

Adsorbent type	Synthesis of adsorbent	Adsorption mechanism	Target pollutant removed	Properties responsible for removal	Industries involved
GO-doped silica aerogels (GO-SA) (Sharma et al., 2023)	GO was prepared using modified Hummer method. GO-SA doping was done by supercritical fluid deposition method	Adsorption mechanism	Removal of dyes like Acid green 25, Crystal Violet, and Sulfamethoxazole from wastewater.	2D structure, Large surface area as a result of their porous and light-weight nature Also, (GO-SA) shows rapid adsorption kinetics	Textile industry wastewater system
Mxene (Ti_3C_2 monolayer) (Liu et al., 2024)	Computational tools like DFT, density functional tight-Binding molecular dynamics (DFTB-MD) are used to investigate the adsorption	Mechanism involved is chemisorption, physi-sorption, and electrostatic attraction.	Removal of dyes like MB, removal of hazardous pollutants	The adsorption behavior is a result of strong interaction properties, chemisorption, physi-sorption, and electrostatic attractions.	Wastewater treatment systems
TiVCTX-MXene-Graphene (Lv et al., 2024)	Hydrothermal self-assembly method followed by freeze drying technique was employed for the preparation of 3D nanosheet aerogels	Adsorption and absorption	Removal of common oils, organic solvents, different pollutants, dyes like MB, rhodamine B, congo red (CR), methyl orange (MO), tetracycline hydrochloride from wastewater	3D structure, combined material composition, photo-thermal heating capacity of the prepared hydrogel enable the removal of organic contaminants from water.	Wastewater treatment systems

Mxene (Pouramini et al., 2023)	A 2-step process is involved where MAX phase precursor preparation is followed by etching and delamination.	Adsorption, photocatalysis	Removal of heavy metals like Pb, Cr (VI), Hg, Cu, Pb, organic and inorganic pollutants.	Mxene possesses specific structural, chemical, electronic properties that make it more flexible for different applications.	Heavy metals evolved from industries like Mining, fertilizers, battery production, metal plating, leather tanning, refractories, electroplating, metallurgy, pharmaceutical industry, Coal and waste burning, cement, and paper production
Mxene—poly melamine formaldehyde composite (Zhang et al., 2024)	Mxene-PMF membrane synthesis	Static adsorption for the elimination of heavy metals	Removal of heavy metals like zinc, lead, organics like phenol and crystal violet dye from wastewater	Presence of active—NH_2 group in PMF particle and hydroxyl group on Mxene sheets enhances the adsorption process.	Wastewater bodies
Zero-valent iron—Mxene intercalated composite (Raza Ul Mustafa et al., 2024)	–	Adsorption process	Removal of polycyclic aromatic hydrocarbons like naphthalene, fluorene, and phenanthrene	Functional groups present on the surface of the MXene facilitated stabilization of zero valent iron into the interlayers and surface of MXene.	Oil & gas industry applications
Surface modified Mxene (Ahmaruzzaman, 2022)	–	Adsorption process induced by chemical affinity among pollutants, interaction with Mxene-surface functional groups.	Removal of toxic heavy metals, radionuclides	Built in Mxene properties like surface area, electron abundance, different surface termination, hydrophilicity	Wastewater systems

(*Continued*)

TABLE 11.4 (Continued)

Adsorbent type	Synthesis of adsorbent	Adsorption mechanism	Target pollutant removed	Properties responsible for removal	Industries involved
LDH-based material (Yang et al., 2016)	–	Adsorption and photocatalysis	Removal of dyes from wastewater	Presence of big interlayer spaces, positively charged surface layer, presence of solvation molecules	Wastewater systems
LDH (El khattabi et al., 2021)	–	Surface adsorption and intercalation between the layers.	Removal tartrazine dye	The layered structure enables efficient adsorption	Wastewater treatment systems, food and beverage industry, textile, pharmaceutical industry
Biopolymer modified double hydroxides (LDH) (R et al., 2020)	one-pot co-precipitation, extrusion method	Adsorption with pseudo-second order model indicating chemisorption	Removal of acid orange II dye	Characteristic microstructural features show efficient adsorption abilities that follow pseudo-second order kinetics, LDH composite also helps in continued stability over a relevant pH range.	Column-based water and wastewater treatment process
Pyrazole-3, 5-dicarboxylate (Hpdc) anions intercalated with LDH (Zheng et al., 2024)	Seeds Embedded-Epitaxial Growth-Ion exchange	Adsorption mechanism	Removal of copper (II) ions	High surface area and abundant exposed active sites	Wastewater treatment systems

Activated boron nitride (Li et al., 2013)	Simple structure-directed method	Adsorption mechanism	Absorption of metal ions like Cr 31, Co 21, Ni 21, Ce 31, Pb 21 Organic pollutants like tetracycline, methyl orange (MO), and congo red (CR) in water Volatile organic compounds (benzene) in air	High surface area, large pore volume, and unique multimodal pore structure	Chemical manufacturing industry
Boron nitride-based nanocomposites (Gadore et al., 2024)	Chemical and mechanical exfoliation (CVD), green synthesis methods were employed for the fabrication of hexagonal BN and its composites	Adsorption and photocatalysis	Remove heavy metals, dyes, pharmaceuticals	Adsorptive and photocatalytic properties of BN enable the removal	Material science manufacturing industry, wastewater treatment systems
Porous BN nanoribbons (Liu et al., 2019)	–	Blend of chelation and direct attraction through chemical bonds on the surface of nanomaterials	Remove cadmium and copper ions from water.	large surface area and small average porosity	Water purification process- wastewater and drinking H_2O
Hexagonal BN nanoparticles (Antipina et al., 2022)	CVD method	Adsorption mechanism involving physicochemical interaction	Absorption of antibiotics from wastewater	Large surface area, distinct physicochemical property, thermal stability, structural attributes.	Wastewater treatment systems and Environmental Science

area, are liable to aggregation, whereas 1D have a moderate surface area and 3D have a lower surface area. Tunable properties of 2D nanoparticles are high such that several nanosheets can be stacked, allowing ultrafast permeability and rejection of the ions. Tunability of 0D is limited by the size of the sheet, whereas it is moderate for 1D and low for 3D nanoparticles. In terms of reusability, 1D nanomaterials are excellent, followed by 1D and 3D where the reusability is moderate; 0D has recovery challenges and suffers poor reusability and recovery (Fatima et al., 2022; Paras et al., 2022).

In terms of adsorption kinetics, 2D nanomaterials followed pseudo 2nd order kinetics for the adsorption of pollutants, indicating that adsorption is determined by the active sites present and the process is managed by chemisorption. With respect to the isotherm models, Langmuir (monolayer) and Freundlich (multilayer) adsorption are generally used, where 2D nanomaterials like GO, Mxene, and BN follow Langmuir behavior, while LDH follows Freundlich behavior as a result of the layered intercalation technique

11.5 Conclusion and perspectives

Micropollutants from diverse environmental origins are regarded as contaminants that cause severe issues to health, the ecosystem, and other living organisms. 2D nanomaterials are a transformational class of adsorbents that have represented great advancement in adsorption technology to remove pollutants from wastewater streams and other environmental bodies. Physicochemical and distinct structural properties enable 2D nanomaterials like GO, MXene, LDH, and BNs to perform better than traditional adsorbents efficiently. Also, the large surface-volume ratio, superior adsorptive performance due to the tunable surface chemistry, kinetics, and versatility to various contaminants make 2D nanomaterials promising for water treatment, environmental remediation, etc. GO facilitates adequate adsorption by the ample O_2 groups, large surface area, etc., while LDH use ion exchange, intra-particle diffusion for adsorption process; Mxenes display excellent adsorption by the inherent hydrophilic, layered structure, plenty surface active sites, etc. Boron nitrides also possess exceptional properties like thermal stability, huge surface area, and chemical resistance, which enable the elimination of pollutants by Lewis acid–base interaction, H bonds, etc. From the 2D nanomaterials listed above, MXenes and their composites have greater absorption capacity by electrostatic and Van der Waals interactions. Furthermore, biocompatible, stable, and magnetic properties present in Mxene and 2D nanomaterial adsorbent materials contribute to water reuse and treatment. In general, 2D nanomaterials are greatly effective for contaminant removal and serve as an excellent alternative for remediation in comparison to conventional treatment techniques.

Although the molecular diversity of 2D nanomaterials provides profound opportunities in the field of water treatment, sometimes there is a difficulty in understanding and precisely predicting the performance of 2D nanomaterials. For 2D nanomaterials in the colloid phase, the complex nature of the structure is intensified by complex physicochemical interaction in aqueous surroundings, in addition to its sensitivity toward chemical degradation. While factors like cost-effective preparation of nanomaterials and stability in different liquid matrices remain critical for industrial applications, economic viability and scale-up of 2D nanomaterial composites still remain unsolved. Hence, it is important to gather deeper information and explore future prospects on features like structural properties of 2D nanomaterials employed in the adsorption of contaminants, on stability of structure, on economical methods in the synthesis of 2D nanomaterials alone, and also by adding composites to the nanomaterials to enhance adsorption, biocompatible nature, recyclability, and efficient performance for water treatment.

References

Ahmaruzzaman, M., 2022. MXenes based advanced next generation materials for sequestration of metals and radionuclides from aqueous stream. Journal of Environmental Chemical Engineering. 10 (5), 108371. Available at https://doi.org/10.1016/j.jece.2022.108371.

Ajith, M.P., James, M.K., 2016. Impact of chamravattam regulator cum bridge on Bharathapuzha river and adjacent areas. Indian Journal of Economics and Development 4.

Ajith, M.P., Priyadarshini, E., Rajamani, P., 2020. Effective and selective removal of heavy metals from industrial effluents using sustainable Si-CD conjugate based column chromatography. Bioresource Technology 314, 123786. Available at https://doi.org/10.1016/j.biortech.2020.123786.

Ajith, M.P., Aswathi, M., Priyadarshini, E., Rajamani, P., 2021. Recent innovations of nanotechnology in water treatment: a comprehensive review. Bioresource Technology 342, 126000. Available at https://doi.org/10.1016/j.biortech.2021.126000.

Alen, S.K., Nam, S., Dastgheib, S.A., 2019. Recent advances in graphene oxide membranes for gas separation applications. International Journal of Molecular Sciences 20 (22), 5609. Available at https://doi.org/10.3390/ijms20225609.

Ali, J., et al., 2023. Aggregation of graphene oxide and its environmental implications in the aquatic environment. Chinese Chemical Letters. 34 (2), 107327. Available at https://doi.org/10.1016/j.cclet.2022.03.050.

Alkhouzaam, A., Qiblawey, H., 2021. Functional GO-based membranes for water treatment and desalination: fabrication methods, performance and advantages. A review. Chemosphere 274, 129853. Available at https://doi.org/10.1016/j.chemosphere.2021.129853.

Alkhudaydi, A.M., Danish, E.Y., Abdel Salam, M., 2024. Magnetite/MXene (Fe3O4/Ti3C2) nanocomposite as a novel adsorbent for environmental remediation of malachite green dye. Molecules (Basel, Switzerland) 29 (6), 1372. Available at https://doi.org/10.3390/molecules29061372.

Amani, A.M., et al., 2025. MXene-based materials for enhanced water quality: advances in remediation strategies. Ecotoxicology and Environmental Safety 291, 117817. Available at https://doi.org/10.1016/j.ecoenv.2025.117817.

Anasori, B., et al., 2015. Two-dimensional, ordered, double transition metals carbides (MXenes). ACS Nano 9 (10), 9507–9516. Available at https://doi.org/10.1021/acsnano.5b03591.

Ang, E.H., Chew, J.W., 2019. Two-dimensional transition-metal dichalcogenide-based membrane for ultrafast solvent permeation. Chemistry of Materials a Publication of the American Chemical Society 31 (24), 10002–10007. Available at https://doi.org/10.1021/acs.chemmater.9b01873.

Anjum, H., et al., 2019. A review on adsorptive removal of oil pollutants (BTEX) from wastewater using carbon nanotubes. Journal of Molecular Liquids 277, 1005–1025. Available at https://doi.org/10.1016/j.molliq.2018.10.105.

Antipina, L.Y., Kotyakova, K.Y., Tregubenko, M.V., Shtansky, D.V., 2022. Experimental and theoretical study of sorption capacity of hexagonal boron nitride nanoparticles: implication for wastewater purification from antibiotics. Nanomaterials 12 (18), 3157. Available at https://doi.org/10.3390/nano12183157.

Aslam, M.K., Xu, M., 2020. A mini-review: MXene composites for sodium/potassium-ion batteries. Nanoscale 12 (30), 15993–16007. Available at https://doi.org/10.1039/D0NR04111D.

Avasarala, S., Sen Gupta, R., Goyal, P., Bose, S., 2024. Functionalized hexagonal boron nitride sheets and charge triggered interpenetrating polymer network based membranes work in tandem towards water remediation. Environmental Science: Water Research & Technology. 10 (2), 408–419. Available at https://doi.org/10.1039/D3EW00712J.

Aziz, M.T., Gill, W.A., Khosa, M.K., Jamil, S., Janjua, M.R.S.A., 2024. Adsorption of molecular hydrogen (H_2) on a fullerene (C_{60}) surface: insights from density functional theory and molecular dynamics simulation. RSC Advances 14 (49), 36546–36556. Available at https://doi.org/10.1039/D4RA06171C.

Baig, N., Ihsanullah, Sajid, M., Saleh, T.A., 2019. Graphene-based adsorbents for the removal of toxic organic pollutants: a review. Journal of Environmental Management 244, 370–382. Available at https://doi.org/10.1016/j.jenvman.2019.05.047.

Baig, N., Kawde, A.-N., Elgamouz, A., 2022. A cost-effective disposable graphene-based sensor for sensitive and selective detection of uric acid in human urine. Biosensors and Bioelectronics: X 11, 100205. Available at https://doi.org/10.1016/j.biosx.2022.100205.

Bao, W., et al., 2025. Engineering the next generation of MXenes: challenges and strategies for scalable production and enhanced performance. Nanoscale 17 (11), 6204–6265. Available at https://doi.org/10.1039/D4NR04560B.

Bhimanapati, G.R., Kozuch, D., Robinson, J.A., 2014. Large-scale synthesis and functionalization of hexagonal boron nitride nanosheets. Nanoscale 6 (20), 11671–11675. Available at https://doi.org/10.1039/C4NR01816H.

Cao, S., 2022. Editorial: one-dimensional nanostructures and their inspired applications in catalysis. Frontiers in Chemistry. 10, 1113117. Available at https://doi.org/10.3389/fchem.2022.1113117.

Carey, M., Barsoum, M.W., 2021. MXene polymer nanocomposites: a review. Materials Today Advances 9, 100120. Available at https://doi.org/10.1016/j.mtadv.2020.100120.

Chen, C., et al., 2018. Functionalized boron nitride membranes with ultrafast solvent transport performance for molecular separation. Nature Communications. 9 (1), 1902. Available at https://doi.org/10.1038/s41467-018-04294-6.

Chen, W., Zhang, D., Yang, K., Luo, M., Yang, P., Zhou, X., 2021. Mxene (Ti3C2T)/cellulose nanofiber/porous carbon film as free-standing electrode for ultrathin and flexible supercapacitors. Chemical Engineering Journal 413, 127524. Available at https://doi.org/10.1016/j.cej.2020.127524.

Chen, X., et al., 2020. Reduced holey graphene oxide membranes for desalination with improved water permeance. ACS Applied Materials & Interfaces 12 (1), 1387–1394. Available at https://doi.org/10.1021/acsami.9b19255.

Chenab, K.K., Sohrabi, B., Jafari, A., Ramakrishna, S., 2020. Water treatment: functional nanomaterials and applications from adsorption to photodegradation. Material Today Chemistry 16, 100262. Available at https://doi.org/10.1016/j.mtchem.2020.100262.

Cheng, L., Wang, X., Gong, F., Liu, T., Liu, Z., 2020. 2D Nanomaterials for cancer theranostic applications. Advanced Materials 32 (13), 1902333. Available at https://doi.org/10.1002/adma.201902333.

Chkhartishvili, L., Sartinska, L., Ramishvili, T., 2016. Adsorption selectivity of boron nitride nanostructures designed for environmental protection. In: Hussain, C.M., Kharisov, B., Hussain, C.M., Kharisov, B. (Eds.), Advanced Environmental Analysis. The Royal Society of Chemistry, pp. 167–192. Available at http://doi.org/10.1039/9781782623625-00167.

Chrystie, R.S.M., 2023. A Review on 1-D nanomaterials: scaling-up with gas-phase synthesis. Chemical Record (New York, N.Y.) 23 (10), e202300087. Available at https://doi.org/10.1002/tcr.202300087.

Coogan, Á., et al., 2023. Layered double hydroxide/boron nitride nanocomposite membranes for efficient separation and photodegradation of water-soluble dyes. Journal of Materials Chemistry. A 11 (23), 12266–12281. Available at https://doi.org/10.1039/D3TA01581E.

Cui, W., Hu, Z.-Y., Unocic, R.R., Van Tendeloo, G., Sang, X., 2021. Atomic defects, functional groups and properties in MXenes. Chinese Chemical Letters 32 (1), 339–344. Available at https://doi.org/10.1016/j.cclet.2020.04.024.

Da Silva, G.H., Ji, J., Maia, M.T., Mattia, D., Martinez, D.S.T., 2025. Exploring the combined toxicity of boron nitride nanosheets, cadmium and natural organic matter on Daphnia magna. Aquatic Toxicology (Amsterdam, Netherlands) 279, 107198. Available at https://doi.org/10.1016/j.aquatox.2024.107198.

Das, A., Yadav, V., Krishnamurthy, C.V., Jaiswal, M., 2023. Percolative proton transport in hexagonal boron nitride membranes with edge-functionalization. Nanoscale Advances 5 (18), 4901–4910. Available at https://doi.org/10.1039/D3NA00524K.

Davoy, X., et al., 2018. High water flux with ions sieving in a desalination 2D sub-nanoporous boron nitride material. ACS Omega 3 (6), 6305–6310. Available at https://doi.org/10.1021/acsomega.8b01076.

Dillon, A.D., et al., 2016. Highly conductive optical quality solution-processed films of 2D titanium Ca. Advanced Functional Materials 26 (23), 4162–4168. Available at https://doi.org/10.1002/adfm.201600357.

Devi, M., 2020. Application of 2D nanomaterials as fluorescent biosensors. In: In: Singh, L., Mahapatra, D.M. (Eds.), ACS Symposium Series, vol. 1353. DC: American Chemical Society, Washington, pp. 117–141. Available at http://doi.org/10.1021/bk-2020-1353.ch006.

Dongre S, S., R, S., Hunsur Ravikumar, C., C, L., Balakrishna, R.G., 2022. Review on 2D arsenene and antimonene: emerging materials for energy, electronic and biological applications. Advanced Materials Interfaces 9 (23), 2200442. Available at https://doi.org/10.1002/admi.202200442.

El khattabi, E.H., Mourid, E.H., Belaaouad, S., Naimi., Y., 2021. "Layered double hydroxide nanomaterial as highly efficient adsorbent and its recycling after removal of a carcinogenic tartrazine dye from wastewater. Biointerface Research Applied Chemistry 12 (6), 7725–7740. Available at https://doi.org/10.33263/BRIAC126.77257740.

Eraković, Z., Stefanović, D., 2022. Purification of contaminated wastewater with the help of graphene composites with hydrogels. Facta Universitatis, Series: Working and Living Environmental Protectiont. 27. Available at https://doi.org/10.22190/FUWLEP2201027E.

Fatima, J., Shah, A.N., Tahir, M.B., Mehmood, T., Shah, A.A., Tanveer, M., et al., 2022. Tunable 2D nanomaterials; Their key roles and mechanisms in water purification and monitoring. Frontiers Environmental Science 10, 766743. Available at https://doi.org/10.3389/fenvs.2022.766743.

Feng, Y., Zhou, Y., Lee, P.-H., Shih, K., 2016. Mineralization of perfluorooctanesulfonate (PFOS) and perfluorodecanoate (PFDA) from aqueous solution by porous hexagonal boron nitride: adsorption followed by simultaneous thermal decomposition and regeneration. RSC Advances 6 (114), 113773–113780. Available at https://doi.org/10.1039/C6RA15564B.

Gadore, V., Mishra, S.R., Singh, A.K., Ahmaruzzaman, M., 2024. Advances in boron nitride-based nanomaterials for environmental remediation and water splitting: a review. RSC Advances 14 (5), 3447–3472. Available at https://doi.org/10.1039/D3RA08323C.

Gai, A., Li, Y., Zhan, F., Zhang, J., Li, R., 2023. Preparation of activated boron nitride and its adsorption characteristics for Zn, Cu, and Cd in flue gas. ACS Omega 8 (30), 27612–27620. Available at https://doi.org/10.1021/acsomega.3c03348.

Gan, X., Lei, D., Ye, R., Zhao, H., Wong, K.-Y., 2021. Transition metal dichalcogenide-based mixed-dimensional heterostructures for visible-light-driven photocatalysis: Dimensionality and interface engineering. Nano Research 14 (6), 2003–2022. Available at https://doi.org/10.1007/s12274-020-2955-x.

Gao, H., et al., 2022. Confined lamellar channels structured by multilayer graphene for high-efficiency desalination. Desalination 530, 115681. Available at https://doi.org/10.1016/j.desal.2022.115681.

Gao, Y., et al., 2021. Pressurized alloying assisted synthesis of high quality antimonene for capacitive deionization. Advanced Functional Materials 31 (34), 2102766. Available at https://doi.org/10.1002/adfm.202102766.

García Doménech, N., et al., 2022. High-performance boron nitride-based membranes for water purification. Nanomaterials 12 (3), 473. Available at https://doi.org/10.3390/nano12030473.

Garnett, E., Mai, L., Yang, P., 2019. Introduction: 1D nanomaterials/nanowires. Chemical Reviews 119 (15), 8955–8957. Available at https://doi.org/10.1021/acs.chemrev.9b00423.

Gaur, N., Narasimhulu, K., Y, P., 2018. Recent advances in the bio-remediation of persistent organic pollutants and its effect on environment. Journal of Cleaner Production 198, 1602–1631. Available at https://doi.org/10.1016/j.jclepro.2018.07.076.

Gehrke, I., Geiser, A., Somborn-Schulz, A., 2015. Innovations in nanotechnology for water treatment. Nanotechnology Science Applied 8, 1–17. Available at https://doi.org/10.2147/nsa.s43773.

Geim, A.K., Novoselov, K.S., 2007. The rise of graphene. Nature Material 6 (3), 183–191. Available at https://doi.org/10.1038/nmat1849.

Ghulam, A.N., Dos Santos, O.A.L., Hazeem, L., Pizzorno Backx, B., Bououdina, M., Bellucci, S., 2022. Graphene oxide (go) materials—applications and toxicity on living organisms and environment. Journal of Functional Biomaterial 13 (2), 77. Available at https://doi.org/10.3390/jfb13020077.

Gogoi, A., Mazumder, P., Tyagi, V., Chaminda G.G, T., An, A., Kumar, M., 2018. Occurrence and fate of emerging contaminants in water environment: a review. Groundwater for Sustainable Development 6. Available at https://doi.org/10.1016/j.gsd.2017.12.009.

Gomes, B.F.M.L., De Araújo, C.M.B., Do Nascimento, B.F., Freire, E.M.P.D.L., Da Motta Sobrinho, M.A., Carvalho, M.N., 2022. Synthesis and application of graphene oxide as a nanoadsorbent to remove Cd (II) and Pb (II) from water: adsorption equilibrium, kinetics, and regeneration. Environmental. Science Pollution Research 29 (12), 17358–17372. Available at https://doi.org/10.1007/s11356-021-16943-3.

Goyal, A., Aggarwal, D., Kapoor, S., Goel, N., Singhal, S., Shukla, J., 2020. A comprehensive experimental and theoretical study on BN nanosheets for the adsorption of pharmaceutical drugs. New Journal of Chemistry. 44 (10), 3985–3997. Available at https://doi.org/10.1039/C9NJ06029D.

Guo, Q., Chen, N., Qu, L., 2020. Two-dimensional materials of group-IVA boosting the development of energy storage and conversion. Carbon Energy 2 (1), 54–71. Available at https://doi.org/10.1002/cey2.34.

Graphene-Info, 2025. Graphene Quantum Dots: Introduction and Market News. Graphene-Info. Available at https://www.graphene-info.com/graphene-quantum-dots.

Han, D., Hong, Q., 2025. Emerging trends in cellulose and lignin-based nanomaterials for water treatment. International Journal of Biological Macromolecules 307, 141936. Available at https://doi.org/10.1016/j.ijbiomac.2025.141936.

He, Y., Fu, X., Li, B., Zhao, H., Yuan, D., Na, B., 2022. Highly efficient organic dyes capture using thiol-functionalized porous organic polymer. ACS Omega 7 (21), 17941–17947. Available at https://doi.org/10.1021/acsomega.2c01250.

Hill, H.M., Rigosi, A.F., Rim, K.T., Flynn, G.W., Heinz, T.F., 2016. Band alignment in MoS2/WS2 transition metal dichalcogenide heterostructures probed by scanning tunneling microscopy and spectroscopy. Nano Letters 16 (8), 4831–4837. Available at https://doi.org/10.1021/acs.nanolett.6b01007.

Hojjati-Najafabadi, A., et al., 2022. Magnetic-MXene-based nanocomposites for water and wastewater treatment: a review. Journal of Water Process Engineering 47, 102696. Available at https://doi.org/10.1016/j.jwpe.2022.102696.

Hu, Z., Wu, Z., Han, C., He, J., Ni, Z., Chen, W., 2018. Two-dimensional transition metal dichalcogenides: interface and defect engineering. Chemical Society Reviews 47 (9), 3100–3128. Available at https://doi.org/10.1039/C8CS00024G.

Huang, L., Ding, L., Wang, H., 2021. MXene-based membranes for separation applications. Small Science 1 (7), 2100013. Available at https://doi.org/10.1002/smsc.202100013.

Huang, T., Jiang, Y., Shen, G., Chen, D., 2020. Recent advances of two-dimensional nanomaterials for electrochemical capacitors. ChemSusChem 13 (6), 1093–1113. Available at https://doi.org/10.1002/cssc.201903260.

Hui, G., 2020. Tuning the electrical conductivity of monolayer transition metal dichalcogenides via chemical vapor deposition synthesis. Cornell University.

Hunge, Y.M., Yadav, A.A., Kang, S.-W., Fujishima, A., Terashima, C., 2024. Transition metal dichalcogenide-based photocatalyst materials for hydrogen production. Book Chapter. Royal Society of Chemistry, pp. 194–221. Available at http://doi.org/10.1039/9781837674664-00194.

Ibragimova, R., Erhart, P., Rinke, P., Komsa, H.-P., 2021. Surface functionalization of 2D MXenes: Trends in distribution, composition, and electronic properties. Journal of Physical Chemistry Letters. 12 (9), 2377–2384. Available at https://doi.org/10.1021/acs.jpclett.0c03710.

Iqbal, M.S., et al., 2023. The potential of functionalized graphene-based composites for removing heavy metals and organic pollutants. Journal of Water Process Engineering. 53, 103809. Available at https://doi.org/10.1016/j.jwpe.2023.103809.

Iravani, S., Varma, R.S., 2022. MXene-based photocatalysts in degradation of organic and pharmaceutical pollutants. Molecules (Basel, Switzerland) 27 (20), 6939. Available at https://doi.org/10.3390/molecules27206939.

Jahan, N., et al., 2022. A comparative study on sorption behavior of graphene oxide and reduced graphene oxide towards methylene blue. Case Studies in Chemical and Environmental Engineering. 6, 100239. Available at https://doi.org/10.1016/j.cscee.2022.100239.

Jatoi, A.H., et al., 2023. Functionalized graphene oxide-based lamellar membranes for organic solvent nanofiltration applications. RSC Advances 13 (19), 12695–12702. Available at https://doi.org/10.1039/D3RA00223C.

Ji, D., Liu, Y., Ma, H., Bai, Z., Qiao, Z., Ji, D., et al., 2022. Study on uranium adsorption property of carbon nanotubes prepared by molten salt electrolysis. ACS Sustainable. Chemistry &. Engineering 10 (36), 11990–11999. Available at https://doi.org/10.1021/acssuschemeng.2c03451.

Ji, J., Choi, J.H., 2022. Recent progress in 2D hybrid heterostructures from transition metal dichalcogenides and organic layers: properties and applications in energy and optoelectronics fields. Nanoscale 14 (30), 10648–10689. Available at https://doi.org/10.1039/D2NR01358D.

Jia, M., Liu, C., Zhao, X., Yu, S., Chen, K., 2022. Efficient adsorption and mechanism of organic dyes on the freeze-dried graphene. SSRN Electronic Journal. Available at https://doi.org/10.2139/ssrn.4215375.

Jiang, D., et al., 2022. Flexible electronics based on 2D transition metal dichalcogenides. Journal of. Materials Chemistry A 10 (1), 89–121. Available at https://doi.org/10.1039/D1TA06741A.

Jiang, H., Zheng, L., Liu, Z., Wang, X., 2020. Two-dimensional materials: from mechanical properties to flexible mechanical sensors. InfoMat 2 (6), 1077–1094. Available at https://doi.org/10.1002/inf2.12072.

Jiao, H., Cui, M., Yuan, S., Dong, B., Xu, Z., 2025. Carbon nanomaterials for co-removal of antibiotics and heavy metals from water systems: An overview. Journal of Hazardous Materials 489, 137566. Available at https://doi.org/10.1016/j.jhazmat.2025.137566.

Jiříček, T., De Schepper, W., Lederer, T., Cauwenberg, P., Genné, I., 2015. Recovery of salts from ion-exchange regeneration streams by a coupled nanofiltration–membrane distillation process. Water Science and Technology a Journal of the International Association on Water Pollution Research 72 (2), 252–259. Available at https://doi.org/10.2166/wst.2015.215.

Joy, J., George, E., Haritha, P., Thomas, S., Anas, S., 2020. An overview of boron nitride based polymer nanocomposites. Journal of Polymer Science 58 (22), 3115–3141. Available at https://doi.org/10.1002/pol.20200507.

Jun, Y., Zhou, Y., Cha, J.J., 2016. Intercalation in two-dimensional transition metal chalcogenides. Inorganic Chemistry Frontiers. 3 (4), 452–463. Available at https://doi.org/10.1039/C5QI00242G.

Kagalkar, A., Dharaskar, S., Chaudhari, N., Vakharia, V., Karri, R.R., 2025. Enhanced metal ion adsorption using ZnO-MXene nanocomposites with machine learning-based performance prediction. Scientific Reports 15 (1), 15563. Available at https://doi.org/10.1038/s41598-025-00324-8.

Kalaitzidou, K., Merachtsaki, D., 2023. Post-use recovery of nanoparticles. In: Simeonidis, K., Mourdikoudis, S. (Eds.), Nanoparticles as Sustainable Environmental Remediation Agents. Royal Society of Chemistry, pp. 120–143. Available at http://doi.org/10.1039/BK9781837670215-00120.

Kallumottakkal, M., Hussein, M.I., Iqbal, M.Z., 2021. Recent progress of 2D nanomaterials for application on microwave absorption: a comprehensive study. Frontiers. Materials. 8, 633079. Available at https://doi.org/10.3389/fmats.2021.633079.

Kamal, C., Ezawa, M., 2015. Arsenene: two-dimensional buckled and puckered honeycomb arsenic systems. Physical Review B 91 (8), 085423. Available at https://doi.org/10.1103/PhysRevB.91.085423.

Kamisan, A.I., et al., 2022. Recent advances on graphene-based materials as cathode materials in lithium-sulfur batteries. International Journal of Hydrogen Energy 47 (13), 8630–8657. Available at https://doi.org/10.1016/j.ijhydene.2021.12.166.

Kang, S., et al., 2020. 2D semiconducting materials for electronic and optoelectronic applications: potential and challenge. 2D Material 7 (2), 022003. Available at https://doi.org/10.1088/2053-1583/ab6267.

Khalid, Z.B., Nasrullah, M., Nayeem, A., Wahid, Z.A., Singh, L., Krishnan, S., 2020. Application of 2D graphene-based nanomaterials for pollutant removal from advanced water and wastewater treatment processes in ACS Symposium Series, 1353, 1353. Adapting 2D Nanomaterials for Advanced Applications 1353. American Chemical Society, pp. 191–217. Available at https://doi.org/10.1021/bk-2020-1353.ch009.

Khaliha, S., et al., 2022. Graphene oxide nanosheets for drinking water purification by tandem adsorption and microfiltration. Separation and Purification Technology. 300, 121826. Available at https://doi.org/10.1016/j.seppur.2022.121826.

Khan, K., et al., 2020. Recent developments in emerging two-dimensional materials and their applications. Journal of Materials Chemistry C 8 (2), 387–440. Available at https://doi.org/10.1039/C9TC04187G.

Khan Musa, M.R., et al., 2020. Insight the process of hydrazine gas adsorption on layered WS_2: a first principle study. Nanotechnology 31 (49), 495703. Available at https://doi.org/10.1088/1361-6528/abb337.

Kimura, J., Ohkubo, T., Nishina, Y., Urita, K., Kuroda, Y., 2021. Adsorption enhancement of nitrogen gas by atomically heterogeneous nanospace of boron nitride. RSC Advance 11 (2), 838–846. Available at https://doi.org/10.1039/D0RA08437A.

Kodali, V., et al., 2022. Understanding toxicity associated with boron nitride nanotubes: Review of toxicity studies, exposure assessment at manufacturing facilities, and read-across. Journal of Materials Research 37 (24), 4620–4638. Available at https://doi.org/10.1557/s43578-022-00796-8.

Kong, J., Wei, Y., Zhou, F., Shi, L., Zhao, S., Wan, M., et al., 2024. Carbon quantum dots: properties, preparation, and applications. Molecules (Basel, Switzerland) 29 (9), 2002. Available at https://doi.org/10.3390/molecules29092002.

Kong, N., et al., 2021. Arsenene-mediated multiple independently targeted reactive oxygen species burst for cancer therapy. Nature Communication 12 (1), 4777. Available at https://doi.org/10.1038/s41467-021-24961-5.

Kotyakova, K.Y., Antipina, L.Y., Sorokin, P.B., Shtansky, D.V., 2022. Efficient and reusable sorbents based on nanostructured BN coatings for water treatment from antibiotics. International Journal of Molecular Sciences 23 (24), 16097. Available at https://doi.org/10.3390/ijms232416097.

Kumar, N., et al., 2023. One-step fragmentation of a 2D MXene across the fine 1D MnO_2 surface and its supercapacitance. CrystEngComm 25 (1), 72–85. Available at https://doi.org/10.1039/D2CE01323A.

Lee, C., Wei, X., Kysar, J.W., Hone, J., 2008. Measurement of the elastic properties and intrinsic strength of monolayer graphene. Science (New York, N.Y.) 321 (5887), 385–388. Available at https://doi.org/10.1126/science.1157996.

Li, D., Müller, M.B., Gilje, S., Kaner, R.B., Wallace, G.G., 2008. Processable aqueous dispersions of graphene nanosheets. Nature Nanotechnology 3 (2), 101–105. Available at https://doi.org/10.1038/nnano.2007.451.

Li, G., et al., 2021. 2D Titanium carbide (MXene) based films: Expanding the frontier of functional film materials. Advanced Functional Materials 31 (46), 2105043. Available at https://doi.org/10.1002/adfm.202105043.

Li, G., et al., 2024. Research progress in graphene-based adsorbents for wastewater treatment: preparation, adsorption properties and mechanisms for inorganic and organic pollutants. C 10 (3), 78. Available at https://doi.org/10.3390/c10030078.

Li, J., et al., 2013. Activated boron nitride as an effective adsorbent for metal ions and organic pollutants. Scientific Reports 3 (1), 3208. Available at https://doi.org/10.1038/srep03208.

Li, J., et al., 2015. Free-standing membranes made of activated boron nitride for efficient water cleaning. RSC Advances 5 (88), 71537–71543. Available at https://doi.org/10.1039/C5RA11899A.

Li, Y., et al., 2019. Emerging two-dimensional noncarbon nanomaterials for flexible lithium-ion batteries: opportunities and challenges. Journal of Material Chemistry A 7 (44), 25227–25246. Available at https://doi.org/10.1039/C9TA09377J.

Li, Y., Li, H., Li, R., Su, X., Shen, S., 2020. Preparation of phosphorus-doped boron nitride and its adsorption of heavy metals from flue gas. Royal Society Open Science 7 (8), 200079. Available at https://doi.org/10.1098/rsos.200079.

Lian, B., et al., 2018. Extraordinary water adsorption characteristics of graphene oxide. Chemical Science 9 (22), 5106–5111. Available at https://doi.org/10.1039/C8SC00545A.

Linda, J., A, G., Suresh, V., Subramanian, B., Menaka, S., 2024. Cost-effective synthesis of mxene cadmium sulfide (CdS) for heavy metal removal. Cureus. Available at https://doi.org/10.7759/cureus.70872.

Liu, H., Wang, X., Wang, R., Xu, H., Yan, W., An, Q., 2024. Computational investigations on the performance of transition metal carbide (MXene) for methylene blue dye adsorption and remediation from wastewater. Material Today Communication 39, 108735. Available at https://doi.org/10.1016/j.mtcomm.2024.108735.

Liu, L., Guo, X., Tallon, R., Huang, X., Chen, J., 2017. Highly porous N-doped graphene nanosheets for rapid removal of heavy metals from water by capacitive deionization. Chemistry Communication 53 (5), 881–884. Available at https://doi.org/10.1039/C6CC08515F.

Liu, T., et al., 2019. Porous boron nitride nanoribbons with large width as superior adsorbents for rapid removal of cadmium and copper ions from water. New Journal of Chemistry. 43 (8), 3280–3290. Available at https://doi.org/10.1039/C8NJ05299A.

Liu, W.X., Song, S., Ye, M.L., Zhu, Y., Zhao, Y.G., Lu, Y., 2022. Nanomaterials with excellent adsorption characteristics for sample pretreatment: a review. Nanomaterials 12 (11), 1845. Available at https://doi.org/10.3390/nano12111845.

Liu, Z., et al., 2018. Novel multifunctional cheese-like 3D carbon-BN as a highly efficient adsorbent for water purification. Scientific Reports 8 (1), 1104. Available at https://doi.org/10.1038/s41598-018-19541-5.

Lu, F., Astruc, D., 2018. Nanomaterials for removal of toxic elements from water. Coordination Chemistry Reviews 356, 147–164. Available at https://doi.org/10.1016/j.ccr.2017.11.003.

Lu, Q., Yang, L., Chelme-Ayala, P., Li, Y., Zhang, X., El-Din, M.G., 2023. "Enhanced photocatalytic degradation of organic contaminants in water by highly tunable surface microlenses. Chemistry (Weinheim an der Bergstrasse, Germany). Available at https://doi.org/10.26434/chemrxiv-2023-tdv80.

Lv, T., et al., 2024. TiVCTX MXene/Graphene nanosheet-based aerogels for removal of organic contaminants from wastewater. ACS Applied Nano Material 7 (7), 7312–7326. Available at https://doi.org/10.1021/acsanm.4c00038.

Lv, Y., Chang, K., Wu, H., Fang, P., Chen, C., Liao, Q., 2021. Highly efficient scavenging of Cr(VI) by two-dimensional titanium carbide nanosheets: kinetics, isotherms and thermodynamics analysis. Water Science and Technology a Journal of the International Association on Water Pollution Research 84 (9), 2446–2456. Available at https://doi.org/10.2166/wst.2021.434.

Manzeli, S., Ovchinnikov, D., Pasquier, D., Yazyev, O.V., Kis, A., 2017. 2D transition metal dichalcogenides. Nature Review Material 2 (8), 17033. Available at https://doi.org/10.1038/natrevmats.2017.33.

Marchesini, S., Wang, X., Petit, C., 2019. Porous boron nitride materials: Influence of structure, chemistry and stability on the adsorption of organics. Frontiers in. Chemistry. 7, 160. Available at https://doi.org/10.3389/fchem.2019.00160.

Marnani, N.N., Tezel, F.H., Basu, O.D., 2024. Adsorptive removal of dyes: A comparison of graphene oxide to granular activated carbon and zeolite NaY. Applied Science 14 (21), 9811. Available at https://doi.org/10.3390/app14219811.

Massoumılari, Ş., Velioğlu, S., 2023. Can MXene be the effective nanomaterial family for the membrane and adsorption technologies to reach a sustainable green world? ACS Omega 8 (33), 29859–29909. Available at https://doi.org/10.1021/acsomega.3c01182.

Matsuzawa, Y., Kimura, Z.-I., Nishimura, Y., Shibayama, M., Hiraishi, A., 2010. Removal of hydrophobic organic contaminants from aqueous solutions by sorption onto biodegradable polyesters. Journal of Water Resource and Protection 02 (03), 214–221. Available at https://doi.org/10.4236/jwarp.2010.23024.

Maurya, M., Sappidi, P.K., Singh, J.K., 2020. Selective separation of CO_2 from flue gas using carbon and boron nitride nanotubes as a membrane. Energy & Fuels an American Chemical Society Journal 34 (6), 7223–7231. Available at https://doi.org/10.1021/acs.energyfuels.0c00311.

Mei, L., et al., 2024. Metallic 1T/1T′ phase TMD nanosheets with enhanced chemisorption sites for ultrahigh-efficiency lead removal. Nature Communication 15 (1), 7770. Available at https://doi.org/10.1038/s41467-024-52078-y.

Michael, J.A., 2021. Covalent surface functionalization of transition metal dichalcogenide thin films via electrochemical methods. The University of Texas at Austin. Available at 10.26153/TSW/48899.

Naguib, M., et al., 2012. Two-dimensional transition metal carbides. ACS Nano 6 (2), 1322–1331. Available at https://doi.org/10.1021/nn204153h.

Nahirniak, S., Ray, A., Saruhan, B., 2023. Challenges and future prospects of the mxene-based materials for energy storage applications. Batteries 9 (2), 126. Available at https://doi.org/10.3390/batteries9020126.

Nam, S.-N., et al., 2023. Review of boron nitride-based membranes in liquid purification/separation applications. Chemical Engineering Journal 453, 139740. Available at https://doi.org/10.1016/j.cej.2022.139740.

Nasar, A., Mashkoor, F., 2019. Application of polyaniline-based adsorbents for dye removal from water and wastewater—a review. Environmental Science and Pollution Research 26 (6), 5333–5356. Available at https://doi.org/10.1007/s11356-018-3990-y.

Novoselov, K.S., Mishchenko, A., Carvalho, A., Castro Neto, A.H., 2016. 2D materials and van der Waals heterostructures. Science (New York, N.Y.) 353 (6298), aac9439. Available at https://doi.org/10.1126/science.aac9439.

Olowojoba, G.B., et al., 2016. In situ thermally reduced graphene oxide/epoxy composites: Thermal and mechanical properties. Applied Nanoscience 6 (7), 1015–1022. Available at https://doi.org/10.1007/s13204-016-0518-y.

Pantano, M.F., Kuljanishvili, I., 2020. Advances in mechanical characterization of 1D and 2D nanomaterials: progress and prospects. Nano Ex 1 (2), 022001. Available at https://doi.org/10.1088/2632-959x/abb43e.

Paras, et al., 2022. A review on low-dimensional nanomaterials: nanofabrication, characterization and applications. Nanomaterials 13 (1), 160. Available at https://doi.org/10.3390/nano13010160.

Park, M., Kim, N., Lee, J., Gu, M., Kim, B.-S., 2021. Versatile graphene oxide nanosheets *via* covalent functionalization and their applications. Materials Chemistry Frontiers 5 (12), 4424–4444. Available at https://doi.org/10.1039/D1QM00066G.

Peng, X., Peng, L., Wu, C., Xie, Y., 2014. Two dimensional nanomaterials for flexible supercapacitors. Chemical Society Reviews 43 (10), 3303. Available at https://doi.org/10.1039/c3cs60407a.

Petukhov, D.I., Kan, A.S., Chumakov, A.P., Konovalov, O.V., Valeev, R.G., Eliseev, A.A., 2021. MXene-based gas separation membranes with sorption type selectivity. Journal of Membrane Science 621, 118994. Available at https://doi.org/10.1016/j.memsci.2020.118994.

Pouramini, Z., et al., 2023. Recent advances in MXene-based nanocomposites for wastewater purification and water treatment: a review. Water 15 (7), 1267. Available at https://doi.org/10.3390/w15071267.

Qamar, M.A., Ali, S.K., 2025. Functionalized MXenes for enhanced visible-light photocatalysis: a focus on surface termination engineering and composite design. Inorganics 13 (2), 45. Available at https://doi.org/10.3390/inorganics13020045.

Qing, Q., et al., 2023. Synchronously enhanced removal ability and stability of MXene through biomimetic modification. Langmuir the ACS Journal of Surfaces and Colloids 39 (27), 9453–9467. Available at https://doi.org/10.1021/acs.langmuir.3c00987.

Qiu, M., et al., 2018. Omnipotent phosphorene: a next-generation, two-dimensional nanoplatform for multidisciplinary biomedical applications. Chemical Society Reviews 47 (15), 5588–5601. Available at https://doi.org/10.1039/C8CS00342D.

R., P., Mandal, S., S.R., B., 2020. Bio-polymer modified layered double hydroxide extrudates: a novel approach towards adsorptive purification of water. Journal of Water Process Engineering 35, 101208. Available at https://doi.org/10.1016/j.jwpe.2020.101208.

Raza Ul Mustafa, M., Khurshid, H., Isa, M.H., 2024. Nanoscale zero-valent iron (NZVI) intercalated MXene ($Ti_3C_2T_x$) for removal of polycyclic aromatic hydrocarbons in wastewater. Journal of Environmental Chemical Engineering 12 (2), 112149. Available at https://doi.org/10.1016/j.jece.2024.112149.

Regis, N.M., da Silva, J., Lima, M.P., 2023. Ab initio investigation of the adsorption properties of molecules on Mos2 pristine and with sulfur vacancy., SSRN. https://doi.org/10.2139/ssrn.4594621.

Ren, X., et al., 2017. Properties, preparation and application of black phosphorus/phosphorene for energy storage: a review. Journal of Materials Science 52 (17), 10364–10386. Available at https://doi.org/10.1007/s10853-017-1194-3.

Saha, D., et al., 2017. Nanoporous Boron nitride as exceptionally thermally stable adsorbent: Role in efficient separation of light hydrocarbons. ACS Applied Materials & Interfaces 9 (16), 14506–14517. Available at https://doi.org/10.1021/acsami.7b01889.

Saini, M., Saini, A., Saroha, B., Kumar, S., 2024. 2D metal chalcogenides for photodegradation and disinfection of wastewater pollutants. Advanced Two-Dimensional Nanomaterials for Environmental and Sensing Applications. CRC Press. Available at https://doi.org/10.1201/9781003436942-10.

Samuel, J.P., Shivaraju, H.P., 2024. 2D layered double hydroxides nanomaterials for the removal of water pollutants. Advanced Two-Dimensional Nanomaterials for Environmental and Sensing Applications. CRC Press. Available at https://doi.org/10.1201/9781003436942-15.

Satyam, S., Patra, S., 2024. Innovations and challenges in adsorption-based wastewater remediation: a comprehensive review. Heliyon 10 (9), e29573. Available at https://doi.org/10.1016/j.heliyon.2024.e29573.

Schwierz, F., Pezoldt, J., Granzner, R., 2015. Two-dimensional materials and their prospects in transistor electronics. Nanoscale 7 (18), 8261–8283. Available at https://doi.org/10.1039/C5NR01052G.

Sharma, A., et al., 2024. Covalently functionalized graphene oxide metal complexes: versatile nanocatalysts for organic transformations. Materials Science and Engineering: B 310, 117671. Available at https://doi.org/10.1016/j.mseb.2024.117671.

Sharma, S.K., Ranjani, P., Mamane, H., Kumar, R., 2023. Preparation of graphene oxide-doped silica aerogel using supercritical method for efficient removal of emerging pollutants from wastewater. Scientific. Reports 13 (1), 16448. Available at https://doi.org/10.1038/s41598-023-43613-w.

Shi, N., Bai, T., Wang, X., Tang, Y., Wang, C., Zhao, L., 2022. Toxicological effects of WS2 nanomaterials on rice plants and associated soil microbes. The Science of the Total Environment 832, 154987. Available at https://doi.org/10.1016/j.scitotenv.2022.154987.

Shukla, S., Kumar, S., Samchetshabam, G., Bharti, V., Kumar, K., Govindarajan, R., 2018. Recent technologies for wastewater treatment: A brief review. In: Jana, B., Mandal, R., Jayasankar, P. (Eds.), Wastewater Management Through Aquaculture. Springer, Singapore, pp. 225–234. Available at https://doi.org/10.1007/978-981-10-7248-2_11.

Shukla, A.K., et al., 2018b. Removal of heavy metal ions using a carboxylated graphene oxide-incorporated polyphenylsulfone nanofiltration membrane. Environmental Science: Water Research & Technology 4 (3), 438–448. Available at https://doi.org/10.1039/C7EW00506G.

Sikri, N., et al., 2025. Recent advancements on 2D nanomaterials as emerging paradigm for the adsorptive removal of microcontaminants. Advances in Colloid and Interface Science 340, 103441. Available at https://doi.org/10.1016/j.cis.2025.103441.

Stoller, M.D., Park, S., Zhu, Y., An, J., Ruoff, R.S., 2008. Graphene-based ultracapacitors. Nano Letters 8 (10), 3498–3502. Available at https://doi.org/10.1021/nl802558y.

Su, S.-K., Chuu, C.-P., Li, M.-Y., Cheng, C.-C., Wong, H.-S.P., Li, L.-J., 2021. Layered semiconducting 2D materials for future transistor applications. Small Structures 2 (5), 2000103. Available at https://doi.org/10.1002/sstr.202000103.

Sultana, N., Degg, A., Upadhyaya, S., Nilges, T., Sen Sarma, N., 2022. Synthesis, modification, and application of black phosphorus, few-layer black phosphorus (FLBP), and phosphorene: a detailed review. Materials in Advances 3 (14), 5557–5574. Available at https://doi.org/10.1039/D1MA01101D.

Sun, L., Huang, H., Peng, X., 2013. Laminar MoS2 membranes for molecule separation. Chemistry Communication 49 (91), 10718. Available at https://doi.org/10.1039/c3cc46136j.

Surwade, S.P., et al., 2015. Water desalination using nanoporous single-layer graphene. Nature Nanotechnology 10 (5), 459–464. Available at https://doi.org/10.1038/nnano.2015.37.

Szuplewska, A., et al., 2022. The 10th anniversary of MXenes: challenges and prospects for their surface modification toward future biotechnological applications. Advanced Drug Delivery Reviews, vol. 182, 114099. Available at https://doi.org/10.1016/j.addr.2021.114099.

Tarhan, T., 2020. Synthesis and characterization of hexagonal boron nitride used for comparison of removal of anionic and cationic hazardous azo-dye: kinetics and equilibrium studies. Turkish Journal of Chemistry 44 (6), 1471–1482. Available at https://doi.org/10.3906/kim-2004-23.

Thines, R.K., Mubarak, N.M., Nizamuddin, S., Sahu, J.N., Abdullah, E.C., Ganesan, P., 2017. Application potential of carbon nanomaterials in water and wastewater treatment: a review. Journal of the Taiwan Institute of Chemical Engineers 72, 116–133. Available at https://doi.org/10.1016/j.jtice.2017.01.018.

Tian, T., et al., 2024. First-principles predictions of MoS_2-WS_2 In-plane heterostructures for sensing dissolved gas species in oil-immersed transformers. ACS Omega 9 (18), 20253–20262. Available at https://doi.org/10.1021/acsomega.4c00681.

Tiwary, S.K., Singh, M., Chavan, S.V., Karim, A., 2024. Graphene oxide-based membranes for water desalination and purification. npj 2D Materials and Applications 8 (1), 27. Available at https://doi.org/10.1038/s41699-024-00462-z.

Tunioli, F., et al., 2023. Adsorption of emerging contaminants by graphene related materials and their alginate composite hydrogels. Journal of Environmental Chemical Engineering 11 (2), 109566. Available at https://doi.org/10.1016/j.jece.2023.109566.

Venkateshalu, S., et al., 2022. Phosphorene, antimonene, silicene and siloxene based novel 2D electrode materials for supercapacitors-a brief review. Journal of Energy Storage 48, 104027. Available at https://doi.org/10.1016/j.est.2022.104027.

Verma, C., Berdimurodov, E., Verma, D.K., Berdimuradov, K., Alfantazi, A., Hussain, C.M., 2023. 3D Nanomaterials: the future of industrial, biological, and environmental applications. Inorganic Chemistry Communications 156, 111163. Available at https://doi.org/10.1016/j.inoche.2023.111163.

Wang, J., et al., 2017. The mechanism of hydrogen adsorption on transition metal dichalcogenides as hydrogen evolution reaction catalyst. Physical Chemistry Chemical Physics PCCP 19 (15), 10125–10132. Available at https://doi.org/10.1039/C7CP00636E.

Wang, Q.H., Kalantar-Zadeh, K., Kis, A., Coleman, J.N., Strano, M.S., 2012. Electronics and optoelectronics of two-dimensional transition metal dichalcogenides. Nature Nanotech 7 (11), 699–712. Available at https://doi.org/10.1038/nnano.2012.193.

Wang, Y., Xue, J., Nie, G., Guo, X., 2020a. Uranium adsorption on two-dimensional irradiation resistant MXenes from first-principles calculations. Chemical Physics Letters 750, 137444. Available at https://doi.org/10.1016/j.cplett.2020.137444.

Wang, T., et al., 2020b. Xenes as an Emerging 2D Monoelemental family: fundamental electrochemistry and energy applications. Advanced in Function Materials 30 (36), 2002885. Available at https://doi.org/10.1002/adfm.202002885.

Wang, Y., Wang, L., Zhang, X., Liang, X., Feng, Y., Feng, W., 2021. Two-dimensional nanomaterials with engineered bandgap: synthesis, properties, applications. Nano Today 37, 101059. Available at https://doi.org/10.1016/j.nantod.2020.101059.

Wu, J., Qiu, X., Chen, S., 2022. "Preparation and characterization of an amine-modified graphene aerogel for enhanced carbon dioxide adsorption. Journal of Materials Science 57 (3), 1727.

Xiao, G., Di, J., Li, H., Wang, J., 2020. Highly thermally conductive, ductile biomimetic boron nitride/aramid nanofiber composite film. Composites Science and Technology 189, 108021. Available at https://doi.org/10.1016/j.compscitech.2020.108021.

Xiao, J., Long, M., Zhang, X., Ouyang, J., Xu, H., Gao, Y., 2015. Theoretical predictions on the electronic structure and charge carrier mobility in 2D Phosphorus sheets. Scientific Reports 5 (1), 9961. Available at https://doi.org/10.1038/srep09961.

Xie, Y., Zhao, D., 2022. 2D materials: a wonderland for physical science. National Science Review 9 (5), nwab202. Available at https://doi.org/10.1093/nsr/nwab202.

Xiong, P., Ma, R., Sakai, N., Bai, X., Li, S., Sasaki, T., 2017. Redox active Cation Intercalation/deintercalation In Two-dimensional Layered MnO_2 nanostructures for high-rate electrochemical energy storage. ACS Applied Materials & Interfaces 9 (7), 6282–6291. Available at https://doi.org/10.1021/acsami.6b14612.

Xu, Y., et al., 2019. Scalable production of high-quality boron nitride nanosheets *via* a recyclable salt-templating method. Green Chemistry an International Journal and Green Chemistry Resource GC 21 (24), 6746–6753. Available at https://doi.org/10.1039/C9GC03285A.

Xuanying, Y., Jili, Y., Haifang, T., Yi, D., Bashir, H., Kai, Y., et al., 2020. Embedding few-layer $Ti_3C_2T_x$ into alkalized g-C_3N_4 nanosheets for efficient photocatalytic degradation. Journal of Colloid and Interface Science 571, 297–306. Available at https://doi.org/10.1016/j.jcis.2020.03.061.

Yang, R., et al., 2023. 2D Transition metal dichalcogenides for photocatalysis. Angewandte Chemie International Edition 62 (13), e202218016. Available at https://doi.org/10.1002/anie.202218016.

Yang, Z., et al., 2016. ChemInform abstract: utilization of LDH-based materials as potential adsorbents and photocatalysts for the decontamination of dyes wastewater: a review. ChemInform 47 (43). Available at https://doi.org/10.1002/chin.201643236.

Yao, J., Li, D.S., Qiu, J., Xu, X., Yang, H.Y., 2023. How boron is adsorbed by oxygen-containing groups functionalized graphene: a density functional theory study. Separation and Purification Technology 330, 125551. Available at https://doi.org/10.1016/j.seppur.2023.125551.

Yousef, A., Thiehmed, Z., Abdul Shakoor, R., Altahtamouni, T., 2022. Recent progress in WS2-based nanomaterials employed for photocatalytic water treatment. Catalysts 12 (10), 1138. Available at https://doi.org/10.3390/catal12101138.

Yu, S., et al., 2018. Boron nitride-based materials for the removal of pollutants from aqueous solutions: a review. Chemical Engineering Journal 333, 343–360. Available at https://doi.org/10.1016/j.cej.2017.09.163.

Yu, W., Sisi, L., Haiyan, Y., Jie, L., 2020. Progress in the functional modification of graphene/graphene oxide: a review. RSC Advances 10 (26), 15328–15345. Available at https://doi.org/10.1039/D0RA01068E.

Zakir, O., et al., 2025. Recent progress in nanomaterials for water treatment: a comprehensive review of adsorption, photocatalytic, and antibacterial applications. Journal of Water Process Engineering 72, 107566. Available at https://doi.org/10.1016/j.jwpe.2025.107566.

Zeng, J., et al., 2018. 3D Graphene fibers grown by thermal chemical vapor deposition. Advanced Materials 30 (12), 1705380. Available at https://doi.org/10.1002/adma.201705380.

Zeng, M., Chen, M., Huang, D., Lei, S., Zhang, X., Wang, L., et al., 2021. Engineered two-dimensional nanomaterials: An emerging paradigm for water purification and monitoring. Materials Horizons 8 (3), 758–802. Available at https://doi.org/10.1039/d0mh01358g.

Zhan, X., Si, C., Zhou, J., Sun, Z., 2020. MXene and MXene-based composites: synthesis, properties and environment-related applications. Nanoscale Horizons 5 (2), 235–258. Available at https://doi.org/10.1039/C9NH00571D.

Zhang, T., Zhang, L.-Z., 2022. Development of a MXene-based membrane with excellent anti-fouling for air humidification-dehumidification type desalination. Journal of Membrane Science 641, 119907. Available at https://doi.org/10.1016/j.memsci.2021.119907.

Zhang, T.-Q., et al., 2024. Exfoliated MXene/poly-melamine-formaldehyde composite membranes for removal of heavy metals and organics from aqueous solutions. Journal of Hazardous Materials 463, 132866. Available at https://doi.org/10.1016/j.jhazmat.2023.132866.

Zhang, W., Matsuda, K., Miyauchi, Y., 2019. Photostability of monolayer transition-metal dichalcogenides in ambient air and acidic/basic aqueous solutions. ACS Omega 4 (6), 10322–10327. Available at https://doi.org/10.1021/acsomega.9b01067.

Zhang, Y., Wang, L., Zhang, N., Zhou, Z., 2018. Adsorptive environmental applications of MXene nanomaterials: a review. RSC Advances 8 (36), 19895–19905. Available at https://doi.org/10.1039/C8RA03077D.

Zhao, Q., et al., 2020a. Adsorption of uremic toxins using $Ti_3C_2T_x$ MXene for dialysate regeneration. ACS Nano 14 (9), 11787–11798. Available at https://doi.org/10.1021/acsnano.0c04546.

Zhao, F., et al., 2020b. Two-dimensional gersiloxenes with tunable bandgap for photocatalytic H2 evolution and CO2 photoreduction to CO. Nature Communications 11 (1), 1443. Available at https://doi.org/10.1038/s41467-020-15262-4.

Zhao, P., Jian, M., Zhang, Q., Xu, R., Liu, R., Zhang, X., et al., 2019. A new paradigm of ultrathin 2D nanomaterial adsorbents in aqueous media: graphene and GO, MoS_2, MXenes, and 2D MOFs. Journal of Materials Chemistry A 7 (28), 16598–16621. Available at https://doi.org/10.1039/C9TA02935D.

Zheng, Y., Sun, F., Han, X., Xu, J., Bu, X., 2020a. Recent progress in 2D metal-organic frameworks for optical applications. Advanced Optical Materials 8 (13), 2000110. Available at https://doi.org/10.1002/adom.202000110.

Zheng, Q., Lee, J., Shen, X., Chen, X., Kim, J.-K., 2020b. Graphene-based wearable piezoresistive physical sensors. Materials Today 36, 158–179. Available at https://doi.org/10.1016/j.mattod.2019.12.004.

Zheng, R., et al., 2024. Pyrazole-3,5-dicarboxylic acid intercalated LDH: A highly selective structured adsorbents for Cu^{2+} ions removal. Separation and Purification Technology 346, 127532. Available at https://doi.org/10.1016/j.seppur.2024.127532.

Zheng, Z., et al., 2021. The oxidation and thermal stability of two-dimensional transition metal carbides and/or carbonitrides (MXenes) and the improvement based on their surface state. Inorganic Chemistry Frontiers 8 (9), 2164–2182. Available at https://doi.org/10.1039/D1QI00041A.

Zhou, X., Liu, B., Chen, Y., Guo, L., Wei, G., 2020. Carbon nanofiber-based three-dimensional nanomaterials for energy and environmental applications. Materials Advances 1 (7), 2163–2181. Available at https://doi.org/10.1039/D0MA00492H.

Zhou, Y., Meng, J., Zhang, M., Chen, S., He, B., Zhao, H., et al., 2019. Which type of pollutants need to be controlled with priority in wastewater treatment plants: Traditional or emerging pollutants? Environment International 131, 104982. Available at https://doi.org/10.1016/j.envint.2019.104982.

CHAPTER

12

Advances of 2D nanomaterials in capacitive deionization

Youfang Zhang[1], *and Jun Jin*[2]

[1]Hubei Key Laboratory of Polymer Materials, Ministry of Education Key Laboratory for Green Preparation and Application of Functional Materials, School of Materials Science and Engineering, Hubei University, Wuhan, Hubei, P.R. China [2]Faculty of Materials Science and Chemistry, China University of Geosciences, Wuhan, Hubei, P.R. China

12.1 Overview of capacitive deionization for water purification

Exploring highly efficient seawater desalination techniques is a key strategy to address the challenges of freshwater scarcity. To tackle this challenge, various desalination techniques have been employed to generate freshwater from relatively abundant seawater or saline water. Seawater desalination technologies can be divided into thermal separation methods and membrane separation methods based on their separation modes (Urban, 2017; Skuse et al., 2021; Olabi et al., 2020; Wang et al., 2024). Thermal separation methods include multistage flash, multieffect distillation, and vapor compression. However, the disadvantages of thermal separation methods are low freshwater output rates, high energy consumption, and the potential for secondary pollution. Membrane separation methods are membrane-based seawater desalination processes, with common techniques including reverse osmosis and electrodialysis. Among these, reverse osmosis is the most effective technology. It is a pressure-driven desalination method that uses hydraulic pressure to push water from a saline solution through a semipermeable membrane, retaining solutes and separating water from salt. Electrodialysis, on the other hand, uses electric power to generate thrust, forcing ions in the solution to move in a specific direction to achieve desalination. The main challenge facing membrane separation methods is the high cost of membrane preparation and their susceptibility to fouling, which necessitates continuous cleaning or periodic replacement of membranes, limiting their large-scale application. Therefore developing a new type of seawater desalination technology that is safe, simple, low cost, low energy, efficient, and free from secondary pollution to obtain freshwater resources from seawater has become a goal pursued by both academia and industry.

Two-Dimensional Nanomaterials for Water Purification
DOI: https://doi.org/10.1016/B978-0-443-44530-9.00020-1

Capacitive deionization (CDI) technology (Jin et al., 2024; Gamaethiralalage et al., 2021; Liu et al., 2021a) is a novel desalination method developed based on the working principle of supercapacitors. As shown in Fig. 12.1, the CDI process Jin et al. (2024) includes two steps. The desalination process and the regeneration process. The desalination process involves applying a constant direct-current voltage or current to the CDI electrodes, creating an electric field between the electrodes. Under the influence of the electric field, anions and cations are driven to their respective electrodes and removed. The regeneration process involves saturating electrodes by short-circuiting the positive and negative terminals (zero-voltage desorption) or applying a reverse voltage (reverse-voltage desorption) to regenerate the electrodes. This approach avoids secondary pollution caused by the use of many chemical additives, and during the desorption process, the CDI module can discharge externally to recover energy. Compared to traditional desalination technologies, CDI features ultralow energy consumption (0.13–0.59 kWh m^{-3}), high efficiency, facile operation for desalination and regeneration processes, environmental friendliness, and recyclability, which provide it with significant cost advantages in desalting brine with medium to low salt concentrations (Liu et al., 2020; Younes et al., 2022; Zhang et al., 2022).

The CDI system (Jin et al., 2024; Xi et al., 2023a, 2022, 2023b) (Fig. 12.1) is composed of a CDI cell, peristaltic pump, raw inlet water, direct-current power supply, and conductivity meter. The desalination process takes place within the CDI cell, and the CDI performance is measured on a conductivity meter. The CDI device has a similar working principle to supercapacitors. The CDI cell comprises two electrodes (anode and cathode) and a separator (e.g., ion exchange membrane). During the CDI adsorption phase, a constant forward voltage (typically less than 1.5 V) is applied to the CDI module using a DC power supply, and the current between the module electrodes and the conductivity of the solution are recorded in real time using a storage multimeter and a conductivity meter. The created

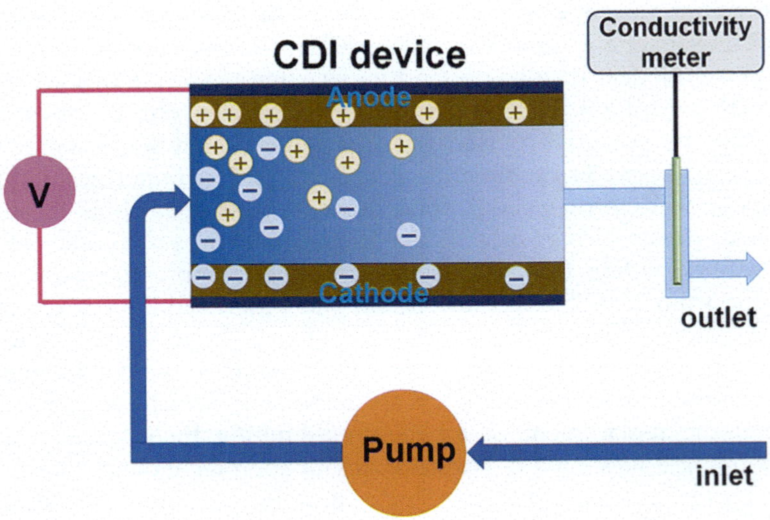

FIGURE 12.1 Schematic illustration of CDI device, including CDI cell, peristaltic pump, raw inlet water, direct-current power supply, and conductivity meter (Jin et al., 2024). *CDI*, capacitive deionization.

electric field forces cations (e.g., Na^+, Li^+, Mg^{2+}) and anions (e.g., Na^+ and Cl^-) to migrate toward the oppositely charged electrodes. The ions are then electro-sorbed or intercalated into the electrodes through electric double-layer absorption or Faradic reactions, leading to the removal of salt from saline water. After a certain period of adsorption, the power supply line is reversed or short-circuited to the CDI module to perform desorption (and regeneration process), while also recording the current and solution conductivity in real-time. In this process, the captured ions are released from the electrodes, forming a concentrated solution. By applying voltage and reversing the connection or creating a short circuit, the cations and anions can be adsorbed and desorbed on or from the electrodes again. Thus cations or anions can be removed from water by using CDI technology. The desalination performance of CDI devices primarily depends on electrode materials.

The CDI technology originated in the early 1960s, first proposed by Blair and Murphy (1960), and at that time, it was referred to as electrochemical demineralization of water. Subsequently, the technology was continuously improved as membrane CDI, particularly by Lee et al. (2006), who assembled a CDI cell by adding anion exchange membranes (AEM) and cation exchange membranes (CEM) on the anode and cathode surfaces, respectively. AEM allows only anions to pass selectively while blocking cations; CEM allows only cations to pass selectively while blocking anions. Membrane capacitive deionization (MCDI) (Biesheuvel and van der Wal, 2010; Choi et al., 2016; Hassanvand et al., 2017) achieves this by adding a layer of anion/cation exchange membrane on the surface of the positive/negative electrodes in traditional CDI, which prevents co-ions from moving from one electrode surface to the other and being adsorbed. This effectively avoids the occurrence of the co-ion effect and improves desalination performance. In addition, due to the presence of the ion exchange membrane, a reverse voltage can be applied during the regeneration process to achieve rapid regeneration. Moreover, the filling of electrode pores by counter ions (i.e., co-ions during the desorption phase) enhances the desalination capacity and adsorption rate for the next cycle. Therefore MCDI has higher energy utilization efficiency, better desalination performance, and faster desalination and regeneration rates compared to traditional CDI. This marked the beginning of research into the impact of device structure on the adsorption performance of CDI.

Generally speaking, MCDI electrodes are solid, and their capacity and volume size are limited. Additionally, their operating mode requires a cyclic process of desalination and regeneration, resulting in a discontinuous desalination process. To address this issue, flow electrode capacitive deionization (FCDI) Liang et al. (2017) was developed based on MCDI, which replaces the solid electrodes with a circulating flow of electrode slurry. This approach enables continuous desalination, improves the rate and efficiency of desalination, and allows for simultaneous deionization and concentration of ion concentration. Therefore, it can also be used as a device to increase salt concentration.Lee et al. (2014) proposed a hybrid capacitive deionization (HCDI) technology. Unlike the carbon-based materials typically used in CDI, they innovatively applied battery materials to the CDI technology. By facilitating reversible oxidation/reduction reactions between the electrode materials and ions in the solution, they achieved chemical adsorption of ions. HCDI demonstrated high ion removal capacity and has been proven to be a viable technology for desalting high-concentration salt solutions. As more and more materials are being used in research within this field, the CDI research in this area is increasingly popular. This lays a foundation for the commercialization of CDI technology.

The working principle of CDI technology can be divided into the principles of double-layer capacitance and pseudo-capacitance (Liu et al., 2021a; Zhang et al., 2022; Li et al., 2020; Suss and Presser, 2018; Feng et al., 2023). The double-layer capacitance principle refers to the process in which, after applying a certain positive and negative voltage to the positive and negative electrodes, the anions and cations in the solution move toward the electrode/electrolyte interface under the influence of the electric field (anions move toward the positive electrode and cations move toward the negative electrode), forming an ion layer to achieve desalination. Once the ion adsorption reaches saturation, the adsorbed ions can be released back into the solution by applying a reverse voltage on the electrode or short-circuiting the electrode, thereby regenerating the electrode. This working principle is also known as non-Faradaic ion storage. Regarding the double-layer capacitance theory for CDI, there are currently three widely accepted classical models: the Helmholtz model, the Gouy–Chapman model, and the Gouy–Chapman–Stern model (Hsieh et al., 2015; Grahame, 1947). The Helmholtz model posits that for every electron moved between the electrodes, the positive and negative electrodes of the CDI device respectively adsorb one anion and one cation, resulting in a charge efficiency of 1. This model is highly idealized and does not account for the effects of ionic thermal motion on the desalination process. The Gouy-Chapman model suggests that the ionic layer in CDI consists of a diffusion layer within a certain distance from the electrode surface, where the adsorbed ions undergo random thermal diffusion. The ion concentration increases as the distance from the electrode surface decreases. Although this model considers Coulombic forces and ionic thermal motion, it does not take into account the size of the ions. In 1924, Stern (1924) proposed the Gouy–Chapman–Stern model based on the Helmholtz and Gouy-Chapman models. This model asserts that the ionic layer is composed of a dense layer of ions (typically hydrated ions) closely packed at the electrode/electrolyte interface, along with a diffusion layer, both of which contribute to the total capacitance. Therefore the Gouy–Chapman–Stern model is currently one of the most recognized theoretical models. Pseudocapacitance theory is a Faradaic process that involves the intercalation of ions into the electrode material and the capture of specific ions through reversible redox reactions. Generally, pseudocapacitance benefits from a large number of fast and reversible redox reactions within the electrode, resulting in higher specific capacitance and greater energy density, leading to high desalination adsorption capacity and rapid adsorption rates.

Based on the introduction to the working principle of CDI, the desalination performance of CDI devices mainly depends on the electrode materials. Ideal CDI electrode materials should be inexpensive, have a high specific surface area and suitable pore structure, high conductivity, good hydrophilicity and surface wettability, a wide and stable electrochemical window, and low resistance. Therefore common supercapacitor electrode materials or battery active materials can generally be used as CDI electrode materials. Currently, common CDI electrode materials mainly include carbon nanomaterials (such as graphene, carbon nanotubes, carbon aerogels), transition metal oxides/sulfides (TMOs/TMDs), MXene, titanates, and layered double hydroxides (LDHs) (Tang et al., 2023; Chang and Hang Hu, 2019; Leong and Yang, 2019; Shen et al., 2020a; Xi et al., 2024; Wang et al., 2022a; Tian et al., 2021). In this chapter, the application of 2D nanomaterials in CDI will be detailed.

12.2 Role of 2D nanomaterials in capacitive deionization systems

CDI is a unique electrochemically driven, material-dependent desalination and separation mode. Its core principle is to control ion migration using an electric field and capture ions with electrode materials. It offers advantages such as low energy consumption, environmental friendliness, safety, modularity, and flexibility, making it a promising technological solution for emerging fields such as selective separation, resource recovery, and distributed water treatment in the future. The development of 2D nanomaterials has elevated CDI to intelligent and precise separation. In CDI electrode materials, commonly used 2D nanomaterials mainly include graphene, MXene, transition metal oxides/sulfides, titanates, and LDHs, and others. In this part, the role of 2D nanomaterials in CDI systems will be discussed.

2D carbon nanomaterials such as graphene and graphene nanoribbons primarily rely on electric double-layer capacitance to adsorb cations or anions from aqueous solutions (Xu et al., 2022; Chang et al., 2021; Guo et al., 2021). Specifically, graphene is a 2D nanomaterial with a honeycomb lattice structure, composed of carbon atoms arranged in a sp^2 hybridization within layers, with a thickness of only 0.34 nm for a single layer. The connections between carbon atoms in graphene are quite flexible, allowing for bending and deformation under external forces, which enables the carbon atoms to adapt without needing to rearrange themselves. This results in remarkable chemical inertness and mechanical stability, making graphene suitable as a supporting material for constructing robust and long-lasting electrodes. Moreover, graphene has a theoretical specific surface area of up to 2630 $m^2\ g^{-1}$. If each carbon atom in a single layer of graphene were to be exposed as a surface site, it could provide a vast number of ion adsorption sites, significantly enhancing desalination performance. The internal electron mobility of graphene can reach up to $4\times10^4\ cm^2\ V^{-1}\ s^{-1}$, with carrier mobility far exceeding that of silicon materials, and its conductivity can be as high as $10^4\ S\ cm^{-1}$. This allows for the construction of a super low-resistance conductive network within electrodes based on graphene, accelerating charge transport and improving desalination performance. As a typical 2D nanomaterial, graphene is easy to functionalize and modify, such as through oxidation, doping, grafting, and assembling into films, creating pores, or constructing three-dimensional structures, which facilitates performance enhancement through precise structural design. Based on the above structural characteristics, graphene plays the following roles in CDI applications. Graphene can expand the electric double layer and is easily modified through functionalization to enhance desalination performance. The intact sp^2 carbon plane of graphene provides rapid ion transport pathways, and its high conductivity allows for efficient charge transport. Therefore the introduction of graphene can endow CDI electrodes with extremely fast desalination kinetics, significantly improving the average desalination rate and achieving rapid response and high-power desalination. Additionally, the interlayer spacing of graphene is adjustable, and its surface properties can be tuned, which can give CDI electrodes high selectivity, transitioning CDI from indiscriminate desalination to selective separation and resource recovery. Furthermore, graphene nanosheets are easy to assemble, enabling the construction of three-dimensional cross-linked networks, such as aerogels and foams, which avoid stacking and provide abundant charge transport pathways, thus imparting high cycling stability to CDI electrodes. Therefore to fully leverage the advantages of graphene as a CDI electrode, future research needs to address the following two questions. How to produce graphene in high quality, in large quantities, and at low cost? How to stably retain the nanoscale performance advantages in macroscopic electrodes through scalable processes (such as coating and 3D printing)?

MXene (Ding et al., 2022; Du et al., 2024; Guo et al., 2018) is a new type of 2D nanomaterial consisting of alternating layers of transition metals and carbides/nitrides. It is typically produced by selectively removing the A layer from MAX phase materials, which have the formula $M_{n+1}AX_n$, where $n = 1, 2,$ or 3. Here, M represents a transition metal, A represents Al or Si, and X denotes C, N, or a combination of C and N. The general formula for MXene, derived from the selective etching of the A phase in MAX phases, is $M_{n+1}XnT_x$, where T_x indicates the terminal functional groups on its surface (-O, -OH, -F, etc.), and the subscript x represents the number of terminal groups, typically satisfying $0 < x < 2$. Taking the most widely studied $Ti_3C_2T_x$ as an example, MXene is composed of alternating layers of transition metal atoms and carbon atoms, with its surface coated with various functional terminal groups. This unique structural composition endows it with advantages such as tunable interlayer spacing, high electrical conductivity, and surface electronegativity and hydrophilicity. Based on the structural characteristics mentioned above, MXene plays the following roles in CDI applications. Similar to graphene, MXene exhibits high electrical conductivity, allowing for rapid charge transfer within the electrode, and features open interlayer ion transport channels. The surface of MXene is rich in functional groups such as -OH, -O, and -F, which not only endow it with excellent electrolyte wettability but also provide a large number of pseudo-capacitance active sites capable of undergoing rapid redox reactions. This enables fast and reversible Faradaic reactions under an electric field, resulting in a high desalination capacity. MXene can precisely tune the interlayer spacing through intercalation, allowing for the sieving of ions of different sizes. Additionally, its rich surface chemistry enables the regulation of surface chemical properties to achieve ion-selective adsorption. Furthermore, like graphene, MXene has high mechanical stability and can easily be combined with other highly active pseudocapacitive electrode materials or assembled into films, aerogels, and other foams.

2D transition metal oxides (TMOs) exhibit high capacity and selectivity due to their unique crystal structures and the ease of undergoing redox reactions. The structure of these materials generally consists of metal-oxygen octahedra forming 2D layered frameworks, with exchangeable cations (such as Na^+, K^+, H^+) and water molecules present between the layers. The interlayer spacing can be adjusted to provide open 2D diffusion pathways for ions. A common structural feature of 2D metal oxides is the presence of abundant redox-active metal centers and vacancies or channels that can accommodate ions within the structure. As a result, 2D metal oxides, based on pseudo-capacitance characteristics, possess a high theoretical specific capacity, with multiple valence states of the same metal allowing for a flexible and reversible charge storage mechanism in such CDI electrodes (Leong and Yang, 2019; Divyapriya et al., 2019; Hosseinzadeh et al., 2022). The crystal form, interlayer spacing, and morphology of 2D metal oxides can be tuned to optimize CDI performance. Therefore similar to MXene, 2D metal oxides exhibit high desalination capacity, high ion selectivity, and rapid reaction kinetics. However, these materials typically have poor conductivity, and their structures are prone to collapse, which generally necessitates compositing with highly conductive nanomaterials such as graphene, carbon nanotubes, and MXene to construct a conductive framework-active material structure.

2D transition metal dichalcogenides (TMDs) are a class of materials formed by the combination of transition metals (groups IV, V, VI, VII, IX, or X) with chalcogen elements (S, Se, or Te) in a specific manner. Taking molybdenum disulfide (MoS_2) as an example, it is a classic 2D layered TMD, where each layer consists of a S-Mo-S sandwich structure. Atoms within the layer are connected by strong covalent bonds, while layers are held together by weak van der Waals forces. This unique structure allows for easy intercalation and exfoliation when used as electrode materials, enabling the storage of more ions. However, this characteristic also leads to significant volume changes during the charge–discharge process. Furthermore, 2D TMDs possess abundant edge unsaturated sites and vacancy defects, serving as highly active adsorption centers. Similar to 2D transition metal oxides, 2D TMDs exhibit tunable interlayer spacing and pseudo-capacitance characteristics, making them suitable for applications in CDI with high desalination capacity, high selectivity, and rapid reaction kinetics (Xing et al., 2017; Zhang et al., 2024; Cai et al., 2021). The drawback, however, is their poor conductivity, necessitating the construction of composite materials with highly conductive nanomaterials.

Titanates are a class of materials formed by the combination of titanate ions with other metal ions, with the chemical formula $A_2Ti_nO_{2n+1}$, where A represents one or more elements from Na, Li, Mg, Ca, K, Cs, H, etc., and $n = 2, 3, 4$. Layered titanates are closely arranged by TiO_6 octahedra through edge-sharing or corner-sharing, with alkali metal cations or hydrogen ions located between the layered framework. Due to the ability of the interlayer alkali metal ions or hydrogen ions to undergo ion exchange with other heavy metal ions, this unique characteristic gives titanates a strong ion adsorption capacity, particularly for heavy metal ions, which has led to extensive research in wastewater treatment (Li et al., 2025). However, when titanates adsorb ions, the ions enter the lattice, causing deformation of the lattice structure. This results in a certain volume change during the charging and discharging processes, and in severe cases, it can lead to the active material detaching from the current collector, which is detrimental to the long-term stability of the material. In addition, titanates have poor conductivity, so they are often compounded with other conductive materials. This not only improves conductivity but also disperses the volume strain, thereby avoiding the issue of stability degradation caused by volume changes.

LDHs are compounds assembled from positively charged brucite-like host layers and interlayer anions through noncovalent interactions. Their structure is similar to that of hydrotalcite, with unit layers formed by MgO_6 octahedra sharing edges. The chemical composition of the host layers and the interlayer spacing can be adjusted, as well as the types and amounts of interlayer guest anions. This material exhibits unique characteristics such as memory effect, excellent anion exchange capacity, chemical structure stability, and environmental friendliness. Therefore LDHs are a unique functional material with designed ionic sieving and memory effects in CDI (Wang et al., 2022a; Hu et al., 2025; Liu et al., 2022; Zhang et al., 2023). However, they have poor intrinsic electronic conductivity, are prone to aggregation, and have poor processability.

2D nanomaterials play an irreplaceable core role in the field of CDI. 2D carbon nanomaterials, MXene, and 2D TMOs, TMD, LDHs, and titanates have complementary advantages. Efficiently combining two or more of these nanomaterials to construct Janus membranes can achieve ion-oriented transport. Developing integrated sensing-desalination electrodes, real-time monitoring and optimization of processes, and deep coupling with renewable energy to build efficient, zero-emission distributed water treatment units is a major research direction for the future.

12.3 Ion capturing mechanisms of capacitive deionization with 2D nanomaterials

To investigate the ion-capturing mechanisms of CDI with 2D nanomaterials and further improve their CDI performances, it is necessary to evaluate the key performance indicators of CDI. The electrochemical performances of CDI devices mainly include CDI capacity, rate, charge efficiency, energy consumption, and cycling performance. CDI capacity refers to the total amount of ions that can be removed per unit mass (or per unit area, or per unit volume) of electrode material. It measures the storage capacity of the electrode material. The higher this value, the more salt the unit electrode material can adsorb, indicating higher performance. This is the primary indicator for evaluating the adsorption capacity of the electrode material itself.

12.3.1 Performance indicators of capacitive deionization electrode materials

The CDI capacity (Q, mg g^{-1}) is calculated according to Eq. (12.1).

$$Q = \frac{(C_0 - C_e) \times V}{m} \tag{12.1}$$

where C_0, C_e, V, and m represent the initial and equilibrium concentration (mg L^{-1}) of NaCl solution, volume of NaCl solution, and mass (g) of active material, respectively.

The average desalination rate refers to the amount of ions that can be removed by a unit mass of electrode material within a unit time. It measures the speed of the desalination process. A higher value indicates faster ion adsorption kinetics of the electrode material and a stronger capacity of the system to treat water. A high-performance CDI system should possess both high specific capacity and high average rate. The average desalination rate (v, mg g^{-1} min^{-1}) was calculated according to Eq. (12.2).

$$v = \frac{Q}{t} \tag{12.2}$$

where t refers to the adsorption time (min).

The charge efficiency refers to (Λ, %) the ratio of the actual charge of adsorbed ions to the total charge input from an external power source. It measures the energy utilization efficiency of the entire CDI process. This value is a key indicator for assessing the energy efficiency of CDI. The higher the value, the greater the proportion of the input electrical energy used for adsorbing ions. In other words, a higher value indicates that the CDI system is more energy-efficient. However, in practical CDI processes, due to side reactions such as water electrolysis, electrode self-discharge, or co-ion effects, the charge efficiency is often much lower than the ideal value of 100%. The charge efficiency (Λ, %) was calculated according to Eqs. (12.3) and (12.4):

$$\Lambda = \frac{Q \times m \times F}{1000 \times M \times \Sigma} \tag{12.3}$$

$$\Sigma = \int I dt \tag{12.4}$$

where Λ, F, M, Σ, and I represent the charge efficiency, Faraday constant (96,485 C mol^{-1}), molar mass of salt, integration (C g^{-1}) of the current during the ion adsorption step, and current (A), respectively.

Energy consumption refers to the amount of electrical energy consumed by the CDI system for removing a unit mass of salt ions from water. It directly determines the operating cost of the technology. The charge efficiency and energy consumption can be used to compare different CDI devices and other desalination techniques. Energy consumption index (ECI) (kWh/kg) was calculated using Eq. (12.5).

$$ECI = \frac{E_{imput}}{M_{salt}} = \frac{I \cdot V \cdot t}{Q \cdot m} \tag{12.5}$$

Where E_{imput} is the total input electric energy (KWh) and M_{salt} (Kg) is the total mass of salt removed. Where I, V, t, m, *and* Q represent the current (A), voltage (V), time (h), mass of active material (mg), and CDI deionization performance (mg g^{-1}), respectively.

Cycling stability refers to the ability of the CDI system to maintain its core performance, such as CDI capacity and charge efficiency at initial values after undergoing multiple consecutive adsorption-desorption cycles. It determines the equipment's lifespan and maintenance costs.

The above CDI performances are mainly determined by the electronic and electrochemical properties of the electrode materials. Thus it is important to investigate ion capturing mechanisms of CDI with 2D nanomaterials. Generally, the ion capture mechanisms in the CDI electrodes can be categorized into electro-sorption (carbon-based electrode materials) and Faradic reaction mechanisms (Faradic electrode materials).

12.3.2 Electro-adsorption mechanisms

Electro-adsorption in the double electric layer is the most fundamental and classic adsorption mechanism in CDI (Fig. 12.2A) (Jin et al., 2024). This theory is mainly applicable to carbon nanomaterials, such as graphene, graphene nanoribbons, active carbon, and carbon aerogels. The mechanism is that when a direct-current voltage lower than the water splitting voltage (typically <1.5 V) is applied to the electrodes, the two electrodes acquire positive and negative charges, respectively. Ions with opposite charges in the solution are electrostatically attracted, rapidly migrating and accumulating at the electrode/solution interface, forming a compact double electric layer. When the external electric field is removed, the charges on the electrode are neutralized, the double electric layer structure dissipates, and the adsorbed ions spontaneously diffuse back into the solution due to the concentration gradient. Alternatively, a reverse voltage can be applied, causing the ions in the double electric layer to be electrostatically repelled. The carbon electrodes have a high

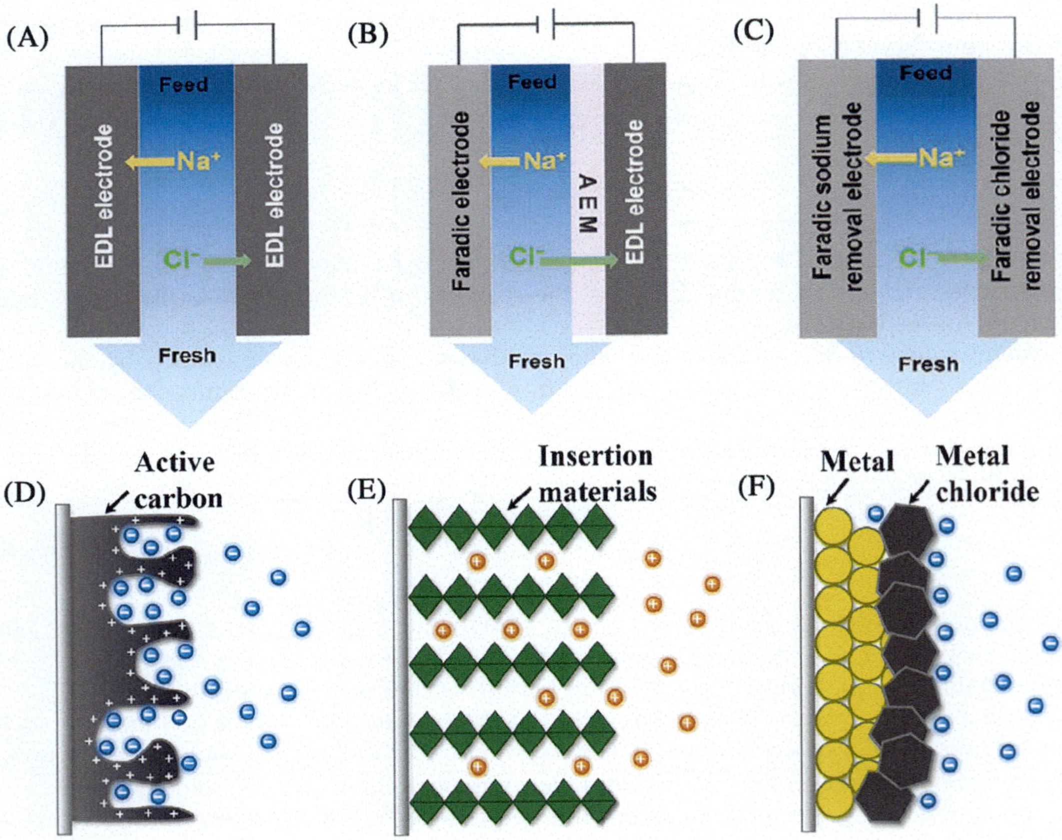

FIGURE 12.2 (A) Conventional CDI composed of EDL electrodes, (B) hybrid CDI composed of EDL electrode and Faradic electrode, and (C) Faradic CDI composed of Faradic electrodes, (D) Schematic illustration of electrosorption reaction, (E) insertion reaction, and (F) conversion reaction (Jin et al., 2024). *CDI*, Capacitive deionization; *EDL*, electric double layer.

specific surface area and fast ion transport, allowing for high adsorption capacity and excellent ion removal rates in CDI devices. However, the biggest obstacle to desalination through electroadsorption mechanisms is the limited adsorption capacity. This can be improved by optimizing charge transport pathways, such as constructing hierarchical porous structures, increasing specific surface area, or efficiently combining with pseudocapacitive electrode materials.

12.3.3 Faradic reaction mechanisms

Faradic reactions in pseudo-capacitance electrode materials mainly include intercalation reaction and conversion reaction (Jin et al., 2024). The intercalation-type electrode materials can capture ions in the crystalline sites or between atomic planes (Fig. 12.2E), while the conversion-type electrode materials undergo chemical transformations to form new compounds (Fig. 12.2F). Most intercalation reaction materials are used to capture cations, while only a few intercalation reaction materials are capable of anion intercalation. Various ion intercalation reaction materials, such as MXene, transition metal oxide/sulfide compounds, layered double hydroxides, sodium titanate, Prussian blue analogues, have been developed as CDI electrode materials. These nanomaterials can provide fast ion diffusion pathways due to their special crystal structure. 2D nanomaterials (e.g., MXene, MoS_2, LDH, NTO) have open 2D nanochannels between their layers, providing fast ionic transport pathways, which are the basis for efficient desalination. Furthermore, the interlayer spacing of 2D nanomaterials can be precisely tuned through intercalation and functional group modification, allowing for the physical sieving of hydrated ions of different sizes (such as the separation of monovalent and divalent ions), which is not achievable with traditional porous carbon.

The conversion reaction mechanism of CDI electrode materials involves the transformation of electrode materials with ions to form new compounds. The conversion-type electrode materials demonstrate high theoretical specific capacities of ion storage. However, the conversion-type electrode materials suffer from issues such as large volume expansion, structure collapse, and poor cycling stability due to the repetitive chemical conversion. To address these issues, it is necessary to construct nanostructures, porous materials, carbon composites, and hierarchical micro-nanostructures for the conversion-type electrode materials in CDI devices.

12.4 Morphology and performance of 2D nanomaterials in capacitive deionization

2D materials possess atomically thin structures and large diameter-length ratios, which give rise to exceptional CDI properties. The morphological categories of 2D nanomaterials (Xi et al., 2023a; Feng et al., 2023; Shen et al., 2020a; Chang et al., 2021; Hosseinzadeh et al., 2022; Li et al., 2023; Liu et al., 2023) mainly include monolayers, such as monolayer graphene, monolayer MoS_2, and monolayer MXene nanosheets; few-layer stacks, such as thin-layer graphene and thin-layer MXene; 3D porous assemblies, such as graphene aerogels, MXene foams, and hollow LDH spheres; as well as vertical array structures, such as vertically aligned graphene and LDH nanosheet arrays. These unique structures collectively endow them with high specific surface area, extremely short ion transport pathways, open and tunable transport channels, rich surface chemistry, pseudo-capacitance characteristics, and ease of assembly. Therefore as CDI electrode materials, 2D nanomaterials theoretically exhibit high desalination capacity and high average adsorption rates. However, common issues associated with these 2D nanomaterials include the tendency of nanosheets to stack and agglomerate, significant differences in conductivity, high-reactivity surfaces that are prone to oxidation, dissolution, or side reactions, susceptibility of layered crystal structures to collapse, and difficulties in cost-effective, scalable preparation. To overcome these limitations and enhance desalination performance, the suggested optimization measures mainly include developing high-quality and low-cost preparation processes, precisely controlling interlayer spacing and surface functional groups, constructing heterostructures through atomic doping, creating hierarchical porous structures to increase specific surface area, building 3D structures assembled from 2D materials, and efficiently combining 2D nanomaterials with various active nanomaterials. By leveraging strengths and avoiding weaknesses, the goal is to design and develop next-generation high-performance, long-lifespan, and low-cost customized CDI electrode materials tailored for specific water treatment scenarios. In this section, the morphology, structure, and desalination performance of these 2D materials (graphene, MXene, TMOs/TMDs, LDHs, and NTO) will be introduced and discussed (Fig. 12.3) (Jin et al., 2024). The CDI performances of 2D nanomaterials are listed in Tables 12.1 and 12.2. We will provide a detailed introduction

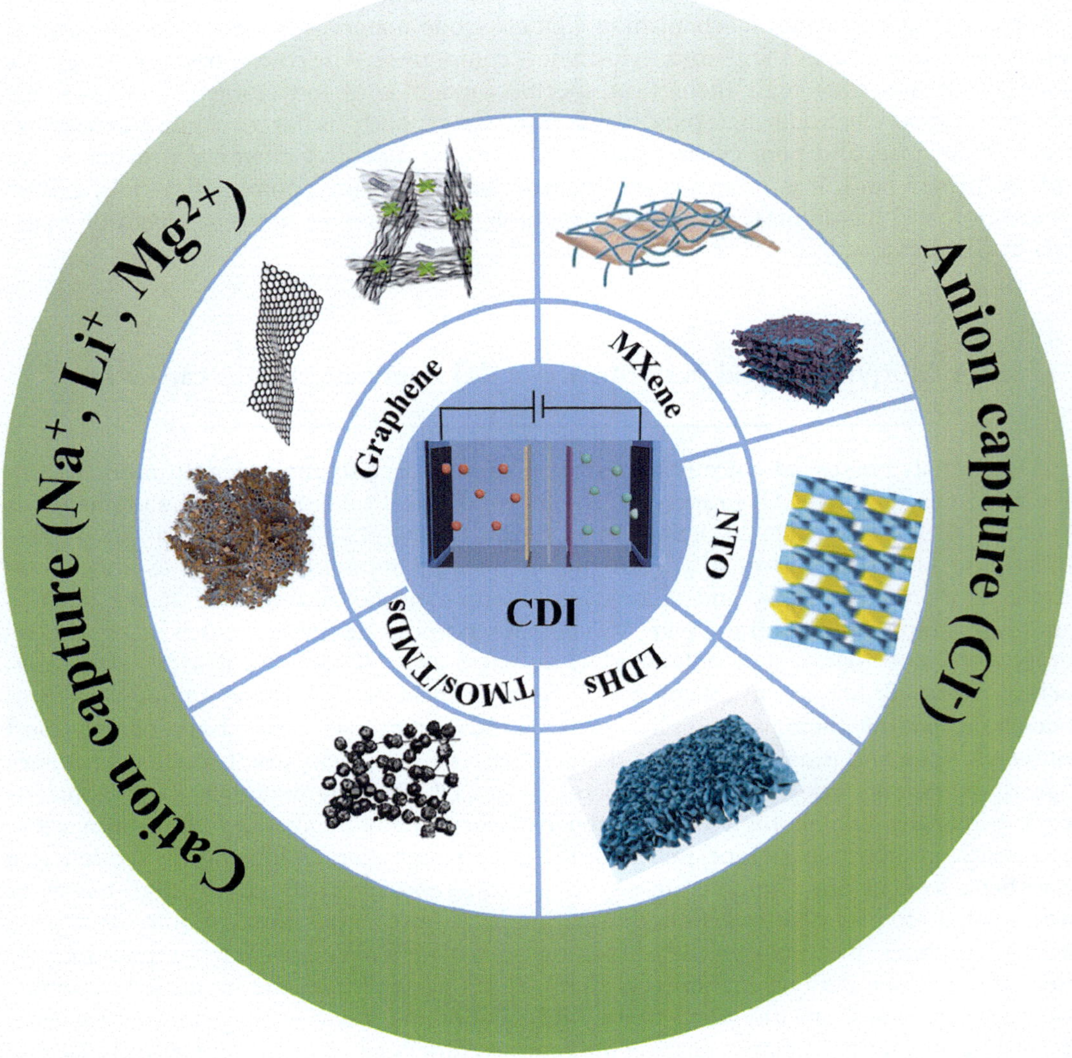

FIGURE 12.3 2D nanomaterials as CDI electrodes applied for cation and anion capture. *CDI*, capacitive deionization

TABLE 12.1 2D nanomaterials as capacitive deionization electrodes to capture cation (Jin et al., 2024; Zhang et al., 2024).

Material	Capturing ions	Desalination capacity (mg g^{-1})	Working potential (V)	Concentration (mg L^{-1})	References
3D rGO	Na^+	8.7	1.2	1000	Leong et al. (2019)
Graphene/Carbon	Na^+, Cl^-	25.5	1.2	500	Feng et al. (2023)
N, P-doped C/graphene	Na^+	39.3	1.2	1000	Guo et al. (2021)
WS_2/rGO-CNT	Na^+	80	10 mA	3000	Vafakhah et al. (2021)
MoS_2/MXene	Na^+	55.8	1.2	500	Zhang et al. (2024)
MoS_2/rGO	Na^+	16.82	1.0	200	Peng et al. (2020)
CFP/3DG/MnO_2	Na^+, Cl^-	21	1.2	1000	Hosseinzadeh et al. (2022)
Defected-MoS_2/rGO	Na^+	25.47	0.8	200	Peng et al. (2021)
NTO-MXene/CNF	Na^+	49.3	1.2	/	Li et al. (2025)
MXene	Na^+	72	1.6	10 mM	Shen et al. (2020b)
CNFs@ MoS_2	Na^+	53.03	1.2	3000	Liu et al. (2021b)
N-doped MXene	Na^+	44.8	1.8	500 ppm	Ding et al. (2022)
Na^+-MXene	Na^+	14.8	1.2	100	Chen et al. (2021a)
MXene/rGO	Na^+	48	1.2	135	Xu et al. (2022)
MXene@carbon dots	Na^+	86.4	1.2	10 mM	Tan et al. (2023)
TiO_2/MXene	Na^+	75.62	1.2	500 mM	Liu et al. (2023)
MoS_2/MXene	Na^+	62.95	0.6	1000	Qie et al. (2022)
MXene/graphene	Na^+	57.57	1.4	30	Shen et al. (2022)
MXene/PANI	Na^+	76	1.2	500	Yang et al. (2022)
MXene/PPy	Na^+	91.2	1.2	500	Wang et al. (2022b)
CoP/Co_3O_4-graphene	Li^+	37	0.6	50	Jin et al. (2021)
GO/Fe_3O_4@C	Li^+	31.24	/	100	Zhao et al. (2021)
MoS_2/CoS_2@TiO_2	Na^+	44.22	1.2	600	Zhao et al. (2022)
MoS_2/C	Na^+	59.9	1.4	2000	Nguyen et al. (2022)
Co-MOF/FeCo-LDH	Na^+	34.2	1.2	600	Li et al. (2021)

TABLE 12.2 2D nanomaterials as capacitive deionization electrodes to capture anion (Jin et al., 2024).

Material	Capturing ions	Desalination capacity (mg g^{-1})	Working potential (V)	Concentration (mg L^{-1})	References
ZnCo-Cl LDH	Cl^-	56.1	1.2	500	Zhang and Li (2021)
CoFeCl-LDH	Cl^-	100.2	1.2	20 mM	Wang et al. (2022a)
CNT@NiCoAl-LDH@ MXene	Cl^-	79.5	1.2	500	Zhang et al. (2023)
MgAl LDHs/G	Cl^-	13.6	1.0	500	Ren et al. (2018)
NiCoAl LDHs/G	Cl^-	24.5	1.4	500	Li et al. (2019)
Bi@MXene	Cl^-	88.2	1.2	50 mM	Gong et al. (2023)
Ag@G/CNT	Cl^-	62.4	1.2	1000 ppm	Rangaraj et al. (2023)
$Ti_3C_2T_x$/Ag	Cl^-	135	20 mA g^{-1}	10 mM	Liang et al. (2020)

to the relationship between the structural morphology of 2D nanomaterials and their desalination performance in CDI.

12.4.1 Graphene-based capacitive deionization electrode materials

Graphene is a carbon nanomaterial with unique sp^2 hybridization, featuring excellent electrical conductivity, high chemical and mechanical stability, high theoretical specific capacitance, and high theoretical specific surface area. It is a widely used electrode material. However, graphene nanosheets are prone to stacking, resulting in a practical specific surface area that is much lower than the theoretical value, which limits ion transport and leads to poor desalination performance. Additionally, the electrochemical desalination mechanism of graphene is based on electrochemical adsorption through double-layer capacitance, which has a limited adsorption capacity. In response to this, researchers have proposed many research solutions.

Li et al. (2009) synthesized graphene using the Hummers method and applied it in CDI. However, due to the self-stacking caused by the π-π conjugation between graphene layers, the specific surface area was relatively small (14.2 m^2 g^{-1}), resulting in an adsorption capacity of only 1.85 mg g^{-1}. To address this limitation, researchers have developed various graphene-based materials with 3D porous networks to increase the specific surface area. This 3D macroscopic interpenetrating network facilitates the transport of ions and charges. Zhu et al. (2020) intentionally designed incomplete hollow spherical graphene shells and successfully constructed a 3D porous graphene structure, significantly enhancing the specific surface area and achieving excellent salt adsorption capacity. Yang et al. (2014) developed sponge-like graphene with in-plane pores through the hydrothermal reduction of graphene oxide sheets, providing a unique interconnected porous structure with a high specific surface area (1261.7 m^2 g^{-1}) that reduced electron transport resistance. Under a voltage of 1.2 V in a 500 mg L^{-1} NaCl solution, the electro-adsorption capacity reached as high as 16.0 mg g^{-1}. Leong et al. (2019) cross-linked GO sheets using polyvinyl alcohol to form 3D macro-porous rGO structure and then activated by KOH. The 3D rGO electrode exhibited a high adsorption capacity of 30 mg g^{-1}. Yang et al. (2023) used O_2-plasma treatment method to regulate the pore structure and surface wettability of graphene sheets. After O_2-plasma treatment, graphene exhibited a mesoporous structure, which could improve ion diffusion and reduce charge transfer resistance in CDI device. The presence of oxygen-containing groups could enhance the salt ions adsorption in the electrode, resulting in a NaCl adsorption capacity of 40.76 mg g^{-1}.

The combination of graphene with porous carbon materials allows for the integration of their respective advantages while preventing the restacking of graphene sheets. Feng et al. (2023) fabricated a 3D graphene-supported N-doped porous carbon using polyacrylonitrile-coated polystyrene spheres as a template. The hollow carbon spheres were evenly embedded in the hierarchically porous graphene sheets. The obtained 3D hierarchically interconnected porous structure could enhance conductivity and diffusion of salt ions, while the N-doped carbon could further improve conductivity and hydrophilicity. Thus the 3D graphene-supported N-doped hierarchically porous carbon electrode exhibited a desalting capacity of 25.5 mg g^{-1}. In addition, Guo et al. (2021) prepared a layered porous N, P-doped carbon/graphene (NPC/rGO) heterostructure by driving the assembly of metal-organic framework nanoparticles on graphene oxide (GO) nanosheets, followed by a stepwise pyrolysis and phosphorization process. This composite material features a high specific surface area, high conductivity, a good meso/microporous structure, and codoping of nitrogen and phosphorus heteroatoms, which is beneficial for ion adsorption and charge transport during the CDI process. The resulting NPC/rGO-based CDI desalination device exhibits ultra-high deionization performance, with a salt adsorption capacity of 39.34 mg g^{-1} in a 1000 mg L^{-1} NaCl solution under conditions of 1.2 V for 30 minutes, and shows good cycling stability over more than 50 cycles.

Efficiently compositing graphene with electrochemically active nanomaterials that exhibit pseudocapacitive characteristics is an important technological approach to address the low desalination capacitance of graphene CDI electrode materials. This section will provide a detailed explanation of the structures and properties of various 2D pseudocapacitive materials.

12.4.2 MXene-based capacitive deionization electrode materials

MXene is composed of alternating layers of transition metal atoms and carbon atoms, with its surface coated by various functional terminal groups. This unique structural composition endows it with advantages such as adjustable interlayer spacing, high conductivity, surface electronegativity, and hydrophilicity. As a result, it serves as an excellent CDI electrode material that integrates pseudo-capacitance, rapid ion transport, and intrinsic high selectivity. The main drawbacks of MXene electrode materials are the high cost of large-scale production of high-quality materials, easy to self-stack and aggregate, and the potential for partial oxidation of some MXene in oxygen-containing aqueous solutions or at positive potentials, which can lead to low desalination capacity and poor cycling stability. To address these issues, researchers have optimized the preparation process to construct thin-layer MXene nanosheets, doping with atoms such as N, P, or metal atoms, effectively combined MXene with carbon nanomaterials and active nanoparticles with pseudocapacitive properties, such as transition metal oxides, to create heterostructured nanocomposites.

Our team has been deeply involved in the applied fundamental research of MXene-based CDI electrode materials. For example, we prepared thin-layer MXene through selective separation using HF and constructed CDI electrodes with this MXene. The effects of MXene electrode thickness and the electrode/electrolyte interface on the CDI behavior of the MXene electrodes was in depth investigated. The study found that reducing the electrode thickness facilitates the penetration of the electrolyte (NaCl solution) into the electrode, providing more contact at the electrode/electrolyte interface and shortening the ion diffusion pathway from the electrolyte to the internal electrode, thereby accelerating the ion diffusion rate and enhancing desalination capability. The optimized MXene‖MXene CDI device has an electrode thickness of 18.41 μm and exhibits a high desalination capacity of 100.9 mg g^{-1} in a 1.2 V NaCl solution (Xi et al., 2023b). Besides, we developed novel MXene/Ni composites using a molten salt etching method. The MXene/Ni composite exhibits a well-preserved accordion-like morphology of MXene, with Ni nanoparticles uniformly distributed on the MXene surface and between layers. The MXene/Ni composite possesses a large interlayer spacing and high specific surface area, which can effectively trap Cl^{-} and Na^{+} ions. Moreover, the MXene/Ni interface can improve electronic conductivity and enhance ion transport. The optimized MXene/Ni‖MXene/Ni CDI symmetry device exhibits a salt removal capacity of 49.4 mg g^{-1} in a 500 μS cm^{-1} NaCl solution at an applied voltage of 1.2 V with a removal rate of 3.00 mg g^{-1} min^{-1} (Xi et al., 2023a). In addition to compounding with metal atoms, we also used MXene as a conductive matrix material to load MoS_2, thereby controllably constructing a hierarchical MoS_2/MXene heterostructure (Zhang et al., 2024). In this work, 2D MXene nanosheets and 2D MoS_2 nanosheets can form a mutually reinforcing structure to prevent aggregation of the nanosheets. MXene nanosheets act as a robust, hydrophilic, and conductive substrate, while MoS_2 nanosheets act as pillars between MXene nanosheets, effectively preventing MXene self-accumulation and providing a pathway for rapid ion transport. Therefore the obtained hierarchical MoS_2/MXene heterostructure CDI electrode demonstrates excellent desalination capacity of 55.8 mg g^{-1} and a fast desalination rate of 13 mg g^{-1} min^{-1}.

The surface property (*e.g.*, -F termination groups, hydrophilicity), morphology (*e.g.*, sheet size), and structure (*e.g.*, porous structure, 2D/3D structure) of MXene can effectively impact the desalination performance. Feng et al. (2023) conducted a study comparing the CDI behavior of two kinds of MXene sheets fabricated through NH_4HF_2 etching and HF etching methods. The NH_4HF_2-etched MXene sheets exhibited a larger interlayer spacing and good hydrophilicity. As a result, the desalination capacity of the NH_4HF_2-etched MXene sheets electrode was 35.5% higher than that of HF-MXene electrode. Qian et al. (2023) synthesized F-free MXene sheets through an electrochemical etching method. The electrochemical-etched MXene sheets were coated with -OH instead of -F groups, resulting in excellent hydrophilicity and fast electrolyte penetration. The -F-free MXene-based CDI device showed excellent desalt removal capacity of 20.27 mg g^{-1} with an adsorption rate of 1.01 mg g^{-1} min^{-1}.

MXene can be used to construct free-standing and binder-free film electrodes owing to their 2D structure, abundant surface functional groups, and interlayer force. For example, we constructed self-supporting MXene/carbon nanofiber (MXene/CNF) films using electrospinning technology, and through in situ growth, we controllably transformed part of the MXene into sodium titanate ($Na_2Ti_3O_7$, NTO), thereby creating NTO-MXene/CNF composite films (Li et al., 2025). The excellent conductivity of both CNFs and MXene enables the NTO-MXene/CNF film to demonstrate an impressive desalination capacity of 49.3 mg g^{-1} and a rapid desalination rate of 9.9 mg g^{-1} min^{-1}. Ma et al. (2020) synthesized a free-standing MXene film using vacuum filtration. The free-standing MXene film could enhance ion transport and improve desalination performance. Moreover, 2D MXene nanosheets could be assembled into 3D porous MXene structure through NaOH-induced crumpling of the MXene nanosheets (Chen et al., 2021b). The 3D porous MXene with 3D interconnected network possessed larger specific surface area and higher electronic conductivity than 2D MXene sheets. As a result, it exhibited a higher desalting capacity of 16.2 mg g^{-1}.

12.4.3 Transition metal oxides/sulfides-based capacitive deionization electrode materials

2D TMOs exhibit pseudo-capacitance characteristics, and 2D layered unique crystal structures with exchangeable cations (such as Na^+, K^+, H^+) between layers. As a result, the interlayer spacing and cations can be adjusted to optimize CDI performance. Thus 2D TMOs exhibit high desalination capacity, high ion selectivity, and rapid reaction kinetics. However, 2D TMOs typically have poor conductivity, and their structures are prone to collapse, which generally necessitates compositing with highly conductive nanomaterials such as graphene, carbon nanotubes, and MXene to construct a conductive framework-active material structure. Bharath et al. (2019) prepared mesoporous peanut shell activated carbon (PSAC) using a low-temperature pyrolysis method at 500 °C, and subsequently developed a novel Fe_3O_4/PSAC nanocomposite through a hydrothermal method at 180 °C. The Fe_3O_4/PSAC exhibits a highly mesoporous structure and a large specific surface area up to 680 m^2 g^{-1}. The optimized Fe_3O_4/PSAC nanocomposite electrode demonstrates a maximum electro-adsorption capacity for Cr (VI) of 24.5 mg g^{-1} at 1.2 V, proving that the CDI system with Fe_3O_4/PSAC electrodes can be used for the removal of toxic metal ions from wastewater. Jin et al. (2021) fabricated oxygen vacancy-rich CoP/Co_3O_4-graphene aerogel (GA/CoP/Co_3O_4) via a hydrothermal reaction. The GA/CoP/ Co_3O_4 composite served as both the anode and cathode in a CDI device for extracting Li^+ ions from salty lake brine. The GA/CoP/ Co_3O_4-based CDI device achieved high electro-adsorption capacity of 37 mg g^{-1}, due to its rich oxygen vacancy, good hydrophilicity, and low charge resistance. Hu et al. (2022) synthesized 3D λ-MnO_2/rGO composite as CDI electrode for Li^+ extraction. The λ-MnO_2/rGO composite exhibited excellent electronic conductivity and fast ion transport. The rGO sheets could prevent the aggregation of λ-MnO_2 particles and increase the active sites for Li^+ electro-adsorption. Consequently, the λ-MnO_2/rGO composite CDI electrode demonstrated a high Li^+ adsorption capacity of 601 μmol g^{-1}. Moreover, the obtained λ-MnO_2/rGO composite electrode also demonstrated excellent adsorption performance for other ions such as Na^+, K^+, Mg^{2+}, and Ca^{2+}.

2D TMDs (e.g., MoS_2, VS_2, WS_2) possess unique 2D structure, large interlayer spacing, and abundant ion adsorption sites. Taking MoS_2 as an example, it is a classic 2D layered TMD, where each layer consists of a S-Mo-S sandwich structure. Atoms within the layer are connected by strong covalent bonds, while layers are held together by weak van der Waals forces. The unique structure allows for ion intercalation and exfoliation when used as electrode materials, enabling the storage of more ions. Moreover, 2D TMDs possess abundant edge unsaturated sites and vacancy defects, serving as highly active adsorption centers. Thus 2D TMDs exhibit tunable interlayer spacing and pseudo-capacitance characteristics, making them suitable for applications as CDI electrode materials. However, TMDs have poor conductivity, easy agglomeration, and structural collapse due to volume expansion during use. They are generally not directly used or used alone.

Jia et al. (2018) studied the desalination performances of pure MoS_2 and defect-rich MoS_2 electrodes for CDI. The defect-rich MoS_2 was fabricated through thermal treatment of MoS2 nanosheets at 300 °C, resulting in abundant negative charges and small inner resistance, which could facilitate electrostatic attraction and electro-sorption of Na^+. Sun et al. (2020) fabricated oxygen-incorporated MoS_2 nanosheets and deeply studied the effects of oxygen incorporation in MoS_2 nanosheets on the CDI performance. The oxygen incorporation could enhance the electronic conductivity, possess more active sites, and reduce the inner resistance. As a result, the oxygen-incorporated MoS_2 nanosheets exhibited a higher desalination capacity of 28.85 mg g^{-1} than that of pure MoS_2 nanosheets.

To address the issues of poor conductivity, easy agglomeration, and volume expansion of TMDs, researchers typically efficiently compounded TMDs with carbon nanomaterials and

MXene to prepare composite electrodes. Cai et al. (2021) developed a 3D conductive architecture by interconnecting MoS_2 sheets with CNT and carbon spheres (CS) (MoS_2@CNT-CS). In this work, MoS_2 nanosheets were connected by CNT, resulting in the 3D architecture, which could prevent MoS_2 sheets from restacking, improve electronic conductivity, and provide continuous charge transfer channel. The MoS_2@CNT-CS composite electrode exhibited a higher desalination capacity of 25.35 mg g^{-1}. Moreover, Cai et al. (2022) constructed a 3D MoS_2/MXene heterostructure. In this work, MXene sheets prevent the aggregation of MoS_2 nanosheets, enhance electronic conductivity, and provide continuous ion transport paths, while MoS_2 nanosheets also prevent MXene sheets restacking and provide more active sites for ion intercalation. Consequently, the MoS_2/MXene heterostructure could demonstrate a desalination capacity of 35.6 mg g^{-1}. Besides, Peng et al. (2021) synthesized defect-containing MoS_2/rGO composite. The defect-containing MoS_2/rGO composite exhibited enhanced hydrophilic properties and displayed more exposed edge sites, which could be beneficial for Na^+ ion diffusion and intercalation. Thus the defect-containing MoS_2/rGO composite achieved a high desalination capacity of 25.47 mg g^{-1} at a voltage of 0.8 V in 200 mg L^{-1} NaCl solution.

12.4.4 Titanate-based capacitive deionization electrode materials

Titanates are a class of layered structural materials formed by the combination of titanate anions and metal cations. They are composed of TiO_6 octahedra closely arranged through edge-sharing or corner-sharing, with metal cations located between the layered frameworks. This unique characteristic gives titanates a strong ion adsorption capacity, particularly for heavy metal ions, making them widely studied in wastewater treatment. However, when titanates adsorb ions, the ions enter the lattice, causing deformation of the lattice structure, which results in certain volume changes during charge and discharge processes. In severe cases, this can lead to the active material detaching from the current collector, negatively impacting the long-term stability of the material. Additionally, titanates have poor conductivity. Thus titanates are often combined with other conductive materials to improve conductivity while dispersing volume strain, thus avoiding stability issues caused by volume changes.

Cech et al. (2017) synthesized sodium titanate with tunnel structure or mixed tunnel/layered structure and conducted charge-discharge studies on $Na_2Ti_6O_{13}$ and $Na_2Ti_6O_{13}/Na_2Ti_3O_7$. The results indicated that the insertion potential of the layered titanate $Na_2Ti_3O_7$ is significantly lower than that of $Na_2Ti_6O_9$, and the specific capacity of the mixed titanate in the first cycle is twice that of pure $Na_2Ti_6O_9$, but its stability is far inferior to that of pure $Na_2Ti_6O_9$. This is because the reaction of $Na_2Ti_6O_{13}$ is primarily controlled by a diffusion process, and the ionic diffusion leads to significant volume changes due to the compression of the $Na_2Ti_6O_{13}$ lattice. In contrast, the samples of $Na_2Ti_3O_7$ incorporate two controlling mechanisms, namely diffusion and capacitive mechanisms, which, to some extent, mitigate the volume changes caused by ionic diffusion. Pan et al. (2022) combined zero-dimensional Co_3O_4 with 2D $Na_2Ti_2O_5$ nanosheets to establish a zero-dimensional/2D heterostructure. The results indicated that compared to NTO and Co_3O_4, the NTO@ Co_3O_4 exhibited superior deionization performance, maintaining a capacity of 927.2 mAh g^{-1} after 150 cycles at 200 mA g^{-1}, and demonstrating a specific capacity of 782.5 mAh g^{-1} after 250 cycles at 1000 mA g^{-1}.

12.4.5 Layered double hydroxides-based capacitive deionization electrode materials

The structure of LDHs is similar to the structure of hydrotalcite. Thus in LDHs crystal structure, the chemical composition of the host layers and the interlayer spacing can be adjusted, as well as the types and amounts of interlayer guest anions. Thus LDHs are often used as CDI for electro-adsorption of Cl^- ions. However, LDHs have poor intrinsic electronic conductivity and processability, easy aggregation.

Zhang and Li (2021) synthesized a ZnCo-Cl LDH with low charge transfer resistance through the coprecipitation approach and Li^+ intercalation. The obtained ZnCo-Cl LDH anode exhibited a salt removal capacity of 56.1 mg g^{-1}. Research shows that LDHs can expand the interlayer spacing through the preintercalation of Cl^{-1}. Wang et al. used Cl^- preintercalated CoFe LDHs (CoFeCl-LDHs) to prepare CDI electrode. The enlarged interlayer spacing of the CoFeCl-LDHs allowed for the fast charge transport and reversible Cl^- intercalation. The CoFeCl-LDHs electrode demonstrated a high desalination capacity of 100.2 mg g^{-1} and fast desalination rate of 0.38 mg g^{-1} s^{-1}. Compositing LDH with conductive nanomaterials can address the issue of poor conductivity in LDH. Liu et al. (2022) coated Cl-preintercalated NiMn LDHs (NiMn-Cl LDHs) on the surface of carbon nanofibers (CNFs@LDHs). The resulting CNFs@LDHs composite possessed a 3D interconnected structure, continuous conductive framework, abundant redox-active sites, and fast charge transport. Moreover, CNFs skeleton can prevent the aggregation of LDHs nanosheets. As a result, the CNFs@LDHs composite achieved a high desalination capacity of 72.04 mg g^{-1} and fast desalination rate. Zhang et al. (2023) fabricated a CNT@NiCoAl-LDH@MXene heterostructure with 3D interconnected network. In this work, MXene sheets acted as a conductive substrate to anchor NiCoAl-LDH nanosheets, while CNT worked as a conductive bridge for fast electron transport. Thus the CNT@NiCoAl-LDH@MXene heterostructure achieved high Cl^- removal capacity up to 79.5 mg g^{-1}.

Moreover, LDHs can be used as a precursor to generate the layered double metal oxides. Hu et al. (2025) synthesized a NiFe-layered double oxides/carbon (NiFe-LDO/C) composite through the thermal treatment of metal–organic framework (MOF) derived NiFe-LDH/MOF, which serves as an effective electrode for Cl^- storage with high desalination capacity and good stability. The NiFe-LDO/C composite electrode demonstrates a high desalination capacity of 43.68 mg g^{-1}, retaining 88.84 % of its initial capacity after 30 cycles. Notably, the NiFe-LDO/C composite can partially convert back into NiFe-LDH during the desalination process, demonstrating a memory effect. Ren et al. (2018) constructed a composite of MgAl layered double metal oxides ($MgAl\text{-}O_x$)/graphene (G) via thermal treatment of MgAl LDHs/G composite. The $MgAl\text{-}O_x$/G composite possessed a high specific surface area, abundant channels for anion, and good electronic conductivity. Thus the optimized $MgAl\text{-}O_x$/G composite CDI electrode exhibited a desalination capacity of 13.6 mg g^{-1}.

12.5 Conclusion and perspectives

2D nanomaterials are widely used as CDI electrode materials due to the following advantages. 2D Materials such as graphene, MXene, and MoS_2 possess a high specific surface area, which effectively increases the electrochemical adsorption active sites and enhances ion adsorption capacity. 2D materials also have excellent electrical conductivity, with graphene's conductivity reaching up to 10^6 S/m, facilitating rapid charge transfer, reducing internal resistance of the electrodes, and improving desalination efficiency. Additionally, 2D materials offer adjustable interlayer spacing and surface properties, allowing for precise control of interlayer distance through intercalation or defect engineering, optimizing the diffusion pathways for ions, and balancing high flux with high selectivity. Their surfaces can be chemically modified with functional groups such as -OH, -O, and -F to adjust hydrophilicity, ion selectivity, and pseudo-capacitance contributions, thereby enhancing the adsorption capacity for specific ions (e.g., Na^+, Cl^-). 2D materials also exhibit mechanical flexibility, enabling the construction of flexible films or 3D porous electrodes to meet complex device designs. However, 2D nanomaterials face several challenges: The nanosheets tend to restack during use or cycling, leading to capacity decay. The 2D nanomaterials loading of electrodes is low, with limited surface density and insufficient active material per unit volume, which restricts the overall desalination performance of the device. The structural collapse or loss of functional groups during use can result in poor cycling stability of CDI electrodes. Moreover, most 2D nanomaterials are in powder form, requiring complex processes to create slurries for coating film electrodes. Additionally, their ion-selective adsorption mechanisms are not well understood, making it difficult to achieve efficient ion-selective separation. To address the above issues, the following solutions are suggested. First, insert carbon nanomaterials into the 2D nanomaterial layers or grow other pseudocapacitive nanoparticles in situ to increase the interlayer spacing of the 2D nanomaterials, suppress the agglomeration of nanosheets, and enhance desalination performance. Second, perform surface modification through plasma treatment and create pores on the nanosheets, or assemble 2D nanomaterials into 3D porous materials to improve electrolyte wettability and promote ion diffusion. Third, utilize 3D printing technology to fabricate self-supporting electrodes while increasing the loading rate of the 2D nanomaterials. In summary, 2D nanomaterials are highly ideal electrode materials for CDI. The development of efficient 2D nanomaterial composites holds promise for addressing the global freshwater crisis.

References

Bharath, G., Rambabu, K., Banat, F., Hai, A., Arangadi, A.F., Ponpandian, N., 2019. Enhanced electrochemical performances of peanut shell derived activated carbon and its Fe_3O_4 nanocomposites for capacitive deionization of Cr(VI) ions. Science of the Total Environment 691, 713–726.

Biesheuvel, P.M., van der Wal, A., 2010. Membrane capacitive deionization. Journal of Membrane Science 346, 256–262.

Blair, J.W., Murphy, G.W., 1960. Electrochemical demineralization of water with porous electrodes of large surface area. Saline Water Conversion: Advances in Chemistry, Vol. 27. American Chemical Society, Washington, DC, pp. 206–223. Available at https://doi.org/10.1021/ba-1960-0027.ch020.

Cai, Y., Wang, Y., Zhang, L., Fang, R., Wang, J., 2022. 3D heterostructure constructed by few-layered MXenes with a MoS_2 layer as the shielding shell for excellent hybrid capacitive deionization and enhanced structural stability. ACS Applied Materials & Interfaces 14, 2833–2847.

Cai, Y., Zhang, W., Fang, R., Zhao, D., Wang, Y., Wang, J., 2021. Well-dispersed few-layered MoS_2 connected with robust 3D conductive architecture for rapid capacitive deionization process and its specific ion selectivity. Desalination 520, 115325.

Cech, O., Castkova, K., Chladil, L., Dohnal, P., Cudek, P., Libich, J., et al., 2017. Synthesis and characterization of $Na_2Ti_6O_{13}$ and $Na_2Ti_6O_{13}/Na_2Ti_3O_7$ sodium titanates with nanorod-like structure as negative electrode materials for sodium-ion batteries. Journal of Energy Storage 14, 391–398.
Chang, L., Hang Hu, Y., 2019. 3D Channel-structured graphene as efficient electrodes for capacitive deionization. Journal of Colloid and Interface Science 538, 420–425.
Chang, L., Fei, Y., Hu, Y.H., 2021. Structurally and chemically engineered graphene for capacitive deionization. Journal of Materials Chemistry A 9, 1429–1455.
Chen, B., Feng, A., Liu, K., Wu, J., Yu, Y., Song, L., 2021b. High-performance capacitive deionization using 3D porous $Ti_3C_2T_x$ with improved conductivity. Journal of Electroanalytical Chemistry 895, 115515.
Chen, B., Feng, A., Liu, K., Wu, J., Yu, Y., Song, L., 2021a. Subsize $Ti_3C_2T_x$ derived from molten-salt synthesized Ti_3AlC_2 for enhanced capacitive deionization. Ceramics International 47, 3665–3670.
Choi, J., Lee, H., Hong, S., 2016. Capacitive deionization (CDI) integrated with monovalent cation selective membrane for producing divalent cation-rich solution. Desalination 400, 38–46.
Ding, Z., Xu, X., Li, J., Li, Y., Wang, K., Lu, T., et al., 2022. Nanoarchitectonics from 2D to 3D: MXenes-derived nitrogen-doped 3D nanofibrous architecture for extraordinarily-fast capacitive deionization. Chemical Engineering Journal 430, 133161.
Divyapriya, G., Vijayakumar, K.K., Nambi, I., 2019. Development of a novel graphene/Co_3O_4 composite for hybrid capacitive deionization system. Desalination 451, 102–110.
Du, P., Xi, W., Zhang, Y., Wang, R., Gong, Y., He, B., et al., 2024. Laser induction of hierarchically micro/nanostructured CoNi-LDH/C composite for high-performance lithium-sulfur batteries. Journal of Alloys and Compounds 985, 174043.
Feng, B., Khan, Z.U., Khan, W.U., 2023. 3D graphene-supported N-doped hierarchically porous carbon for capacitive deionization of saline water. Environmental Science: Nano 10, 1163–1176.
Gamaethiralalage, J.G., Singh, K., Sahin, S., Yoon, J., Elimelech, M., Suss, M.E., et al., 2021. Recent advances in ion selectivity with capacitive deionization. Energy & Environmental Science 14, 1095–1120.
Gong, S., Liu, H., Zhao, F., Zhang, Y., Xu, H., Li, M., et al., 2023. Vertically aligned bismuthene nanosheets on MXene for high-performance capacitive deionization. ACS Nano 17, 4843–4853.
Grahame, D.C., 1947. The electrical double layer and the theory of electrocapillarity. Chemical Reviews 41, 441–501.
Guo, J., Xu, X., Hill, J.P., Wang, L., Dang, J., Kang, Y., et al., 2021. Graphene–carbon 2D heterostructures with hierarchically-porous P,N-doped layered architecture for capacitive deionization. Chemical Science 12, 10334–10340.
Guo, L., Wang, X., Leong, Z.Y., Mo, R., Sun, L., Yang, H.Y., 2018. Ar plasma modification of 2D MXene $Ti_3C_2T_x$ nanosheets for efficient capacitive desalination. FlatChem 8, 17–24.
Hassanvand, A., Wei, K., Talebi, S., Chen, G., Kentish, S., 2017. The role of ion exchange membranes in membrane capacitive deionisation. Membranes 7, 54.
Hosseinzadeh, M., Mozaffari, S.A., Ebrahimi, F., 2022. Porous 3D-graphene functionalized with MnO_2 nanospheres and NiO nanoparticles as highly efficient electrodes for asymmetric capacitive deionization: evaluation by impedance-derived capacitance spectroscopy. Electrochimica Acta 427, 140844.
Hsieh, W., Horng, T.L.A., Huang, H.C., Teng, H.S., 2015. Facile simulation of carbon with wide pore size distribution for electric double-layer capacitance based on Helmholtz models. Journal of Materials Chemistry A 3, 16535–16543.
Hu, B., Shang, X., Nie, P., Zhang, B., Yang, J., Liu, J., 2022. Lithium ion sieve modified three-dimensional graphene electrode for selective extraction of lithium by capacitive deionization. Journal of Colloid and Interface Science 612, 392–400.
Hu, J., Xi, W., Zhang, Y., Wang, R., Wang, H., Gong, Y., et al., 2025. Rational design of LDH-derived NiFe layered double oxides as capacitive deionization anode for efficient chlorine ion storage with a "memory effect" Applied Surface Science 687, 162289.
Jia, F., Sun, K., Yang, B., Zhang, X., Wang, Q., Song, S., 2018. Defect-rich molybdenum disulfide as electrode for enhanced capacitive deionization from water. Desalination 446, 21–30.
Jin, J., Xi, W., Li, Z., Hu, J., Wang, R., Gong, Y., et al., 2024. Recent advances of 2D materials in capacitive deionization. Desalination 578, 117468.
Jin, W., Hu, M., Sun, Z., Huang, C.H., Zhao, H., 2021. Simultaneous and precise recovery of lithium and boron from salt lake brine by capacitive deionization with oxygen vacancy-rich CoP/Co_3O_4-graphene aerogel. Chemical Engineering Journal 420, 127661.

Lee, J., Kim, S., Kim, C., Yoon, J., 2014. Hybrid capacitive deionization to enhance the desalination performance of capacitive techniques. Energy & Environmental Science 7, 3683–3689.

Lee, J.B., Park, K.K., Eum, H.M., Lee, C.W., 2006. Desalination of a thermal power plant wastewater by membrane capacitive deionization. Desalination 196, 125–134.

Leong, Z.Y., Yang, H.Y., 2019. A study of MnO_2 with different crystalline forms for pseudocapacitive desalination. ACS Applied Materials & Interfaces 11, 13176–13184.

Leong, Z.Y., Lu, G., Yang, H.Y., 2019. Three-dimensional graphene oxide and polyvinyl alcohol composites as structured activated carbons for capacitive desalination. Desalination 451, 172–181.

Li, B., Sun, K., Xu, W., Liu, X., Wang, A., Boles, S., et al., 2023. Tailoring interlayer spacing in MXene cathodes to boost the desalination performance of hybrid capacitive deionization systems. Nano Research 16, 6039–6047.

Li, D., Wang, S., Wang, G., Li, C., Che, X., Wang, S., et al., 2019. Facile fabrication of NiCoAl-layered metal oxide/graphene nanosheets for efficient capacitive deionization defluorination. ACS Applied Materials & Interfaces 11, 31200–31209.

Li, H.B., Lu, T., Pan, L.K., Zhang, Y.P., Sun, Z., 2009. Electrosorption behavior of graphene in NaCl solutions. Journal of Materials Chemistry 19, 6773–6779.

Li, Q., Zheng, Y., Xiao, D., Or, T., Gao, R., Li, Z., et al., 2020. Faradaic electrodes open a new era for capacitive deionization. Advanced Science 7, 2002213.

Li, X., Fu, X., Zhang, Y., Zhang, Q., Wei, Z., Wang, K., et al., 2025. Electrospun sodium titanate-MXene/carbon nanofibers as binder-free electrode for enhanced hybrid capacitive deionization. Chemical Engineering Journal 511, 162040.

Li, Z., Mao, S., Yang, Y., Sun, Z., Zhao, R., 2021. Controllable synthesis of a hollow core-shell Co-Fe layered double hydroxide derived from Co-MOF and its application in capacitive deionization. Journal of Colloid and Interface Science 585, 85–94.

Liang, M., Wang, L., Presser, V., Dai, X., Yu, F., Ma, J., 2020. Combining battery-type and pseudocapacitive charge storage in $Ag/Ti_3C_2T_x$ MXene electrode for capturing chloride ions with high capacitance and fast ion transport. Advanced Science 7, 2000621.

Liang, P., Sun, X., Bian, Y., Zhang, H., Yang, X., Jiang, Y., et al., 2017. Optimized desalination performance of high voltage flow-electrode capacitive deionization by adding carbon black in flow-electrode. Desalination 420, 63–69.

Liu, N., Yu, L., Liu, B., Yu, F., Li, L., Xiao, Y., et al., 2023. Ti_3C_2-MXene partially derived hierarchical 1D/2D TiO_2/Ti_3C_2 heterostructure electrode for high-performance capacitive deionization. Advanced Science 10, 2204041.

Liu, Y., Gao, X., Wang, K., Dou, X., Zhu, H., Yuan, X., et al., 2020. Rocking-chair capacitive deionization with flow-through electrodes. Journal of Materials Chemistry A 8, 8476–8484.

Liu, Y., Du, X., Wang, Z.P., Zhang, L.Y., Chen, Q.L., Wang, L.H., et al., 2021b. MoS_2 nanoflakes-coated electrospun carbon nanofibers for "rocking-chair" capacitive deionization. Desalination 520, 115376.

Liu, Y., Wang, K., Xu, X., Eid, K., Abdullah, A.M., Pan, L., et al., 2021a. Recent advances in faradic electrochemical deionization: system architectures versus electrode materials. ACS Nano 15, 13924–13942.

Liu, Y., Du, X., Wang, Z., Wang, L., Liu, Z., Shi, W., et al., 2022. Layered double hydroxide coated electrospun carbon nanofibers as the chloride capturing electrode for ultrafast electrochemical deionization. Journal of Colloid and Interface Science 609, 289–296.

Ma, J., Cheng, Y., Wang, L., Dai, X., Yu, F., 2020. Free-standing $Ti_3C_2T_x$ MXene film as binder-free electrode in capacitive deionization with an ultrahigh desalination capacity. Chemical Engineering Journal 384, 123329.

Nguyen, T.K.A., Wang, T.-H., Doong, R.-a, 2022. Architectures of flower-like MoS_2 nanosheet coated N-doped carbon sphere electrode materials for enhanced capacitive deionization. Desalination 540, 115979.

Olabi, A.G., Elsaid, K., Rabaia, M.K.H., Askalany, A.A., Abdelkareem, M.A., 2020. Waste heat-driven desalination systems: perspective. Energy 209, 118373.

Pan, Y.Y., Li, Y.X., Song, J., Xiang, Y.X., Sui, Z.Y., Tian, Q.H., 2022. Co_3O_4 nanoparticles anchored on stable $Na_2Ti_2O_5$ nanobelts for improved lithium storage. Applied Surface Science 601.

Peng, W., Wang, W., Qi, M., Miao, Y., Huang, Y., Yu, F., 2021. Enhanced capacitive deionization of defect-containing MoS_2/graphene composites through introducing appropriate MoS_2 defect. Electrochimica Acta 383, 138363.

Peng, W.J., Wang, W., Han, G.H., Huang, Y.K., Zhang, Y.S., 2020. Fabrication of 3D flower-like MoS_2/graphene composite as high-performance electrode for capacitive deionization. Desalination 473.

Qian, A., Wu, H., Wang, G., Sun, N., Cheng, H., Zhang, K., et al., 2023. Freeing fluoride termination of $Ti_3C_2T_x$ via electrochemical etching for high-performance capacitive deionization. ACS Applied Materials & Interfaces 15, 9203–9211.

Qie, H., Liu, M., Fu, X., Tan, X., Zhang, Y., Ren, M., et al., 2022. Interfacial charge-modulated multifunctional MoS_2/$Ti_3C_2T_x$ penetrating electrode for high-efficiency freshwater production. ACS Nano 16, 18898–18909.

Rangaraj, V.M., Yoo, J.I., Song, J.K., Mittal, V., 2023. MOF-derived 3D MnO_2@graphene/CNT and Ag@ graphene/CNT hybrid electrode materials for dual-ion selective pseudocapacitive deionization. Desalination 550, 116369.

Ren, Q., Wang, G., Wu, T., He, X., Wang, J., Yang, J., et al., 2018. Calcined MgAl-layered double hydroxide/ graphene hybrids for capacitive deionization. Industrial & Engineering Chemistry Research 57, 6417–6425.

Shen, X., Xiong, Y., Hai, R., Yu, F., Ma, J., 2020b. All-MXene-based integrated membrane electrode constructed using $Ti_3C_2T_x$ as an intercalating agent for high-performance desalination. Environmental Science & Technology 54, 4554–4563.

Shen, X., Li, L., Xiong, Y., Yu, F., Ma, J., 2022. Graphene-assisted Ti_3C_2 MXene-derived ultrathin sodium titanate for capacitive deionization with excellent rate performance and long cycling stability. Journal of Materials Chemistry A 10, 10192–10200.

Shen, X., Hai, R., Wang, X., Li, Y., Wang, Y., Yu, F., et al., 2020a. Free-standing 3D alkalized $Ti_3C_2T_x$/$Ti_3C_2T_x$ nanosheet membrane electrode for highly efficient and stable desalination in hybrid capacitive deionization. Journal of Materials Chemistry A 8, 19309–19318.

Skuse, C., Gallego-Schmid, A., Azapagic, A., Gorgojo, P., 2021. Can emerging membrane-based desalination technologies replace reverse osmosis? Desalination 500, 114844.

Stern, O., 1924. The theory of the electrolytic double shift. Zeitschrift Fur Elektrochemie Undangewandte Physikalische Chemie 30, 508–516.

Sun, K., Yao, X., Yang, B., Jia, F., Song, S., 2020. Oxygen-incorporated molybdenum disulfide nanosheets as electrode for enhanced capacitive deionization. Desalination 496, 114758.

Suss, M.E., Presser, V., 2018. Water desalination with energy storage electrode materials. Joule 2, 10–15.

Tan, Z., Wang, W., Zhu, M., Liu, Y., Yang, Y., Ji, X., et al., 2023. $Ti_3C_2T_x$ MXene@carbon dots hybrid microflowers as a binder-free electrode material toward high capacity capacitive deionization. Desalination 548, 116267.

Tang, Y., Ding, J., Zhou, W., Cao, S., Yang, F., Sun, Y., et al., 2023. Design of uniform hollow carbon nanoarchitectures: different capacitive deionization between the hollow shell thickness and cavity size. Advanced Science 10, 2206960.

Tian, S., Zhang, X., Zhang, Z., 2021. Novel MoS_2/NOMC electrodes with enhanced capacitive deionization performances. Chemical Engineering Journal 409, 128200.

Urban, J.J., 2017. Emerging scientific and engineering opportunities within the water-energy nexus. Joule 1, 665–688.

Vafakhah, S., Saeedikhani, M., Huang, S.Z., Yan, D., Leong, Z.Y., Wang, Y., et al., 2021. Tungsten disulfide-reduced GO/CNT aerogel: a tuned interlayer spacing anode for efficient water desalination. Journal of Materials Chemistry A 9, 10758–10768.

Wang, K., Liu, Y., Ding, Z., Chen, Z., Xu, X., Wang, M., et al., 2022a. Chloride pre-intercalated CoFe-layered double hydroxide as chloride ion capturing electrode for capacitive deionization. Chemical Engineering Journal 433, 133578.

Wang, X.-R., Wang, X., Nian, H.-e, Chen, T., Zhang, L., Song, S., et al., 2022b. Hierarchical MXene/Polypyrrole-decorated carbon nanofibers for asymmetrical capacitive deionization. ACS Applied Materials & Interfaces 14, 53150–53164.

Wang, Y., Zhang, X., Liang, S., Wang, C., Sun, L., Deng, W., et al., 2024. Exploring MXene's role in capacitive deionization: advances, challenges, and future directions. Chemical Engineering Journal 496, 154130.

Xi, W., Jin, J., Zhang, Y., Wang, R., Gong, Y., He, B., et al., 2022. Hierarchical MXene/transition metal oxide heterostructures for rechargeable batteries, capacitors, and capacitive deionization. Nanoscale 14, 11923–11944.

Xi, W., Zhang, Y., Wang, R., Gong, Y., He, B., Wang, H., et al., 2023b. The effect of electrode thickness and electrode/electrolyte interface on the capacitive deionization behavior of the $Ti_3C_2T_x$ MXene electrodes. Journal of Alloys and Compounds 947, 169701.

Xi, W., Guo, T., Xie, Z., Zhang, Y., Wang, R., Gong, Y., et al., 2023a. Molten salt etching synthesis of $Ti_3C_2T_x$/Ni composites for highly efficient capacitive deionization. Desalination 574, 117241.

Xi, W., Zhang, Y., Hu, J., Li, Z., Wang, R., Wang, H., et al., 2024. In-situ growth of Ni, Co-Prussian blue nanocubes on molten salt etched lamellar $Ti_3C_2T_x$/Ni for enhanced hybrid capacitive deionization. Chemical Engineering Journal 499, 155941.

Xing, F., Li, T., Li, J., Zhu, H., Wang, N., Cao, X., 2017. Chemically exfoliated MoS_2 for capacitive deionization of saline water. Nano Energy 31, 590–595.

Xu, H., Li, M., Gong, S., Zhao, F., Zhao, Y., Li, C., et al., 2022. Constructing titanium carbide MXene/reduced graphene oxide superlattice heterostructure via electrostatic self-assembly for high-performance capacitive deionization. Journal of Colloid and Interface Science 624, 233–241.

Yang, T., Zhang, H., Guo, L., Wang, J., Guo, Z., Du, Y., et al., 2023. Mesopore-enhanced graphene electrodes with modified hydrophilicity for ultrahigh capacitive deionization. Desalination 567, 116984.

Yang, X., Jia, Z., Zhang, W., Ying, G., Wang, Z., Lu, Z., et al., 2022. Facile fabrication of intercalation-type pseudocapacitive S-$Ti_3C_2T_x$/PANI/F-$Ti_3C_2T_x$ cathode for asymmetric capacitive deionization. Desalination 535, 115816.

Yang, Z.Y., Jin, L.J., Lu, G.Q., Xiao, Q.Q., Zhang, Y.X., Jing, L., et al., 2014. Sponge-templated preparation of high surface area graphene with ultrahigh capacitive deionization performance. Advanced Functional Materials 24, 3917–3925.

Younes, H., Lou, D., Rahman, M.M., Choi, D., Hong, H., Zou, L., 2022. Review on 2D MXene and graphene electrodes in capacitive deionization. Environmental Technology & Innovation 28, 102858.

Zhang, B., Boretti, A., Castelletto, S., 2022. Mxene pseudocapacitive electrode material for capacitive deionization. Chemical Engineering Journal 435, 134959.

Zhang, L., Wang, Y., Cai, Y., Fang, R., Huang, S., Zhao, Y., et al., 2023. Heterostructure of NiCoAl-layered double hydroxide nanosheet arrays assembled on MXene coupled with CNT as conductive bridge for enhanced capacitive deionization. Chemical Engineering Journal 478, 147270.

Zhang, Y., Li, X., Xi, W., Zhang, Q., Ge, X., You, J., et al., 2024. Constructing a hierarchical MoS_2/MXene heterostructure for efficient capacitive deionization of saline water. Chemical Synthesis 4, 19.

Zhang, Z., Li, H., 2021. Promoting the uptake of chloride ions by ZnCo-Cl layered double hydroxide electrodes for enhanced capacitive deionization. Environmental Science: Nano 8, 1886–1895.

Zhao, H., Liang, Q., Yang, Y., Liu, W., Liu, X., 2021. Magnetic graphene oxide surface lithium ion-imprinted material towards lithium extraction from salt lake. Separation and Purification Technology 265, 118513.

Zhao, Z., Zhao, J., Sun, Y., Ye, M., Wen, X., 2022. In-situ construction of 3D hierarchical MoS_2/CoS_2@TiO_2 nanotube hybrid electrodes with superior capacitive performance toward water treatment. Chemical Engineering Journal 429, 132582.

Zhu, Y.S., Zhang, G.J., Xu, C., Wang, L.Z., 2020. Interconnected graphene hollow shells for high-performance capacitive deionization. ACS Applied Materials & Interfaces 12, 29706–29716.

Index

Note: Page numbers followed by "*f*" and "*t*" refer to figures and tables, respectively.

Q

R

S

V

W

X

Z

Printed and bound by CPI Group (UK) Ltd, Croydon, CR0 4YY
25/09/2026
02223046-0002